POLYMER SCIENCE AND TECHNOLOGY

POLYPROPYLENE

SYNTHESIS, APPLICATIONS AND ENVIRONMENTAL CONCERNS

POLYMER SCIENCE AND TECHNOLOGY

Additional books in this series can be found on Nova's website
under the Series tab.

Additional E-books in this series can be found on Nova's website
under the E-book tab.

MATERIALS SCIENCE AND TECHNOLOGIES

Additional books in this series can be found on Nova's website
under the Series tab.

Additional E-books in this series can be found on Nova's website
under the E-book tab.

POLYMER SCIENCE AND TECHNOLOGY

POLYPROPYLENE

SYNTHESIS, APPLICATIONS AND ENVIRONMENTAL CONCERNS

LUCIANO PAULINO SILVA
AND
EDUARDO FERNANDES BARBOSA
EDITORS

nova publishers
New York

For permission to use material from this book please contact us:
Telephone 631-231-7269; Fax 631-231-8175
Web Site: http://www.novapublishers.com

Additional color graphics may be available in the e-book version of this book.

Library of Congress Cataloging-in-Publication Data

ISBN: 978-1-62417-142-0

Published by Nova Science Publishers, Inc. † New York

CONTENTS

PREFACE

In an almost poetic way, the discovery of synthetic polymeric plastics can be compared to the "Big Bang" event. Both resulted in a vast universe of novel opportunities. After this, the formed substances have been passing through a real evolution. In this book, it is possible to verify that this chemical evolution of thermoplastics was possible and guided by Ziegler-Natta and metallocene catalyzers. The properties of the obtained materials could be understood considering each crystal structure obtained, which is a direct consequence of catalyzer influence on polymer formation. Crystal morphology guides the particular physical and chemical material behaviours. These intrinsic features also guide how these substances can interact with other materials, forming blends or composites. Finally, topographic morphology of the nanometric arrangement and surface physical properties showed in this book is able to help us to understand a little more about the “nanoworld” of some plastic surfaces. Have a nice trip!

In: Polypropylene
Editors: L. P. Silva and E. F. Barbosa
ISBN: 978-1-62417-142-0

Chapter 1

AN INTRODUCTION OF STRUCTURE, SYNTHESIS AND SAFETY CONCERNING POLYPROPYLENE APPLICATIONS

Huimin Sun*[*] *and Langui Xie
The National Institutes of Food and Drug Control,
Beijing, People's Republic of China

ABSTRACT

Polypropylene, a thermoplastic polymer, can be made from the monomer propylene by Ziegler-Natta polymerization and by metallocene catalysis polymerization. According to three-dimensional structure or tacticity, polypropylene can be classified into three types: isotactic, syndiotactic, and atactic. All the methyl groups in isotactic propylene are on the same side of the chain, methyl groups in syndiotactic propylene are positioned alternately, and the methyl groups in atactic polypropylene are placed randomly on both sides of the chain. The relative orientation of methyl groups has a strong effect on the polymer's ability to form crystals, which affects the polymer's physical properties and its application. Isotactic polypropylene, which has high crystallinity, is rugged and unusually resistant to many chemical solvents, bases and acids. It is used in a wide variety of applications including packaging, textiles, furniture, stationery, reusable containers, appliances, automotive components, and construction materials. Syndiotactic polypropylene with low crystallinity has a promising application prospect in films, medical tubing, and medical adhesive. Atactic polypropylene is rubbery and widely used as hot melt adhesives, sealing materials, emulsifiers, paint and etc. Since polypropylene contains only carbon and hydrogen atoms, it does not poison the environment after its disposal, but deterioration time in the land field will be long. The best way to limit PP disposal is to recycle the PP wastes, or incinerate the wastes especially when they are seriously polluted. Thus, from the aspect of environmental protection and public health assurance, PP wastes should be recycled as much as possible. However, the recycled PP products should not be used as pharmaceutical packaging materials due to technical and hygienic reasons.

[*] E-mail address: sunhm@126.com

INTRODUCTION

Polypropylene (PP) is made from the polymerization of propylene and is the largest-volume commodity thermoplastic. As one important member of the polyolefin resin family, PP is molded or extruded into many plastic products that require toughness, flexibility, light weight, and heat resistance.

The structure of PP is linear, and its molecular weight ranges from 100,000 to more than 500,000. Tacticity, also known as stereoregularity, is a special feature in the PP structure that must be considered for understanding its properties. Tacticity generally refers to the relative stereochemistry of adjacent chiral centers within a macromolecule, and in the case of PP it refers to the relative orientation of each methyl group with regard to the methyl groups in neighboring monomer units. PP can be classified accordingly into three types: isotactic, syndiotactic, and atactic. The methyl groups in isotactic PP (i-PP) are all on the same side of the chain, whereas in syndiotactic PP they are positioned alternately and in atactic PP (a-PP) they are oriented randomly on both sides of the chain. The relative orientation of methyl groups has a strong effect on the polymer's ability to form crystals. Both isotactic and syndiotactic PP are crystalline, whereas atactic PP, which can also be prepared with special Ziegler–Natta catalysts as a byproduct, is an amorphous, soft and tacky material used in adhesives, sealants, and caulk products. The tacticity of the PP is determined by the catalyst employed in the synthesis and affects the polymer's physical properties as well as application.

SYNTHESIS OF PP

Polymerization of propylene monomer to form PP is thermodynamically favorable. However, for a very long time, the polymerization of propylene monomers failed to generate high molecular weight polymers and could only form branched low-molecular weight oil that was of no interest, mainly owing to the lack of proper initiator and the dynamic infeasibility. With financial support from Montecatini Company (now Montedison SpA), a large Italian chemical company, the Italian chemist Giulio Natta and his assistant Paolo Chini at Polytechnic Institute of Milan extended the discovery of the German chemist Karl Ziegler at the Max Planck Institute for coal chemistry to prepare isotactic PP (with 175°C melting point) in 1954. The Nobel Prize in Chemistry in 1963 was awarded jointly to Karl Ziegler and Giulio Natta "for their discoveries in the field of the chemistry and technology of high polymers". In the U.S., meanwhile, Paul Hogan and Robert Banks at Phillips Petroleum (now Chevron Phillips) were experimenting with high-octane gasoline when they discovered high-melting crystalline PP in 1952. Practically useful PP was discovered nearly at the same time in the U.S. and Europe. Large-scale commercial production of PP by Montecatini in Italy, Hercules Inc. in the United States, and Hoechst AG in West Germany (now in Germany) began in 1957. Since the early 1980s, production and consumption have increased significantly, owing to the invention of more efficient catalyst systems by Montedison and the Japanese Mitsui Petrochemical Industries, Ltd.

The invented Ziegler–Natta catalyst restricts the propylene monomer to adopt a specific orientation to produce linear crystalline PP, either isotactic or syndiotactic. Currently, two main types of catalysts are mostly employed in the synthesis of isotactic and syndiotactic PP.

The first class of modern Ti-based catalysts is derived from titanium chlorides. Ziegler discovered that the combination of $TiCl_4$ and Al $C_2H_5)_2Cl$ was suitable for polymerization of ethylene into a linear, highly crystalline polymer. Natta further developed this catalyst and used crystalline α-$TiCl_3$ in combination with $Al(C_2H5)_3$ to produce the first isotactic PP (with 40% crystalline). The Ziegler–Natta catalyst later evolved into numerous variations of formulae consisted of the combinations of IVB–VIIIB transition metal compounds and IA–IIIA organometallic compounds. Moreover, further addition of electron donor and catalyst support to the Ziegler–Natta catalyst can improve the reactivity and tacticity in the polymerization reaction.

Thus, all modern supported Ziegler–Natta catalysts designed for preparing isotactic PP use $TiCl_4$ as the active ingredient, organoaluminum compounds such as triethylaluminium $Al(C_2H_5)_3$ as cocatalysts, an ester of an aromatic diacid or a diether as the organic modifier, and $MgCl_2$ as the support. Two most important technological characteristics of all supported catalysts are high productivity and a high fraction of crystalline isotactic polymer, when produced at 70–80°C under standard polymerization conditions.

The elemental reactions in the coordination polymerization of propylene consist of chain initiation, chain propagation, chain transfer and chain termination, which are detailed as follows:

[1] Chain initiation: the titanium and aluminum components of the catalyst react to form an active species $©^{\delta+}—^{-\delta}R$ (abbreviated as ©-R), on whose surface the chain initiation takes place :

$©\text{-R} + CH_2\text{=}CHCH_3 \rightarrow ©\text{-}CH_2\text{-}CH(CH_3)\text{-R}$

[2] Chain propagation: the monomer inserts between the transition metal–carbon bond, thus propagating the chain :

$©\text{-}CH_2\text{-}CH(CH_3)\text{-R} + nCH_2\text{=}CHCH_3 \rightarrow ©\text{-}CH_2\text{-}CH(CH_3)\text{-}[CH_2\text{-}CH(CH_3)]_n\text{-R}$

[3] Chain transfer: the active polymer chain can transfer to alkyl aluminum or propylene, although the transfer constant is usually small. During production, it is necessary to add hydrogen as the chain transfer agent to control the molecular weight of the prepared polymer. Chain transfer to the alkyl aluminum:

$©\text{-}CH_2\text{-}CH(CH_3)\text{-}[CH_2\text{-}CH(CH_3)]_n\text{-R} + Al\ (CH_2\text{-}CH_3)_3 \rightarrow$

$©\text{-}\ CH_2\text{-}CH_3 + Al\ (CH_2\text{-}CH_3)_2\text{-}CH_2\text{-}CH(CH_3)\text{-}[CH_2\text{-}CH(CH_3)]_n\text{-R}$

Chain transfer to the propylene monomer:

$©\text{-}CH_2\text{-}CH(Mt)\text{-}[CH_2\text{-}CH(Mt)]_n\text{-R} + C_3H_6 \rightarrow ©\text{-}C_3H_7 + CH_2\text{=}CH(Mt)\text{-}[CH_2\text{-}CH(Mt)]_n\text{-R}$

Chain transfer to hydrogen:

©-CH_2-CH(CH_3)-[CH_2-CH(CH_3)]$_n$-R + H_2 →

©-H + CH_3-CH (CH_3)-[CH_2-CH(CH_3)]$_n$-R

[4] Chain termination: coordination polymerization is difficult to terminate and may self-terminate after a long time of reaction by transferring to the β-H within the molecular chain. Many mechanisms, including monometallic and bimetallic mechanisms, have been proposed for the tacticity in the coordination polymerization of propylene. The monometallic mechanism is currently more widely accepted, which is proposed by P. Cossee and E.J. Arlman in 1960 arguing that the active catalytic species consists of a single transition metal (Ti) and the monomer coordinates to Ti before insertion into the Ti–C bond, thus achieving chain propagation.

After alkyl exchange, the $TiCl_4$ and alkyl aluminum form an active catalytic species, in which the transition metal rests in the center of the octahedral configuration and is attached to four chlorine atoms and one alkyl ligand, leaving one vacant site. Propylene is adsorbed to the surface of this active species and coordinated with the alkylated Ti^{3+}. Due to its electron donating property, the π electrons in the double bond activate the Ti–C bond and a four-membered ring transition state is formed. Afterwards, the CH_2 or alkyl group is transferred from the transition metal to the alkene and addition reaction takes place. Alternatively, the alkene may insert between the Ti–C bonds to expose the vacant site again but at different position. The vacant site must return to the original position in order to achieve isotactic propagation of propylene chain, otherwise the syntactic structure will form.

The second class of catalysts is metallocenes. They are soluble in aromatic hydrocarbons and usually contain two components, a metallocene complex and a special organometallic cocatalyst, methylalumoxane (MAO, [–O–Al(CH_3)–]$_n$). The ideal metallocene catalysts have the composition Cp_2MCl_2 (M = Ti, Zr, Hf) such as titanocene dichloride. Typically, the organic ligands are derivatives of cyclopentadienyl. In some complexes, the two Cp rings are linked with bridges, like –CH_2–CH_2–. Depending on the type of their cyclopentadienyl ligands, metallocene catalysts can produce either isotactic or syndiotactic polymers of propylene. Commercial synthesis of syndiotactic PP is mostly carried out with the use of a special class of metallocene catalysts. They employ bridged bis-metallocene complexes of the type (Cp_1) (Cp_2)$ZrCl_2$, where the first Cp ligand is the cyclopentadienyl group, the second Cp ligand is the fluorenyl group, and the bridge between the two Cp ligands is -CH_2-CH_2-. These complexes are converted to polymerization catalysts by activating them with MAO.

The polymerization mechanism of metallocene catalysts is similar to that of Ziegler–Natta catalysts: propylene molecules coordinate with transition metal complexes and then insert between the chained segments and the metal to form the long chain polymer.
In both cases atactic PP is produced as a by-product. It is an amorphous rubbery material, and the methyl groups in its PP chain exhibit no preferred orientation.

Two process techniques are used for the production of PP: Liquid Propylene Pool Process (LIPP) and gas phase polymerization. In LIPP, the polymer is produced in a liquid pool of PP by the inclusion of a Ziegler-Natta catalyst, which forces the molecules of the monomer to align and join in a certain manner to enable long chain formation, or polymerization, into PP.

When long chain formation is sufficient, growth of the molecule is stopped via hydrogen addition. Following the termination of reaction, the catalyst is deactivated and resultant polymer is separated from the liquid propylene. In gas phase polymerization, a gas phase reactor is essentially a fluidized bed of dry polymer particles maintained either by stirring or by passing the propylene monomer gas at a high flow rate through it. As in LIPP, high activity specialties catalysts are used in the polymerization of the gaseous monomer into PP. Pressures are usually relatively low at ~2MPa and temperatures are usually in the range of 60 - 80°C.

APPLICATION OF PP

The low density characteristics of PP, with a density of only 0.9 g/cm^3, is the key factor enabling its wide applications in lighter weighting products comparing with the higher density materials. Besides, PP has outstanding chemical resistance to organic chemicals, which is the best of all thermoplastics. In fact, there is no solvent that can dissolve PP at room temperature, although PP may swell in some solvents. PP can be flammable and ignite at the temperature of about 350°C. Although its burning rate is slow when tested to ASTM D635, it can be made flame-retardant by incorporating suitable flame-retardant agent. When used in applications involving extended UV and outdoor exposure, PP will require the addition of UV stabilizing additives. Therefore, PP has found its way in many applications that require low density, high melting point, long-term durability, and resistance to stress, cracking and chemical corrosion. This section will illustrate the wide use of PP. In some application areas, PP is used from the beginning, and cannot be replaced by other resins. While in some other application areas, PP is proven to have more economic benefits or higher performance than other resins. The application of PP will be described according to its market share [1].

According to Phillip Townsend Associates, [1] fibers represent 30% of PP consumption on a global scale. Fibers used in industrial and household textiles include carpets, slit tape, nonwoven fabrics, apparel, cordage, netting, rugs and mats. PP began as an outdoor carpet fiber thanks to its resistance to moisture and mildew, and then new advances in its look and feel have made it popular for the house. PP shows several advantages for use as carpet face yarn: it possesses excellent stain resistance, static electricity resistance and lower specific gravity which means more fiber volume (for bulk as it is known in the industry) per given weight of carpet. Throughout the world, PP has replaced natural fibers (primarily jute) as the material in almost all carpet backing. The usage of PP in the slit tape market exceeds its use in the bulk continuous filament carpet market. Fabric woven from slit tape are usually used in tarpaulins, erosion abatement screens, sand bags, fertilizer and feed bags, swimming pool covers, and other similar heavy-duty products. The largest single fiber application of PP is in nonwoven fabrics, especially used as diapers and hygiene products. Disposable diapers, made from PP nonwoven fabric, have provided a convenient alternative to cloth diapers, and can improve the quality of life of infants and parents, and are universally replacing the cloth diapers for infants. Disposable sterile surgical or therapeutic garments made from PP nonwoven fabrics protect medical personnel, which are also soft, breathable, lightweight, comfortable to wear, odorless, non-toxic and can be disposed of easily and safely [1]. PP apparel such as lightweight Jackets made from PP fiber textile is waterproof and windproof, and usually used as sports apparel and military clothing. PP cordages prepared from PP single fiber are mostly used for marine application, with the benefit of lightweight, rigidity and good resistance to moisture absorption.

PP is also widely used in manufacturing injection molded consumer products, including appliances (e.g., dish-washing machines, washing machines, and coffee makers), general consumer products (e.g., luggage carriers, outdoor furniture, and portable refrigerator), medical applications (e.g., syringe, medical non-woven fabrics), rigid and flexible packaging (e.g., yogurt containers, syrup bottles), etc. As we known, the environment of a dishwasher interior is very hostile, which requires cycling through hot water, aggressive detergents, dry heat, and physical abuse from dropped dishes. PP material could provide the necessary chemical resistance, rigidity, appearance, and durability to endure the hostile impact, last the expected life of the product, and allow its production at sufficiently low cost to render the competitive price of the products. For the same reasons, PP is extensively used in the production of washing machines. Glass reinforced PP, which has very high effect on mechanical properties, is usually used for the outer drum of washing machines. Most of the small electronic products are also made from PP materials. Advantages such as lightweight, rigidity, easy molding, dyeability and excellent insulation can promote the quality of these products, which lead to the good quality and affordable final appliance. Obviously, the use of PP to mold outdoor furniture has become more and more popular. The high-stiffness of PP products allows the substantial reduction and even elimination of fillers to the PP base for outdoor furniture, thus improving the surface appearance of the final article and the color retention when scratched or suffered surface abrasion. Other advantages include the lower material density, which results in lighter weight of the item and easier recycling operations, owning to the mono-material composition of the final products. These PP furniture exhibits excellent resistances to humidity, warm, acid rain, ultraviolet ray, and other adverse environmental conditions, which could cause damage in metal and wood furniture. But they must be properly stabilized and formulated to avoid polymer degradation.

PP becomes the preferred resin for most of the previously discussed general injection molded applications because of its light weight and low cost. Several medical applications, however, take advantage of not only these two important characteristics but also other unique properties. It is possible, for example, to formulate PP molding compounds to undergo steam or radiation sterilization after molding and still retain sufficient intended physical properties. The most notable such application is the disposable syringe. [1 Due to their availability in sterilized condition, ready to use, and cost effectiveness, disposable syringes are fast replacing the age-old glass syringes. Rigid PP packaging is also used to protect various items, such as foods and drugs. There are several advantages to using PP for food and drug packaging. Fluid and moldable, PP can be made into sheets, shapes, and structures, offering considerable design flexibility [2]. The use of PP in food packaging has continued to increase due to the low cost of materials and functional advantages (such as thermosealability, microwavability, optical properties, and unlimited sizes and shapes) over traditional materials such as glass and tinplate [3]. The other reason for the development of PP rigid packaging is the pass of the environmental laws, particularly in the U.S. and Europe, which has raised more interest in PP. In some U.S. states, for example, the use of packaging materials is being restricted by legislative decree. The package must be recycled at a specified recovery rate, reduced in total weight, or have 25% post-consumption content in order to reduce the load on available landfills. The strength of PP enables the manufacturer to achieve a source reduction of approximately 10 to 30%. Recycling ability is volume sensitive; the larger the content of PP in packaging, the more economical it can be recycled. Therefore, more PP is better and it is exactly what PP has been enjoying in rigid packaging. The stiffness of PP allows thinner walls, thereby reducing the total package weight, a feature that not only reduces the cost of shipping but also reduces the amount of packaging material mandated for recycling.

In addition, PP is used to prepare piping systems that need high-purity, strength and rigidity (e.g., those intended for use in potable plumbing, hydronic heating and cooling, and reclaimed water). PP is selected for such applications due to its resistance to corrosion and chemical leaching, its resilience against most forms of physical damage (including impact and freezing), its environmental benefits, and its ability to be joined by heat fusion rather than gluing.

It is not surprising that the global automotive sector is the largest market when it comes to the application of PP. Compared with other plastics, such as phenolic resin, polystyrene, and cellulose acetate butyrate, PP shows better mechanical properties (e.g., low density and good impact/rigidity profile). Not only many inner components but also many external components in the automobile are made from PP.

Another important application of PP is in film. Biaxially Oriented PP (BOPP) is a good example. It has become a popular, high growth film on the world market because of a unique combination of properties such as resistant to grease, oil and moisture permeation, high tensile strength even at elevated temperatures, good impact and flexibility at very low temperatures, good shrinkage, stiffness on the converting line, high seal strength, twist retention and high barrier. Aesthetically, PP film displays excellent transparency and glossiness, and has replaced glass papers (former popular packaging materials) as the transparent packaging film globally. BOPP can be made via two distinctly different processes: tubular and tenter frame. Both processes were used in the early commercial production of BOPP film, and both are still in use today. Both processes yield exceptionally clear films that can be coated, heat sealed, or metalized to accommodate today's sophisticated converting lines. BOPP is usually used to pack fast food, breads and candies. The rapid chilling of extruded PP film produces an un-oriented film with excellent clarity. The film produced in this manner lacks the stiffness of BOPP and, while clear, is not as clear as BOPP. However, it is less costly to make and it can serve certain competitive markets where soft clear films are desired. Stationery products have used much of the cast film manufactured today, which include clear overlays, dividers, and ancillary film products such as photo albums and baseball card protector pages that proliferate in today's homes and offices. The environmental concerns surrounding PVC have accelerated the use of cast PP in these applications, many of which were previously served by PVC film [1].

Many plastic items for medical or laboratory use can be made from PP which is autoclavable so that it can withstand the heat in an autoclave. Moreover, PP has been used in a variety of pharmaceutical packages such as infusion bottles, multilayer polymer infusion bags and bottles for oral drugs, due to their low density, high tenacity, easy manufacturing, relatively low cost and resistance to bacterial growth. PP also finds its way in miscellaneous applications such as the Australian banknotes.

ENVIRONMENTAL CONCERNS OF PP

PP has been widely used in an array of industries and has become an indispensable ingredient of human life. The safety, environmental, and health concerns about PP have been continuously raised within industry and from consumers [3]. Without any additive, PP is not deemed as a hazardous material by any known definition, regulation or law, although it can be construed under U.S. OSHA HAZCOM that the use of additives may make PP resins hazardous. The PP polymers can become hazardous because of the ingredients used in formulating the final product. For instance, the PP mixture that blends >1% fiberglass is considered a hazardous material under U.S. OSHA HAZCOM regulation, because fiberglass is a skin irritant and is considered a hazardous material. Other ingredients, such as flame-retardants and mineral fillers like talc, can also make PP formulations hazardous. If a PP mixture containing a hazardous ingredient is tested and exhibit no adverse health, safety, or environmental effects from its use, the PP mixture can be considered as a non-hazardous substance [4].

Other environmental concerns of PP polymer come from its production process. Operations of the traditional solvent slurry process lead to the relatively high emission of volatile organic compounds (hydrocarbons) into the atmosphere. Moreover, the use of alcohol and water for catalyst washing in the solvent slurry process also presents environmental problems with regard to the disposal of the extracted catalyst residue solids (titanium and aluminum complexes) and the malfunctioning of the alcohol recovery systems. The extracted solid catalyst residues are allowed to settle in on-site ponds, many of which are still in existence today. In addition, the malfunctioning of the alcohol recovery systems usually results in high concentrations of alcohol draining with the supposedly clean water, which greatly threatens the environment. In addition to the air emissions in solvent recovery, there is also a need to remove solids from the solvent. This process results in atactic PP being removed, collected and, in some cases, placed in landfills. With the large volumes of solvents and alcohol stored on-site, there is also a greater risk of spill, causing problems related to air, water, and soil contamination. As the manufacture of propylene evolved and the use of propylene as the polymerization medium increased, manufacturing processes have become more environmentally friendly. Environmental issues have not been completely eliminated with the new technologies; however, the emissions to air, water, and land have been greatly reduced with their use [1].

Another environmental concern of PP polymer focuses on the problems of the PP wastes. Rapid growth of the world population and economy led to increased demand for PP. Due to its non-biodegradability, short life and vast applications in many fields, PP forms a large fraction of municipal waste, which is one of the major environmental concerns in the world. According to a 2003 Environment Agency report, 80% of post-consumer plastic waste is sent to landfill, 8% is incinerated and only 7% is recycled [5]. Recycling plastic can reduce the amount of plastics waste that requires disposal. It also has several other advantages such as conservation of non-renewable fossil fuels (plastic production uses 8% of the world's oil production, 4% as feedstock and 4% during manufacture), reduced energy consumption, reduction of solid waste to send to landfill, and reduced emissions of carbon dioxide (CO_2), nitrogen oxide (NO) and sulfur dioxide (SO_2). Although PP like other thermoplastics is capable of being recycled, for a multitude of reasons some of it will end up in landfill.

Landfill is the least preferred end result for a variety of reasons. However, in isolated areas or where the costs and energies required to transport and recycle waste outweigh the benefits, landfill may be a better option than the existing recycling methods. PP is inert under landfill conditions, it does not contribute to leaching in landfill sites, it does not generate methane gas, and it is stable within the landfill [1].

There is a variety of safe ways to manage the end of life PP products, depending on the circumstances. If large and relatively clean streams are available and can be collected in an economically viable way, recycling is the preferred option. PP can be readily recycled mechanically several times. Studies conducted on the properties of recycled PP show a slight loss in physical properties with each recovery cycle. The retained properties are adequate to enable reuse in existing and established myriad of applications. The reuse of the PP wastes is considered as the best environmental alternative for solving the problem of PP disposal. Legislation in many industrialized countries includes a number of provisions concerning waste reduction and recycling in all material categories.

Products made from recycled PP have many areas of potential applications, which include auto parts, new automotive battery cases, furniture, pails, water meter boxes, bag dispensers, carpets, refuse and recycling containers, grocery cart handles, and industrial fibers. Examples of PP recycling include:

- Automotive battery cases where the batteries are collected at the end of their life, the lead and acid are recovered, and the PP battery case material is recycled. This recycled material typically goes back into new automotive battery cases and can also be used in other applications.
- Australian banknotes are made from PP and upon their withdrawal from use are collected from Banking Institutions and granulated into small pieces. These pieces are then re-melted and formed into pellets ready to be used as a raw material for a range of new products.
- Municipal curbside collected packaging items such as used ice cream, yoghurt and margarine containers, spreads/dip containers, cordial bottles, personal care containers, laundry product containers and more.
- Materials handling applications such as pallets, crates, containers and tote boxes. These multiple use durable applications are sought after by the recycling industry as a source of feedstock for recycled PP. Recycling companies will typically form collection deals with major sources of such items [6].

In the packaging area, very often the PP bottles can be recycled and turned into reusable pallets or transportation containers, but the post-consumer PP closures cannot be used in the production of plastic packaging containers that directly contact with drugs. The Federal Food, Drug, and Cosmetic Act stated that a drug was deemed to be adulterated "if its container is composed, in whole or in part, of any poisonous or deleterious substance which may render the contents injurious to health...." Not only FDA but also SFDA in China, European Commission in Europe, and ministry of health, labor and welfare in Japan hold the same view. Nevertheless, FDA and European Commission allow some recycled plastics products to be used in the manufacturing of plastic packaging containers for food if 1) the remaining food and contaminants in the plastics intended for recycling was effectively washed way, and 2) the result of subsequent analysis of the processed polymer for the surrogate contaminants shows great efficacy in the recycling process, and 3) conformed to the same special standards as for the virgin plastics, especially the migration of the components in the recycled products from the packaging to the food. Thus, from the aspect of environmental protection and public health assurance, PP wastes should be recycled effectively and the recycled PP products should not be applied as pharmaceutical packaging materials due to technical and hygienic reasons. Many analytical methods have been developed to detect pharmaceutical packaging materials adulteration, such as the adulteration of PP bottles for Oral Drug Package [7].

CONCLUSION

The discovery of Ziegler–Natta catalyst leads to the commercial production of practically useful PP. As one of the world's most exciting and promising manufacturing materials, PP's outstanding property profile renders its broad application, which has been fully used by the manufacturing industry. Moreover, PP is highly safe for human use, unlikely to be leached out even when exposed to warm or hot water, and approved by the FDA for use with food, beverage and drug storage. As an environmentally safe and eco-friendly plastic, PP can be readily recycled several times. In order to protect the environment and save energy, the post-consumption PP wastes should be recycled as much as possible. However, it is important to note that the recycled PP products should not be used as pharmaceutical packaging materials due to technical and hygienic reasons.

REFERENCES

[1] N. Pasquini, *Polypropylene handbook*, Hanser Gardner, United States (2005)

[2] K. Marsh, B. Bugusu. *J. Food Sci.* 72, 37 (2007).

[3] Lopez-Rubio, E. Almenar, P. Hernandez-Muñoz, J. M. Lagarón, R. Catalá, Rafael Gavara, *Food Reviews International*, 20, 357 (2004).

[4] H. Zweifel, R. D. Maier, M. Schiller, *Plastics Additives Handbook*, Hanser Verlag, Germany (2009).

[5] R. Siddique, J. Khatib et al, *Waste Manag.*, 28(10): 1835-1852. (2008).

[6] L. Basell Australia Pty Ltd. Environmental Information Document Australian Manufacture POLYPROPYLENE [pdf]. 2012 at *http://www.lyondellbasell.com/NR/rdonlyres/C2ED0A47-6430-45FA*-87A4-D4018108814D/0/AusPPEnvirostatementJan12final.pdf.

[7] X. Lan-Gui, S. Hui-Min, J. Shao-Hong, *Anal. Chimica Acta*, 706, 312 (2011).

In: Polypropylene
Editors: L. P. Silva and E. F. Barbosa

ISBN: 978-1-62417-142-0

Chapter 2

POLYPROPYLENE FABRICS: PREPARATION, MODIFICATION, AND APPLICATIONS

Xiang-Yu Ye and Zhi-Kang Xu*

MOE Key Laboratory of Macromolecular Synthesis and Functionalization,
Department of Polymer Science and Engineering, Zhejiang University, Hangzhou, China

ABSTRACT

Polypropylene is extremely versatile as a fiber-forming material, and can be used in a wide range of fibrous forms, such as monofilament, yarns, and non-woven meshes. Nowadays, the polypropylene fabric industry occupies more than 30% of incessantly expanding global polypropylene market share. The most common industrial manufacturing processes are melt spinning, spun-bond and melt-blown. However, the diameters of commercial polypropylene fabrics are usually about 10 to 20 μm or more, which greatly limit their performance. Therefore, untiring laboratory efforts have been made to develop novel methods for fabricating polypropylene nanofibers. Some preliminary successes have been achieved so far, including melt electrospinning/high- temperature solution electrospinning, direct melt extrusion of immiscible blend and subsequent extraction removal of matrix (phase separation), and template synthesis, which open up a new avenue for high-performance utilization of polypropylene. The unique characteristics of polypropylene fabrics such as excellent mechanical properties, high thermal and chemical stability, easy processability and comparatively low-cost, make them exhibit high potentials for comprehensive applications in fiber-based reinforcement and especially in the nonwovens industry. Despite the advantages of polypropylene nonwoven fabrics, like a random network of overlapped fibers, multiple connected pores and overall high porosity, the inherent hydrophobic nature of polypropylene greatly limits their usage as separation media. It is widely recognized that the surface properties play a crucial role, therefore various methods have been proposed to modulate the surface hydrophilicity of polypropylene nonwoven membrane or further achieve surface functionalization (ion-exchange groups, antibacterial and affinity ligands), e.g. in-situ modification, dip-coating, layer deposition, chemical treatment, thermal/plasma/UV/radiation induced graft polymerization, biocatalysis or the combination of different methods. The

* E-mail address: xuzk@zju.edu.cn. Fax: + 86 571 8795 1773.

modified polypropylene nonwoven membrane can be extensively applied in several fields of protective cloth, wastewater treatment, affinity membrane, and biomedical textile.

INTRODUCTION

Polypropylene, as one of the most common plastic materials, is eminent for its light-weight, low cost, good mechanical strength, excellent moisture, heat and chemical resistances, favorable processing and recycle characteristics. Global PP capacity is forecasted to increase from 52 million tones/year in 2008 to 69.1 million tones/year in 2013. Due to its unique properties, polypropylene has found extensive use in both packaging and engineering applications, especially textile and can be applied in a wide range of fibrous forms, such as monofilament, yarns, and non-woven meshes. Nowadays, more than 30% of incessantly expanding global polypropylene market share goes into fiber and fiber-related products.

Polypropylene fabric industry, which mainly includes polypropylene fibers and polypropylene non-woven meshes, has become a global industry and always been increasing exponentially. Despite the great achievements, there are still two remaining challenges to overcome for further progress: 1) It is better to reduce the fiber size to a nanometer range. The diameters of conventional textile fibers made by typical industrial manufacturing processes are usually about 10 to 20 μm or more. Nanofiber means larger surface area to volume ratio and flexibility in surface functionalities, which makes the nanofibers to be optimal candidates for intriguing applications in many areas [1]; 2) Polypropylene is well known as a hydrophobic polymer and without polar functional groups. These futures greatly limit the end-use performance of polypropylene fabrics, especially in non-woven mesh applications. It is of significant importance to develop efficient methods for producing polypropylene nanofibers and modifying polypropylene fabrics for both academic research and practical application.

Herein, firstly we intend to provide a brief discuss of the preparation methods of polypropylene fabrics, with focus on the cutting edge technology developed for producing nanofibers including electrospinning, direct melt extrusion of immiscible blend and subsequent extraction removal of matrix (phase separation), and template synthesis. Then, we list a comprehensive account of the modification methods including *in-situ* modification during the manufacturing process and post modification (especially surface modification). Their industrial prospects are fully demonstrated. Further, a range of interesting applications are assessed based on the fibrous structures, functionalities, and specific properties. Finally we give concluding remarks and a personal outlook of the polypropylene fabric industry.

PREPARATION OF POLYPROPYLENE FABRICS

Fibrous forms of polypropylene fabrics include staple, bicomponent staple, monofilament, multifilament, slit film yarns, slit-fibrillated film yarns, woven meshes, spun-bond (S), melt-blown (M) nonwovens, or a combination of spun-bond and melt-blown (SMS), extruded nettings, and synthetic pulps. Among them, the most two common industrial products are fibers and nonwoven meshes.

Polypropylene Fibers

Melt Spinning

Melt spinning is the preferred manufacturing process for the formation of synthetic fibers. Three of the major commercial fibers, nylon, olefin and polyester, are produced with this process. A typical melt spinning process is described as follows (Figure 1). Small polypropylene chips are injected into a screw extruder, melted and pumped through a spinneret. The spinneret is, in effect, a plate with numerous holes (one to thousands), one for each filament and in different cross-sectional shapes, including round, trilobal, pentagonal, octagonal, hollow, and others. The molten fibers are cooled by cold air, solidified and collected on a take-up wheel. Finally the fibers in the molten or solid states are stretched for orientation of the polymer chains to enhance the physical-mechanical properties, creating a considerably strong yarn. The processability of polypropylene fibers is influenced by the rheological properties of polypropylene chips and operating parameters like throughput rate, spinneret geometry, temperature, cooling rate, suction air speed, take-up speed, draw ratio, and other factors. Many different types of morphologies such as spherulites, shish-kebad, lamellar and fibrillar structures can be obtained by adjusting the processing conditions [2]. Up to date, various models have already been proposed; we are able to predict the physical properties of spun fibers by correlating them with polypropylene chips and operating parameters [3].

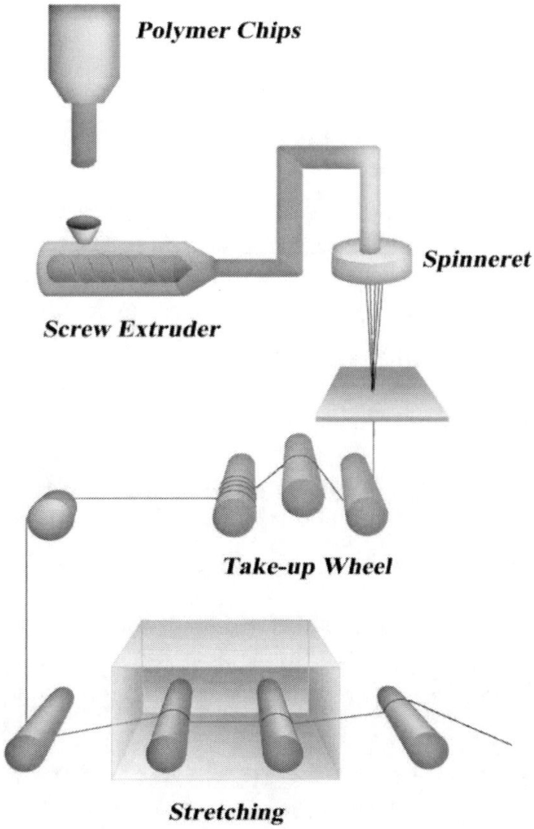

Figure 1. Schematic representation of a typical melt spinning process.

Typically, the industrial melt spinning production line adopts electrical heating system ever since 1950s. More recently, a carbon dioxide (CO_2) laser thinning apparatus has been developed in laboratory, which was used to welding, cutting and cladding for ceramics and metals, etching of polymer [4]. The advantage of laser heating method is that the polypropylene chip can be locally and instantaneously melted so that the molten polymer is not exposed to long-term thermal degradation and thermal dispersion is minimized [5]. Under optimal conditions, the CO_2 laser-thinning apparatus could wind up 3.5 μm fibers in the range of 100 m/min to 2500 m/min without requiring large-scale equipment compared with conventional techniques, which makes it have great industrial prospect [6].

However, the surface properties of polypropylene fibers have to be adjusted to meet the requirements for the end-use applications. It is a facile yet effective method by blending of additives prior to or during the fiber melt spinning process, or rather *in-situ* modification. For example, pigments mixed with the melt can be extruded together to produce pigmented polypropylene fibers for indoor/outdoor carpeting [7]. Various additives, such as polar polymer, aluminosiloxane, flame retardant, nanoclay, montmorillonite, carbon nanotube, nucleating agent, and silver powder, stabilizer, or just another grade polypropylene chip, were incorporated to improve the dyeability, hydrophilicity, flame retardancy, antistatic ability, antibacterial activity, thermal stability, crystallinity, and mechanical property [8-16]. It is also feasible to functionalize polypropylene with initiator and monomer during melt extrusion, a process called melt reactive extrusion [17,18]. Even we can simply modify the surface of polypropylene fibers by passing through a spin finish aqueous solution or on-line corona treatment line after the melt spinning process [19,20].

Phase Separation

Phase behavior is one of the most important characteristics for a polymer blend with different components. Usually, polymer blends are immiscible and exhibit phase separation. Separated phases are found in the forms of sphere, lamellar, cylinder and bicontinuous structures, which depend on the composition of the polymer blends as well as their processing condition. One form, named 'microfibrilar reinforced blend' [21], consists of a thermo-plastic polymer as matrix and thermoplastic polymer microfibrils as *in-situ* reinforcement. It provides a promising method to prepare micro or nanofibers when the matrix was selectively removed [22-25]. Isotactic polypropylene (iPP) nanofibers were prepared by this method using cellulose acetate butyrate (CAB) as the polymer matrix [26]. The nanofibers were in continuous and uniform yarns with an average diameter of 215 nm. Figure 2 depicts the schematic diagram of iPP nanofiber fabrication. In this case, the fiber size depends on major factors such as the dispersion of iPP in CAB and the shear and elongational deformation of iPP in the blend. Nanofibers with diameters varying from 100 nm to micrometer size were well controlled by modulating the blend ratio and the condition of extrusion process. Similarly, iPP nanofibers were also obtained from blends with polyethylene terephthalate (PET) [27], nylon 6 [28], and polylactide [29]. This phase separation method makes the mass production of polypropylene nanofibers possible and can be readily applied in industrial practice.

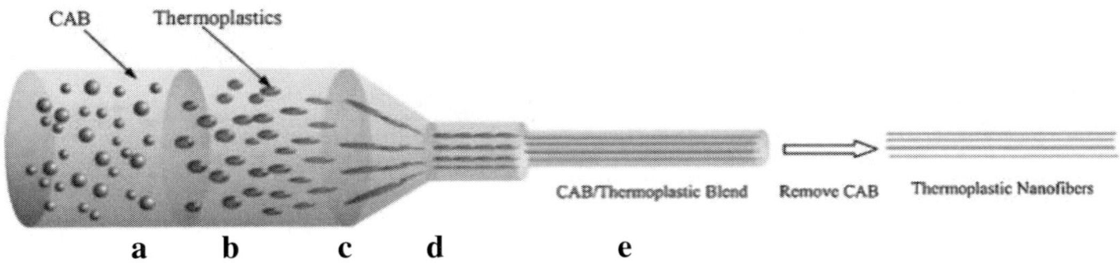

Figure 2. Schematic diagram of thermoplastic nanofiber fabrication. (a) Dispersion of thermoplastics in CAB matrix into microsized micelles, (b) deformation and elongation of dispersed thermoplastic micelles into ellipsoids, (c) elongation and coalescence of thermoplastic ellipsoids, (d) CAB/thermoplastics composite fibers, (e) thermoplastic nanofibers after the removal of CAB matrix. Regions of (a–b) inside the extruder, (c) inside the die, and (d–e) in the air. Reprinted from [26]. Copyright WILEY-VCH Verlag GmbH & Co. KGaA, Weinheim, 2007.

Template Synthesis

The template synthesis method has been widely used for preparing micro and nanostructured materials. One typical example is the synthesis of nanofibrils with uniform diameter within the pores of a membrane template [30]. Anodic aluminum oxide (AAO) membranes are well-established templates because they have straight and separated cylindrical pores with monodisperse diameters in nanometer scale. For example, polypropylene melts was pressed into the pores by using AAO membrane as the mold for the 'template melt extrusion' process [31]. Nanofiber arrays on the bulk feeding film were reserved after removal of the AAO membrane. At last, isolated nanofibers with diameters ranging from 150 to 400 nm were got by ultrasonic in ethanol for 5 min. However, the viscosity of molten polymer is too high to enter the pores completely, and therefore the length and thinnest diameter of nanofibers are greatly restricted.

Figure 3. (a) A 2×2 cm^2 microfibre array patch holding two 200 g weights (400 g) on a vertical smooth glass slide without normal load. (b) A scanning electron microscopy (SEM) image of a microfibre array. Average length and diameter of the microfibres are 18 mm and 0.6 mm, respectively. Reprinted from [35]. Copyright The Royal Society, 2008.

A very interesting phenomenon is the vertically aligned arrays of high-density polypropylene nanofibers with a gecko-inspired structure. These arrays were fabricated by casting one layer of polypropylene film into a 20 μm thick polycarbonate track-etched membrane filter in a vacuum oven at 200°C for 25 min, and then dissolving in methylene chloride [32-34]. Friction experiments demonstrated that the fibrillar polymer surface exhibits directional adhesion, and is able to hold heavy weights, as shown in Figure 3 [35]. Nevertheless, this template synthesis method needs exquisite operation and is not so suitable for industrial production.

Polypropylene Non-Woven Meshes

Structurally, a non-woven mesh is a highly porous fabric and composed of random network of overlapped fibers which are bonded mechanically, thermally or chemically. Generally, there are three categories of non-woven as dry laid, wet laid, and polymer laid [36]. Spun-bond and melt-blown are high profile technologies often used in well-known consumer and industrial businesses. Meanwhile, untiring laboratory efforts have been made to fabricate polypropylene nanofibrous non-wovens by electrospinning.

Spun-Bond Fabrics and Melt-Blown Non-Woven Meshes

Spun-bond fabrics are usually produced by depositing extruded, spun filaments onto a collecting belt in a uniform random manner followed by bonding the fibers. Many methods can be used to bond the fibers including mechanical needling, and thermal or chemical treatment. The last two may bond large regions (area bonding) or small regions (point bonding) of the web by fusion or adhesion of fibers. Currently, thermal point bonding has mostly been employed for webs of low to medium basis weight [37]. The benefits of this technique include both energy saving and environmentally friendliness. Melt spinning is a basis technique for spun-bond fabrics. Fiber spinning and web formation can fit together to form a moving production line by placing the bonding device in line with spinning. The filaments pass through a constricted area of low pressure but high velocity as in a venturi tube, causing the filaments to fan out and randomize, which are then laid on a moving collector belt. The collecting surface is usually perforated to prevent the air stream from deflecting, and suction fans below remove the air and help in the lay-down. The bonding conditions greatly influence the structure and properties of the spun-bond fabrics. Statistical analysis shows that the bonding temperature has a major effect, and the mechanical strength increases with bonding area and bonding size [38]. The tensile properties can also be enhanced by applying *in-situ* additive-modified polypropylene fibers [39]. For many applications, solid polypropylene fibers can be replaced by hollow polypropylene fibers without suffering the loss of compression-recovery and opacity behaviors, which can substantially reduce the material cost [40]. Today, spun-bond polypropylene fabrics are widely used as packing material and baby diapers, both of them include markets that are usually large and constant in demand. The current growth rate for spun-bond nonwovens is on the order of 6-8% per year.

Melt-blown is a single step process for producing non-woven meshes by extruding a polymer melt through an orifice die and drawing down the extrudate with two convergent jets of hot air (typically at the same temperature as the molten polymer) [41]. Bicomponent PP/PET melt-blown microfiber can also be fabricated by using side-by-side spinneret [42]. This environmentally benign processing method was first developed under U.S. government sponsorship in the 1950s. The Naval Research Laboratory initiated this work with the goal of producing sub-micron fibers for the collection of radioactive particles in the upper atmosphere. As early as in 1994, melt-blown was used to produce about 93 million kg of nonwoven webs worldwide, and it is estimated over 90% of them are made from polypropylene [43]. A majority of the research has been conducted on the relationship of process parameters and mean fiber diameter in order to understand how to produce fibers with small size and high quality [44]. Process variables that have already been studied can be divided into three groups: operational conditions, off-line parameters, and material characteristics. Various models were proposed on the basis of surface response methodology [45], single/thin-filament model [46], and computer simulation [47]. However, the melt-blown process is a complex one that involves turbulence making it hard to work out theoretical or empirical models for industrial applications.

In general, melt-blown microfibers have diameters in the range of 2 to 4 μm, smaller than melt spinning fibers. On-line measurements of fiber diameters using high-speed flash photography reveal that most of the fiber attenuation, more than 96% in some cases, is found to occur within 1.5 cm from the die [48]. Further efforts to produce nanofibers by melt-blown process resulted in surface tension driven fiber breakup instabilities and led spherical particles to disperse among the non-woven mesh. These instabilities may present the onset of an underlying fundamental limit to the process [41].

The spun-bond and melt-blown processes are somewhat identical from an equipment and operator's point of view. With advanced composite technology on the rise, three-ply system has been developed by consisting of spun-bond/melt-blown/spun-bond (SMS). In this case, it supplements the fiber and web properties of the melt blown with the strength and toughness of the spun-bond surface fabrics.

Electrospinning

Electrospinning is a simple and versatile method for fabricating continuous nano- and/or micrometer fibers. The process consists of three stages: jet initiation and the extension of the jet along a straight line; jet thinning by bending instability and solvent evaporation; jet solidification into fibers [49]. Up to now, more than 50 types of polymers have already been successfully electrospun into fabrics. However, electrospinning of polyolefins is challenging due to their poor solubility in polar solvent. Polyolefin solutions from non-polar solvents usually have low dielectric constant and poor conductivity, which in turn result in poor electro-spinnability [50]. In the case of isotactic polypropylene (iPP) especially, there is no proper solvent to prepare spinning solution at room temperature. Therefore, electrospinning of iPP requires either dissolving it into a solvent at elevated temperature or electrospun directly at molten state. Related researches are still in primary stages and only few results have been published.

Syndiotactic polypropylene (sPP) can be dissolved in a mixture of cyclohexane, acetone, and dimethylformamide (80/10/10 by weight ratio) at 60 °C to form a homogeneous solution and is stable after cooling to room temperature. sPP fiber with average diameter of 650 nm can be electrospun, however, twisted ribbon-shaped rather cylinder fibers are frequently observed and there are irregular microholes on the sPP microfibers [51]. Using the same sPP solution, a more straightforward method, which is termed novel handspinning, is reported for producing uniaxially aligned sPP nanofibers [52]. As shown in Figure 4, sPP nanofibers of 420 nm can be easily obtained simply by the pull/push motions of thumb and index fingers without an applied electric field. The easy and simple processing is the most crucial advantage in this handspinning method. Compared with electrospun sPP fibers, the handspun nanofibers exhibit smoother surface morphologies, smaller diameters and increased mechanical strength.

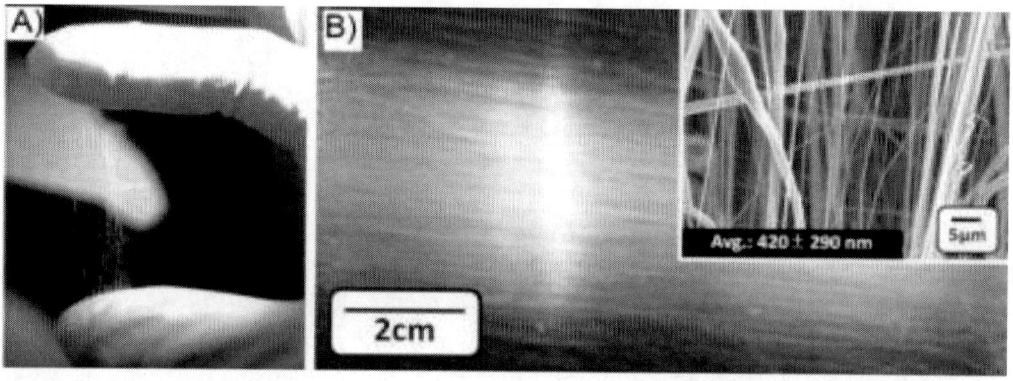

Figure 4. (A) Simple and easy process of handspinning. Rapid pressing and pulling away process of polymer solution by using thumb and index fingers can produce the uniaxially aligned nanofibers. (B) Digital photo of uniaxially aligned sPP nanofibers, and typical SEM image (inset) of the corresponding handspun sPP nanofibers. Reprinted from [52]. Copyright WILEY-VCH Verlag GmbH & Co. KGaA, Weinheim, 2011.

However, solution electrospinning is inevitably required at elevated temperature for iPP. Figure 5 shows a schematic setup of the high-temperature electrospinning apparatus [53]. A jacket-type heat exchanger is designed to maintain the temperature of polymer solution at a constant value by circulating heated silicone oil, where the temperature can be adjusted up to 150°C. To resolve the gel formation problem in the long and thin needle, high-energy laser light is used to locally control the temperature of the needle effectively as high as 120°C. In addition, an infrared emitter is used to modulate the environmental temperature by its throughput and distance to the spinning apparatus. In this experiment, o-dichlorobenzene was used as a solvent and tetra-n-butyl ammonium perchlorate soluble salt was added to increase the solution conductivity. A stable cone-jet electrospinning mode can be maintained at a temperature of up to 100°C. Compared to conventional room temperature electrospinning, it is very interesting to find that iPP nanofibers of 300 nm were successfully electrospun from 1 wt% solution, which is much lower than the entanglement concentration calculated for the present iPP species, possibly due to the rapid evaporation of solvent. Another study reveals that there is only a narrow operational window available (from 140 to 170°C) to electrospin iPP/decalin solution [54].

For solution electrospinning, the use of organic solvent would result in serious environmental and health problems, low productivity (typically solution concentration <20%), and high cost. Thus, melt electrospinning becomes a subject of great interest [55-57]. The advantages of this process include: elimination of organic solvent; high throughput; suitable for polymer blends that have no common solvent; the potential to produce hollow and noncircular cross-sectional fibers by changing the nozzle shape. The first patent about melt electrospinning was approved in 1936 [58], and the first trial of melt electrospinning of polypropylene was published in 1981 [59-61]. However, little progress has been made in recent decades [62]. The underlying difficulties include high viscosity and low conductivity of polypropylene melt, thermal degradation, high electrical field strength needed, interference between heating system and high voltage. A major drawback of melt electrospinning is that the resulting fibers are thicker than those of solution electrospinning mainly because the bending stability is restricted by the rapid cooling of polypropylene melt jet and lack of solvent evaporation.

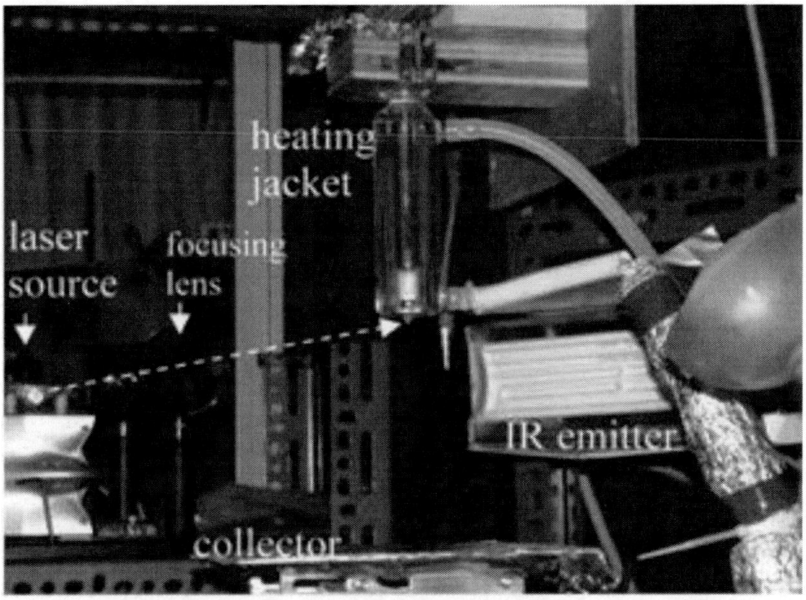

Figure 5. Schematic setup of the high-temperature electrospinning apparatus. The dashed line shows the path of laser light. Reprinted from [53]. Copyright American Chemical Society, 2010.

Melt electrospinning equipment can be readily manufactured from melt spinning extruder [63]. The molecular weight of polypropylene is the predominant factor in determining the fiber diameter. High tacticity tends to produce smaller fiber, and the other factors like electrical field strength, supplied volume, needle diameter, collection distance show similar effects as solution electrospinning process [64,65]. The diameter of polypropylene fiber is usually thicker than 2 μm. Efforts to produce thin fiber have achieved certain success aiming at reducing the jet solidification speed by electrospinning in vacuum [66] and further heating the spin line [67]. A major step is made by using a viscosity-reducing additive, the diameter of polypropylene fibers can be reduced to 840 ± 190 nm [68]. An experimental line of continuous melt electrospinning on filter media has already been established in laboratory (Prof. Yoog Lak Joo's group, Cornell University), however, there is still a long way to accomplish industrialization. It is of great promise to modify the melt-blown device into melt-blown/melt electrospinning combination machine. Table 1 summarizes of different preparation methods for iPP fabrics.

Table 1. Comparison of different preparation methods for polypropylene fabrics

Preparation Method	Fiber Diameter (μm)	Industrial Prospect
Melt spinning	10~20	Strong
Melt-blown	2~4	Strong
Phase separation	0.1~1	Medium
Template synthesis	0.1~0.8	Low
Solution electrospinning	0.3~2.5	Low
Melt electrospinning	0.7~10	Medium

POST MODIFICATION OF POLYPROPYLENE FABRICS

The unique characteristics of polypropylene fabrics make them exhibit high potentials for comprehensive applications in fiber-based reinforcement and especially in the nonwovens industry as separation membrane. Nevertheless, the inherent hydrophobic nature of polypropylene greatly limits their in-service performance, and thus post modification of polypropylene fabrics is essential. It is widely recognized that the surface properties play a crucial role, and our group puts forward a new concept, so called 'Surface Engineering of Polymer Membranes [69]. The purpose may be to minimize fouling, modulate hydrophilicity/hydrophobicity, enhance biocompatibility, construct a diffusion barrier, provide bio or chemical functionalities, mimic a biomembrane, fabricate nanostructures or simply improve the aesthetic appearance of the membrane surface. Various surface modification methods have already been developed, and usually two or more methods are adopted at a time.

Physical Treatment

Coating

Coating is the simplest modification method. Spin finishes are frequently used in industry by a so called 'pad-dry' process. For example, polypropylene nonwoven mesh was padded with chitosan oligomer solution to give 100% pick-up (wetting agent was added in the finishing formulation to improve the penetration of finishing solution), then the padded nonwoven mesh was dried at 100°C for 3 min, and the as-prepared fabrics showed high antimicrobial activities [70]. Dip-coating is a similar way to prepare thin and dense skin layers on fabric surface. Polyvinyl alcohol (PVA) was usually selected, for it is well-known as nonfouling material and spontaneously adsorb onto hydrophobic substrate, such as polypropylene. After dip-coating, PVA was further lightly cross-linked by glutaraldehyde to get a stable hydrogel layer [71]. It was even possible to modify the pore walls by filtering PVA copolymer solution through the polypropylene nonwoven mesh [72]. Or using polypropylene nonwoven mesh as collector, one can simply deposited a thin and uniform layer of electrospun PVA [73] or chitosan [74] nanofibers on the mesh surface to endow it with enhanced flux, effluent quality and antifouling property.

Another popular way is electrostatic-driven self-assembly by means of alternate adsorption of linear polycations and polyanions. Typically, the first acidic layer is produced by grafting of acrylic acid onto the nonwoven mesh. However, the deposition of succeeding polyelectrolyte layers failed to provide a complete coverage of the modified surface, and the surface properties depended exclusively on the type of the top layer [75]. SiO_2 nanoparticles can be readily deposited onto plasma pretreated polypropylene nonwoven mesh (charged substrate), while adhere weakly to untreated one [76]. The zeta potential of the nanoparticles was critical for the deposition behavior. Deposition of nanoparticles with ξ=36.4 mV resulted in a monolayer coating with a patchy coverage, while a multilayer coating was obtained when nanoparticles with ξ=21.5 mV were used. Accordingly, the water contact angles changed from 120° to 45° and then superhydrophilicity, which the authors believe is the first demonstration of a superhydrophilic polypropylene nonwoven mesh. Another great advantage of this approach is that the coatings are virtually unchanged even after 1 h of sonication in the solvent. This method also works well for other nanoparticles like TiO_2 [77].

Deposition

Types of deposition include physical vapor deposition (PVD), chemical vapor deposition (CVD), plasma deposition, sputtering, atomic layer deposition (ALD), and molecular layer deposition (MLD). Gold/platinum sputter deposition is a common way to prepare samples for SEM observation. The greatest benefit of this method is the uniformity. Copper [78,79], aluminum [73], carbon [80], zinc oxide [81], organo-silver compound [82], or even polymer [81] have been used as the target material for polypropylene nonwoven mesh modification.

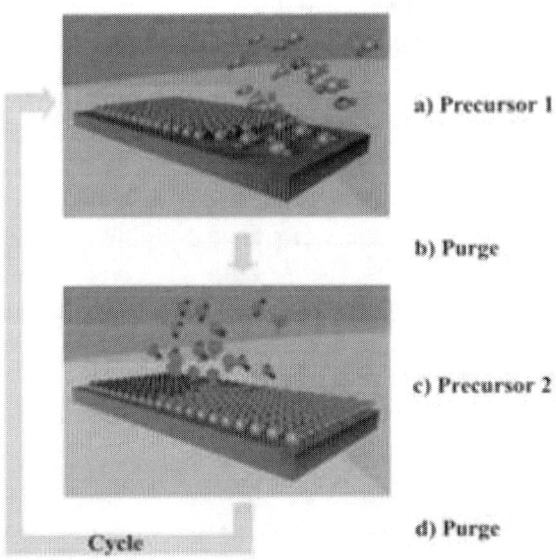

Figure 6. Schematic of an ALD process. One ALD cycle consists of four separate steps. In step a) the substrate is exposed to precursor molecules (precursor 1) which adsorb ideally as a monolayer on the surface. In step b) the excess of pre- cursor 1 in the gas phase is removed by inert gas purging. In step c) the substrate is exposed to precursor 2 which reacts with the adsorbed precursor 1 to form a layer of the desired material. In step d) the excess of precursor 2 and the reaction by-products are removed by purging. This cycle is repeated until the desired thickness of the deposit is obtained. Reprinted from [83]. Copyright WILEY-VCH Verlag GmbH & Co. KGaA, Weinheim, 2007.

Atomic layer deposition is particularly attractive [83], because it is based on successive, self-limiting reactions from the gas phase to produce thin film and overlayers on complex surface geometries, where the thickness is controlled in the nanometer range with perfect conformality by the number of ALD reaction cycles (Figure 6). Upon deposition by ALD aluminum oxide, nonwetting hydrophobic polypropylene fibers transition to either a metastable hydrophobic or a fully wetting hydrophilic state, consistent with well-known Cassie-Baxter and Wenzel models of surface wetting of roughened surfaces [84]. The as-formed Al_2O_3 coating further provides active sites for covalently grafting organic ligands such NH_2, COOH with high surface density, making inert polypropylene fibers chemically responsive [85]. Now, ALD is already the method-of-choice in the semiconductor industry, but for polypropylene nonwoven industry, there is some distance to reach the commercial stage.

Chemical Treatment

Chemical modification of polypropylene fabrics includes chlorination, bromination, nitration, sulfonation and phosphonylation [86-88]. Among them, sulfonation is the most commonly studied, for polypropylene fabrics with sulfonic acid groups will find applications not only in routine ion-exchange and adsorption applications, but also in advanced technical fields such as a solid support for catalysts [89,90]. Different sulfonation reactants like fuming H_2SO_4, gaseous SO_3 and chlorosulfonic acid have been investigated. Molecular orbital calculations suggest that the sulfonation proceeds via electrophilic addition of SO_3 to the side CH_3 group rather than the main chain of PP, resulting from 'the conformational change restricted selectivity (CCRS)' [91]. However, a more recent detailed study reveals the presence of a complex mechanism involving sulfonation, oxidation, decomposition, and aromatization to form carbonaceous structures [92].

To further increase the density of the sulfonic acid groups, two other approaches can be used by radiation-chemical technologies: one is grafting of glycidyl methacrylate (GMA) [93] or styrene [89, 94] as a precursor-monomer with subsequent sulfonation of grafted chains; the other is single-step grafting of a monomer with a sulfonic acid functional group. Nevertheless, it is unable to directly graft sodium styrene sulfonate monomers in aqueous solutions, only copolymerization with acrylic acid can we get desired results [95].

Surface Grafting Polymerization

Graft polymerization is a universal modification method to fabricate 'tailored' fabric surface with desired functions. Techniques including ultraviolet (UV), plasma, radiation, chemical and biocatalysis have been presented.

Ultraviolet-Induced Grafting Polymerization

Surface graft polymerization induced by UV irradiation exhibits some advantages, e.g., fast reaction rate, low cost of processing, simple equipment, easy industrialization, and versatile for various vinyl monomers with desirable functionality [96]. Benzophenone (BP) is the most widely used photoinitiator for the modification of polypropylene fabrics. The UV-excited triplet states of BP facilitate hydrogen abstraction to form tertiary carbon radicals and benzpinacol radicals on the fabric surface, thus graft polymerization can be initiated without prior modification. Different initiating types have been reported, which include BP soaking, adsorption, entrapment, and sequential living grafting [97]. Intriguingly, a dynamic UV process was developed where the initiator and the monomer solution were circulated separately by a peristaltic pump [98]. This method can be used to modify places difficult to be irradiated directly like the inner part of melt-blown polypropylene filter element.

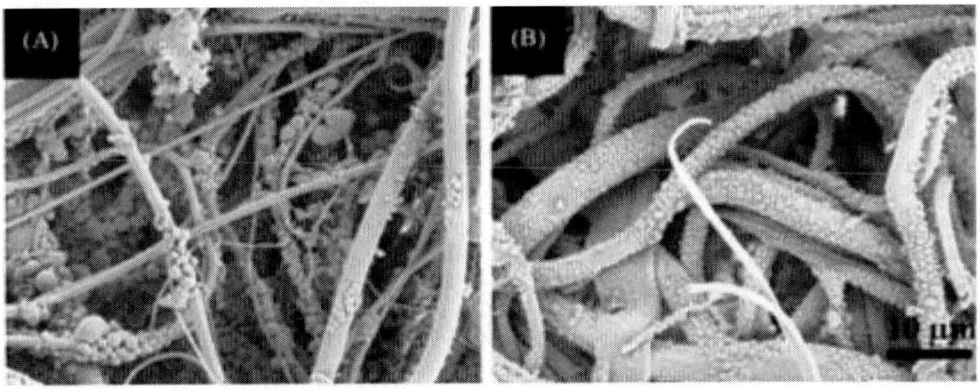

Figure 7. Grafting with 20% GMA solution for 30 min (A) without pretreatment; (B) with UV pretreatment of 15 min. Reprinted from [100]. Copyright Elsevier, 2010.

One important issue has to take into consideration is the spatial grafting uniformity for the UV grafting onto porous and thick polypropylene nonwoven mesh, because UV light decays continuously as passing through the nonwoven mesh [99]. Nonetheless, UV pretreatment, which can significantly improve the adsorption of benzophenone to the fiber surface, results in uniform and conformal grafts on the fiber surfaces [100](Figure 7). The spatial variation effects can be minimized further by careful optimization of the mesh thickness, the UV light intensity, and the reaction times as well as the initiator concentration.

Continuous UV grafting under nitrogen atmosphere or in the absence of inert gases were conducted to simulate the commercial textile manufacturing process by combination of previous mentioned 'pad-dry' method [101]. For example, polypropylene nonwoven was impregnated with a grafting solution of HEMA, BP, and 1 g/L Triton X-100 and the padded mesh was subsequently squeezed to have 89±0.5 % wet pick up. Continuous UV irradiation was then carried out. However, homopolymer formation was inevitable and grafting efficiency was 50~60% when grafting yield exceeded 10%. The grafting efficiency could be increased to 90% or greater by the addition of trimethylolpropane triacrylate (TMPTA, cross-linker) and mohr's salt (Fe^{2+}, free radical trapper) to monomer solution. This improved process provides huge industrial potential [102].

Plasma-Induced Grafting Polymerization

Plasma can be considered to be a gaseous mixture of oppositely-charged particles with a roughly zero net electrical charge. It is estimated that 99% of the matter in the observable universe is in the plasma state [103]. Polypropylene fabrics were modified by low-pressure/atmospheric pressure, non-equilbrium plasma (cold plasma) with various gases such as oxygen [104], argon [105], nitrogen [77], and helium [106]. Development of dielectric-barrier discharge [107] makes the application of plasma modification in a large industrial scale possible, due to its simple device and very short treatment time (~ 1 s), [108]. Surface modification by plasma can be mainly divided into two classes: plasma treatment and plasma-induced grafting polymerization.

Plasma treatment introduces polar groups such as C-O, C=O and O-C=O onto the fiber surface and usually results in increased surface waviness or roughness (Figure 8), both of them contribute to the improved hydrophilicity [106]. Conversely, highly hydrophobic surface can be obtained by using CF_4 gas to form a thin fluorinated layer [109]. However, there is still one major drawback termed aging or hydrophobic recovery, and the plasma treated surfaces can only preserve their hydrophilicity for about one month [110]. This behavior is believed to be caused by the re-orientation of polar groups from the top surface into the bulk of polypropylene [111]. In contrast, plasma-induced grafting polymerization can endow the fabrics with a permanent effect [112].

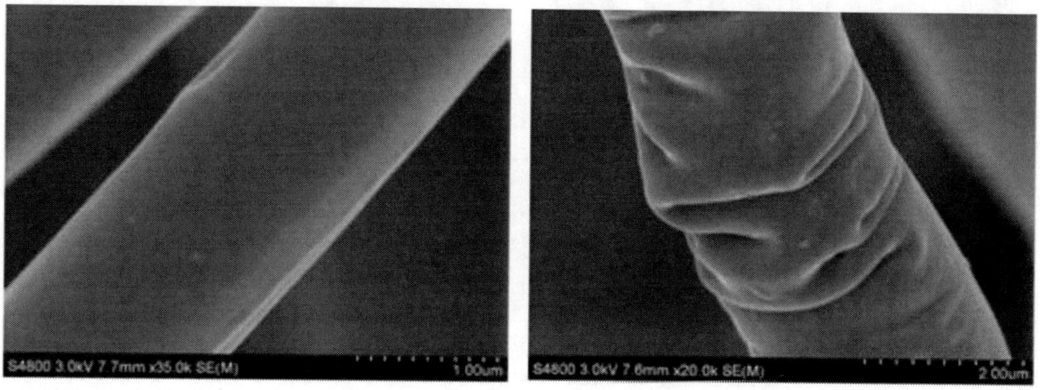

Figure 8. Surface morphology of untreated (Left) and plasma-treated (right) polypropylene nonwoven fabric.

Plasma treatment generates a large number of peroxide on the material surfaces, and the maximum concentration can reach about 6.0×10^{-7} mol/cm^2 (determined by the 1,1-diphenyl-2-picrylhydrazyl (DPPH) method with good accuracy) [112]. Long plasma exposure time results in decomposition (etching effect) and cross-linking reaction on the polypropylene fiber surface which would decrease the grafting degree. The peroxide group can then initiate the graft polymerization of various functional monomers like acrylic acid, glycidyl methacrylate, GMA-IDA and zwitterions [113-116]. In particular, in the plasma-grafting of acrylic acid, XPS analysis suggests the grafting process may follow the concept of 'front mechanism', which means the initial grafting takes place on the surface and then moves further into the surface by progressive diffusion of monomer deep across the initially grafted layer [117].

High-Energy Radiation-Induced Grafting Polymerization

At present, electron beam (EB) and γ-ray radiation are mostly used high-energy radiations, however, caution should be taken for both of them are harmful, even in low amounts. Generally, high-energy radiation-induced grafting polymerization for polypropylene fabrics proceeds in two ways: pre-irradiation and mutual irradiation technique. It has been found that the free radicals and peroxides created by irradiation can be kept for a long time, depending on the storage condition. No statistical variation of grating activity was observed when the pre-irradiated polypropylene fabrics were storage at -130°C even after 30 days [118]. The absorbed irradiation dose is the most prominent factor, typically mutual irradiation technique has much higher grafting efficiency over the pre-irradiation one [119].

Glycidyl methacrylate (GMA) and acrylonitrile (AN) are the most common monomers used for the irradiation grafting modification of polypropylene nonwoven mesh. The epoxide group of GMA can further react with different reactants like phosphoric acid [120], cyclodextrin [121], amines [122] and dipyridylamine [123] (PDA, a bidentate ligand with high metal ions coordination capacity) to achieve desired functionalities, while the nitrile group of AN is usually conversed to amidoxime group for metal recovery with high sorption rate and capacity [124-127]. It should be especially noted that in the case of radiation grafting of styrene methanol solution, the grafting degree increased sharply to a peak with monomer concentration due to a trommsdorff effect brought by the insolubility of the growing polystyrene chains in methanol [128].

Chemical Grafting Polymerization

In chemical grafting polymerization, the initiator plays a very important role as it determines the path of the grafting process. Comparison was conducted on the grafting efficiency with water-soluble, oil-soluble, and redox initiators for polypropylene fiber. Results demonstrated the oil-soluble one might be more suitable [129]. Benzoyl peroxide (BPO) has been widely used, however, homopolymer formation is usually serious [130]. To overcome this problem, BPO-coated [131] and BPO pre-thermal treated methods [132] are presented making it possible to obtain high grafting degrees with a minimal quantity of homopolymer as a side product.

The development of surface-initiated atom transfer radical polymerization (SI-ATRP) makes the grafted polymer with well define chain length and chain density possible. C-Br can be introduced onto the polypropylene surface by directly UV irradiation of 2-benzophenonyl bromoisobutyrate (BP derivative) deposited sample [133]. A two-step procedure was also proposed: first the hydroxyl groups were introduced on the polypropylene nonwoven mesh by ozone treatment and deoxidizing by potassium iodide (KI); then the hydroxyl groups were reacted with 2-bromoisobutyryl bromide (BIBB) to obtain the alkyl bromide-functionalized surface [134]. Poly(N-vinyl-2-pyrrolidone) (PVP) can be successfully grafted in a controlled manner and the modified mesh has high permeation fluxes, low flux loss, protein resistance and depressed bacterial adhesion.

Biocatalysis

A quite new method is grafting modification with enzymatic biocatalysis. Firstly, polypropylene nonwoven mesh was pre-modified with 2-aminoethyl-methylacrylate (AEMA) by plasma treatment. Then laccase was used to catalyze the covalent coupling reaction between guaiacol sulfonic acid (GSA) and primary amine [135]. Successful grafting of phenolic compound onto the surface was approved by HPLC-MS. The advantage of biocatalysis technology is that the covalent binding of functional groups can be suitably tailored by enzymes possessing substrate specificity and regional selectivity.

Other Methods

Besides the previous mentioned modification methods, ozone treatment and corona charging have also been used. Ozone can be readily produced in laboratory by commercial ozone generator. Ozone pretreatment generates carbonyl and carboxyl groups on the polypropylene nonwoven mesh surface, which facilitates the next chitosan dip-coating step and increase the structural stability of the composite mesh [136]. Corona discharge on meltblown nonwoven polypropylene increases the surface charge density and improve its filtration efficiency of indoor air [137]. Surface potential decay measurement suggests that the use of a triode-type electrode arrangement may improve the uniformity of corona charging in the case of nonwoven media [138].

In summary, different modification methods for polypropylene fabrics have been introduced above, and the selection of optimal process is based on the cost consideration and end-use applications.

APPLICATIONS OF POLYPROPYLENE FABRICS

It is widely acknowledged that the structure and property determine the application in the field of material science. The unique structure and the easily modulated surface functionality by modification make polypropylene fabrics suitable candidates for various applications.

Composite reinforcement

One of the most important applications of polypropylene fiber is to be used as reinforcements in composite developments including concrete [139], soil [140], and mortar [141]. The interfacial bonding between the fiber and matrix has a key effect on the mechanical performance of the composites, and it is found that both oxyfluorination [142] and γ-ray radiation treatment [143] result in improved mechanical properties. Polypropylene nonwoven mesh can also be employed as reinforcement for hydrogel material, which shows an increase of 800% in tensile strength, without changes in the swelling and cytotoxicity [144].

Protective Cloth

Polypropylene nonwoven mesh can be used in three types of sanitary protection products: feminine sanitary napkin, spunbond-meltblown diaper top sheet, and disposable adult incontinence absorbent products. By simply depositing a thin layer of carbon or aluminum on the mesh surface as mentioned above, it possess electromagnetic field shielding and thermal insulation properties accordingly [73,80]. It can also be used as a mask or ventilation/air-conditioning filter media for its high dust particle trap efficiency [137]. However, for aerosol particles whose sizes lie between 100 and 500 nm (most penetrating particle size), there exists a local minimum collection efficiency for commercial melt-blown polypropylene mesh.

Theoretical calculations predict that this efficiency may be significantly increased using nanofibrous media [145].

For agricultural workers, obviously, the most important feature in protective clothing material is the effectiveness of material as a barrier to pesticides, while it should be thermal comfort. Melt electrospun polypropylene mesh exhibits a combination of high protection performance and an acceptable level of air/vapor transport properties [146]. Further, mesh spun by solution with thin diameter shows superhydrophobicity and enhanced barrier performance even towards pesticide mixtures with low surface tension without losing permeability [54]. In short, for protective cloth application, it is always 'the thinner the better'. Therefore, it has great market prospect to figure out an optimal process for the mass-production of polypropylene nanofibers.

Wastewater Treatment

Polypropylene nonwoven fabrics are especially appropriate for wastewater treatment due to the extraordinary characteristics such as random network of overlapped fibers, multiple connected pores, and overall high porosity. Polypropylene nonwoven mesh is hydrophobic and show high affinity to oil substance. For this reason, unmodified nonwoven mesh can be directly packaged in parallel as a sorption bed and efficiently absorb oil from oil-in-water emulsion [147].

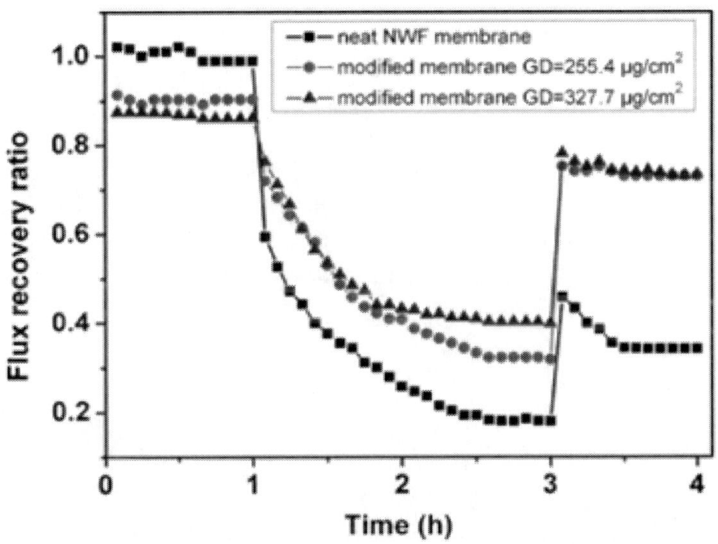

Figure 9. Time-dependent flux recovery ratio (Frr) of neat and modified NWF membranes during the BSA solution microfiltration process. Reprinted from [116]. Copyright Elsevier, 2011.

Typically, in wastewater treatment, nonwoven mesh was predominantly used as a prefilter and the cost is about 50% lower than conventional pretreatment method [36]. It is considered suitable for use in membrane bioreactor (MBR) process for municipal wastewater treatment [148]. For example, melt-blown polypropylene nonwoven mesh can reduce COD, NH_3-N values of reclaimed water by over 80% and 95% [149]. However, membrane fouling is a serious problem for polypropylene nonwoven mesh because of the hydrophobic interaction between bacteria, proteins and the membrane surface or the deposition of suspended floc particles of activated sludge [150]. Membrane fouling causes a large energy use, a high cleaning frequency and a short life span of the membrane. As a result, modification of polypropylene fabrics is very necessary. Surface modification with hydrophilic polymers is regard as an efficient way for reducing fouling and improving biocompatibility. To this end, conventional water-soluble polymers including PVA, PAA, chitosan, PHEMA, PVP have been used as mentioned above [70-74, 101, 102, 113, 116, 117, 134]. More recently, zwitterionic materials have drawn great attention for its outstanding antifouling performance. For example, [3-(methacryloylamino)propyl]-dimethyl(3-sulfo-propyl) ammonium hydroxide (MPDSAH) was grafted onto polypropylene mesh via plasma pretreatment and UV irradiation [116]. Water flux recovery ratio can be achieved as high as 90% for the modified membrane with respect to only about 30% for the neat NWF membrane (Figure 9). Catalytic MBR was fabricated by depositing TiO_2 layer [151] or grafting metal phthalocyanine derivative [152] for the degradation of organic pollutants in the wastewater. Or some phenolic organic pollutants were adsorbed by cyclodextrins-grafted mesh through host guest inclusion interactions [153]. Further, enzymes such as lipase [154] and urease [155] were also immobilized on the mesh for simultaneous products synthesis and separation.

Heavy metal ion like mercury is highly toxic and the drinking water criterion for mercury established by USEPA is 2 μg/L. Polypropylene mesh with aminated chelating groups shows very high adsorption capacity for mercury ions (the equilibrium adsorption amount could be as high as 657.9 mg/g), which results from the large surface area of mesh and the strong chelating interaction between amine groups and mercury ions. The residual concentration can achieve less than 1 μg/L for feed concentration of mercury lower than 1 mg/L after absorption, indicating the great application potential in drinking water purification [156]. However, most of the published methods for removal of heavy metals ions depend on electrostatic interaction, which lacks specificity [157]. A novel Cu^{2+}-imprinted polypropylene fiber was prepared by cross-linking polyethylenimine (PEI) pre-grafted fiber in the present of Cu^{2+}. Besides having a high adsorption capacity for Cu^{2+} (120mg/g), the Cu^{2+}-imprinted polypropylene fiber shows a high selectivity towards Cu^{2+}. The selectivity coefficient of Cu^{2+}-imprinted fiber for Cu^{2+} with respect to Zn^{2+} could reach 76.4, the capacity and selectivity can be well maintained after 10 regeneration cycles [158].

Bioseparation

Membrane chromatography is increasingly being used for protein bioseparation. For polypropylene nonwoven mesh modified with ligands and spacers [100], the porous structure and improved hydrophilicity make it possible to achieve high binding capacities without severely increasing the pressure drop by stacking multiple layers. It also helps create sharp chromatographic elution profiles [159]. Further attempts to improve the binding capacity led to the creation of polypropylene nonwoven mesh supported/ macroporous gel based 3D binding domain composite membrane [160] (Figure 10). The as-prepared gel-filled membrane was used for the separation of human plasma proteins HSA and HIgG, and the equilibrium HSA binding capacity was in the range of 290–300 mg/mL under optimum conditions. This value is significantly higher than other chromatographic membranes, even higher than conventional gel based chromatographic media. Therefore, polypropylene nonwoven mesh is well suitable for affinity membrane application.

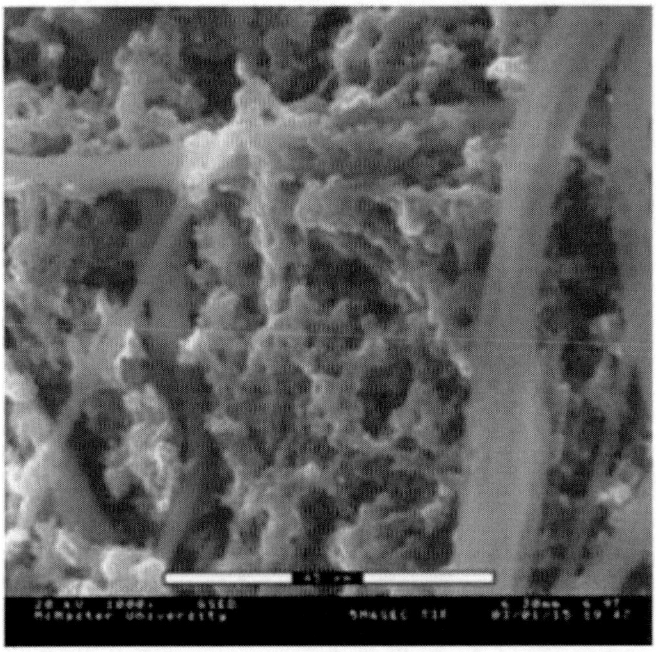

Figure 10. ESEM picture of macroporous gel-filled membrane. Reprinted from [160]. Copyright Elsevier, 2007.

Biomedical Textile

Polypropylene nonwoven mesh can be used as wounding dressing for burn injury, but the infection and colonisation problem caused by bacteria must be taken into consideration. Therefore, various approaches have been developed to endow the mesh with antibacterial activity. Silver nanoparticle were readily formed on the polypropylene fiber surface with Ag^+ chelating groups by reduction with UV light and formaldehyde solution [115]. However, such method may have leaking problem and run the risk of the development of bacterial resistance. An alternative strategy is to employ polycations with intrinsic antibacterial activity to create permanent antibacterial surfaces including polypyridinium salt [72], polyquaternary ammonium salt [161], and chitosan. Chitosan is currently the most widely used antibactrial material for polypropylene nonwoven mesh. It could be easily coated on the mesh surface by pad-dry process [70] or grafted via coupling reaction between the amino group of chitosan and pre-modified carboxyl group [162]. Another feature is the NH^{3+} sites on the immobilized chitosan exhibit biofunctional in anticoagulation and in antibacterial property [113]. Collagen was also incorporated to improve the biocompatibility for it is responsible for the functional integrity of skins [130,163-164]. A novel auto-stripped tri-layer wound dressing was fabricated using thermosensitive poly(N-isopropylacrylamide) as the second layer [165,166]. When the wound dressing needs removed, just let the temperature of the substrate below the

LCST to make PNIPAAm swelling and then it would be easily stripped-off from the wound surface (Figure 11).

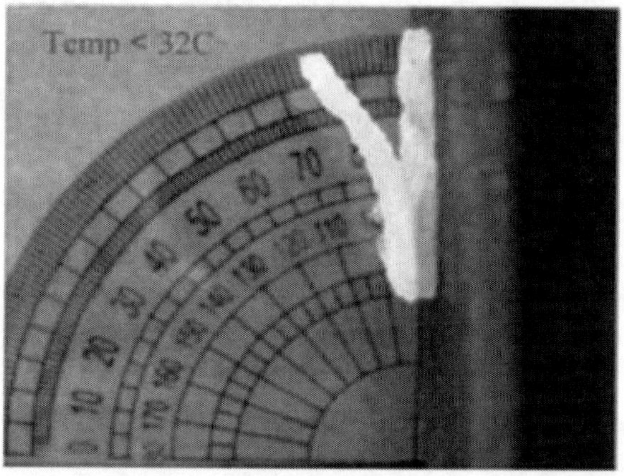

Figure 11. Photo of auto stripped-off behavior of tri-layer wound dressing. Reprinted from [166]. Copyright Elsevier, 2005.

Conclusions

The past 50 year has witnessed tremendous achievements in the area of polypropylene fabrics industry. With years of concerted efforts, it has become a global industry and various theoretical or empirical models have been proposed for practical guidance. Detailed studies of the effects of modification methods and operation parameters on the functionalization of polypropylene fabrics are conducted to improve their in-service performance and satisfy mounting demand. Herein, we provide a simple review of the preparation, modification methods of polypropylene fabrics and their application in various fields.

Melt spinning, spun-bond and melt-blown technologies have already been commercialized for the preparation of polypropylene fabrics, however, the diameters of as-prepared fibers are usually about 10 to 20 μm or more, which greatly limit their performance. Novel methods, such as phase separation, template synthesis and electrospinning, have been developed for fabricating polypropylene nanofibers in laboratory scale. Some preliminary successes have been achieved so far, yet there is still a long way to accomplish industrialization. It is of great promise to modify the melt-blown device into melt-blown/melt electrospinning combination machine for mass production of polypropylene nanofibrous mesh.

Different modification processes have been demonstrated to modulate the surface hydrophilicity or further achieve surface functionalization for high-performance utilization of polypropylene fabrics. These processes include *in-situ* modification, physical treatment, chemical treatment, surface grafting polymerization, biocatalysis and so on. However, more efforts need to make to incorporate modification process with well defined functionality into the productive line.

A primary concern in the near future is to extend the application scope of polypropylene fabrics to find new market rising points. With the mutual involvement of a bigger academic and engineering community, we can anticipate even great development in polypropylene fabrics industry and there is plenty of room!

ACKNOWLEDGMENTS

Financial supports are acknowledged to the National Natural Science Foundation of China in a grant number 50933006 and Zhejiang Provincial Innovative Research Team in a grant number 2009R50004.

REFERENCES

[1] Z. M. Huang, Y. Z. Zhang, M. Kotaki, M., S. Ramakrishna, *Compos. Sci. Technol.,* 63, 2223 (2003).
[2] D. Rovere, R. L. Shambaugh, E. A. Orear, *J. Appl. Polym. Sci.,* 77, 1921 (2000).
[3] H. H. George, *ACS. Symp. Ser.*, 260, 355(1984).
[4] Suzuki, S. Narusue, *J. Appl. Polym. Sci.*, 92, 1534 (2004).
[5] N. Ogata, S. Yamaguchi, N. Shimada, G. Lu, T. Iwata, K. Nakane, T. Ogihara, *J. Appl. Polym. Sci.*, 104, 1640 (2007).
[6] Suzuki, S. Narusue, *J. Appl. Polym. Sci.*, 99, 27 (2006).
[7] J. Broda, J. *Polymer.*, 44, 1619 (2003).
[8] C. S. Yoo, D. S. Ji, *Fiber. Polym.*, 4, 210 (2003).
[9] Vroman, S. Giraud, F. Salaun, F., S. Bourbigot, *Polym. Degrad. Stab.*, 95, 1716 (2010).
[10] G. Smart, B. K. Kandola, A. R. Horrocks, S. Nazare, D. Marney, *Polym. Adv. Technol.*, 19, 658 (2008).
[11] Z. Strecka, A. Ujhelyiova, E. Bolhova, P. Alexy, E. Borsing, *J. Text. I.*, 101, 315 (2010).
[12] Ujhelyiova, Z. Strecka, E. Bolhova, M. Dulikova, P. Bugaj, *Fibers. Text. East. Eur.*, 15, 37 (2007).
[13] M. Kristofic, I. Vassova, A. Ujhelyiova, J. Ryba, *Fibers. Text. East. Eur.*, 19, 18 (2011).
[14] D. G. Svetec, *J. Appl. Polym. Sci.*, 100, 1067 (2006).
[15] C. S. Li, T. X. Liang, W. Z. Lu, C. H. Tang, X. Q. Hu, M. S. Cao, J. Liang, *Compos. Sci. Technol.*, 64, 2089 (2004).
[16] S. Y. Yeo, S. H. Jeong, *Polym. Int.*, 52, 1053 (2003).
[17] M. R. Badrossamay, G. Sun, *Polym. Int.*, 59, 155 (2010).
[18] D. Shi, J. H. Yang, Z. H. Yao, Y. Wang, H. L. Huang, W. Jing, J. H. Yin, G. Costa, *Polymer.*, 42, 5549 (2001).
[19] Yeo, B. Hong, H. Byun, *J. Ind. Eng. Chem*, 11, 732 (2005).
[20] Hautojarvi, S. Laaksonen, *Textile. Res. J.*, 70, 391 (2000).
[21] X. Yi, L. Xu, Y. L. Wang, G. J. Zhong, X. Ji, Z. M. Li, *Eur. Polym. J.*, 46, 719 (2010).
[22] D. Wang, G. Sun, B. S. Chiou, J. P. Hinestroz, *Polym. Eng. Sci.*, 47, 1865 (2007).
[23] C. H. Xue D. Wang, B. Xiang, B. S. Chiou, G. Sun, *J. Polym. Res.*, 18, 1947 (2011).

[24] C. H. Xue, D. Wang, B. Xiang, B. S. Chiou, G. Sun, *J. Polym. Sci., Part B: Polym. Phys.*, 48, 921 (2010).
[25] D. Wang, G. Sun, *Eur. Polym. J.*, 43, 3587 (2007).
[26] D. Wang, G. Sun, B. S. Chiou, *Macromol. Mater. Eng.*, 292, 407 (2007).
[27] W. S. Lyoo, Y. G. Choi, J. H. Choi, W. S. Ha, B. C. Kim, *Int. Polym. Proceess.*, 15, 369 (2000).
[28] E. Fallahi, M. Barmar, M. H. Kish, M.H. *J. Appl. Polym. Sci.*, 108, 1473 (2008).
[29] P. Liu, Y. Ouyang, R. Xiao, *J. Appl. Polym. Sci.*, 123, 2859 (2012).
[30] Z. Y. Yang, J. G. C. Veinot, *J. Mater. Chem.*, 21, 16505 (2011).
[31] H. Y. Li, Y. C. Ke, Y. L. Hu, *J. Appl. Polym. Sci.*, 99, 1018 (2006).
[32] C. Majidi, R. E. Groff, Y. Maeno, B. Schubert, S. Baek, B. Bush, R. Maboudian, N. Gravish, M. Wilkinson, K. Autumn, R. S. Fearing, *Phys. Rev. Lett.*, 97, 076103 (2006).
[33] J. Lee, R. S. Fearing, K. Komvopoulos, *Appl. Phys. Lett.*, 93, 191910 (2008).
[34] J. Lee, R. S. Fearing, *Langmuir.*, 24, 10587 (2008).
[35] J. Lee, C. Majidi, B. Schubert, R. S. Fearing, *J. R. Soc. Interface.*, 5, 835 (2008).
[36] S. Kaur, R. Gopal, W. J. Ng, S. Ramakrishna, T. Matsuura, *MRS. Bull.*, 33, 21 (2008).
[37] R. Nanjundappa, G. S. Bhat, *J. Appl. Polym. Sci.*, 98, 2355 (2005).
[38] G. S. Bhat, P. K. Jangala, J. E. Spruiell, *J. Appl. Polym. Sci.*, 92, 3593 (2004).
[39] V. Caldas, G. R. Brown, R. S. Nohr, J. G. Macdonald, L. E. Raboin, *J. Appl. Polym. Sci.*, 65, 1759 (1997).
[40] D. Rovere, R. L. Shambaugh, *Ind. Eng. Chem. Res.*, 40, 176 (2001).
[41] J. Ellison, A. Phatak, D. W. Giles, C. W. Macosko, F. S. Bates, *Polymer.*, 48, 3306 (2007).
[42] R. Zhao, L. C. Wadsworth, *Polym. Eng. Sci.*, 43, 463 (2003).
[43] R. Zhao, L. C. Wadsworth, C. Sun, D. Zhang, *Polym. Int.*, 52, 133 (2003).
[44] C. Dutton, *Scholarly JTATM.*, 6, 1 (2008).
[45] D. Zhang, C. Sun, L. C. Wadsworth, *Textile. RES. J.*, 71, 301 (2001).
[46] Jarecki, A. Ziabicki, *Fibers. Text. East. Eur.*, 16, 17 (2008).
[47] L. Jarecki, A. Ziabicki, Z. Lewandowski, A. Blim, *J. Appl. Polym. Sci.*, 119, 53 (2011).
[48] V. Bansal, R. L. Shambaugh, *Ind. Eng. Chem. Res.*, 37, 1799 (1998).
[49] D. Li, Y. N. Xia, *Adv. Mater.*, 16, 1151 (2004).
[50] S. R. Givens, K. H. Gardner, J. F. Rabolt, D. B. Chase, 40, 608 (2007).
[51] K. H. Lee, O. Ohsawa, K. Watanabe, I. S. Kim, S. R. Givens, D. B. Chase, J. F. Rabolt, *Macromolecules.*, 42, 5215 (2009).
[52] K. Watanabe, B. S. Kim, Y. J. Enomoto, I. S. Kim, *Macromol. Mater. Eng.*, 296, 568 (2011).
[53] C. Wang, T. C. Hsieh, Y. W. Cheng, *Macromolecules.*, 43, 9022 (2010).
[54] D. W. Cho, H. J. Zhou, Y. J. Cho, D. Audus, Y. L. Joo, *Polymer.*, 51, 6005 (2010).
[55] J. Lyons, J. F. Ko, *Polym. News.*, 30, 1 (2005).
[56] D. H. Reneker, A. L. Yarin, *Polymer.*, 49, 2387 (2008).
[57] Gora, R. Sahay, V. Thavasi, S. Ramakrishna, *Polym. Rev.*, 51, 265 (2011).
[58] C. L. Norton, *US. patent.* 2048651
[59] L. Larrondo, R. S. J. Manley, *J. Polym. Sci., Part B: Polym. Phys.*, 19, 909 (1981).
[60] L. Larrondo, R. S. J. Manley, *J. Polym. Sci., Part B: Polym. Phys.*, 19, 921 (1981).
[61] L. Larrondo, R. S. J. Manley, *J. Polym. Sci., Part B: Polym. Phys.*, 19, 933 (1981).
[62] D. W. Hutmacher, P. D. Dalton, *Chem. Asian J.*, 6, 44 (2011).

[63] C. Erisken, D. M. Kalyon, H. J. Wang, *Nanotechnology.*, 19, 165302 (2008).
[64] J. Lyons, C. Li, F. ko, *Polymer.*, 45, 7597 (2004).
[65] Y. Kadomae, Y. Maruyama, M. Sugimoto, T. Taniguchi, K. Koyama, *Fiber. Polym.*, 10, 275 (2009).
[66] R. Rangkupan, D. H. Reneker, *J. Met. Mat. Miner.*, 12, 81 (2003).
[67] C. S. Kong, K. J. Jo, N. K. Jo, H. S. Kim, *Polym. Eng. Sci.*, 49, 391 (2009).
[68] P. D. Dalton, D. Grafahrend, K. Klinkhammer, D. Klee, M. Moller, *Polymer.*, 48, 6823 (2007).
[69] Z. K. Xu, X. J. Huang, L. S. Wan, Surface Engineering of Polymer Membranes. Zhejiang university press and Springer. (2009).
[70] Y. S. Shin, D. I. Yoo, K. Min, *J. Appl. Polym. Sci.*, 74, 2911 (1999).
[71] C. H. Zhang, F. L. Yang, W. J. Wang, B. Chen, *Sep. Purif. Technol.*, 61, 276 (2008).
[72] C. C. Wang, F. L. Yang, L. F. Liu, Z. M. Fu, Y. Xue, *J. Membr. Sci.*, 345, 223 (2009).
[73] H. J. Wu, J. T. Fan, X. H. Qin, S. Mo, Hinestroza. *J. Appl. Polym. Sci.*, 110, 2525 (2008).
[74] K. Desai, K. Kit, J. J. Li, P. M. Davidson, S. Zivanovic, H. Meyer, *Polymer.*, 50, 3661 (2009).
[75] D. Stawski, C. Bellmann, *Colloids. Surf. A. Phys. Eng. Aspects.*, 345, 191 (2009).
[76] J. Fang, A. Kelarakis, L. Estevez, Y. Wang, R. Rodriguez, E. P. Giannelis, *J. Mater. Chem.*, 20, 1651 (2010).
[77] R. Szabova, L. Cernakova, M. Wolfova, M. Cernak, *Acta. Chim. Slov.*, 2, 70 (2009).
[78] Stefecka, M. Kando, H. Matsuo, Y. Nakashima, M. Koyanagi, T. Kamiya, M. Cernak, *J. Mater. Sci.*, 39, 2215 (2004).
[79] Q. F. Wei, X. L. Xiao, D. Y. Hou, H. Ye, F. L. Huang, *Surf. Coat. Technol.*, 202, 2535 (2008).
[80] M. Jaroszewski, P. J. Ziaia, *J. Non-Cryst. Solids.*, 356, 625 (2010).
[81] Q. F. Wei, L.Y. Yu, D. Y. Hou, F. L. Huang, *J. Appl. Polym. Sci.*, 107, 132 (2008).
[82] Poulter, X. N. Berbel, A. L. Johnson, A. J. Dowling, N. Waterfield, A. T. Jenkins, *Chem. Comm.*, 45, 7312 (2009).
[83] M. Knez, K. Nielsch, L. Niinisto, *Adv. Mater.*, 19, 3425 (2007).
[84] G. K. Hyde, G. Scarel, J. Spagnola, Q. Peng, K. Lee, G. N. Parsons, *Langmuir.*, 26, 2550 (2010).
[85] Peng, B. Gong, G. N. Parsons, *Nanotechnology.*, 22, 155601 (2011).
[86] C. D. Shan, D. K. Jain, *Textile. Res. J.*, 54, 742 (1984).
[87] J. M. Bouknight, P. A. Depalma, J. R. Cobb, S. W. Shalaby, *J. Appl. Polym. Sci.*, 76, 93 (2000).
[88] H. Kim, J. S. Bae, *Fiber. Polym.*, 10, 320 (2009).
[89] H. K. Cho, J. S. Park, D. H. Han, I. Bondar, *Nucl. Instrum. Methods Phys. Res., Sect. B.*, 269, 636 (2011).
[90] K. Stakne, M. S. Smole, K. S. Kleinschek, *J. Mater. Sci.*, 38, 2167 (2003).
[91] H. Tada, S. Ito, *Langmuir.*, 13, 3982 (1997).
[92] J. P. Matos, M. T. C. Sansiviero, R. M. Lago, *J. Appl. Polym. Sci.*, 115, 3586 (2010).
[93] Y. Bondar, H. J. Kim, S. H. Yoo, Y. J. Lim, *React. Funct. Polym.*, 58, 43 (2004).
[94] Vahdat, H. Bahrami, N. Ansari, F. Ziaie, *Radiat. Phys. Chem.*, 76, 787 (2007).
[95] S. W. Lee, Y. Bondar, D. H. Han, *React. Funct. Polym.*, 68, 474 (2008).
[96] J. P. Deng, L. F. Wang, L. Y. Liu, W. T. Yang, *Prog. Polym. Sci.*, 34, 156 (2009).

[97] D. M. He, H. Susanto, M. Ulbricht, *Prog. Polym. Sci.*, 34, 62 (2009).
[98] L. L. Wu, G. X. Wu, S. G. Xu, H. Zhong, Y. J. Shen, H. Z. Liu, *J. Environ. Sci.*, 19, 1387 (2007).
[99] Y. Zheng, P. V. Gurgel, R. G. Carbonell, *Ind. Eng. Chem. Res.*, 50, 6115 (2011).
[100] Y. Zheng, Y. H. Liu, P. V. Gurgel, R. G. Carbonell, *J. Membr. Sci.*, 364, 362 (2010).
[101] J. H. Jang, W. S. Go, *Fiber. Polym.*, 9, 375 (2008).
[102] K. C. Na, H. J. Park, *J. Appl. Polym. Sci.*, 114, 387 (2009).
[103] F. S. Denes, S. Manolache, *Prog. Polym. Sci.*, 29, 815 (2004).
[104] W. T. Ren, C. Z. Cheng, R. M. Wang, X. Li, *J. Appl. Polym. Sci.*, 116, 2480 (2010).
[105] N. Yaman, E. Ozdogan, N. Seventekin, *J. Textile. Inst.*, 101, 746 (2010).
[106] Y. J. Hwang, M. G. Mccord, J. S. An, B. C. Kang, S. W. Park, *Textile Res. J.*, 75, 771 (2005).
[107] U. Kogelschatz, U. *Plasma. Chem. Plasma. Process.*, 23, 1 (2003).
[108] J. Rahel, M. Simor, M. Cernak, M. Stefecka, Y. Imahori, M. Kando, M. *Surf. Coat. Technol.*, 169, 604 (2003).
[109] Y. A. Kwon, *Fiber. Polym.*, 3, 174 (2002).
[110] E. Masaeli, M. Morshed, H. Tavanai, F. Ashrafizadeh, *Fiber. Polym.*, 9, 461 (2008).
[111] R. Morent, N. D. Geyter, C. Leys, L. Gengembre, E. Payen, *Textile Res. J.*, 77, 471 (2007).
[112] K. S. Chen, H. R. Lin, S. C. Chen, J. C. Tsai, Y. A. Ku, *Polym. J.*, 38, 905 (2006).
[113] J. D. Liao, S. P. Lin, Y. T. Wu, *Biomacromolecules.*, 6, 392 (2005).
[114] N. Yaman, E. Ozdogan, N. Seventekin, *Fiber. Polym.*, 12, 35 (2011).
[115] C. H. Tseng, C. C. Wang, C. Y. Chen, *J. Phys. Chem. B.*, 110, 4020 (2006).
[116] J. Zhao, Q. Shi, S. F. Luan, L. J. Song, H. W. Yang, X. L. Li, *J. Membr. Sci.*, 369, 5 (2011).
[117] S. Saxena, A. R. Ray, J. Mindemart, J. Hilborn, B. Gupta, *Plasma Process. Polym.*, 7, 610 (2010).
[118] J. S. Park, J. H. Kim, Y. C. Nho, O. H. Kwon, *J. Appl. Polym. Sci.*, 69, 2213 (1998).
[119] L. C. Lopergolo, L. H. Catalani, L. D. B. Machado, P. R. Rela, A. B. Lugao, *Radiat. Phys. Chem.*, 57, 451 (2000).
[120] S. H. Choi, Y. C. Nho, *Korean. J. Chem. Eng.*, 16, 241 (1999).
[121] P. L. Thuaut, B. Martel, G. Grini, U. Maschke, X. Cooueret, M. Morcellet, *J. Appl. Polym. Sci.*, 77, 2118-2125.
[122] Y. C. Nho, J. S. Spark, J. H. Jin, *J. Macro. Sci, Part A.*, 34, 831 (1997).
[123] M. Barsbay, P. A. Kavakli, O. Guven, *Phys. Chem.*, 79, 227 (2010).
[124] S. Das, A. K. Pandey, A. Athawale, V. Kumar, Y. K. Bhardwaj, S. Sabharwal, V. K. Manchanda, *Desalination.*, 232, 243 (2008).
[125] T. L. Prasad, P. K. Tewari, D. Sathiyamoorthy, *Ind. Eng. Chem. Res.*, 49, 6559 (2010).
[126] J. H. Zu, S. Q. He, Y. C. Nho, J. J. Pyo, F. Yan, *J. Appl. Polym. Sci.*, 116, 1414 (2010).
[127] J. H. Zu, Y. C. Nho, J. J. Pyo, R. Q. Liu, L. Liu, F. Yan, M. Xia, *J. Appl. Polym. Sci.*, 119, 3483 (2011).
[128] Y. C. Nho, J. Chen, J. H. Jin, (1999) *Radiat. Phys. Chem.*, 54, 317 (2011).
[129] W. Wang, L. Wang, C. L. Wang, T. X. Sun, H. J. Yu, T. Chen, *J. Appl. Polym. Sci.*, 99, 734 (2006).
[130] C. Wang, C. C. Chen, *J. Appl. Polym. Sci.*, 98, 391 (2005).
[131] Y. F. Duann, Y. C. Chen, J. T. Shen, Y. H. Lin, *Polymer.*, 45, 6839 (2004).

[132] J. Buchenska, *J. Appl. Polym. Sci.*, 83, 2295 (2002).
[133] J.Y. Huang, H. Murata, R. R. Koepsel, A. J. Russell, K. Matyjaszewski, *Biomacromolecules.*, 8, 1396 (2007).
[134] C. C. Wang,; R. R. Feng, F. L. Yang, *J. Colloid Interface Sci.*, 357, 273 (2011).
[135] M. Schroeder, E. Fatarella, J. Kovac, G. M. Guebitz, V. Kokol, *Biomacromolecules.*, 9, 2735 (2008).
[136] C. C. Wang, F. L. Yang, H. M. Zhang, *Sep. Purif. Tech.*, 75, 358 (2010).
[137] Z. Z. Yang, J. H. Lin, I. S. Tsai, T.Y. Kuo, *Textile Res. J.*, 72, 1099 (2002).
[138] B. Tabti, M. R. Mekidenche, M. C. Plopeanu, L. M. Dumitran, L. Henrous, L. Dascalescu, *IEEE. T. Ind. Appl.*, 46, 634 (2010).
[139] S. Cattaneo, L. Biolzi, *J. Mater. Civ. Eng.*, 22, 836 (2010).
[140] H. T. Jiang, Y. Cai, J. Liu, *J. Mater. Civ. Eng.*, 22, 1315 (2010).
[141] C. P. Ostertag, C. K. Yi, *Mater. Struc.*, 40, 679 (2007).
[142] L. Tu, D. Kruger, J. B. Wangener, P. A. B. Carstens, *J. Adhesion.*, 62, 187 (1997).
[143] G. M. Barrera, F. U. Nunez, O. Gencel, W. Brostow, *Composites. Part. A.*, 42, 567 (2011).
[144] L. C. Lopergolo, A. B. Lugao, L. H. Catalainni, *J. Appl. Polym. Sci.*, 86, 662 (2002).
[145] A. Podgorski, A. Balazy, L. Gradon, *Chem. Eng. Sci.*, 61, 6804 (2006).
[146] S. Lee, S. K. Obendorf, *J. Appl. Polym. Sci.*, 102, 3430 (2006).
[147] J.Kaluzka, M. Lebiedowski, *Fibers. Text. East. Eur.*, 11, 64 (2003).
[148] L. S. Chang, M. Gander, B. Jefferson, S. J. Judd, *Trans IChemE.*, 79, (2001).
[149] L. Liu, Z. H. Xu, C.Y. Song, Q. B. Gu, Y. M. Sang, G. L. Lu, H. L. Hu, F. S. Li, *Desalination.*, 201, 198 (2006).
[150] M. C. Chang, R. Y. Horng, H. Shao, Y. J. Hu, *Desalination.*, 191, 8 (2006).
[151] H. Han, R. B. Bai, *Sep. Purif. Tech.*, 73, 142 (2010).
[152] D. L. Cho, C. N. Choi, H. J. Kim, A. K. Kim, J. H. Go, *J. Appl. Polym. Sci.*, 82, 839 (2001).
[153] B. Martel, P. L. Thuaut, S. Bertini, G. Grini, M. BacQuet, G. Torri, M. Morcellet, *J. Appl. Polym. Sci.*, 85, 1771 (2002).
[154] Li, W.N.; Chen, B.Q. and Tan, T.W. (2011) Comparative study of the properties of lipase immobilized on nonwoven fabric membranes by six methods. *Process. Biochem.*, 46, 1358-1365.
[155] K. H. Yeon, R. M. Lueptow, *J. Chem. Technol. Biotechnol.*, 81, 940 (2006).
[156] N. F. Ma, Y. Yang, S. X. Chen, Q. K. Zhang, *J. Hazard. Mater.*, 171, 288 (2009).
[157] T. B. Mostafa, *J. Appl. Polym. Sci.*, 111, 11 (2009).
[158] T. Y. Li, S. X. Chen, H. C. Li, L. Wu, *Langmuir.*, 27, 6753 (2011).
[159] T. J. Menkhaus, H. Varadaraju, L. F. Zhang, S. Schneiderman, S. Bjustrom, L. Liu, H. Fong, *Chem. Commun.*, 46, 3720 (2010).
[160] D. M. Kanani, E. Komkova, T. Wong, A. Mika, R. F. Childs, R. Ghosh, *Biochem. Eng. J.*, 35, 295 (2007).
[161] S. Mosleh, S. M. Gawish, F. H. Khalil, R. F. Bieniek, *J. Appl. Polym. Sci.*, 98, 2373 (2005).
[162] J. M. Yang, H. T. Lin, T. H. Wu, C. C. Chen, *J. Appl. Polym. Sci.*, 90, 1331 (2003).
[163] C. C. Wang, J. P. Chen, C. C. Chen, *Mater. Sci. Eng., C.*, 29, 1133 (2009).
[164] C. C. Wang, C. H. Su, J. P Chen, C. C. Chen, *Mater. Sci. Eng., C.*, 29, 1715 (2009).

[165] F. H. Lin, J. C. Tsai, T. M. Chen, K. S. Chen, J. M. Yang, P. L. Kang, T. H. Wu, *Mater. Chem. Phys.*, 102, 152 (2007).
[166] K. S. Chen, Y. A. Ku, C. C. Lee, H. R. Lin, F. H. Lin, T. M. Chen, *Mater. Sci. Eng., C.*, 25, 472 (2005).

In: Polypropylene
Editors: L. P. Silva and E. F. Barbosa

ISBN: 978-1-62417-142-0

Chapter 3

CRYSTALLIZATION IN MISCIBLE AND IMMISCIBLE POLYMER BLENDS

Petr Svoboda

Centre of Polymer Systems, Department of Polymer Engineering,
Tomas Bata University in Zlin, Zlin, Czech Republic

ABSTRACT

Crystalline morphology influences greatly mechanical properties. Firstly the influence of crystallization temperature on elastic behavior of miscible polypropylene (PP)/hydrogenated polybutadiene (hBR) blends is discussed in terms of residual strain values after pre-defined stretching. The structure was directly observed in nano-scale by transmission electron microscopy (TEM) and quantitatively evaluated by differential scanning calorimetry (DSC). PP/hBR blends serve as a model for explanation of the elastic behaviour of polyolefinic TPE containing PP. There is a big difference in structure and properties for neat PP and PP/hBR blend. Regardless on crystallization temperature, PP shows always yielding point, i.e. plastic deformation after stretching even to small deformation. In neat PP, the crosshatch lamellar crystals were always formed. The situation for PP/hBR blend was completely different. The elastic properties are highly affected by crystallization temperature. The optimum in temperature of cooling medium was found to be around room temperature. In this case the formation of long crosshatch lamellae was prevented and the small fragmented crystal lamellae were formed. The fragmented crystallites play a role of tie points for amorphous chains to render good elastomeric behaviour. There is no yielding point in stress-strain curves for those blends. When the blend crystallizes at high temperature, big crosshatch lamellae were formed and the blend showed very poor elastic properties. Secondly, isothermal crystallization of immiscible PP/ ethylene–octene copolymer (EOC) blends was studied by DSC and optical microscopy at various temperatures (123-140°C). By DSC it was found that small amounts of EOC (10-30%) increase the bulk crystallization kinetics; in some cases even 80% increase was observed. Larger amounts of EOC (50-80%) have caused a decrease in crystallization kinetics. Optical microscopy has revealed that crystallization kinetics of a single spherulite is decreased by the presence of EOC, and at the same time there are more spherulites in the blend. EOC acts as a nucleation agent increasing the bulk crystallization rate and at the same time it acts as an obstacle to the growing front of a spherulite when the PP lamellae have to go around the EOC obstacle. At low PP content (20%) PP forms very small isolated particles with size smaller than 1 micrometer. At 40% of PP a structure resembling a co-continuous one was found. When the PP is a majority phase (20-40% of EOC) rather large EOC particles were found.

INTRODUCTION

One can prepare the polyolefinic thermoplastic elastomer (TPE) by dynamic vulcanization of the blend of polypropylene (PP) with ethylene-propylene rubber (EPDM); e.g. by melt-mixing a 40/60 PP/EPDM blend in the presence of curatives, such as sulphur, accelerators and peroxides [1, 2]. This process yields two-phase material in which cured EPDM particles with a diameter of a few μm are dispersed in PP matrix. There have been several approaches to explain the elastic behaviour of this blend. One was the analysis of deformation mechanism of the two-phase system by the finite element method (FEM) [3]. In this FEM analysis, the matrix was implicitly assumed to consist of neat PP. However, one has to be suspicious about it; i.e., the matrix might be a mixture of PP and EPDM. The background is as follows.

Most pairs of dissimilar polymer are immiscible at quiescent state. However, it was reported that at high shear rates the LCST (lower critical solution temperature) was shifted to higher temperatures and thus one-phase mixture was achieved at high shear rates [4-6]. In polyolefinic blends, Sano *et al.* found a regularly phase-separated structure with periodic distance of about 150 nm in injection-moulded blend of PP with high-density polyethylene (PP/HDPE 60/40). This structure can be achieved only by spinodal decomposition mechanism (under zero shear rate in mould) from a single-phase mixture attained (by UCST depression or LCST elevation) in high shear fields in an injection moulding machine (UCST – upper critical solution temperature) [7].

Yang *et al.* found that the matrix in commercial polyolefinic TPE, Santoprene®, is probably composed of PP/EPDM blend. It implies that the matrix is less ductile and more elastomeric than neat PP. In rubber particles the lamellae of PP were found. Thus, the partial miscibility between those two components seems to be caused by high shear field during the dynamic vulcanization [8].

Otsuka *et al.* found that hydrogenated poly(styrene-co-butadiene) is miscible with PP above the melting point of PP even at quiescent condition. This could be a good model blend for the matrix of commercial TPE. However the UCST was found at around 100°C. Then a two-phase structure was formed with periodic distance about 40 nm. This blend shows very good elastomeric properties and can be called as a new type of TPE. However, the structure is still rather complicated to understand deeply the elastic properties of polyolefinic TPE, because the crystallization is probably affected by spinodal decomposition [9, 10].

Recently it was found that PP is completely miscible with hydrogenated polybutadiene (hBR) above the melting point of PP and the UCST or LCST type of phase boundary does not exist. Then the situation is simpler; i.e., the single-phase melt can be crystallized without liquid-liquid phase separation and the crystallized blend could be a nice model system for the matrix of TPE. The aim of this research was at first to find the effect of hBR (high molecular weight impurity) on mechanical properties for blends crystallized at different crystallization temperatures and second to find the relationship between mechanical properties and the crystalline morphology. The elastic behaviour was tested at tensile test machine by measuring the residual strain after stretching to desired strain. For the morphology analysis, transmission electron microscopy (TEM), wide-angle X-ray diffraction (WAXD) and differential scanning calorimetry (DSC) were used.

Mechanical properties are closely related to the crystalline structure. For example Xu et al. found out that smaller size of the spherulites (in pure PP) may be responsible for the increase in impact strength [11]. Avella et al. discovered that the connectivity of the spherulitic boundaries is the major factor in controlling the fracture toughness [12]. Zhang et al. found out that decrease in spherulite size caused by addition of nucleating agent in PP/EOC blends has caused tremendous increase in impact strength and moderate increases in tensile strength and flexural strength [13]. All these examples show the importance of controlling the crystalline morphology in relationship with mechanical properties.

Polypropylene (PP) is one of the most versatile low cost commodity polymers. It has good chemical and moisture resistance, good ductility and stiffness and easy processability. Although PP has seen widespread application, its limited impact strength, especially at lower temperature due to its relatively high T_g is an obstacle to broader utilization as an engineering plastic. The impact properties of PP can be considerably improved by incorporation of a rubbery phase. Accordingly, rubber-toughened PP blends with various impact modifiers have been studied, including ethylene-propylene rubber (EPR), ethylene-propylene-diene rubber (EPDM), and ethylene-propylene-styrene rubber (SEBS). Recently, impact modification of PP, using metallocene-catalyzed ethylene-octene copolymer (EOC) has attracted attention. EOC provides better efficiency of impact modification than EPR, and is more cost effective than EPDM [14].

In 1993, DuPont Dow Elastomers has introduced POEs under the brand name ENGAGE®. They are ethylene-octene copolymers produced via advanced INSITETM catalyst and process technology designed to be processed like thermoplastic but can be also compounded like elastomers. The exceptional performance of ENGAGE® is attributed to extraordinary control over polymer structure, molecular weight distribution, uniform comonomer composition and rheology. They are being considered for use in diverse applications such as cushioning agents, gaskets, and particularly good alternative for sealing application due to their structural regularity and non-toxic composition.

The PP/EOC blend has been investigated from different aspects such as crystallization and melting behavior [15], mechanical properties[16, 17], morphology and rheology [18, 19]. Moreover, there have been studies on thermal expansion [20], adhesion [21, 22], and non-isothermal crystallization [23-27]. However, no comprehensive study has been done on the isothermal crystallization behavior of PP/EOC blends with high EOC content and in wide-range of crystallization temperatures and blend compositions.

The main objective of this research was to analyze quantitatively the influence of blend composition on isothermal crystallization kinetics of PP/EOC blends. The morphology of the blends was observed by transmission electron microscopy (TEM). The crystallization behavior of PP in the blends was investigated by differential scanning calorimetry (DSC). The crystallization kinetics was also measured by optical microscopy. Crystallization kinetics parameters based on the isothermal crystallization were analyzed according to by the Avrami's equation [28, 29] and by Hoffman-Lauritzen equation [30, 31].

Materials and Methods

The isotactic polypropylene (PP) was commercial polymer supplied by Mitsui Chemicals Inc. (J3HG, M_w=3.5 x 10^5 g mol^{-1} and M_n=5x10^4 g mol^{-1}). Hydrogenated polybutadiene (hBR) was a special sample prepared by JSR Co. Ltd. (M_w = 3 x 10^5 g mol^{-1}; ethylene/ butene molar ratio = 20/80, degree of hydrogenation ≥ 97%).

The PP and hBR were melt-mixed (charge 0.7 g) at 200°C for 7 min at 100 rpm in a miniature mixer, Mini-Max Moulder (model CS-183MMX, Custom Scientific Instruments, Inc.). Three blend ratios were used: PP/hBR 75/25, 50/50 and 25/75 in wt. ratio. The melt-mixed blend was extruded and compression moulded to a thin sheet of 0.2 mm thickness at 200°C. The moulded sheet was quenched in various ways: dry ice-methanol (-73°C), ice-water (0°C), water (26°C), or put into annealing box set at different crystallization temperatures (60~140°C). The neat PP specimens with the same thermal histories were also prepared as control samples.

The tensile stress-stain curves were measured at room temperature with a tensile testing machine, Tensilon UTM-II-20 (Toyo Baldwin Co. Ltd.) with a crosshead speed of 2 mm/min. After the pre-set strain was attained, the crosshead returned at the same speed as stretching. After the strain reached zero, the sample was released from the clamps and the residual strain was measured at certain intervals.

For transmission electron microscopy (TEM) observations, the specimens were stained with ruthenium tetroxide (RuO_4) vapour for 30 min at 50°C. The stained specimen was microtomed to obtain ultrathin sections of ca. 70 nm thickness using the ultra-cryomicrotome, Ultracut N (Reichert-Nissei) at -70°C. The structure in section was observed with a JEOL electron microscope, JEM-100CX.

In a differential scanning calorimeter (Seiko Instruments – EXSTAR 6000), the specimens were heated in nitrogen atmosphere. In case of linear temperature increase the rate was 10°C min^{-1}. The melting temperature was obtained from the minimum and the enthalpy of fusion from the area of the endothermic peak. The crystallinity, X_c, was calculated by $X_c = \Delta H^* / \Delta H^0_{PP}$, where ΔH^* is the enthalpy of fusion per gram of PP or that in the blend and ΔH^0_{PP} is the heat of fusion per gram of 100% crystalline PP (=209 J g^{-1})[32]. In case of t.m.d.s.c. (temperature modulated differential scanning calorimetry) experiment [33], the temperature increased in sinus function; $T=T_0+bt+B\sin(\omega t)$, where T_0 is the starting temperature (100°C), t is the time, b is the heating rate (2°C min^{-1}), ω is the frequency (0.04Hz) and B is the amplitude (0.5°C). Fourier transform treatment of the measured response was used for data deconvolution.

Ethylene-octene copolymer was a special sample prepared by Dow Chemicals. The octene content is 40 wt% (or 14.29 mol %) which means that there are about 6 ethylene units per 1 octene unit. Molecular weight M_n was 2.28 x 10^5 g mol^{-1}.

The PP and EOC were melt-mixed (charge 5 g) at 200°C for 5 min at 50 rpm in a miniature HAAKE MiniLab mixer, two screws being in counter-rotating mode. Blend ratio was gradually varied across the whole blend composition. The melt-mixed blend was extruded and quenched.

For the isothermal crystallization of the PP/EOC samples by DSC, the samples were heated up to 200°C (at 100°C min^{-1} of heating rate) and then cooled (at 50°C min^{-1}) to the isothermal crystallization temperatures (124, 126, 128, 130, 132, 134, and 136°C). In all cases, sample was held at 200°C for 1 min to eliminate previous thermal history.

The WAXD pattern was observed by Rigaku Denki RU-200 X-ray diffraction apparatus using scintillation counter system. The radiation from the Cu anode was reflected from a graphite monochromator to obtain monochromatic Cu Kα radiation with a wavelength of 0.1541 nm. The generator was operated at 50 kV and 180 mA.

For the optical microscopy observations, the specimen was melt-pressed for 1 min between two cover glasses on a hot stage at 200°C. The melted specimen was then placed onto a LINKAM hot stage of the microscope set to temperatures in range 130-140°C. Structural development during the isothermal annealing was observed under the optical microscope (LMU 406 SP) equipped with a video recording system [34, 35].

RESULTS AND DISCUSSION

Miscible Blends

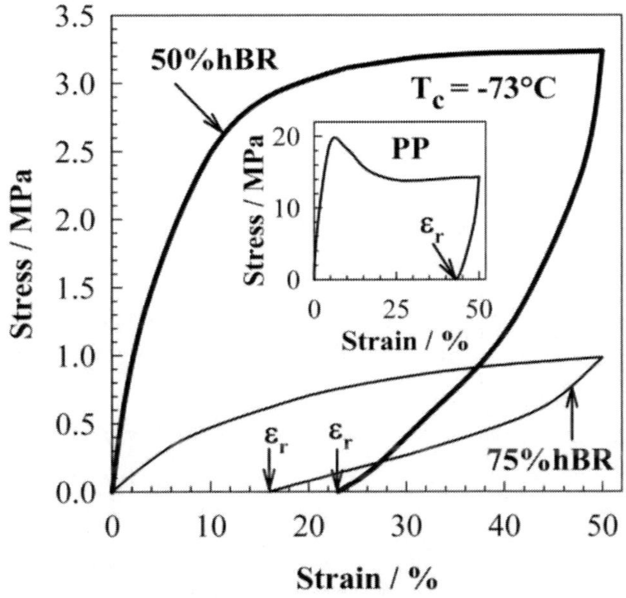

Figure 1. Typical stress-strain curves for quenched blends in comparison with pure PP.

The Figure 1 shows the stress-strain behaviour of the 50/50 and 25/75 PP/hBR blends. The blends were stretched to 50% elongation and then the movement of clamp was reversed at the same speed. The inserted small picture shows the curve for neat PP. Even though the temperature of quenching medium was very low (-73°C), the curve for PP shows a clear yielding behaviour with plastic deformation starting at about 8%-elongation. The residual strain of the neat PP is very high (43%), suggesting a big plastic deformation. For immiscible blends with PP, similar curves with yielding point were obtained [36]. In contrast, for the blends with miscible hBR, there is no yielding point and the residual strain ε_r is much smaller compared to neat PP. Furthermore, the residual strain decreases with resting time at room temperature, as shown in Figure 2.

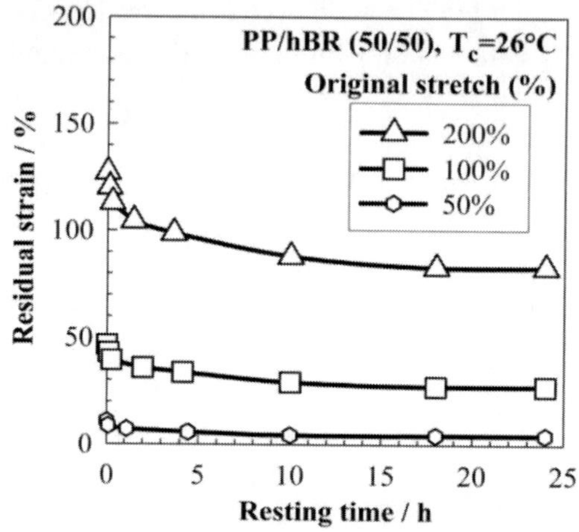

Figure 2. Residual strain after release from 200, 100 or 50%-elongation as a function of resting time.

In Figure 2 the residual strain is plotted as it was measured at certain time intervals. The steep recovery occurs during the first hour of resting, then the recovery speed decreases. After 24 h, the value did not change any more. Hence, the value of residual strain after 24 h should be compared for various blend compositions, applied strains and crystallization temperatures. Note that the recovery of neat PP was quite small (from 175 to 170% during 24h, after stretching to 200%).

The influence of hBR content on residual strain is shown in Figure 3. The lowest residual strain values were found for blends containing 50 and 75% hBR after stretching to 50%-elongation. The values ε_r were about 4 and 2%, respectively.

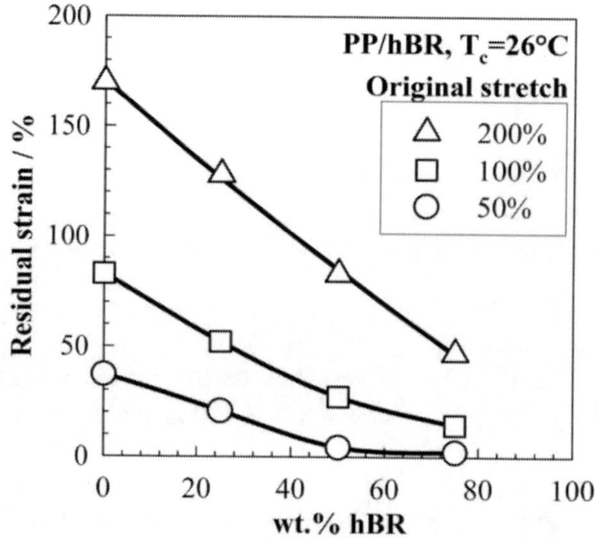

Figure 3. Residual strain vs. blend composition.

The values of residual strain are plotted as a function of applied strain for blends with different blend compositions in Figure 4.

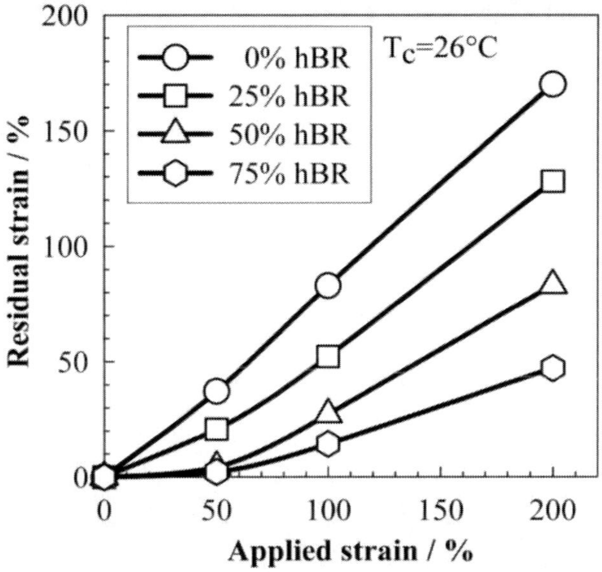

Figure 4. Applied strain vs. residual strain for samples quenched to 26°C.

In Figure 5 one can find the optimum crystallization condition in order to get the best elastic properties. The results of two blends stretched to 50%-elongation are shown here. The optimum temperature of cooling medium was found to be around room temperature. Crystallization at high temperatures is very harmful for elasticity. When the blend is crystallized at very low temperatures (-73°C), the elasticity gets a little bit worse, but the blend still remains elastic.

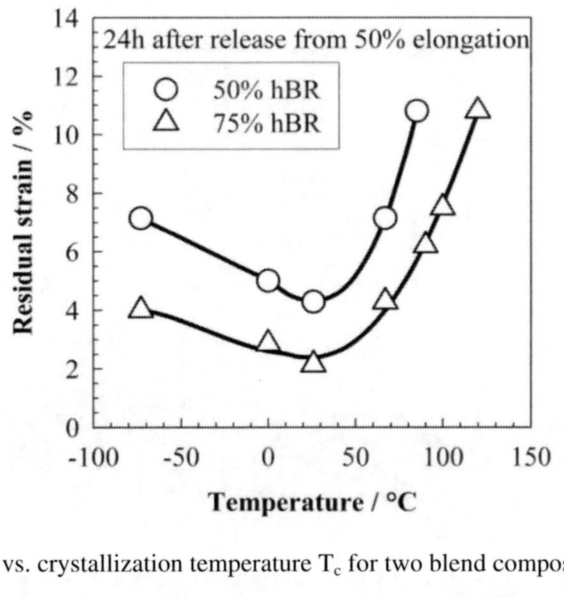

Figure 5. Residual strain vs. crystallization temperature T_c for two blend compositions.

It is well known that the PP spherulite consists of crosshatched lamellae; i.e., subsidiary lamellae grow tangentially to the radiating primary lamellae [37]. The fraction of subsidiary lamellae in PP spherulite is high at a low crystallization temperature T_C, and it decreases with increasing T_C [38].

In Figure 6 are shown three TEM micrographs (a~c) of blend containing 75% of hBR crystallized at high temperatures. The crosshatch lamellar structure is seen. The lamellae are very long and their thickness was about 10, 12 and 14 nm for crystallization temperatures 100, 120 and 140°C, respectively. For comparison, TEM of neat PP crystallized at 140°C is shown in Figure 6 (d). In contrast to the blends, the lamellae prefer the parallel alignment and they are very close together. The well-developed crosshatch lamellar structure may be harmful for good elastic properties, because by bulk deformation a plastic deformation of the lamellae would occur.

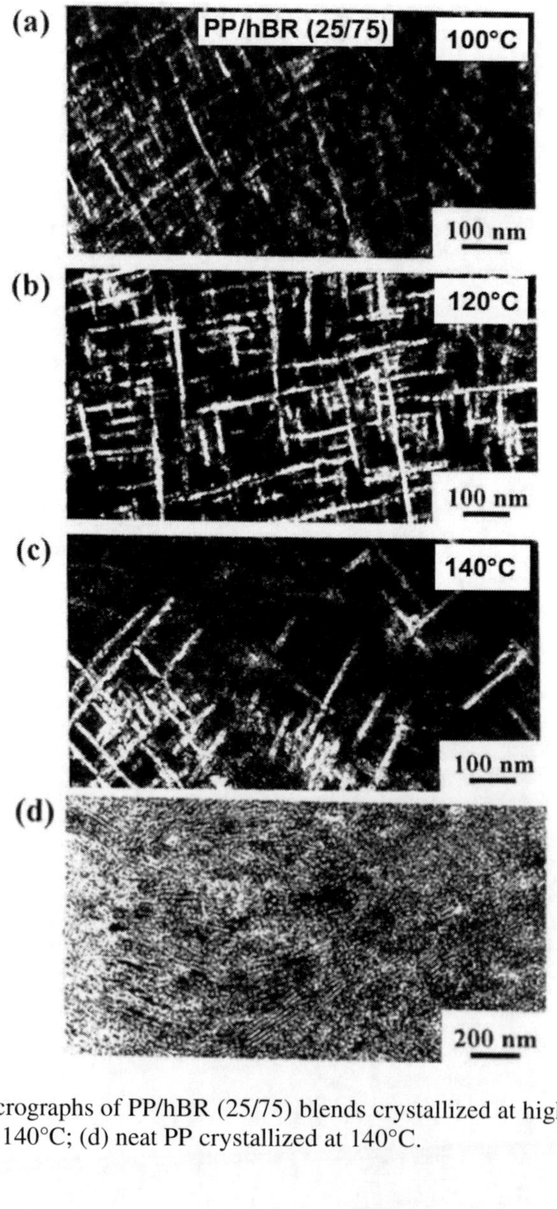

Figure 6. (a)-(c) TEM micrographs of PP/hBR (25/75) blends crystallized at high temperatures: (a) 100°C, (b) 120°C and (c) 140°C; (d) neat PP crystallized at 140°C.

Figure 7 shows three TEM micrographs of blends containing 50% of hBR quenched to low temperatures. There are no long crosshatch lamellae. One sees only round particles (bright domains). The round particles could be assigned to PP crystallites, since the crystallites should be less stained by RuO_4. Diffraction peaks of WAXD from the blend (see Figure 11) appeared at exactly same diffraction angles as for neat PP. The melting temperatures T_m by DSC were very close to T_m of neat PP (Table 2). These results clearly support that even in the blends quenched to very low temperatures, PP crystals do exist, and they are observed as the round particles under TEM. From the broadness of WAXD profile, one can estimate the apparent crystal size (D) by the Scherrer equation [39]. The crystal size in the perpendicular direction to the [110] plane was estimated from a peak at $2\theta = 14.16°$. The apparent crystal size in the blend quenched to low temperature (–73°C and 0°C) is quite small. This is consistent with the TEM observation (Figures 7a and 7b). Then, we call them "fragmented lamellar crystallites", because the crystal growth to big lamellar crystals would be suppressed by the presence of high molecular weight impurity (hBR).

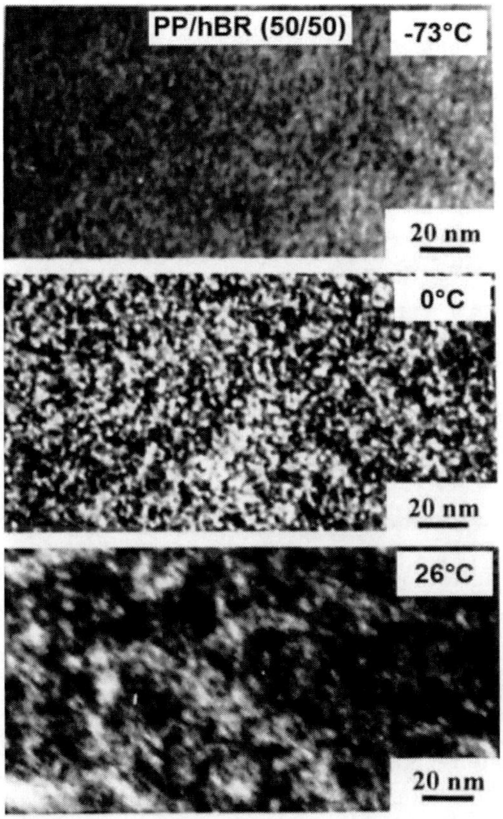

Figure 7. TEM micrographs of PP/hBR (50/50) blends quenched to low temperatures: (a) -73°C, (b) 0°C and (c) 26°C.

The structure of fragmented lamellar crystallites may be depicted as shown in Figure 8. The fragmented lamellar crystallites may play a role of tie points for amorphous chains to provide a nice elastic behaviour.

Figure 8. Schematic picture of the fragmented lamellar crystallites.

When the blend was quenched to room temperature (Figure 7c), the crystal size is slightly bigger. Note that for this blend the optimum of elastic properties was found. Combining the results in Figures 5 ~ 7 and Table 1, one can conclude that there is an appropriate crystal size to achieve good elastic property; i.e., when the crystals are too small or too long, it is not good for nice elastic behaviour.

Table 1. Melting point T_m, crystallinity X_c, apparent crystal size D and morphology as a function of crystallization temperature T_c; obtained by DSC, WAXD and TEM analyses

T_c /°C	T_m /°C	X_c /%	D / nm	Morphology
-73	159.8	15.3	5.4	fragmented
0	161.3	18.3	10.0	fragmented
26	161.5	21.8	13.9	fragmented
140	164.2	28.0	15.0	crosshatch
0°C +annealing at 140°C	161.7	22.8	12.7	fragmented

Finally, it is interesting to compare elastic behaviour of PP/hBR blend with that of the PP/hSBR blend which contains non-crosslinked rubber particles of size 20 nm, and with that of a commercial TPE Santoprene®, which contains crosslinked rubber particles 1 ~ 3 μm. The comparison is shown in Figure 9. Both PP/hSBR blend and Santoprene® have better elastic behaviour than PP/hBR; i.e., by the presence of rubber particles the elastic behaviour seems to be improved. The favourable contribution of rubber particles to the elastic behaviour can be interpreted in terms of the volumetric strain of the rubber phase with high Poisson's ratio (~ 0.5; very small volumetric change with bulk deformation), as has been discussed by FEM analysis [40]. However, one can understand that the contribution is minor, if one compares with the strain recovery of neat PP. Major contribution is by the elastic nature of matrix in PP/hSBR and Santoprene®, which originates from the characteristic morphology of "fragmented PP crystallites" as in the case of PP/hBR blend.

As has been discussed, the more fragmented crystallites formed by quenching to very low temperatures (0°C, -73°C) render the less elastic behaviour (see Figure 5 and Table 1). It may imply that the less ordered crystallites cannot play the role of nice tie points, since they are rather easily subjected to plastic-deformation by bulk deformation. Then, if the crystallites get more ordered state by thermal annealing, the elastic property could be improved.

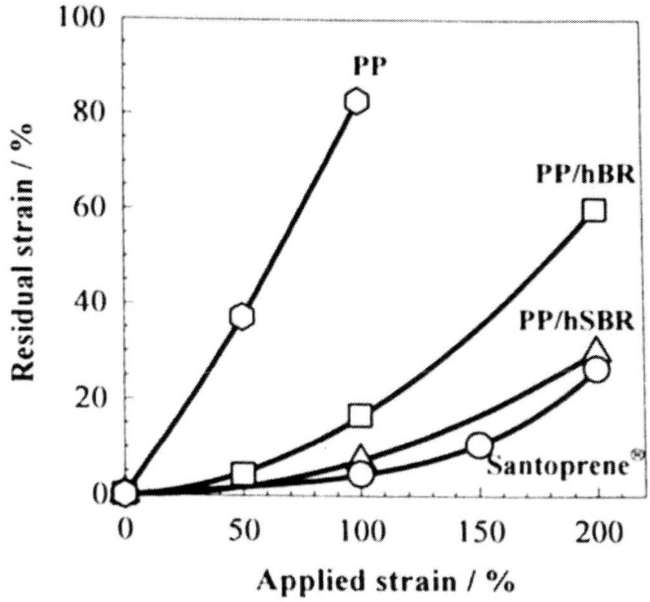

Figure 9. Comparison of elastic behaviour of Santoprene®[8], 50/50 PP/hSBR blend[10], 50/50 PP/hBR blend and neat PP.

Figure 10 shows the results of t.m.d.s.c. analysis for samples crystallized at 140°C and –73°C. Conventional DSC showed curves similar to TOTAL curves in this Figure. In the TOTAL curves, there is only difference in heat of fusion (area below the peak). However, at NONREVERSE curve one can see the exothermic peak (shown by arrow) for blend quenched to –73°C, which suggests that when the quenched blend was heated above 120°C the recrystallization took place. The reference sample crystallized at 140°C does not show any exothermic peak, implying that the well-crystallized lamellae had been formed. The recrystallization in the blend quenched to –73°C may involve a healing of defects and an increase in crystal size.

To investigate the effect of recrystallization on crystal structure and elastomeric property, a 50/50 blend was first quenched to 0°C, and then it was annealed for 17h at 140°C. In Figure 11 the WAXD curves before and after the annealing are shown. With annealing, the crystallinity increased from 18.3 to 22.8% and the apparent crystal size increased from 10.0 to 12.7 nm. The structure of the blend after annealing by TEM is shown in Figure 12. At the upper right part there is a white stripe of the size approximately 7 x 35 nm (shown by arrow), showing clearly that the size of fragmented lamellae increased with annealing. TEM also shows that fragmented character of lamellae still remains even after annealing. It implies that the fragmented lamellae cannot grow up to the crosshatch lamellae as in the case of neat PP.

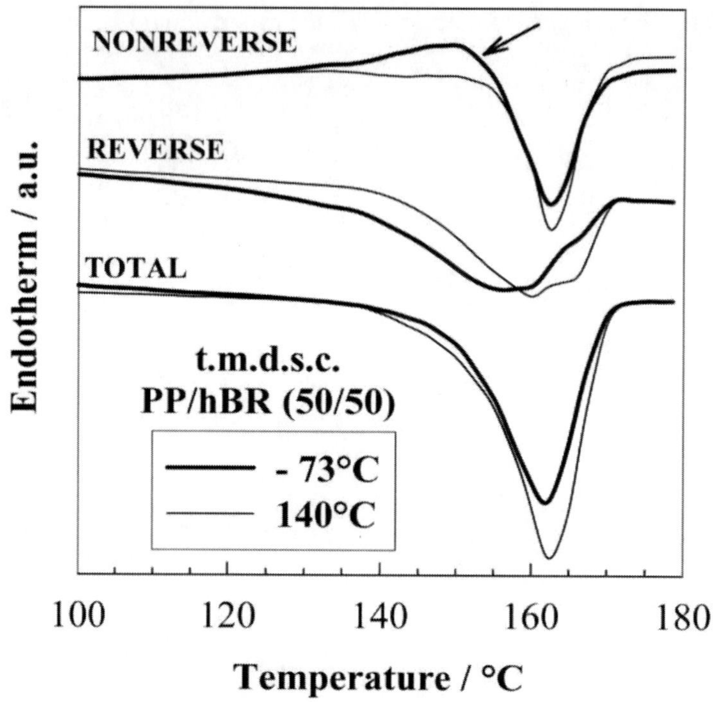

Figure 10. t.m.d.s.c. of PP/hBR (50/50) blends crystallized at –73°C and 140°C.

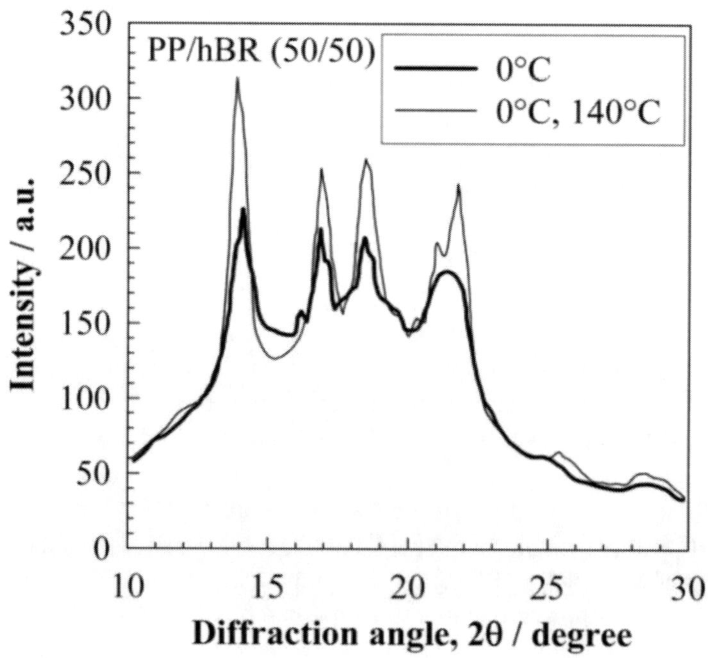

Figure 11. WAXD curves of PP/hBR (50/50) blends.

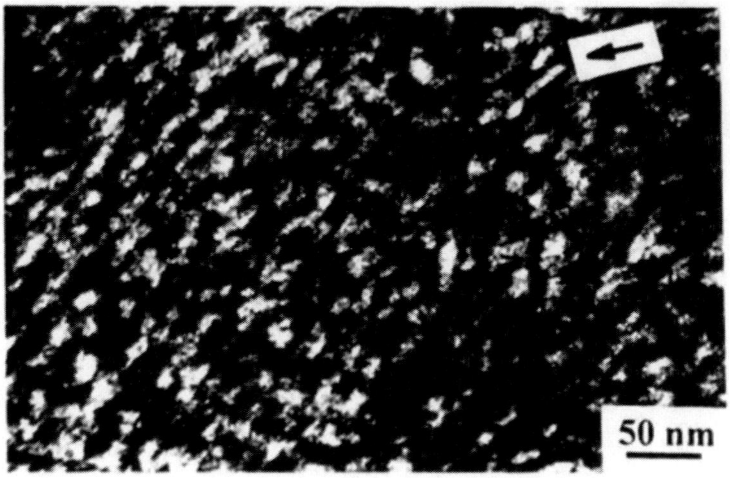

Figure 12. TEM micrograph of PP/hBR (50/50) blend quenched to 0°C and then annealed 17h at 140°C.

Figure 13 shows the strain recovery behaviour of the blend before and after annealing. There is a small improvement. The improvement seems to be caused by ordering in crystal structure.

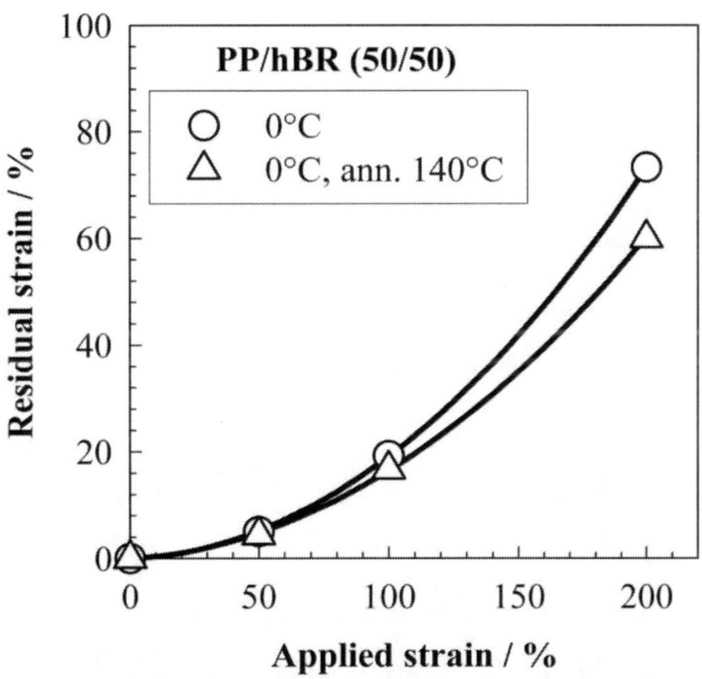

Figure 13. Residual strain vs. applied strain of PP/hBR (50/50) blends of blend quenched to 0°C in comparison with blend quenched to 0°C and then annealed to 140°C.

Crystallization – Theoretical Background

Whenever a polymer crystallizes, the extent of the phase transformation depends upon the crystallizing species and the experimental conditions. High molecular weight polymers do not crystallize completely because of topological constraints that lower crystallinity considerably. The classical isothermal transformation kinetics, initially formulated by Kolmogorov and Goler et al. were extended later by the Avrami theory that was initially formulated for metals and later modified, for example, by Evans and others for polymers [32].

The fraction X_t is obtained from the area of the exothermic peak in DSC isothermal crystallization analysis at a crystallization time t divided by the total area under the exothermic peak [41]:

$$X_t = \frac{\int_0^t \left(\frac{dH}{dt}\right) dt}{\int_0^\infty \left(\frac{dH}{dt}\right) dt} \tag{1}$$

Where the numerator is the heat generated at time t and the denominator is the total heat generator up to the complete crystallization.

The crystallization kinetics of polymers is analyzed using a classical Avrami equation as given in Eq. (2):

$$1 - X_t = exp(-kt^n) \tag{2}$$

Where the k value is the Avrami rate constant and the n value is the Avrami exponent. Both k and n depend on the nucleation and growth mechanisms of spherulites. In order to deal conveniently with the operation, Eq. (2) is usually rewritten as the double logarithmic form as follows:

$$ln[-ln(1 - X_t)] = \ln k + n \ln t \tag{3}$$

The k and n values could be directly obtained using Eq. (3) from the slope and intercept of the best-fit line.

The crystallization behavior of the polymers was also studied according to the relationship between chain folded crystal growth rates and undercooling proposed by Hoffman and Lauritzen [30, 31]:

$$G = G_0\, exp\left[\frac{-U^*}{R(T_c - T_\infty)} - \frac{K_g}{T_c(\Delta T)f}\right] \tag{4}$$

where G is the crystal growth rate, U^* is a constant characteristic of the activation energy for repetitive chain motion and is equal to 1500 cal mol^{-1} [42, 43], R is the gas constant, T_c is the crystallization temperature (K), $T_\infty = T_g - 30K$ and (for PP the glass transition temperature $T_g = 270K$ [44]), $\Delta T = T_m^0 - T_c$, T_m^0 is the equilibrium melting temperature of an infinitely thick crystal, K_g is the nucleation constant, f is a correction factor and equals to $2T_c/(T_m^0 + T_c)$ and G_0 is a pre-exponential factor. In order to perform nonlinear regression by software SigmaPlot

12 it was necessary to rewrite the equation into a computer language. The sequence of simplification is shown by Eqs. 5 to 7 and example of equation for pure PP where T_∞=233 K and $T_m^{\ 0}$=459 K is given by Eq. 8.

$$G = G_0 e^{\left[\frac{-U^*}{R(T_c - T_\infty)} - \frac{K_g}{T_c(T_m^0 - T_c)\frac{2T_c}{(T_m^0 + T_c)}}\right]} \tag{5}$$

$$G = G_0 e^{\left[\frac{-U^*}{R(T_c - T_\infty)} - \frac{K_g(T_m^0 + T_c)}{2T_c^2(T_m^0 - T_c)}\right]} \tag{6}$$

$$y = ae^{\left[\frac{-b}{8.314*(x - T_\infty)} - \frac{c(T_m^0 + x)}{2x^2(T_m^0 - x)}\right]} \tag{7}$$

$$y = a * \exp((-b)/(8.314 * (x - 233)) - c * (459 + x)/(2 * x^2 * (459 - \mathrm{x}))) \tag{8}$$

For linear data analysis and estimation of initial parameters a, b and c for nonlinear regression it is convenient to use a logarithmic version of Eq. (4) shown by Eq. (9). This equation was used for optical microscopy evaluation.

$$ln(G) + \frac{U^*}{R(T_c - T_\infty)} = ln\, G_0 - \frac{K_g}{T_c \Delta T f} \tag{9}$$

For the evaluation of DSC results we have replaced G by $1/\tau_{1/2}$.

$$ln\left(\frac{1}{\tau_{1/2}}\right) + \frac{U^*}{R(T_c - T_\infty)} = ln\, G_0 - \frac{K_g}{T_c \Delta T f} \tag{10}$$

A major extension of the theory involved the recognition that the deposition of a single critical nucleus may not always occur and that multiple nucleation generate different situation. The situation is handled best in general conceptual terms by considering it to be a competitive situation between the rate at which critical nuclei are deposited on the surface and the rate at which the chains deposit laterally to complete the growth step. This leads to three distinct situations or regimes; regime I the classical situation in which the rate of secondary nucleation is slowest, regime II a situation in which the rates of secondary nucleation and lateral spreading are comparable, and regime III a situation in which the rate of secondary nucleation is the fastest. These three situations occur naturally in many polymers as the crystallization temperature is reduced. The vast majority of polymers studied show regimes II and III, whereas few show regime I which is the classical situation.

Basically, the diffusion process has been described as consisting of two elementary processes: the deposition of the first stem on the growth front (secondary nucleation process) and the attachment of following stems in the chain on the crystal surface (surface spreading process). According to the Hoffman-Lauritzen theory, *G* is mostly governed by the rate of secondary nucleation, *i*, in regimes I and III, while it is governed by both *i* and the rate of surface spreading, *g*, in regime II:

$$G \propto i \qquad for\frac{i}{g} \ll 1\ (regime\ I) \tag{11}$$

$$G \propto (ig)^{\frac{1}{2}} \qquad for\frac{i}{g} \sim 1\ (regime\ II) \tag{12}$$

$$G \propto i \qquad for\frac{i}{g} \gg 1\ (regime\ III) \tag{13}$$

Where *i* consists of both β_g and $exp[-K_g/T_c(\Delta T)f]$ and g consists of only β_g. The diffusion coefficients in the surface nucleation process and the substrate completion process are defined as D_M and D_S, respectively. Assuming that β_g is proportional to the diffusion coefficient, i and g may be given by:

$$i \propto D_M \exp\left[-\frac{K_g}{T_c \Delta T f}\right] \tag{14}$$

$$g \propto D_S \tag{15}$$

From equations (11-15) one can obtain:

$$\beta_g \propto \phi_1 D_M \qquad (regime\ I\ and\ III) \tag{16}$$

$$\beta_g \propto \phi_1^{\frac{1}{2}} (D_M D_S)^{\frac{1}{2}} \qquad (regime\ II) \tag{17}$$

where a prefactor ϕ_1 is introduced, since *i* is proportional to the number of crystallizable molecules at the crystal surface, which is proportional to the volume fraction of crystalline polymer ϕ_1 [46].

Immiscible Blends

After 5 min in Haake mixer the blends were extruded to air and pressed between two cold metal plates. Figure 14 shows the TEM micrographs (at 1000x magnification) of blends with increasing EOC content. For 30 and 40% of EOC the matrix is composed of PP and the dispersed phase is EOC. There seems to be an increase in EOC particle size from about 2 μm to 5 μm when the blend composition is changed from 30 to 40% of EOC. At 60% of EOC the structure resembles a co-continuous one. At 80% of EOC the situation is reversed: EOC became a matrix and PP forms very small droplets with the size smaller than 1 μm. One could expect that the crystallization will be affected quite significantly for these very different morphologies.

While the observation of morphology at 1000x magnification is important for seeing which polymer forms the matrix and which one is the dispersed phase, TEM enables us also to look much deeper at magnification e.g. 30000x when one can see the lamellar structure.

These structures at much higher magnification are shown in Figure 15. RuO_4 stains preferably EOC, so the bright phase is PP. In our previous study [47] we have shown the existence of partial miscibility between PP and EOC based on the T_g shift for EOC being 3.5°C and for PP 2.3°C. The crystallization of PP is most likely affected by the presence of small amount of EOC in PP-rich phase.

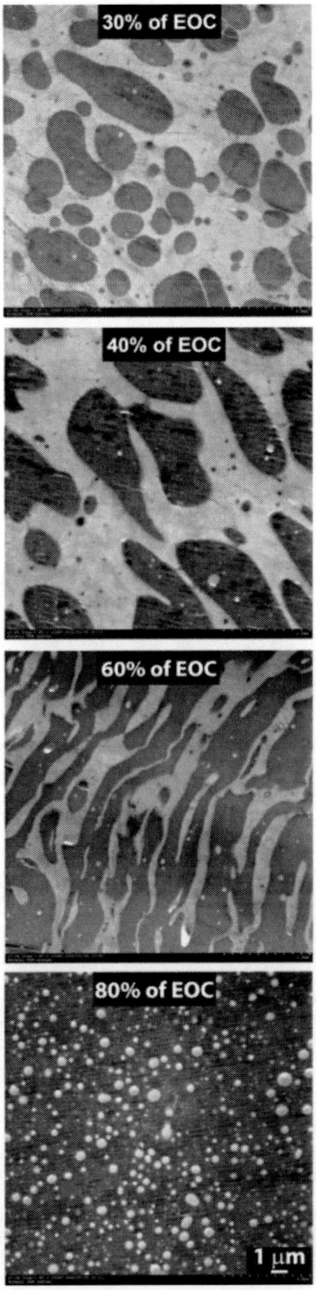

Figure 14. TEM micrograph of PP/EOC blends after mixing at 50 rpm for 5 min at 200°C.

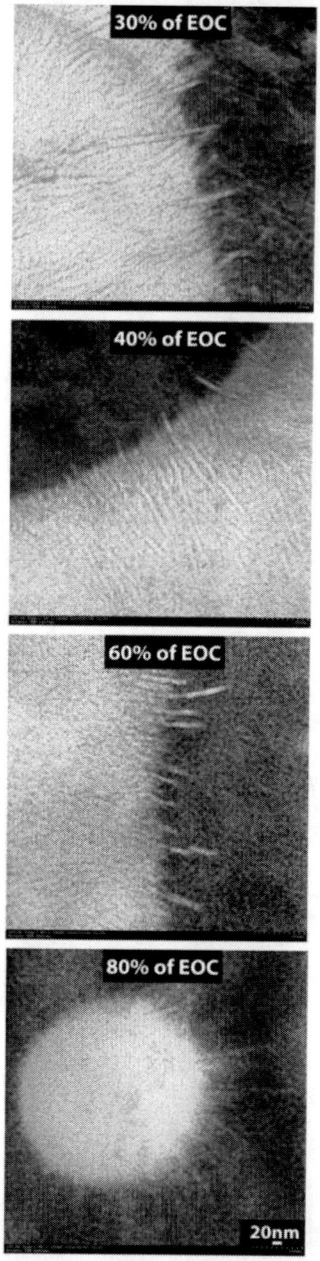

Figure 15. TEM micrograph of PP/EOC blends after mixing at 50 rpm for 5 min at 200°C, higher magnification.

In addition to that there are PP lamellae present also in EOC-rich phase. Crystallization of these lamellae could have quite different kinetics compared to lamellae in PP-rich phase. As one can observe the situation is quite complex and we will try to understand it and explain it in following section.

The crystallization was studied at first by DSC and then also by optical microscopy. Figure 16 illustrates the results from DSC experiments. Initially the samples with weight about 10-15 mg were heated from room temperature to 200°C at rate 100°C/min, then the temperature was kept at 200°C for 1 min to fully melt the PP crystals (T_m of PP is about 165°C) and the EOC crystals as well (T_m of EOC is about 48°C). The next step was fast cooling to desired isothermal crystallization temperature. This cooling at rate 50°C/min was attained by a cooling unit capable of cooling to -30°C. The last step was isothermal crystallization at desired temperature.

When the heat flow curve reaches minimum value and starts to grow to form an exothermal peak we have assigned this time to be 0. By the integration of heat flow curve one can get a relative crystallinity curve shown on the inserted picture of Figure 16. When the crystallinity reaches 0.5 (or 50%) one gets $\tau_{1/2}$.

Temperature affects the crystallization heat flow curves in a way shown in Figure 16. While at 124°C crystallization takes few minutes, at 134°C crystallization takes almost an hour. One cannot go beyond the 123-136°C temperature range with DSC experiment very far. When we set the crystallization temperature lower than 123°C, crystallization starts already during cooling to desired temperature. On the other hand if we choose temperature higher than 136°C, the heat flow curves have large fluctuations and noises and the $\tau_{1/2}$ evaluation becomes less accurate. For very slowly crystallizing blends this experimental window was limited only to 123-130°C.

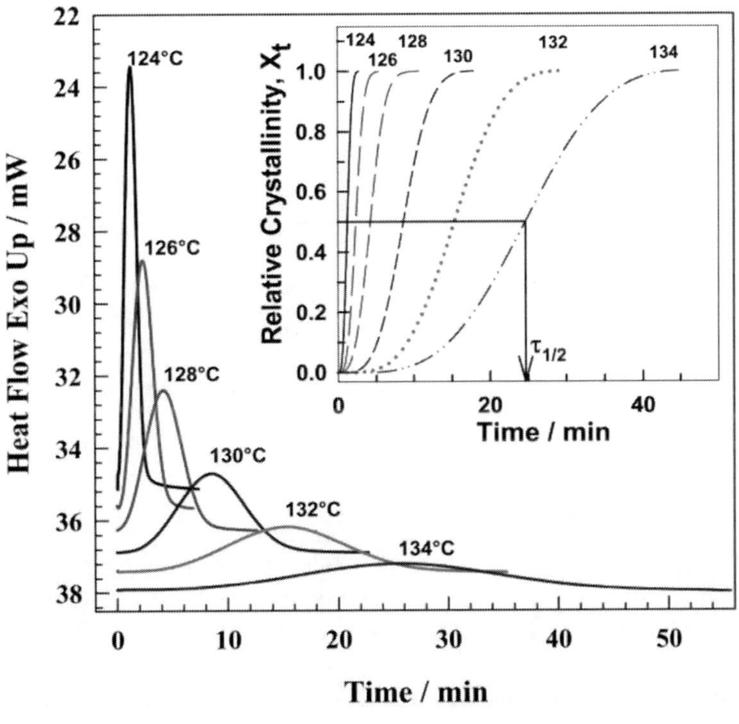

Figure 16. Isothermal crystallization peaks of PP/EOC (70/30) blend by DSC at various crystallization temperatures.

This is shown on the small inserted picture of Figure 17 for example for PP/EOC (20/80) blend when we had to stop experiments at about 129°C since the crystallization was extremely slow $\tau_{1/2}$ being more than 3000s. It is interesting that pure PP's curve lays somewhere in the middle. As we have mentioned there are blends much slower than pure PP, such as above mentioned PP/EOC (20/80) blend, but there are also blends faster than pure PP, such as PP/EOC (90/10 and 60/40), as shown in the small inserted picture of Figure 17. There is a non-linear increase in $\tau_{1/2}$ as a function of temperature.

More direct way to see the crystallization kinetics is plotting of $\tau_{1/2}^{-1}$ as a function of temperature as shown in main Figure 17. While there is only a moderate increase in kinetics with decreasing temperature in temperature range 135-129°C, this increase is much more pronounced in temperature range 129-123°C. From the composition point of view the kinetics decreases with increasing EOC content 20, 30, 50 and 75%.

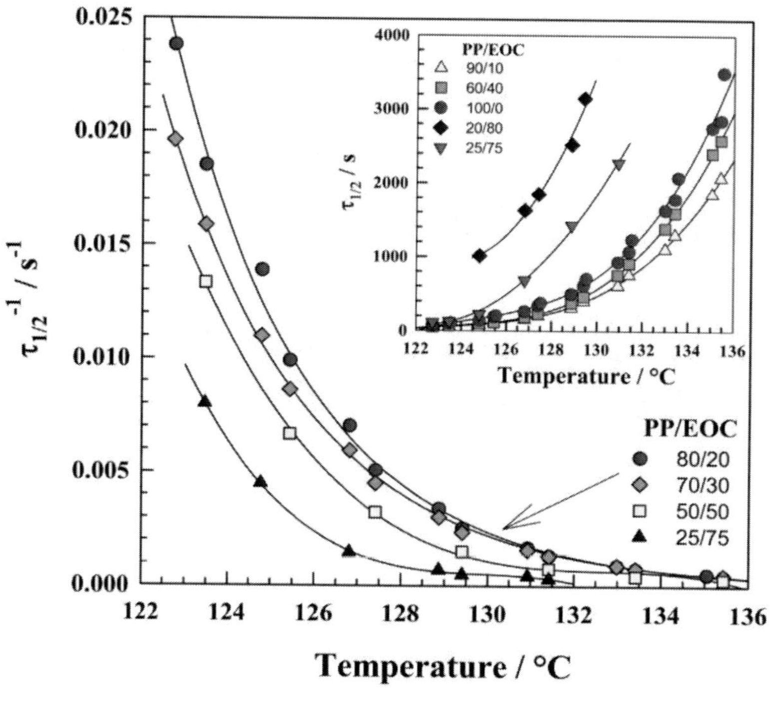

Figure 17. Crystallization kinetic ($\tau_{1/2}^{-1}$) and half time ($\tau_{1/2}$) from DSC as a function of temperature for PP/EOC blends.

To see the effect of EOC content and the morphology on the crystallization kinetics represented by $\tau_{1/2}^{-1}$ we have focused on lower temperatures (124-128°C) in Figure 18. For example there is about 80% increase in crystallization kinetics from pure PP to 20% of EOC, then the kinetics decreases with increasing EOC content. How the morphology affects the kinetics it is possible to see in the inserted TEM pictures. Absolutely slowest crystallization was found for 80% of EOC when PP exists in very small droplets (smaller than 1μm). For 90% of EOC we were not able to detect PP crystallization by this isothermal DSC method. Müller et al. [48] presented similar shape curves for blends of copolymers of ethylene/butene and ethylene/hexene initially heated above UCST curve (upper critical solution temperature) and then crystallized below UCST where in 0-15 and 85-100% partial miscibility still existed. This could be similar to our case.

To analyze the crystallization behavior further we have carried out an Avrami and Hoffman-Lauritzen analyses as follows.

Avrami equation is describing relative crystallinity as a function of time (see Eq. 2). One has to do the integration step (shown in Figure 16) to get the relative crystallinity curve. Convenient way to evaluate the constants from Avrami equation is using the double logarithmic form of this equation (see Eq. 3). The comparison of the pure PP and PP/EOC 70/30 blend at 132°C is shown on Figure 19.

When we plotted the data from the whole experiments we got a curve rather than a single straight line. It is possible to divide the curve into two straight regions. To understand the problem deeper we have connected the Avrami curve with the original heat flow curve as shown on the inserted picture in Figure 19. The Avrami curve's straight regions can be limited by points shown as AB and BD. At the heat flow curve also the maximum is an interesting point (called C in Figure 19). The conclusion of this analysis is this. The majority of the crystallization process happened between points B and D while only very early stage of crystallization happened between points A and B. This is not obvious when one looks only at Avrami curve when the AB section seems to have the same length like BD section. It is clear that the Avrami curves for pure PP and PP/EOC 70/30 are different.

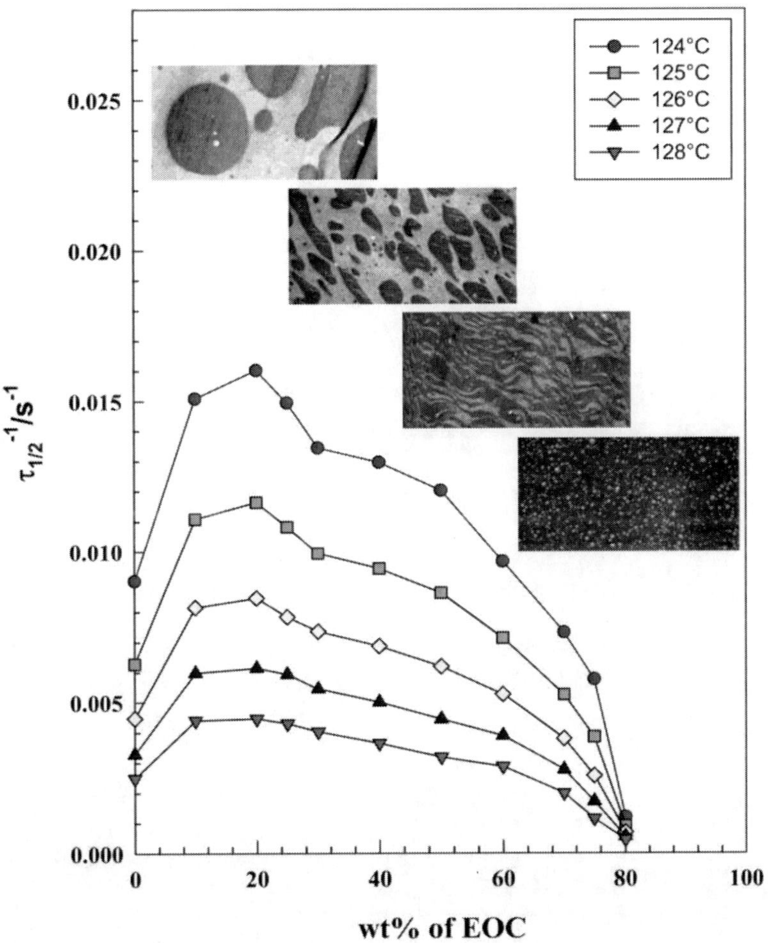

Figure 18. Crystallization kinetics ($\tau_{1/2}^{-1}$) from DSC as a function of EOC content in PP/EOC blends. The structure of the four corresponding blends (20, 40, 60, and 80% of EOC) is shown by 4 inserted TEM pictures.

Table 2 summarizes the Avrami exponents n and rate constants k for selected blend compositions and crystallization temperatures. At 124°C the exponent n for pure PP was found to be 2.12 which is close to value 1.97 mentioned by Huang [43] and corresponds to two-dimensional spherulite growth. In contrast the blend containing 10% of EOC had the exponent n=2.99 which is very close to 3, and can be attributed to heterogeneous nucleation of a three-dimensional growth in a spherical form. With increasing EOC content this n value gradually decreased up to 2.36 for 80% of EOC.

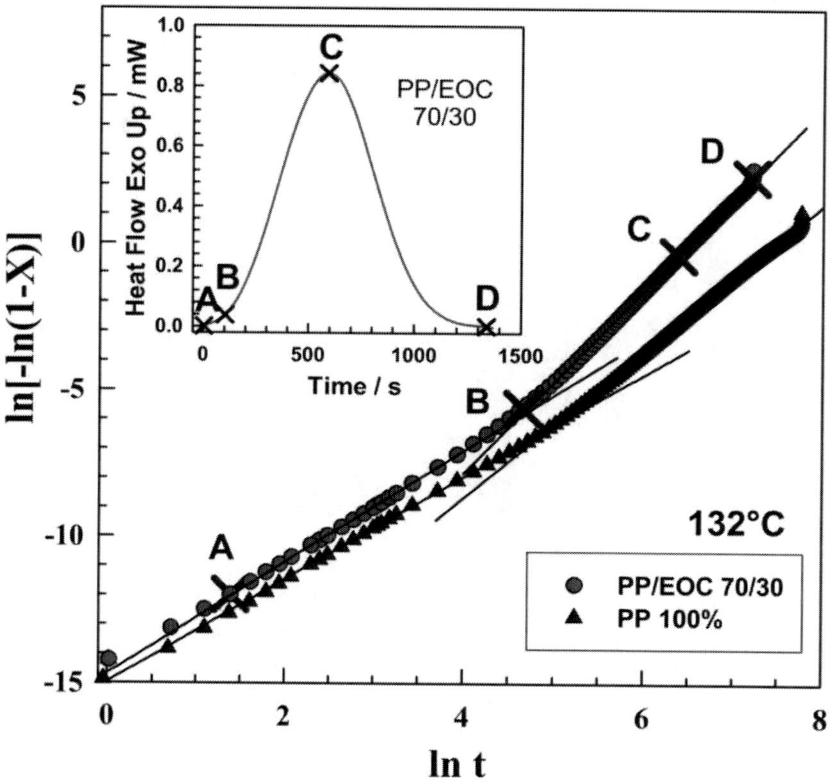

Figure 19. Avrami plots for pure PP and PP/EOC (70/30) blend isothermally crystallized at 132°C.

Table 2. Avrami parameters from linear regression

Wt.% of EOC	124°C		126°C		128°C		130°C		132°C		134°C	
	n	k (min^{-1})	n	k (min^{-1})	n	k (min^{-1})	n	k (min^{-1})	n	k (min^{-1})	n	k (min^{-1})
0	2.12	0.11798	2.26	0.02842	2.34	0.00756	2.53	0.00113	2.52	0.000316	2.52	0.000094
10	2.98	0.76721	2.80	0.26405	2.77	0.04294	2.83	0.00292	2.76	0.001121	2.64	0.000211
30	2.77	0.66801	2.76	0.16869	2.75	0.02317	3.01	0.00206	2.85	0.000596	3.06	0.000147
40	2.67	0.53584	2.74	0.11433	2.87	0.01416	2.83	0.00194	2.78	0.000336	2.71	0.000117
60	2.70	0.26101	2.75	0.04678	2.69	0.00914	2.50	0.00103	2.43	0.000282	2.31	0.000090
75	2.31	0.02222	2.26	0.00309	2.19	0.00082	2.14	0.00047	–	–	–	–
80	2.35	0.00251	2.28	0.00073	2.21	0.00066	2.26	0.00013	–	–	–	–

The tendency for changes in n value is due to the transition between heterogeneous and homogeneous nucleation. This tendency was observed for all crystallization temperatures. The n values are not seriously affected by crystallization a temperature which is not a case for rate constant k which is extremely influence by crystallization temperatures. For given crystallization temperatures, the dependence of k on EOC content is very similar to $\tau_{1/2}^{-1}$ dependence shown in Figure 18 and the dependence of k on crystallization temperature is very similar to the one shown in Figure 17. In connection with TEM observations the mechanism of crystallization should be very much different for blends with 40% and 80% of EOC. In case of 40% of EOC once nucleus is formed the spherulite can grow through the continuous PP phase. However in case of 80% of EOC nucleus has to be formed in each small PP droplet which influenced not only the kinetics (k) but also the Avrami exponent n.

Next analysis was done according to Hoffman-Lauritzen equation (Eq. 4). For this treatment we also needed to obtain equilibrium melting points T_m^0 according to Hoffman-Weeks [49] plot. The T_m^0 for pure PP was found to be 183°C and for the PP/EOC blends in range 177-181°C. PP/EOC blends have only slightly lower T_m^0 than PP. This observation goes well with Nishi and Wang [50] theory and the fact of only partial miscibility between the two polymers.

According to Eq. 10 the Hoffman-Lauritzen plots are shown on Figure 20a and 20b some of the blends exhibit two slopes representing regime II and III of crystallization with transition being around 130°C. Nucleation constant K_g values and pre-exponential factor G_0 values for selected blends are listed in Table 3. Initially the pure PP and PP/EOC (70/30) blend are compared on Figure 20a. They show similar behavior with a small shift in absolute values affecting the $\ln G_0$ values. These values correspond with different kinetics. The K_g values are very similar. The slope of the line changed quite significantly from about 5.9 to 3.8. In literature it is usually interpreted as different crystallization regime. At the lowest temperature (123-130°C), so called regime III, there is a multiple nucleation on the surface of lamellae and the crystallization is the fastest. K_g value was found to be about 5.9 x10^5 for pure PP and also for the blend. Then at 130°C there is transition from regime III to regime II where the rates of secondary nucleation and lateral spreading are comparable. The K_g was found to be about 3.8x10^5 again for pure PP and the blend. In literature we found values K_g 7.76x10^5 for regime III and 4.46x10^5 for regime II [51], K_g 3.33x10^5 for regime III and 1.68x10^5 for regime II [52], K_g 4.805x10^5 for regime III [53], K_g 5.2x10^5 for regime III [54]. Our values lay in the range reported by other authors for iPP.

How the EOC content influences the Hoffman-Lauritzen plot is shown on Figure20b. The curves for 10, 40 and 60 wt. % of EOC show very similar trend and are only little bit shifted to lower values with increasing EOC content. The situation is completely different for 75 and 80% of EOC. Both the K_g and $\ln G_0$ values are much lower (see Table 3). As shown by TEM pictures (Figure 14) in case of high amount of EOC the PP is located in small islands and therefore the crystallization kinetics is dramatically reduced (which is reflected by pre-exponential factor G_0) and also nucleation parameter K_g is reduced from about 6 to 4 x10^{-5} K^2.

In Hoffman-Lauritzen equation some researchers use a constant value for activation energy U*=6300 J/mol [52]. This is very convenient in first evaluation (or estimation) of HL parameters K_g and $\ln G_0$ according to logarithmic Hoffman-Lauritzen Eq. 10 which is shown on Figure 20a and b. Many researchers are calculating the activation energy U* by nonlinear regression because it is the most important material constant and it is changing with blend/composites composition [55-60]. The nonlinear regression was done by SigmaPlot 12 software using Eqs. 7-8.

Table 3. Hoffman-Lauritzen parameters from linear regression

Wt.% of EOC	Regime II		Regime III		Transition (°C)
	$K_g \times 10^{-5}$ (K^2)	ln G_0	$K_g \times 10^{-5}$ (K^2)	ln G_0	
0	3.731	15.47	5.916	25.82	130.0
10	3.798	16.21	5.863	26.13	130.5
30	3.811	16.24	5.897	26.16	131.0
40	3.747	15.72	5.929	26.28	131.5
60	3.698	15.22	5.988	26.24	132.5
75	–	–	4.275	17.71	–
80	–	–	4.005	15.31	–

The activation energy U* calculated in this way is shown by Figure 21. The activation energy for the blends was found higher than for the pure PP. The meaning of this observation is this. EOC acts as an obstacle (barrier) for crystallization of PP and therefore more energy is required for crystallization. Similar increase in activation energy was discovered by Huang [43] who has concluded that foreign particles act not only as nucleating agent but also as a hindrance to retard the crystallization.

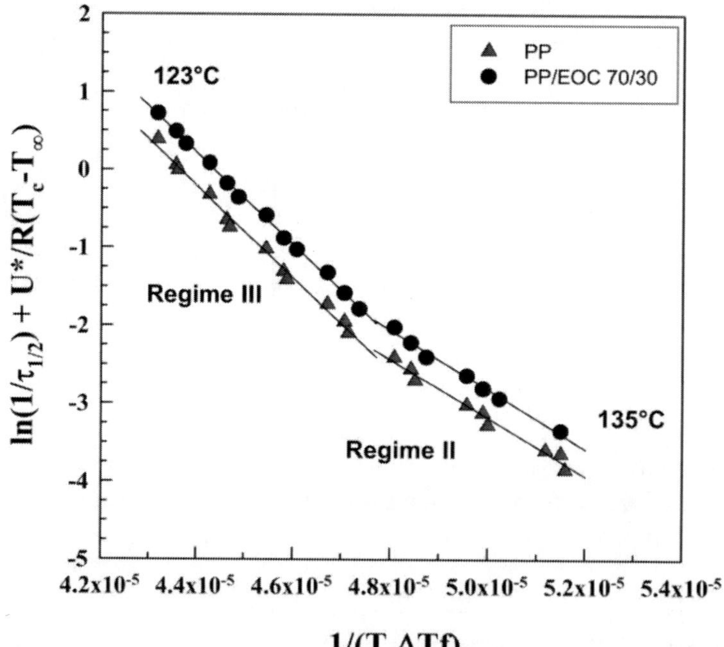

(A)

Figure 20 (Continued).

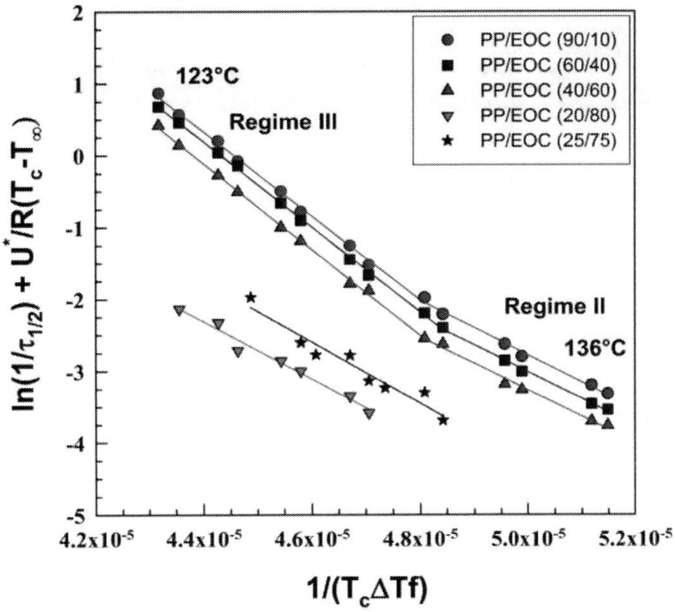

(B)

Figure 20 (a-b). Hoffman-Lauritzen plots for various PP/EOC blends from DSC.

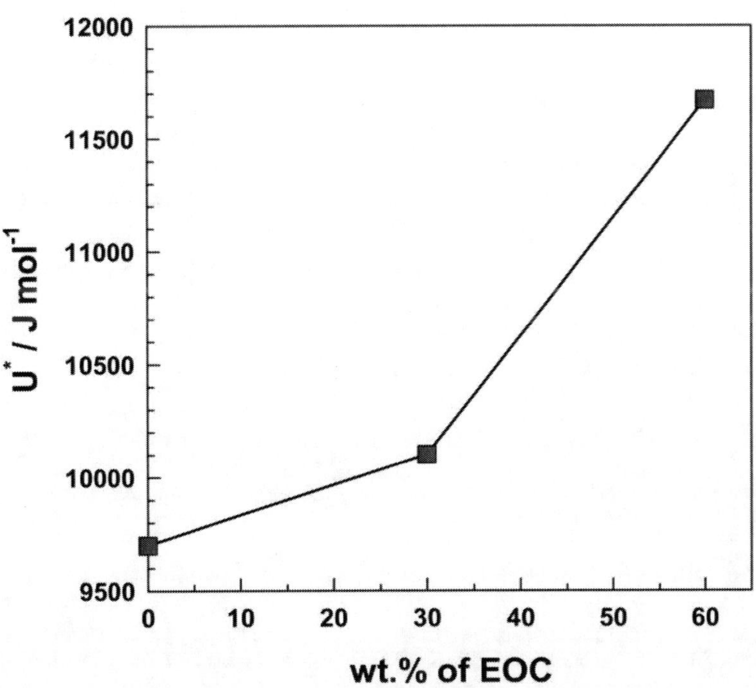

Figure 21. Activation energy (U^*) as a fuction of EOC content according to nonlinear regression of Hoffman-Lauritzen equation.

Both of these models, Avrami and Hoffman-Lauritzen, attempt to describe the crystallization behavior. While Avrami is focused on the time dependence of crystallinity at given temperature, Hoffman-Lauritzen is capturing crystallization rate as a function of temperature. We have observed relation of Avrami exponent n with HL K_g values and also Avrami k rate constant and HL lnG_0 values (decrease in 40-80 wt% of EOC range).

To understand this phenomenon better we have carried out another independent experiment optical microscopy. We have recorded the structure development in the computer equipped with a TV video card for later precise analysis.

Examples of pictures taken from the video at certain time intervals are shown in Figure 22. The spherulites are growing in time in both cases, in pure PP and in PP/EOC (70/30) blend. While the PP spherulites are clear, the spherulites growing in blend are quite coarse. The EOC does not prevent spherulite growth at this concentration. There are more spherulites in case of the blend. And the volume fills with them faster in case of the blend. Most likely EOC acts as nucleating agent for PP. This is quite surprising fact bearing in mind that at the measurement temperature (135°C) the EOC is a liquid (T_m of EOC is 48°C). Nevertheless it is a foreign subject in PP and causes faster bulk crystallization.

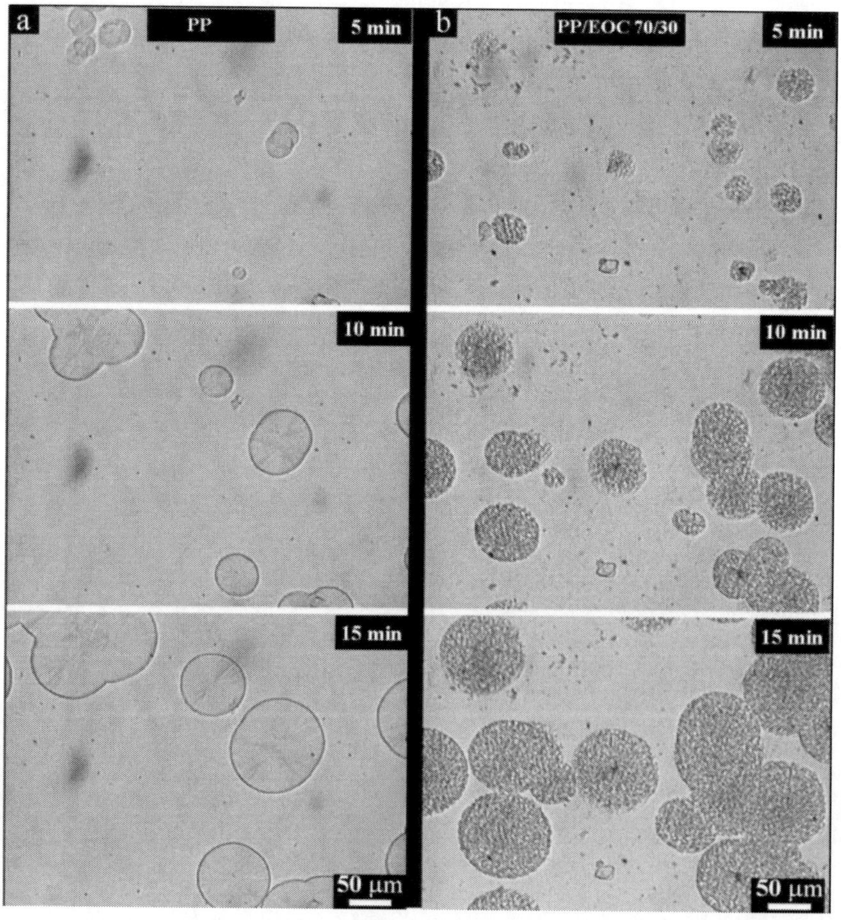

Figure 22. Growth of spherulites in time at T_c= 135°C in pure PP and PP/EOC (70/30) blend by hot-stage optical microscopy after 1 min pre-heating at 200°C.

We have also analyzed the growth rate of individual spherulites to see if EOC affects the crystallization kinetics of individual spherulite (see Figure 23). In contrast with DSC observations we have found the crystallization kinetics of the blend to be slower than that of pure PP. This behavior is quite understandable. Growing PP lamella hits an obstacle in the form of EOC and has to go around this obstacle. Then the overall growth rate of spherulite is slower.

In conclusion the EOC is slowing down the growth of individual spherulites but also acts as nucleation agent causing faster bulk crystallization. This result is in good agreement with above mentioned increase in activation energy.

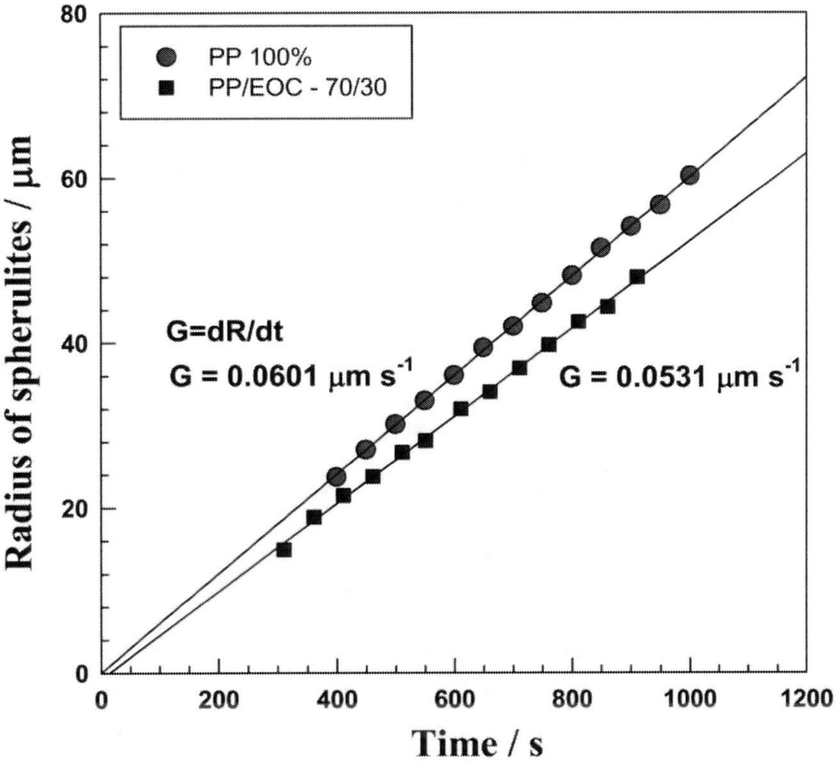

Figure 23. Two examples of spherulite growth rate evaluation from the plot of radius of spherulite vs. time for pure PP and PP/EOC 70/30 blend at T_c = 135°C.

The PP/EOC 70/30 blend was investigated in the temperature range 130-140°C by optical microscopy. Figure 24 shows that the growth was linear for all temperatures.

Again we have used the Hoffman-Lauritzen theory to plot the data according to Eq. 5 as shown in Figure 25. The K_g value in the temperature range 130-136°C is very similar to that one received by DSC (3.7x10^5) and corresponds to regime II crystallization. Then at the highest temperatures (136-140°C) the crystallization is very slow, the K_g value is 1.7x10^5 and the crystallization proceeds in regime I when the secondary nucleation is the slowest.

As mentioned earlier partial miscibility between PP and EOC does exist. During mixing at high shear PP gets into EOC regions and EOC penetrates into PP. Thus we have to use "PP-rich" and "EOC-rich" terms. When mixing stops and blend is quickly quenched the structure with fine morphology and partial miscibility is preserved. However when the sample is heated again for the crystallization experiment, this time there is no shear, the blend is in quiescent state. The blend is thermodynamically unstable and the polymers tend to phase separate.

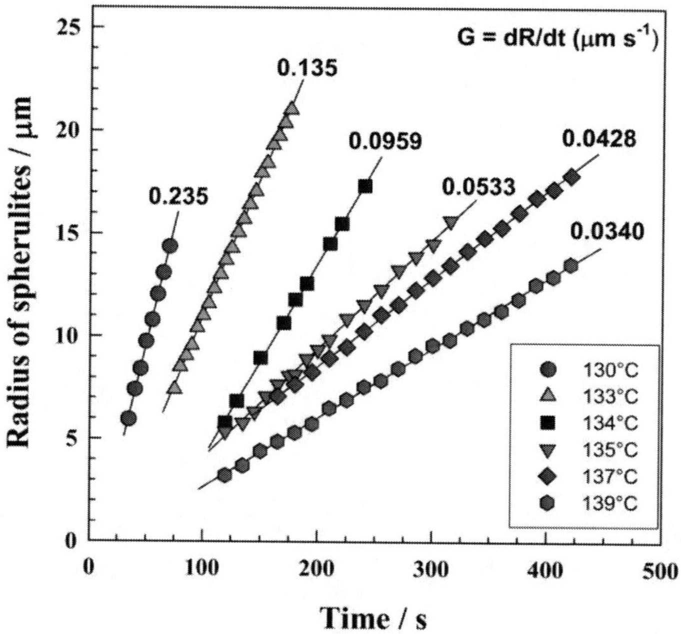

Figure 24. Spherulite growth rate evaluation for PP/EOC (70/30) blend from optical microscopy.

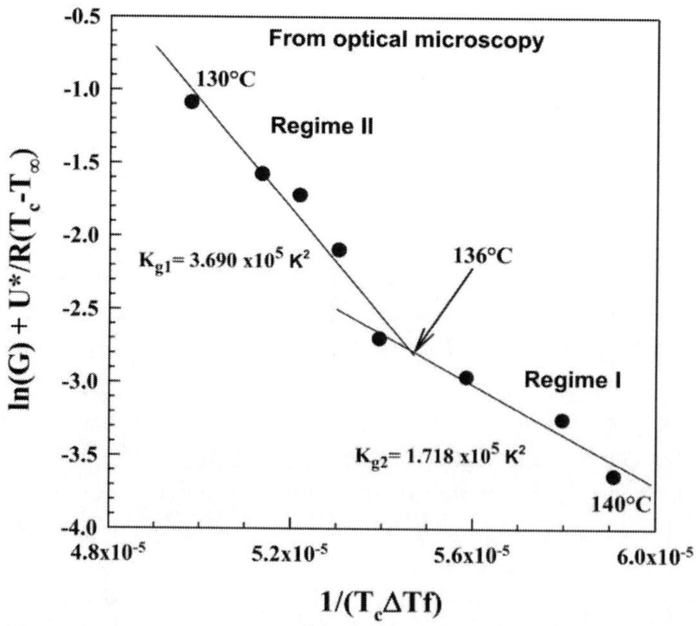

Figure 25. Hoffman-Lauritzen plots for PP/EOC (70/30) blend from optical microscopy.

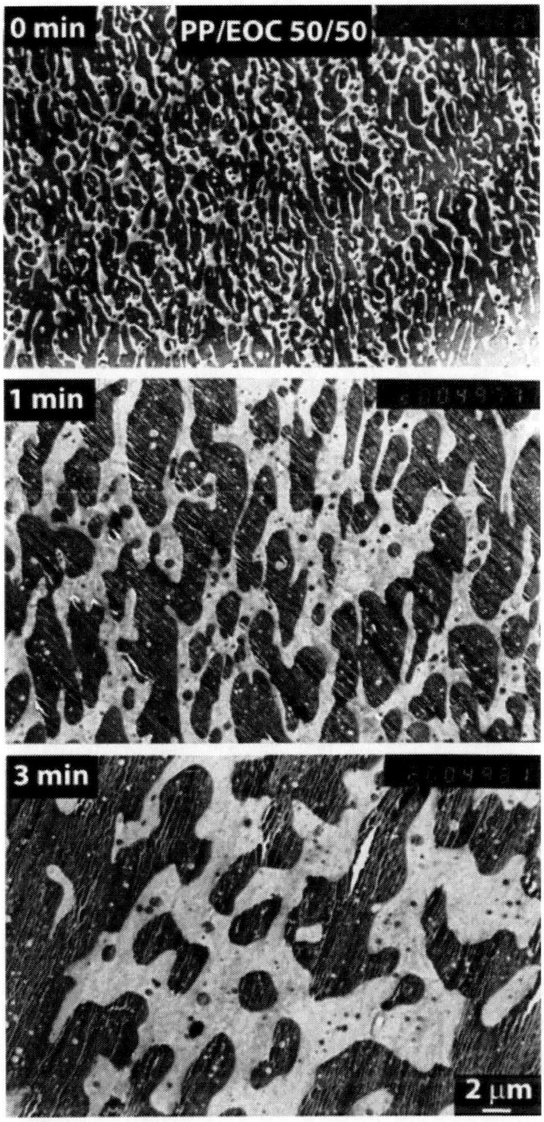

Figure 26. Coarsening of PP/EOC (50/50) blend at 200°C by TEM.

The example of such phase separation at 200°C is shown in Figure 26 [47]. Clearly a phase coarsening is happening at 200°C. Because of the shift in thermodynamics conditions (high shear→zero shear), also the miscibility is probably affected. This has been shown in past by shifting of LCST (UCST) due to the presence of high shear. The term "flow-induced mixing and demixing" was established [61]. On this basis we can assume that in quiescent state at high temperature where both PP and EOC are liquids, there is a strong thermodynamic driving force that causes movement of EOC molecules out of PP-rich phase and PP molecules out of EOC-rich phase.

Formally we have mentioned that EOC most likely acts in PP as nucleating agent causing faster bulk crystallization. This was observed in highest extent (e.g. for PP/EOC 80/20 blend) at low crystallization temperatures (122-126°C) when the crystallization takes just few minutes. It is clearly visible in Figure 27 where we compare the kinetics of a blend to that of pure PP. The ratio for PP/EOC 80/20 blend at 126°C has 1.8 value, meaning 80% increases in crystallization kinetics. However at high crystallization temperatures (132-136°C) when the crystallization takes more than 1h the ratio has decreased to about 1.2. The difference between the blend and pure PP is not that high. Most likely the EOC had enough time to move from PP-rich phase back to EOC-rich phase causing decreased number of nucleation centers in PP-rich phase. Then the bulk crystallization kinetics of the blend is just slightly higher (20%) than that of pure PP. The 50/50 blend has very interesting behavior. In 124-128°C temperature range the crystallization of the blend is faster than that of pure PP. Then in the temperature range 128-134°C the crystallization of pure PP is faster than that of the 50/50 blend. Finally for PP/EOC (20/80) blend the crystallization kinetics is always much slower than that of pure PP (the ratio is about 0.19).

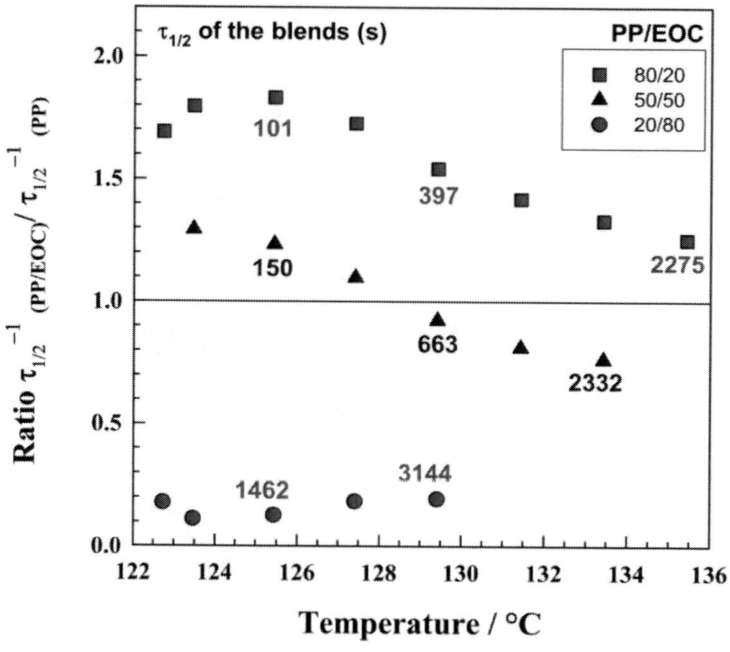

Figure 27. Ratio of crystallization kinetics ($\tau_{1/2}^{-1}$) of PP/EOC blends over neat PP for various temperatures.

Conclusion

Miscible PP/hBR blends served as a model for explanation of the elastic behaviour of polyolefinic TPE containing PP. There is a big difference in structure and properties for neat PP and PP/hBR blend. Regardless on crystallization temperature, PP shows always yielding point, i.e. plastic deformation after stretching even to small deformation. In neat PP, the crosshatch lamellar crystals were always formed. The situation for PP/hBR blend was completely different. The elastic properties are highly affected by crystallization temperature. The optimum in temperature of cooling medium was found to be around room temperature. In this case the formation of long crosshatch lamellae was prevented and the small fragmented crystal lamellae were formed. The fragmented crystallites play a role of tie points for amorphous chains to render good elastomeric behaviour. There is no yielding point in stress-strain curves for those blends. The elastic behaviour can be improved by annealing of the blend at higher temperature (after quenching), when the crystal size increased, but the fragmented character remained. When the blend crystallizes at high temperature, big crosshatch lamellae were formed and the blend showed very poor elastic properties.

Concerning the immiscible blends the crystallization kinetics is greatly affected by the blend composition. As it was found by DSC, small amounts of EOC (10-30%) increased the bulk crystallization kinetics. On the other hand, larger amounts of EOC (50-80%) caused the decrease in kinetics. Optical microscopy revealed a decrease in a single spherulite growth rate caused by EOC obstacles, and at the same time larger number of spherulites. EOC in small amount acts as a nucleation agent for PP and increases the bulk crystallization rate.

Crystallization kinetics was tremendously decreased when the EOC content was very high (such as 70-80%). As it was shown by TEM, in this case PP formed small islands in EOC matrix and crystallization in these isolated domains had to wait till nucleation happened in each of them. This is very slow process. We were not able to measure the isothermal kinetics for blends with only 10% of PP.

The analysis according to Avrami's equation revealed that rate constant k and also the Avrami exponent n follow the same trend like $\tau_{1/2}^{-1}$, initially increase and then decrease with increasing EOC content. The Hoffman-Lauritzen equation was at first analyzed in linear form which produced parameters K_g and lnG_0 that decreased significantly when the EOC content was 75-80%. Initial parameters K_g and lnG_0 were then used in nonlinear regression that revealed increase in activation energy U* caused by addition of EOC into PP. EOC acts as nucleating agent which increases the bulk crystallization rate, but also acts as hindrance to crystallization which decreases the growth rate of individual PP spherulite. This conclusion was confirmed by direct observation of spherulites growth on the hot stage in the polarized optical microscope. The analysis according to Hoffman-Lauritzen theory revealed also interesting transitions in crystallization regimes at 130 and 136°C.

Comparison of crystallization of the blend to pure PP revealed an existence of competition of crystallization with phase separation. At lower temperatures the crystallization proceeds fast and is not affected by phase separation that much. At high temperatures when the crystallization kinetics is very slow the phase separation has enough time to proceed before crystallization which results in crystallization kinetics of the blend being close to that one of pure PP (for PP/EOC 80/20).

ACKNOWLEDGMENT

This work has been supported by Operational Programme Research and Development for Innovations co-funded by the European Regional Development Fund (ERDF) and national budget of Czech Republic within the framework of the Centre of Polymer Systems project (reg.number: CZ.1.05/2.1.00/03.0111).

REFERENCES

[1] A. Y. Coran, *Handbook of Elastomers-New Developments and Technology*, Marcel Dekker, New York (1988).

[2] A. Y. Coran and R. Patel, *Rubber Chem Technol*,53, 141 (1980).

[3] Y. Kikuchi, T. Fukui, T. Okada and T. Inoue, *Polymer Engineering and Science*,31, 1029 (1991).

[4] J. D. Katsaros, M. F. Malone and H. H. Winter, *Polymer Engineering and Science*,29, 1434 (1989).

[5] F. B. C. Larbi, M. F. Malone, H. H. Winter, J. L. Halary, M. H. Leviet and L. Monnerie, *Macromolecules*,21, 3532 (1988).

[6] L. P. Rector, K. A. Mazich and S. H. Carr, *J Macromol Sci-Phys*,B27, 421 (1988).

[7] H. Sano, H. Yui, H. G. Li and T. Inoue, *Polymer*,39, 5265 (1998).

[8] Y. Yang, T. Chiba, H. Saito and T. Inoue, *Polymer*,39, 3365 (1998).

[9] N. Otsuka, Y. Yang, H. Saito, T. Inoue and Y. Takemura, *Polymer*,39, 1533 (1998).

[10] Y. Yang, N. Otsuka, H. Saito, T. Inoue and Y. Takemura, *Polymer*,40, 559 (1999).

[11] T. Xu, J. Yu and Z. Jin, *Materials and Design*,22, 27 (2001).

[12] M. Avella, R. Dellerba, E. Martuscelli and G. Ragosta, *Polymer*,34, 2951 (1993).

[13] X. Zhang, F. Xie, Z. Pen, Y. Zhang, Y. Zhang and W. Zhou, *European Polymer Journal*,38, 1 (2002).

[14] S. Hom, A. R. Bhattacharyya, R. A. Khare, A. R. Kulkarni, M. Saroop and A. Biswas, *Polymer Engineering and Science*,49, 1502 (2009).

[15] R. Adhikari, R. Godehardt, W. Lebek and G. H. Michler, *Journal of Applied Polymer Science*,103, 1887 (2007).

[16] R. R. Babu, N. K. Singha and K. Naskar, *Journal of Applied Polymer Science*,113, 1836 (2009).

[17] S. Mohanty and S. K. Nayak, *Journal of Applied Polymer Science*,104, 3137 (2007).

[18] J. W. Lim, A. Hassan, A. R. Rahmat and M. U. Wahit, *Polymer International*,55, 204 (2006).

[19] A. L. N. D. Silva, M. C. G. Rocha, F. M. B. Coutinhoa, R. Bretas and C. Scuracchiod, *Polymer Testing*,19, 363 (2000).

[20] M. Onoa, J. Washiyamaa, K. Nakajimab and T. Nishi, *Polymer*,46, 4899 (2005).

[21] P. Dias, Y. J. Lin, B. Poon, H. Y. Chen, A. Hiltner and E. Baer, *Polymer*,49, 2937 (2008).

[22] L. Godail and D. E. Packham, *J Adhesion Sci Technol*,15, 1285 (2001).

[23] G. Bogoeva-Gaceva, A. Janevski and A. Grozdanov, *Journal of Applied Polymer Science*,67, 395 (1998).

[24] J.-W. Huang, Y.-L. Wen, C.-C. Kang, M.-Y. Yeh and S.-B. Wen, *Journal of Applied Polymer Science*,109, 3070 (2008).

[25] H. Lu, Y. Hu, M. Li and L. Song, *Polymer Composites*,29, 1358 (2008).

[26] D. Rana, K. Cho, T. Woo, B. H. Lee and S. Choe, *Journal of Applied Polymer Science*,74, 1169 (1999).

[27] J.-R. Ying, S.-P. Liu, F. Guo, X.-P. Zhou and X.-L. Xie, *Journal of Thermal Analysis and Calorimetry*,91, 723 (2008).

[28] M. Avrami, *The Journal of Chemical Physics*,7, 1103 (1939).

[29] M. Avrami and J. B. Little, *Journal of Applied Physics*,13, 255 (1942).
[30] J. D. Hoffman, L. J. Frolen, G. S. Ross and J. I. J. Lauritzen, *J Res Natl Bur Stand Sect A Phys Chem*,79 A, 671 (1975).
[31] J. I. J. Lauritzen and J. D. Hoffman, *Journal of Applied Physics*,44, 4340 (1973).
[32] J. Brandrup, E. H. Immergut, E. A. Grulke, A. Abe and D. R. Bloch, *Polymer Handbook*. 4 ed, John Wiley & Sons., (2005).
[33] M. Reading, D. Elliott and V. L. Hill, *J Therm Anal*,40, 949 (1993).
[34] P. Svoboda, H. Saito, T. Chiba, T. Inoue and Y. Takemura, *Polymer Journal*,32, 915 (2000).
[35] P. Svoboda, K. Trivedi, D. Svobodova, P. Mokrejs, V. Vasek, K. Mori, T. Ougizawa, T. Inoue, *Materials Chemistry and Physics*,131, 84 (2011).
[36] M. T. Payne and C. P. Rader, *Elastomer Technology Handbook*, CRC Press, Boca Raton (1993).
[37] F. J. Padden and H. D. Keith, *Journal of Applied Physics*,30, 1479 (1959).
[38] D. R. Norton and A. Keller, *Polymer*,26, 704 (1985).
[39] B. Wunderlich, *Macromolecular Physics*, Academic Press, New York, N.Y. (1973).
[40] M. Okamoto, K. Shiomi and T. Inoue, *Polymer*,35, 4618 (1994).
[41] P. D. Hong, W. T. Chung and C. F. Hsu, *Polymer*,43, 3335 (2002).
[42] J. D. Hoffman, G. T. Davis and J. I. J. Lauritzen. Treatise on solid state chemistry. In: Hannay NB, editor. New York: Plenum press; 1976. p. 497.
[43] J.-W. Huang, *Polymer Engineering and Science*,49, 1855 (2009).
[44] G. Z. Papageorgiou, D. S. Achilias, D. N. Bikiaris and G. P. Karayannidis, *Thermochimica Acta*,427, 117 (2005).
[45] J. D. Hoffman and R. L. Miller, *Polymer*,38, 3151 (1997).
[46] M. Okamato and T. Inoue, *Polymer*,36, 2739 (1995).
[47] P. Svoboda, D. Svobodova, P. Slobodian, T. Ougizawa and T. Inoue, *European Polymer Journal*,45, 1485 (2009).
[48] A. J. Muller, M. L. Arnal, A. L. Spinelli, E. Canizales, C. C. Puig and H. Wang, *Macromol Chem Phys*,204, 1497 (2003).
[49] J. D. Hoffman and J. J. Weeks, *Journal of Chemical Physics*,37, 1723 (1962).
[50] T. Nishi and T. T. Wang, *Macromolecules*,8, 909 (1975).
[51] Q. L. Ni, J. Q. Fan and J. Y. Dong, *Journal of Applied Polymer Science*,114, 2180 (2009).
[52] Y. G. Shangguan, Y. H. Song and Q. Zheng, *Polymer*,48, 4567 (2007).
[53] V. Goodarzi, S. H. Jafari, H. A. Khonakdar, S. A. Monemian, R. Hassler and D. Jehnichen, *J Polym Sci Pt B-Polym Phys*,47, 674 (2009).
[54] Y. Feng, X. Jin and J. N. Hay, *Journal of Applied Polymer Science*,69, 2089 (1998).
[55] W. Leelapornpisit, M. T. Ton-That, F. Perrin-Sarazin, K. C. Cole, J. Denault and B. Simard, *J Polym Sci Pt B-Polym Phys*,43, 2445 (2005).
[56] S. Y. Liu, Y. N. Yu, Y. Cui, H. F. Zhang and Z. S. Mo, *Journal of Applied Polymer Science*,70, 2371 (1998).
[57] F. Marguerat, P. J. Carreau and A. Michel, *Polymer Engineering and Science*,42, 1941 (2002).
[58] S. Nandi and A. K. Ghosh, *J Polym Res*,14, 387 (2007).
[59] Y. Seo, J. Kim, K. U. Kim and Y. C. Kim, *Polymer*,41, 2639 (2000).
[60] S. Vyazovkin and N. Sbirrazzuoli, *Macromol Rapid Commun*,27, 1515 (2006).

[61] I. A. Hindawi, J. S. Higgins and R. A. Weiss, *Polymer*,33, 2522 (1992).

In: Polypropylene
Editors: L. P. Silva and E. F. Barbosa

ISBN: 978-1-62417-142-0

Chapter 4

EFFECT OF BETA NUCLEATING AGENTS ON THE CRYSTALLIZATION, MORPHOLOGY, MECHANICAL PROPERTIES AND HEAT RESISTANCE OF INJECTION MOLDED ISOTATIC POLYPROPYLENE

Qiang Dou*
College of Materials Science and Engineering,
Nanjing University of Technology, Nanjing, China

ABSTRACT

Injection molded β-isotactic polypropylene (β-iPP) was prepared with two commercial β nucleating agents, NT-A and NT-C, respectively. The effect of β nucleating agents on the crystallization, morphology, mechanical properties and heat resistance of iPP was investigated by means of differential scanning calorimeter (DSC), wide angle X-ray diffractometer (WAXD), polarized light microscope (PLM), scanning electron microscope (SEM), mechanical and heat deflection tests. DSC and WAXD analysis show that β crystal content of nucleated iPP is higher than that of pure iPP, and β crystal content of the core is higher than that of the skin in the central part of injection molded specimen. PLM and SEM observations show that injection molded iPP specimen have obvious skin-core structure. β nucleating agents induce abundant β crystals and make spherulites small, which obviously improve Izod notched impact strength. Heat deflection temperatures of the specimens are also improved by the β nucleating agents.

INTRODUCTION

Isotactic polypropylene (iPP) is one of the most common commercial plastics, which is widely used in many fields such as packaging, housewares, automotive industries, owing to its low cost, facile processing, chemical resistance and good mechanical properties.

* E-mail address: douqiang.njut@163.com, College of Materials Science and Engineering, Nanjing University of Technology, Nanjing, 210009, China

It is well known that iPP has five crystallographic forms: α, β, γ, δ and smectic phases. Monoclinic α is the most common and stable one, which can be obtained easily under normal processing conditions. Other forms will transform to α form under certain conditions. Some methods such as isothermal crystallization at a certain temperature [1], temperature gradient [2] and shear [3] will induce β-phase, but the content is low. The best way to obtain high purity β polypropylene is the addition of β nucleating agents. Fujiyama [4] prepared β-PP with γ-quinacridone and quinacridonequinone as β-crystal nucleating agents, and found that β-crystal nucleating ability of quinacridonequinone is superior to that of γ-quinacridone. Obadal and coworkers [5] used N, N'-dicyclohexylnaphthalene-2, 6-dicarboxamide to prepare β-iPP. Varga and Menyhard [6] found N, N'-dicyclohexyl-2, 6-naphthalenedicarboxamide is an efficient nucleating agent to prepare iPP samples rich in β-form, but not a selective one. Dou et al. found that N,N'-diphenyl glutaramide (DPG) [7], N, N'-diphenyl adipamide (DPA) [8] and N, N'-dicyclohexylsuccinamide (DCS) [9] can induce β crystals, and the maximum proportion of β crystal is 0.86, 0.97 and 0.79 with 0.2wt% DPG, 0.1wt% DPA and 0.05wt% DCS, respectively. Tjong and partners [10, 11] prepared β-phase PP with a bi-component β nucleating agent (a mixture of equal amount of pimelic acid and calcium stearate) and studied the morphology and mechanical behavior of injection molded specimens. They found that the content of β crystal in the core is 92%, β-spherulite exhibits a sheaf-like structure and no clear boundaries exist between the spherulites. Li et al. [12] used pimelic acid/calcium stearate mixture to produce β-iPP, but composition ratio of the components is 1/1.5. Juhász and coworkers [13] reported that calcium suberate is a selective β nucleating agent for iPP with exclusive β-form. Meng and Dou [14] found that pimelic acid modified wollastonite can induce more β crystals and makes the spherulites size of PP small. Feng et al. [15] added nineteen kinds of minerals, alone and in conjunction with LaC (a mixed ternary complex of trivalent lanthanum stearate and stearin) to iPP. They found that the mineral or LaC alone could not induce β-form, whereas the bicomponent mixtures containing a calcium compound could.

Injection molding is a common processing method for polypropylene. During injection molding, the melt will suffer complex thermal and flow conditions, which will significantly affect the structural development of polypropylene. The hot melt contacts the cold wall of the mould, which undergoes high shear and cooling rate, and oriented skin forms. While the melt away from the cold wall undergoes low shear and cooling rate, leading to the relaxation of the molecular chain, so the spherulite has enough time to grow. Kantz and colleagues [16] reported that three layers (skin, shear and core layers) exit in injection molded polypropylene for the first time. While Menges et al. [17] divided shear layer into three layers. Fitchmum and Mencik [18] found skin, transition and core layers and divided skin layer into three layers. Karger-Kocsis and Friedrich [19] observed six microstructural zones; five of them grouped in the skin region (including two shear zones) and one was defined as the core region. Fujiyama [20] reported that clear skin/core structures were observed in PP injection moldings without or with γ-quinacridone. Obadal, Cermak and colleague [5, 21] also found skin-core stuctures in injection-moulded polypropylenes and such stucture was strongly influenced by varying mould temperatures.

In this chapter β-iPPs were prepared with two commercial β nucleating agents, NT-A and NT-C, respectively. The melting and crystallization behavior, crystal morphology, mechanical properties and heat deflection temperatures (HDT) of β-iPPs are investigated. The aim of the work is to elucidate the effect of contents of β nucleating agents on the structures and properties of iPP.

MATERIAL AND METHODOS

Materials

iPP pellets, Grade F401, MFR=1.4 g/10min at 2.16 Kg and 230 °C, produced by Sinopec Yangzi Petrochemical Company Ltd. (Nanjing, China); β nucleating agent, NT-A and NT-C, supplied by Nanjing Chengkuan Trade Company Ltd. (Nanjing, China); and liquid paraffin, supplied by Shanghai Lingfeng Chemical Reagent Company Ltd. (Shanghai, China) were used in this work.

Preparation of Samples

iPP pellets were premixed with 0.5 wt% liquid paraffin (based on iPP weight) and variable amounts of NT-A or NT-C. The mixtures were compounded in a twin-screw extruder (TE-20, D=20mm, L/D=32/1, Coperion Keya Machinery Co., Ltd., Nanjing, China). The rates of feeding screw and main screw were 40 rpm and 200 rpm, respectively. The temperature of the extruder from the hopper to the die was 200 °C, 210 °C, 210 °C, and 200 °C, respectively. The melt was cooled and pelletized. The pellets were injection-molded with a reciprocating-screw injection molding machine (CJ80M3V, Chen De Plastics Machinery Co., Ltd., Foshan, China), and standard test specimens for mechanical and heat resistance tests were obtained. The injection temperature range was 200-205 °C and the injection pressure was 70 MPa.

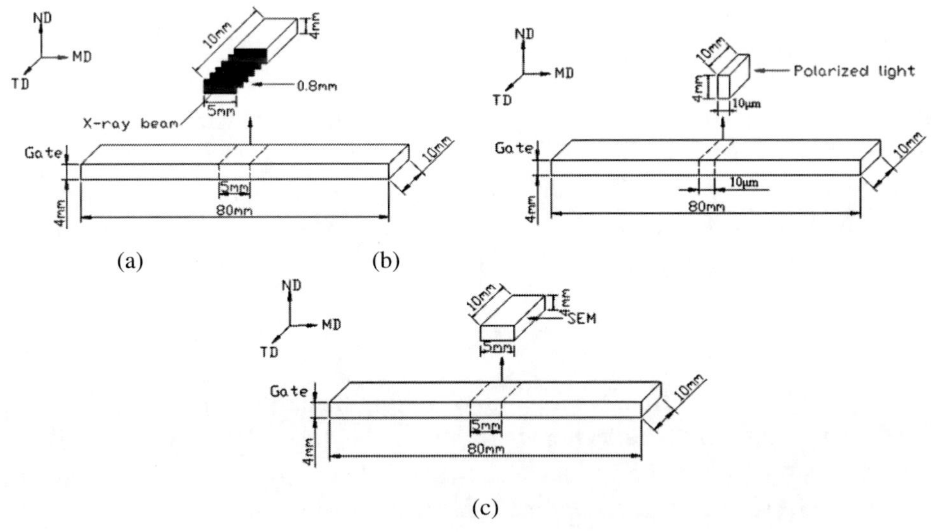

Figure 1. Schematics of the samples for characterization. (a) DSC and WAXD (b) PLM (c) SEM.

The samples for DSC, WAXD analysis and PLM, SEM observations were thin sections which were sliced from the central parts of the injection molded impact specimens with a microtome, as shown in Figure 1.

DSC Analysis

The melting and crystallization behaviors of the samples were studied with a DSC apparatus (ZF-DSC-D2, Shanghai Zufa Industry Co., Ltd., Shanghai, China) in a dry nitrogen atmosphere. The heat flow and temperature were calibrated with indium, stannum and zinc. Each sample of 5-6 mg was heated from room temperature to 210 °C and held for 5min. In order to avoid β-α transformation, which will affect the calculated β form content, the sample was then cooled to 100 °C, and held for 5min. Finally, the sample was reheated to 210 °C. Both the heating and cooling rates were 10 °C/min. The melting enthalpies of α-iPP ($H_{m,\alpha}$) and β-iPP ($H_{m,\beta}$) were obtained from the separation of the area of α and β melting peaks. The onset crystallization temperature (T_{con}) and the peak crystallization temperature (T_{cp}) were obtained from the cooling curves. The crystallinities of α-iPP (X_α) and β-iPP (X_β) and the content of β crystalline form (K_{DSC}) were calculated as follows.

$$X_\alpha = (H_{m,\alpha} / H^\circ_{m,\alpha}) \times 100\% \qquad (1)$$

$$X_\beta = (H_{m,\beta} / H^\circ_{m,\beta}) \times 100\% \qquad (2)$$

$$K_{DSC} = [H_{m,\beta}/(H_{m,\alpha} + H_{m,\beta})] \times 100\% \qquad (3)$$

Where $H^\circ_{m,\alpha}$ and $H^\circ_{m,\beta}$ are the enthalpies of fusion of α-iPP with 100% crystallinity (177 J/g) and β-iPP with 100% crystallinity (168.5 J/g) [12], respectively.

WAXD Characterization

WAXD characterization was carried out in an X-ray diffractometer (ARL X'TRA, Thermo Electron Corporation, USA). Radial scans of intensity versus diffraction angle (2θ) were recorded in the range of 10-30° with a scanning rate of 10°/min and a step size of 0.02°.

The relative content of β crystalline form was measured in terms of the K_{WAXD} proposed by Turner Jones et al. [22].

$$K_{WAXD} = [H_{\beta,1}/(H_{\beta,1} + H_{\alpha,1} + H_{\alpha,2} + H_{\alpha,3})] \times 100\% \qquad (4)$$

Where $H_{\beta,1}$ is the height (from the top of the peak to the background curve) of the peak at 2θ=16°, belongs to the (300) reflection of β-crystalline form. $H_{\alpha,1}$, $H_{\alpha,2}$ and $H_{\alpha,3}$ are the heights of three strong diffraction peaks at 2θ=14.1°,16.9° and 18.8°, corresponding to the (110), (040) and (130) plane of α-crystalline form, respectively.

PLM Observation

Crystalline textures were observed on a polarized light microscope (LW-200-4JS, Shanghai LW Scientific Co., Ltd., Shanghai, China) with a cross polar and a CCD camera. PLM pictures were captured and stored in a computer.

SEM Observation

Because the samples could not be etched effectively by the permanganate solution suggested in the literature [23]. The undermentioned procedure was used in this work. First, the samples were immersed in turpentine at 90 °C for 30min, and were washed with distilled water. Then the samples were etched in chromium trioxide solution (20 ml concentrated sulfuric acid, 80 ml distilled water and 50 g chromium trioxide) in 90 °C for 30 min, and were washed with plenty of distilled water. The dry etched samples were coated with gold. The morphologies were observed on a scanning electron microscope (SSX-550, Shimadzu, Japan).

Mechanical Test

The tensile and flexural tests were performed on an electromechanical universal testing machine (SANS5254, Shenzhen SANS Testing Machine Co., Ltd., Shenzhen, China) according to ISO 527-2/1BA/50 and ISO178, respectively. And the test speeds are 50 mm/min and 2 mm/min, respectively. The Izod notched impact strength was measured on a pendulum impact testing machine (MZ2056, Jiangdu Mingzhu Testing Machine Factory, Jiangdu, China) according to ISO 180/A.

HDT Test

Heat deflection temperatures (HDT) were measured according to ISO 75 with a HDT-VICAT tester (ZWK1302-B, Shenzhen SANS Testing Machine Co., Ltd., Shenzhen, China). The flexural stress was 0.45 MPa and the heating rate was 120 °C/h.

RESULTS AND DISCUSSION

DSC Analysis

Nucleated iPP Pellets

Figure 2 (a) shows the second melting curves of iPP pellets nucleated by NT-A. For pure iPP, a small peak at 155.6 °C and a big peak at 168.2 °C exist, which represent melting peaks of β- and α-phase, respectively. The β melting peak becomes big when NT-A is added. And the β melting peak is larger than α peak when NT-A content is less than 0.075%. After that, the α peak is larger than β peak gradually. Relevant DSC parameters are described in Figure 2 (b). It is shown than X_α decreases dramatically when 0.01% NT-A is added, after that, it increases gradually. But the change of X_β is on the contrary. The K_{DSC} attains the maximum at 0.01% content, and then it declines gradually.

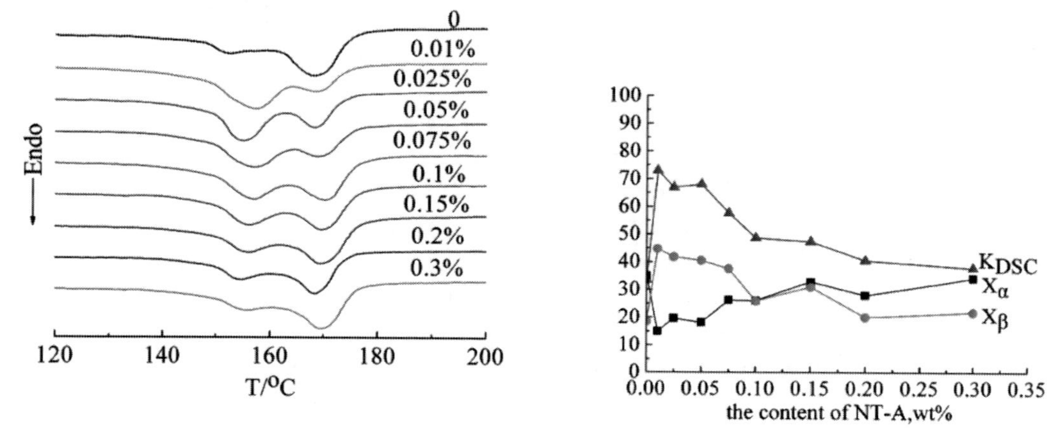

(a) (b)

Figure 2. DSC melting curves and relevant parameters of NT-A nucleated iPP pellets. (a) the second melting curves (b) X_α, X_β and K_{DSC}.

Figure 3 (a) shows the cooling curves of iPP pellets nucleated by NT-A. Relevant crystallization temperatures are illustrated in Figure 3 (b). T_c^{on} and T_c^{p} increase with the content of NT-A. The curve is leveled off when NT-A content is more than 0.15%.

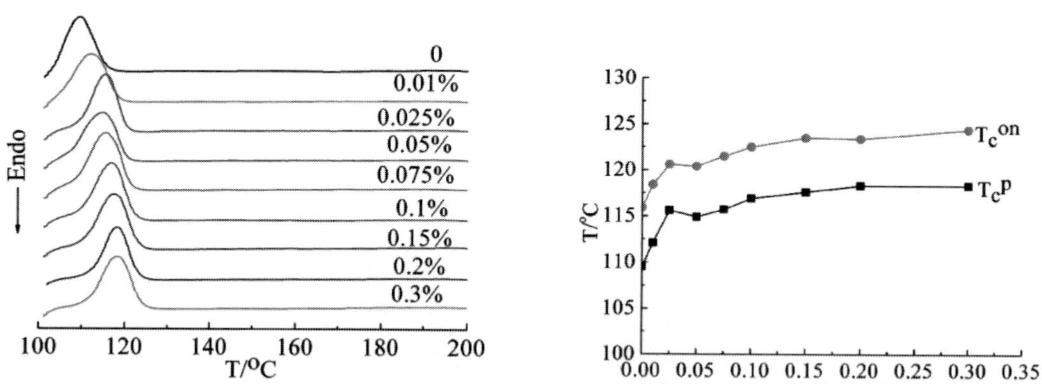

(a) (b)

Figure 3. DSC cooling curves and relevant parameters of NT-A nucleated iPP pellets. (a) the cooling curves (b) T_c^{on} and T_c^{p}.

Figure 5 (a) shows the cooling curves of iPP pellets nucleated by NT-C. The peak of crystallization shifts to higher temperature with the increasing content of NT-C. Relevant crystallization temperatures are illustrated in Figure 5 (b). T_c^{on} and T_c^{p} increase with the content of NT-C. With the addition of NT-C, a large number of nuclei are introduced into iPP melt, making the density of nuclei increase, so the nucleated iPP can crystallize in a higher temperature, and T_c^p and T_c^{on} increase.

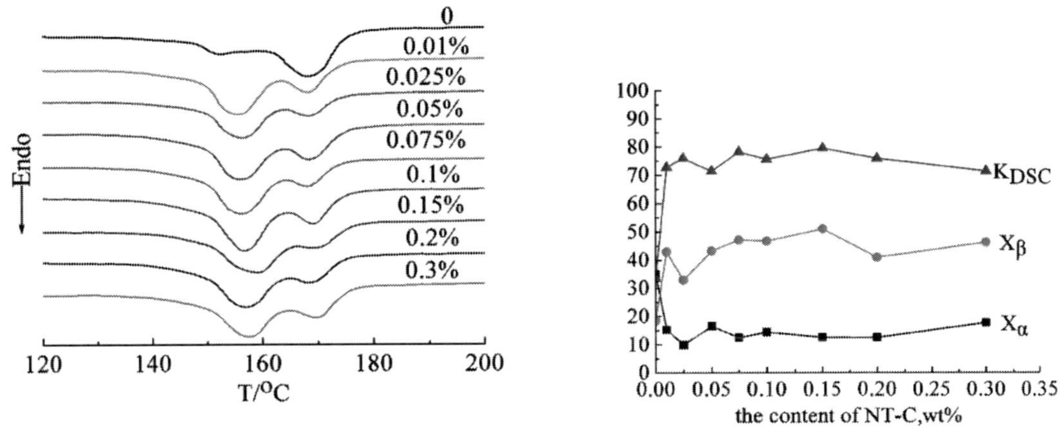

(a) (b)

Figure 4. DSC melting curves and relevant parameters of NT-C nucleated iPP pellets. (a) the second melting curves (b) Xα, Xβ and KDSC.

Figure 4 (a) shows the second melting curves of iPP pellets nucleated by NT-C. For the nucleated iPP, the melting peak of β-phase becomes large, while the melting peak of α-phase becomes small. Relevant DSC parameters are described in Figure 4 (b). X_α of nucleated iPP is lower than that of pure iPP, while X_β and K_{DSC} of nucleated iPP is higher than that of pure iPP. When only 0.01wt% NT-C is added, X_α decreases from 35.01% to 15.28%, X_β increases from 18.46% to 43.02%, and K_{DSC} increases from 33.42% to 72.82%.

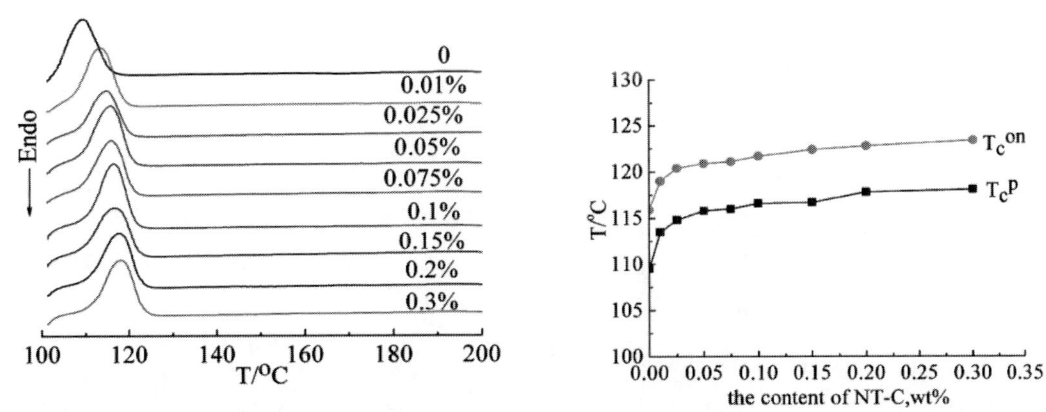

(a) (b)

Figure 5. DSC cooling curves and relevant parameters of NT-C nucleated iPP pellets. (a) the cooling curves (b) T_c^{on} and T_c^{p}..

Injection molded specimens

Figure 6 shows the first melting curves of injection molded specimens nucleated by NT-A. For pure iPP, at the surface, there is only one big peak at about 167.5 °C, which belongs to the melting peak of α-phase; away from the surface, besides the big α melting peak, there are two small peaks located between 145 °C and 155 °C, belonging to β-phase. For 0.05wt% NT-A nucleated iPP, the double small melting peaks of β-phase are more obvious than those of pure iPP, and the double melting peaks of β-iPP of the samples away from the surface are obvious than those of the surface. For 0.2wt% NT-A nucleated iPP, the samples at the distances ≤ 1.6 mm from the surface, their curves are similar to those of 0.05wt% NT-A nucleated iPP; while the distance is beyond 1.6 mm, there is only one melting peak of β-phase.

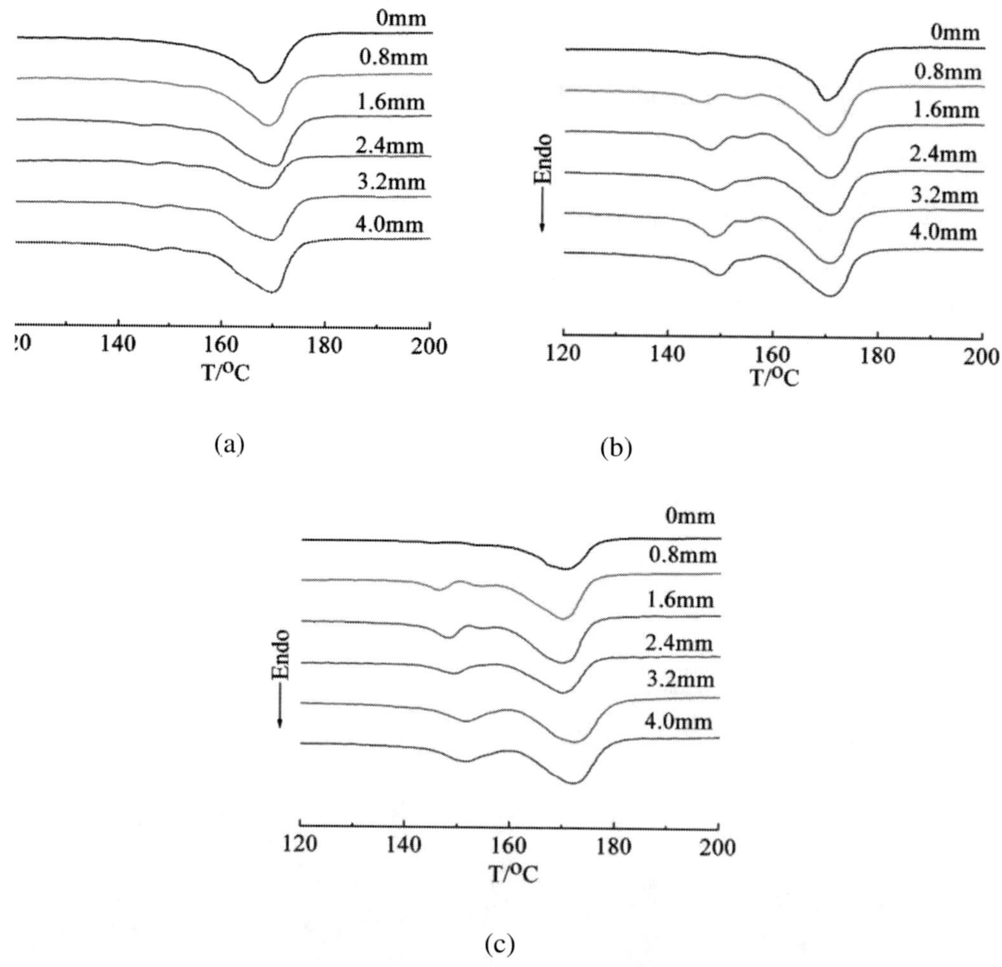

Figure 6. The first melting curves of the injection molded specimens nucleated by NT-A. The distances from the surfaces are indicated in the figures. (a) pure iPP (b) 0.05wt% NT-A (c) 0.2wt% NT-A.

Figure 7 shows the second melting curves of the injection molded specimens nucleated by NT-A. For pure iPP, two small melting peaks of β-phase exist. And away from the surface, the melting peaks of β-phase are more obvious than those of the first melting. For nucleated iPP, besides the small α peak, there is only one β melting peak, which is more obvious than those of the first melting. Compared with pure iPP, nucleated iPP has larger β melting peak.

X_α , X_β and K_{DSC} of the fist melting of injection molded specimens nucleated by NT-A are described in Figure 8. X_α of the sample at the surface is higher than that of the sample away from the surface. While X_β and K_{DSC} of the sample at the surface are lower than those of the sample away from the surface. In addition, X_α is higher than X_β for all samples at the same NT-A content and the same distance from the surface. It reflects the thermal and shear history of the specimen during the injection molding.

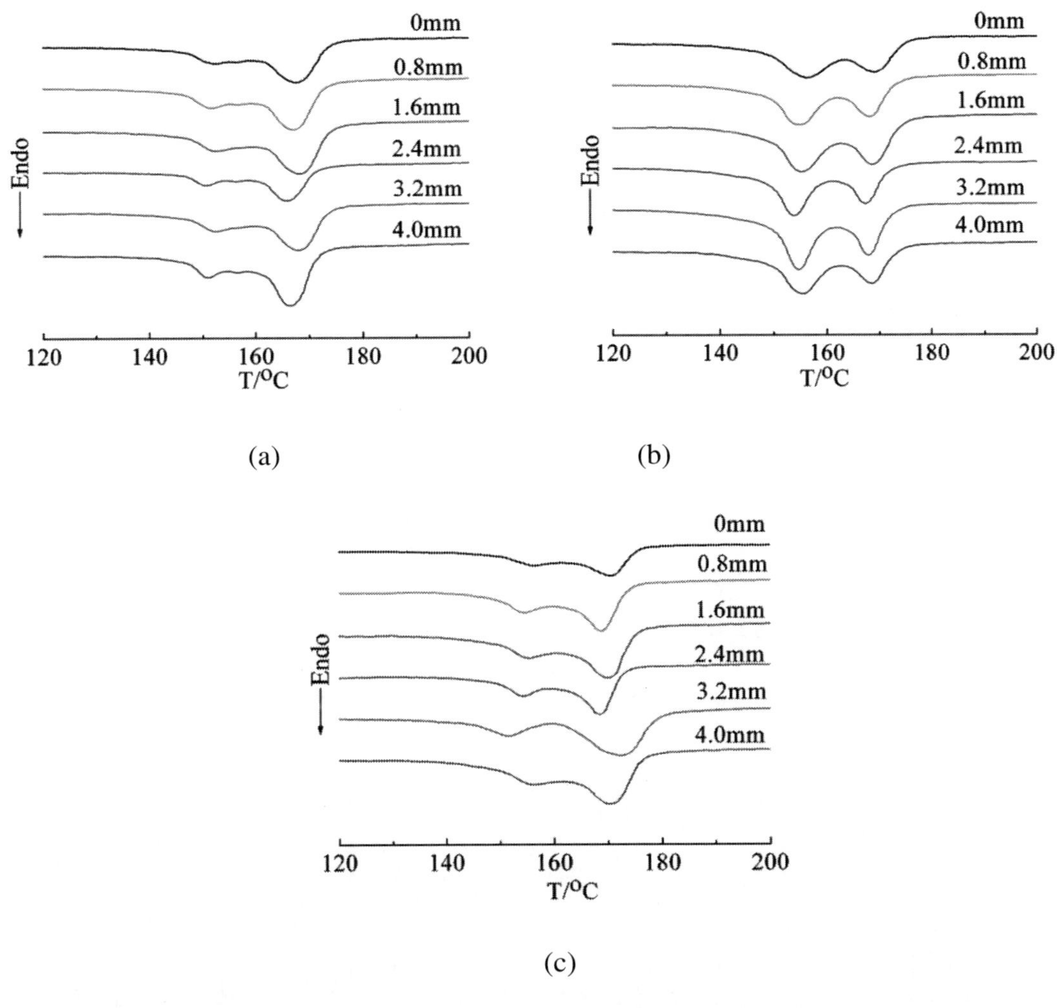

Figure 7. The second melting curves of the injection molded specimens nucleated by NT-A. The distances from the surfaces are indicated in the figures. (a) pure iPP (b) 0.05wt% NT-A (c) 0.2wt% NT-A.

X_α , X_β and K_{DSC} of the second melting of injection molded specimens nucleated by NT-A are described in Figure 9. The differences of X_α, X_β and K_{DSC} between the skins and the cores are marginal.

Comparing Figure 8 with Figure 9, it is found that X_α of the first melting is higher than that of the second melting, while X_β and K_{DSC} of the second melting are higher than those of the first melting. According to ISO 11357, DSC measurements on polymers are greatly affected by the thermal history and morphology of the sample and the test specimen. The first melting evaluates the original properties of injection specimens and removes thermal history. The second melting evaluates the properties of the samples which suffer nonisothermal crystallization in DSC. Samples suffer different thermal history in injection molding and in DSC measurement.

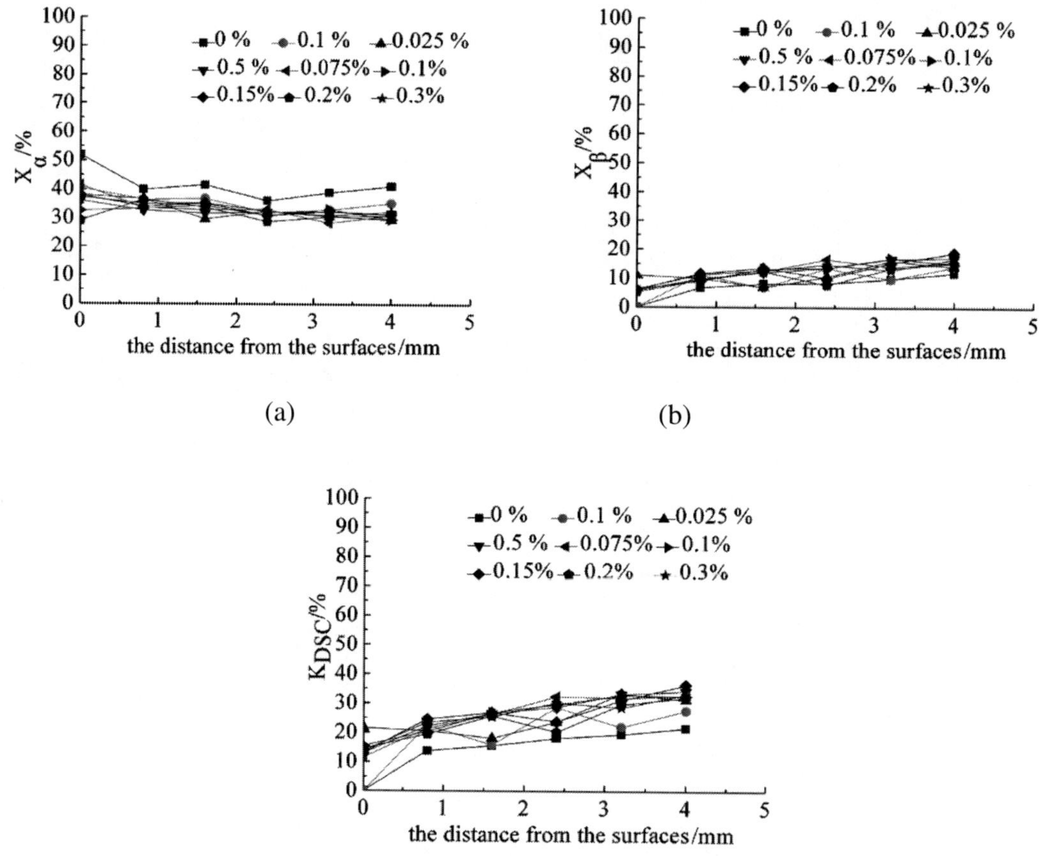

Figure 8. X_α, X_β and K_{DSC} of the fist melting of injection molded specimens nucleated by NT-A. The NT-A contents are indicated in the figures. (a) X_α (b) X_β (c) K_{DSC}.

Crystallization temperatures of injection specimens nucleated by NT-A are shown in Figure 10. T_c^p and T_c^{on} increase with the content of NT-A. With addition of NT-A, a large number of heteropical nuclei are introduced into iPP melt, making the density of nuclei increase, so iPP can crystallize in a higher temperature, and T_c^p and T_c^{on} increase. The difference of crystallization temperatures among different layers of the same specimen is marginal.

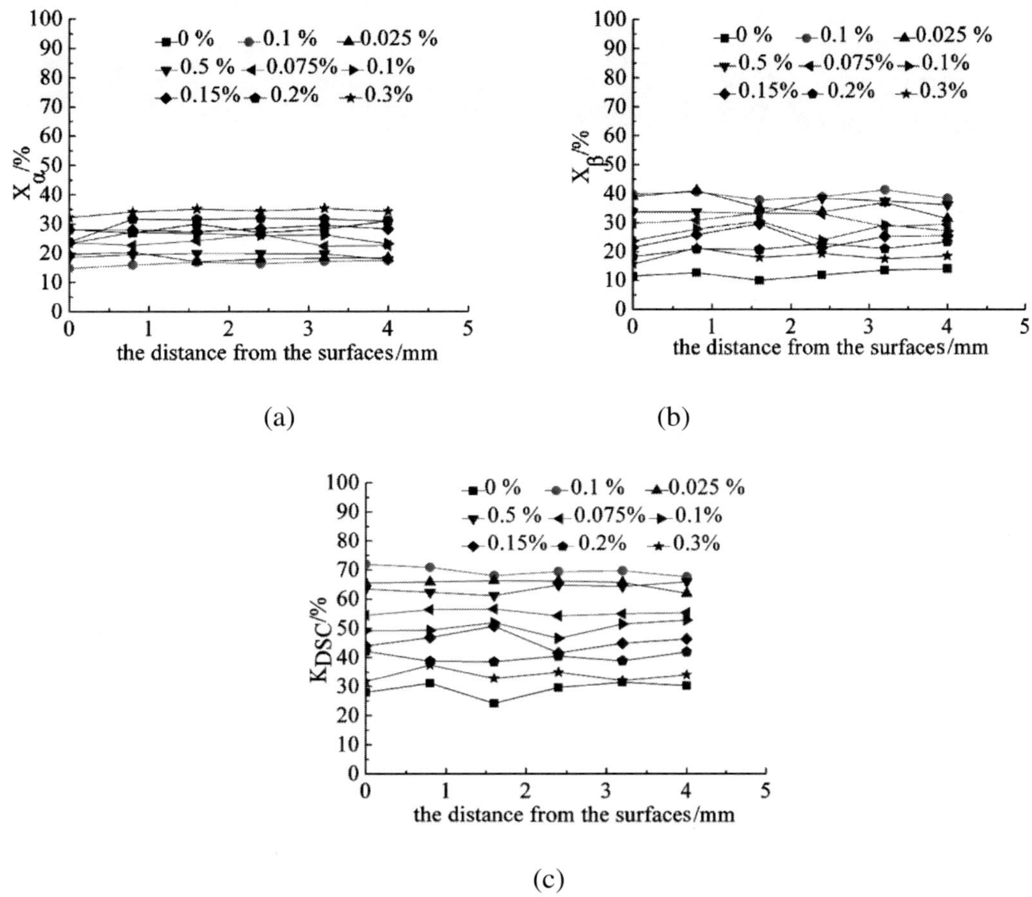

Figure 9. X_α, X_β and K_{DSC} of the second melting of injection molded specimens nucleated by NT-A. The NT-A contents are indicated in the figures. (a) X_α (b) X_β (c) K_{DSC}.

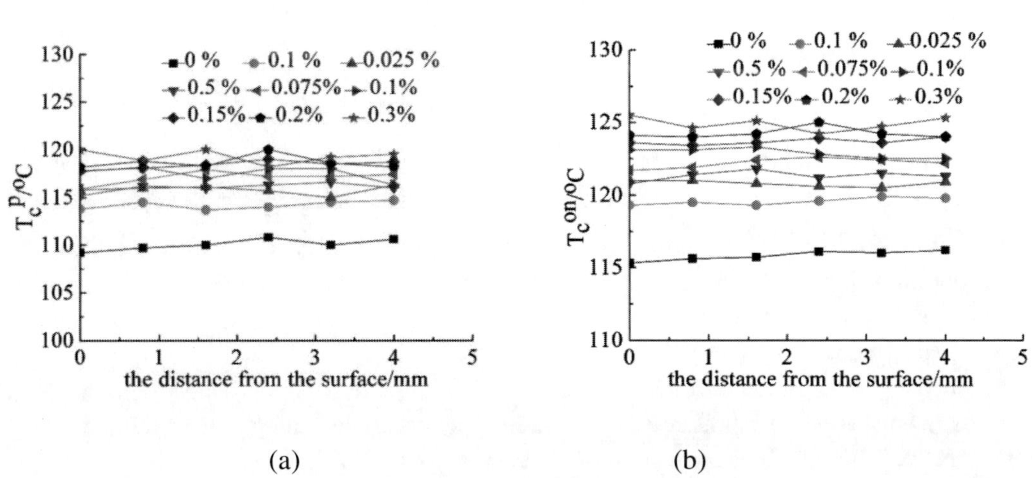

Figure 10. crystallization temperature of injection molded specimens nucleated by NT-A. The NT-A contents are indicated in the figures. (a) T_c^p (b) T_c^{on}.

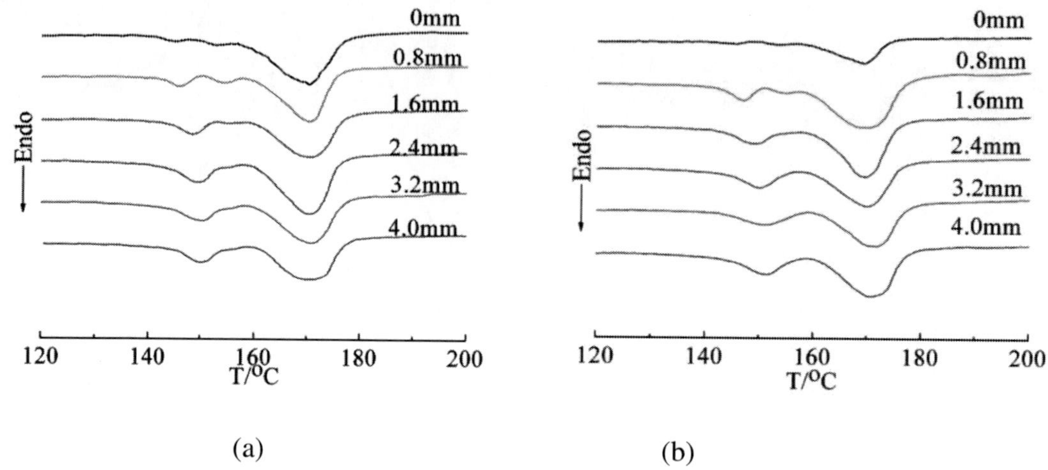

Figure 11. The first melting curves of the injection molded specimens nucleated by NT-C. The distances from the surfaces are indicated in the figures. (a) 0.05wt% NT-C (b) 0.2wt% NT-C.

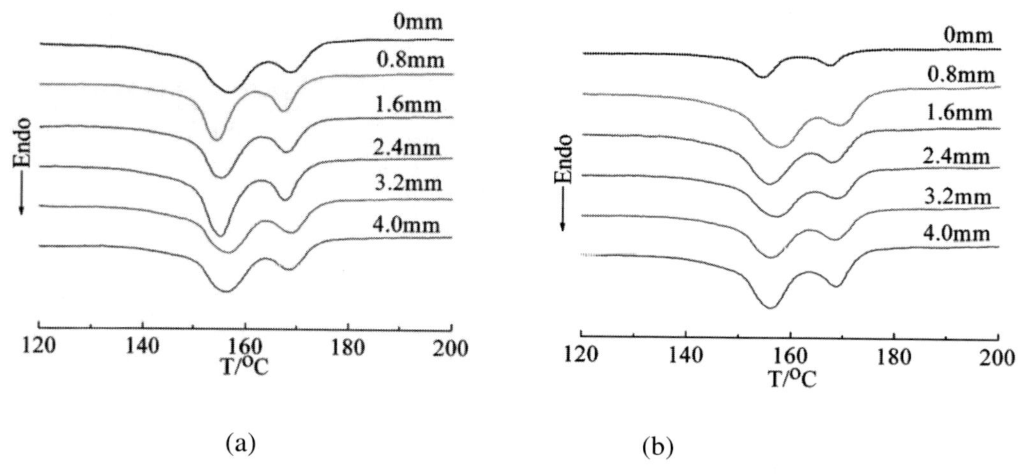

Figure 12. The second melting curves of the injection molded specimens nucleated by NT-C. The distances from the surfaces are indicated in the figures. (a) 0.05wt% NT-C (b) 0.2wt% NT-C.

Figure 11 shows the first melting curves of the injection molded specimens nucleated by NT-C. Double β melting peaks are present for 0.05% NT-C nucleated iPP specimen. But if the depth is larger than 0.8 mm, only one small β melting peak present for 0.05% NT-C nucleated specimen.

Figure 12 shows the second melting curves of the injection molded specimens nucleated by NT-C. The β melting peaks are larger than α peaks of nucleated iPP specimen at each depth. It reflects high β nucleation ability of NT-C.

Figure 13 shows X_{α} , X_{β} and K_{DSC} of the fist melting of injection molded specimens nucleated by NT-C. Generally speaking, X_{α} of the skin is higher than that of the core. While X_{β} and K_{DSC} of the skin are lower than those of the core. Figure 14 shows X_{α}, X_{β} and K_{DSC} of the second melting of injection molded specimens nucleated by NT-C. The differences of X_α, X_β and K_{DSC} between the skin and the core are marginal.

Comparing Figure 13 with Figure 14, we find that X_α of the second melting is lower than that of the first melting, while X_β and K_{DSC} of the second melting are higher than those of the first melting.

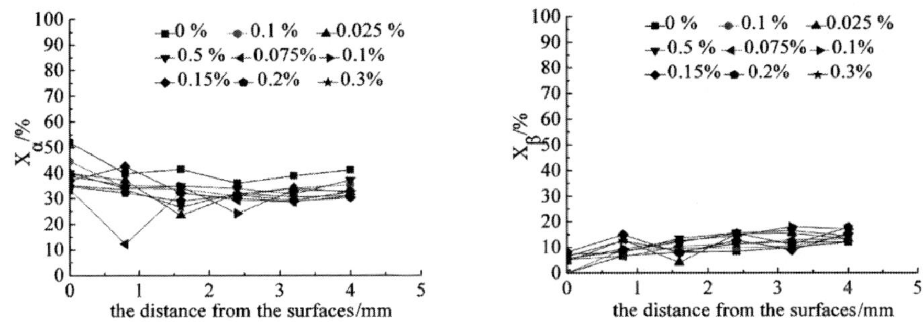

(a) (b)

Figure 13. X_α, X_β and K_{DSC} of the fist melting of injection molded specimens nucleated by NT-C. The NT-C contents are indicated in the figures. (a) X_α (b) X_β (c) K_{DSC}.

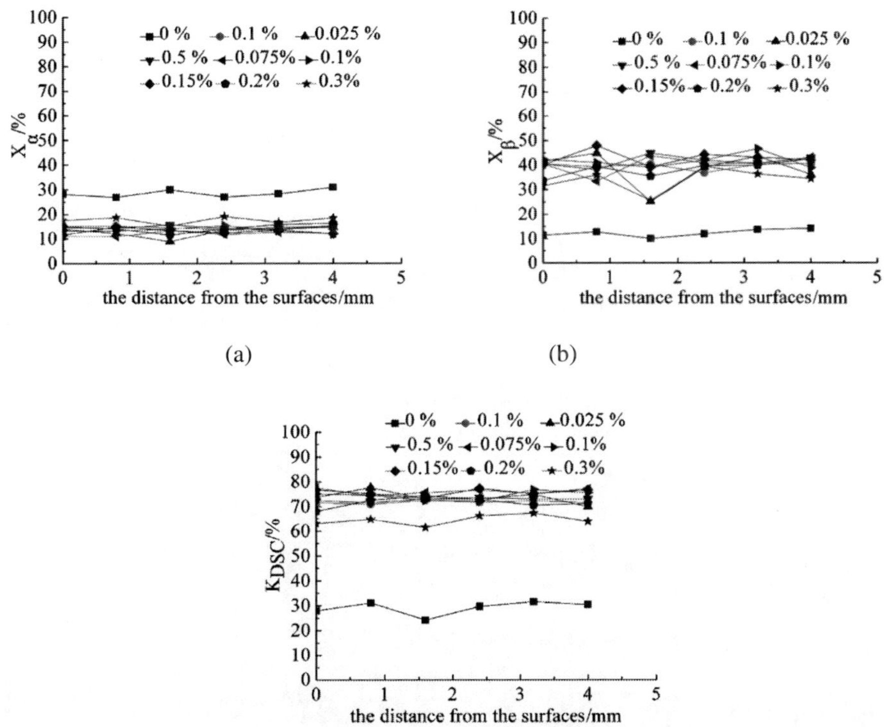

(a) (b)

(c)

Figure 14. X_α, X_β and K_{DSC} of the second melting of injection molded specimens nucleated by NT-C. The NT-C contents are indicated in the figures. (a) X_α (b) X_β (c) K_{DSC}.

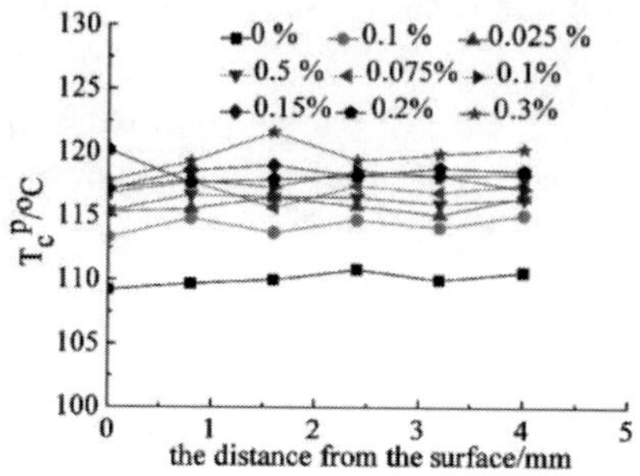

(a)

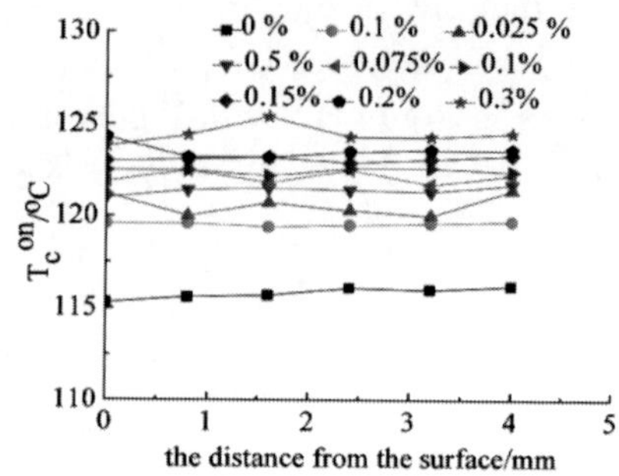

(b)

Figure 15. Crystallization temperatures of injection molded specimens nucleated by NT-C. The NT-C contents are indicated in the figures. (a) T_c^p (b) T_c^{on}.

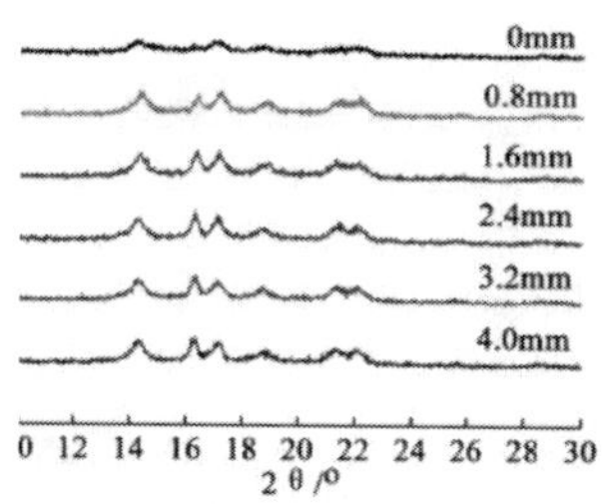

(a)

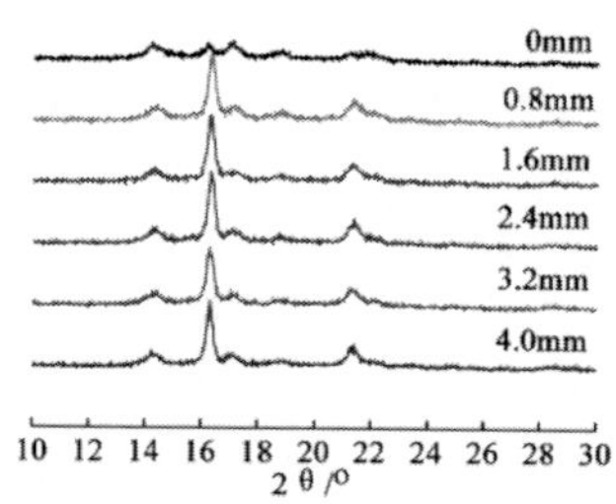

(b)

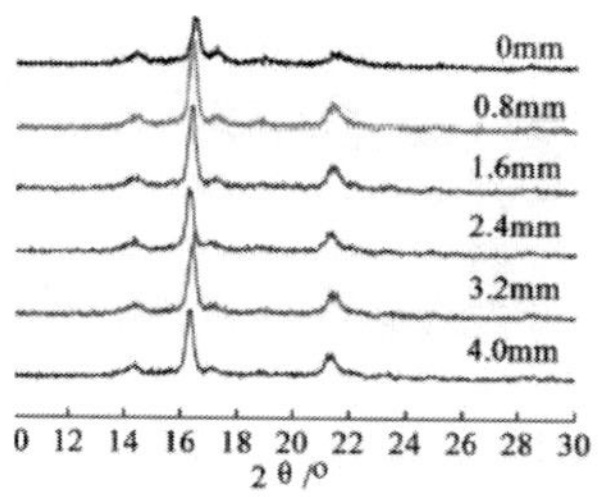

(c)

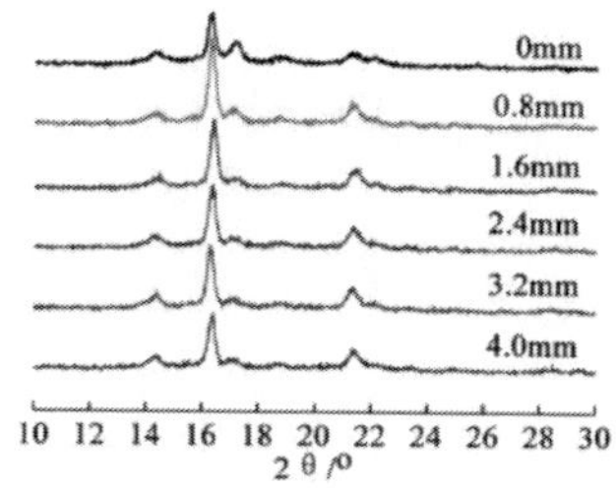

(d)

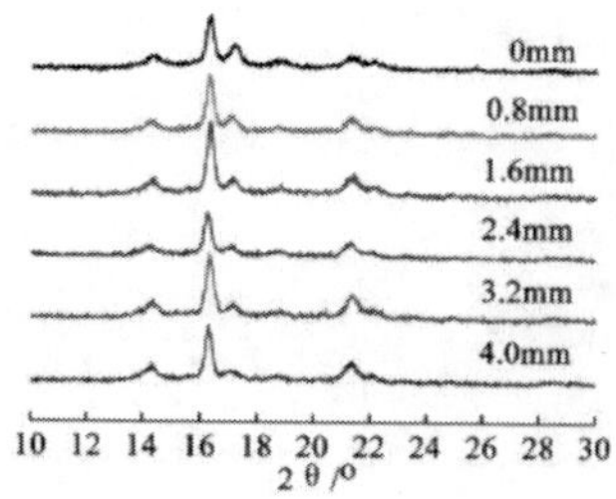

(e)

Figure 16. WAXD patterns of injection molded specimens nucleated by NT-A. The distances from the surfaces are indicated in the figures. (a) pure iPP (b) 0.01wt% NT-A (c) 0.05wt% NT-A (d) 0.1wt% NT-A (e) 0.2wt% NT-A.

Crystallization temperatures of injection molded specimens nucleated by NT-C are shown in Figure 15. T_c^p and T_c^{on} increase with the content of NT-C, which is in accordance with the results of DSC analysis of nucleated iPP pellets. The difference of crystallization temperature between skin and core is marginal.

WAXD Characterization

WAXD patterns of injection molded specimens nucleated by NT-A are shown in Figure 16. For pure iPP, in the surface, three characteristic peaks are observed at the diffraction angle of $2\theta=14.1^o$, 16.9^o, 18.8^o, corresponding to the reflection plane of (110), (040), and (130), respectively, which belong to α crystal form; while away from surface, besides characteristic peaks of α crystal form, the peak at the diffraction angle of $2\theta=16.2^o$, which belongs to (300) plane of β crystal form, is also observed.

For iPP nucleated by NT-A, there are characteristic peaks of α and β crystal form in both skin and core, and the characteristic peak of β crystal form of the core is more obvious than that of the skin, which suggests that content of β crystal in the core is higher than that in the skin.

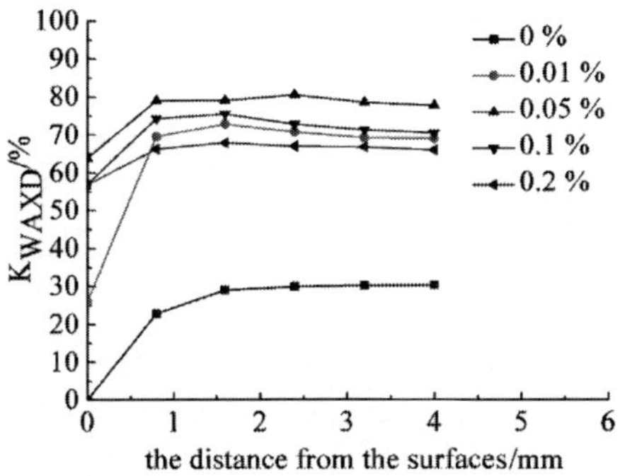

Figure 17. K_{WAXD} of injection molded specimens nucleated by NT-A. The NT-A contents are indicated in the figures.

The relative contents of β crystal, in terms of K_{WAXD}, are illustrated in Figure 17. K_{WAXD} attains the maximum when the content of NT-A is 0.05wt%, and then K_{WAXD} decreases somewhat with the content of NT-A. Fujiyama [20] also found that the content of β crystals shows a maximum at a γ-quinacridone content of about 100 ppm. Whether pure or nucleated iPP, K_{WAXD} increases largely from the surface to 0.8 mm away from the surface; after that, changes of K_{WAXD} are small. For the injection molded PP with a bi-component β nucleating agent, K_{WAXD} in the core layer (0.92) is higher than that in the skin layer (0.54), which was reported by Tjong et al [10]. They explained that polymers are subjected to a complex environment of stress and thermal gradients during injection molding; the cooling rate differs dramatically from the mold walls to the center of the molding, resulting in development of skin-core morphologies.

WAXD patterns of injection molded specimens nucleated by NT-C are described in Figure 18. For nucleated specimens, there are characteristic peaks of α and β crystal forms in both skins and cores, and the characteristic peaks of β crystal forms of the cores are larger than those of the skins, which show that the content of β crystal in the core is higher than that in the skin. Moreover, the characteristic peak of β crystal form of nucleated specimens is also larger than that of pure iPP.

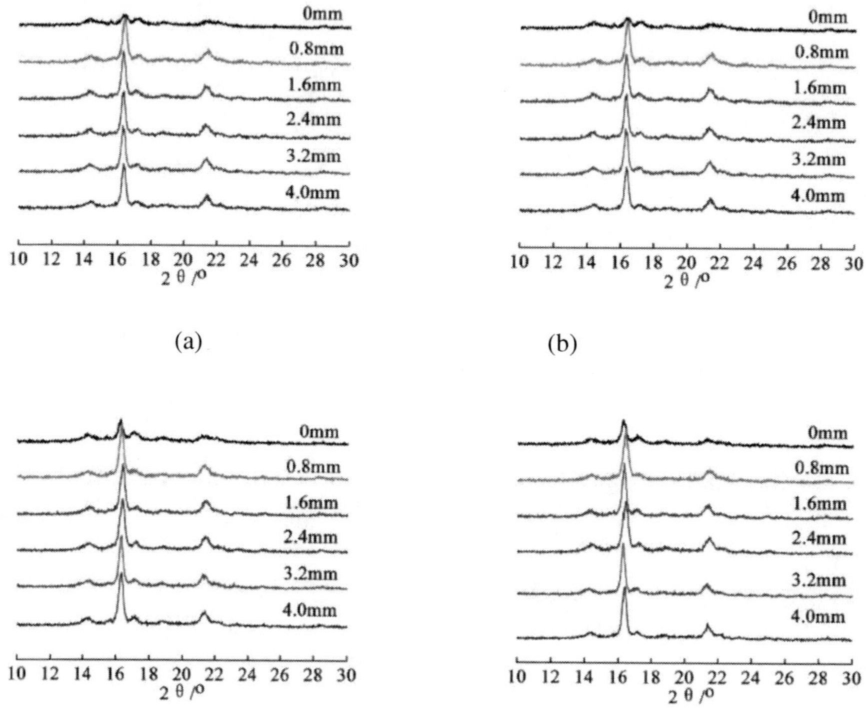

(a) (b)

(c) (d)

Figure 18. WAXD patterns of injection molded specimens nucleated by NT-C. The distances from the surfaces are indicated in the figures. (a) 0.01wt% NT-C (b) 0.05wt% NT-C (c) 0.1wt% NT-C (d) 0.2wt% NT-C.

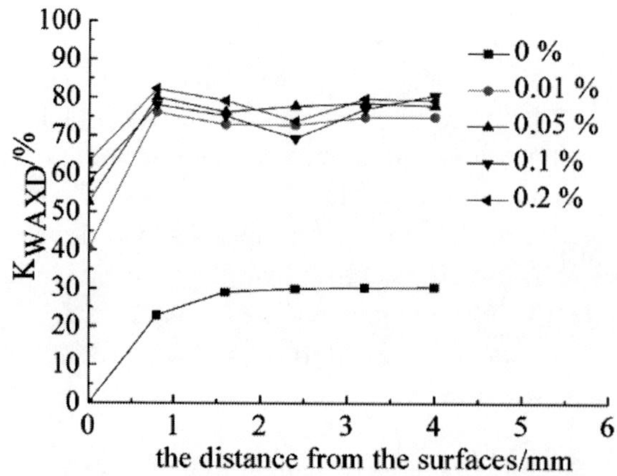

Figure 19. K_{WAXD} of injection molded specimens nucleated by NT-C. The NT-C contents are indicated in the figures.

β crystal content, in terms of K_{WAXD}, is illustrated in Figure 19. K_{WAXD} increases with the content of NT-C. Whether pure or nucleated iPP, K_{WAXD} increases largely from the surface to 0.8 mm away from the surface; after that, the changes of K_{WAXD} are small.

PLM Observation

Figure 20 shows PLM pictures of injection molded specimens nucleated by NT-A at low magnification. Skin-core structures are observed in both pure iPP and NT-A nucleated iPP. For pure iPP, there is a clear transition layer between the skin and the core, and the size of spherulites becomes larger gradually from the transition layer to the core. With addition of NT-A, the size of spherulites of the core becomes smaller. In addition, for nucleated iPP, there is not any clear transition layer between the skin and the core.

PLM pictures of skin and core of Figure 20 at high magnification are shown in Figure 21. Whether pure iPP or nucleated iPP, no spherulites exist in the skins. But there are spherulites in the cores. For pure iPP, β crystals are few and the size is large. For nucleated iPP, β crystals are abundant and the size is small.

PLM pictures of injection molded specimens nucleated by NT-C at low magnification are shown in Figure 22. With the addition of NT-C, the size of spherulites of the core becomes smaller. In addition, for the nucleated iPP, the difference between the skin and the core is small.

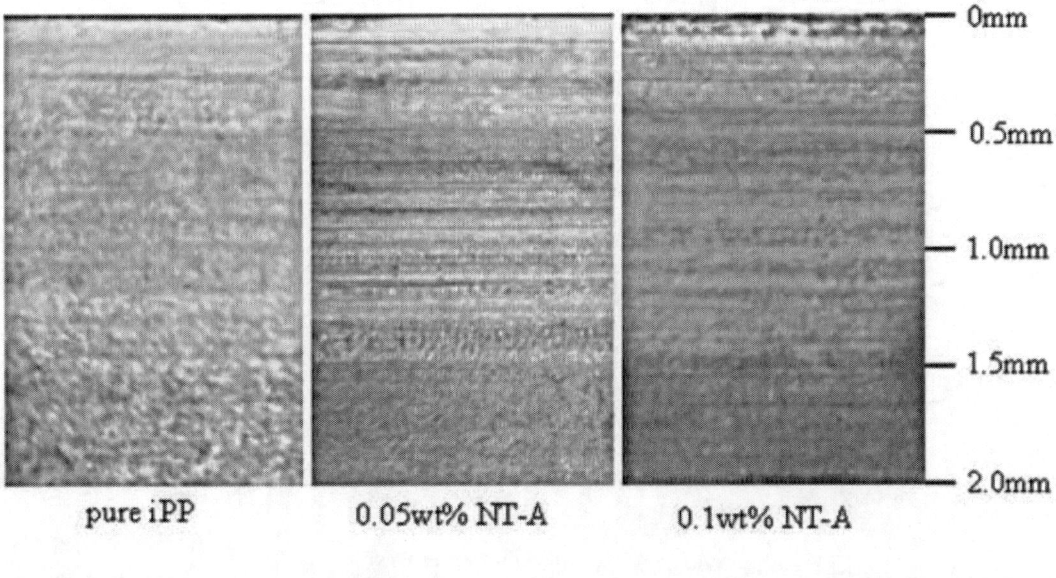

Figure 20. PLM pictures of injection molded specimens nucleated by NT-A at low magnification. The distance from the surface is indicated in the figure.

Figure 23 shows PLM pictures of skin and core of Figure 22 at high magnification. No spherulites exist in the skin of pure iPP or nucleated iPP, while there are spherulites in the core. For pure iPP, the bright β crystals are few and the size is large. For nucleated iPP, small β spherulites are abundant.

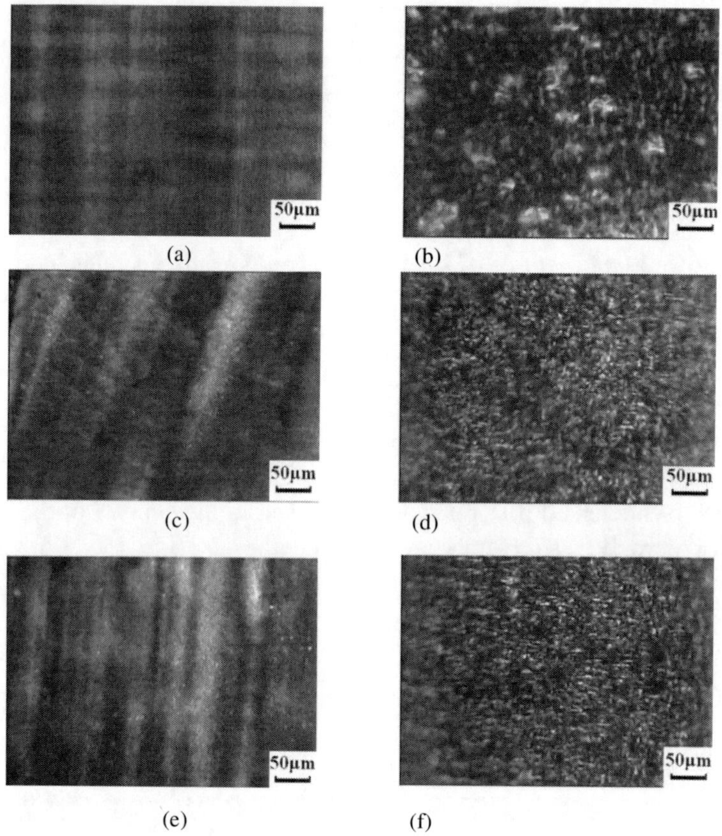

Figure 21. PLM pictures of skins and cores of injection molded specimens nucleated by NT-A. (a) pure iPP (skin) (b) pure iPP (core) (c) 0.05wt% NT-A (skin) (d) 0.05wt% NT-A (core) (e) 0.1wt% NT-A (skin) (f) 0.1wt% NT-A (core).

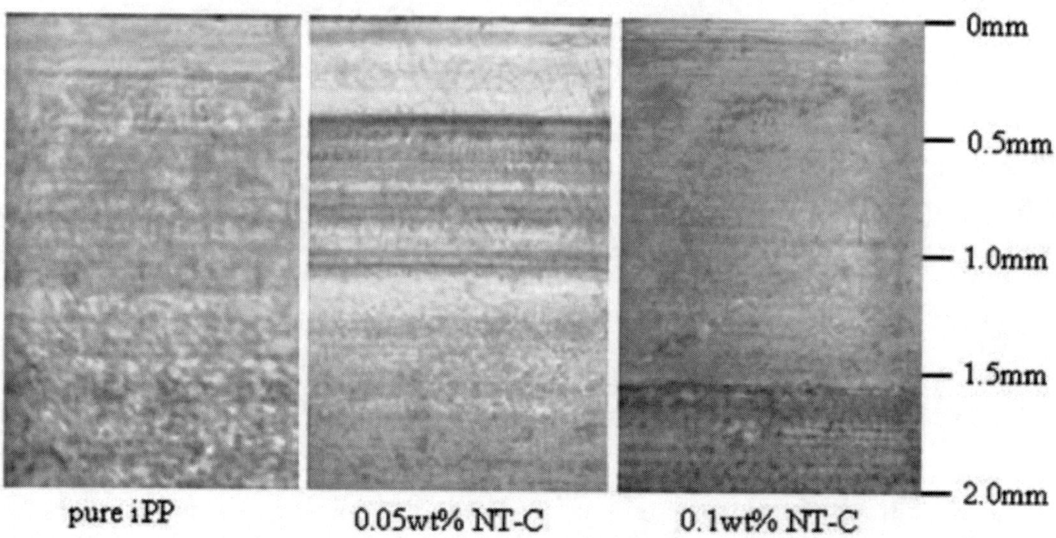

Figure 22. PLM pictures of injection molded specimens nucleated by NT-C at low magnification. The distance from the surface is indicated in the figure.

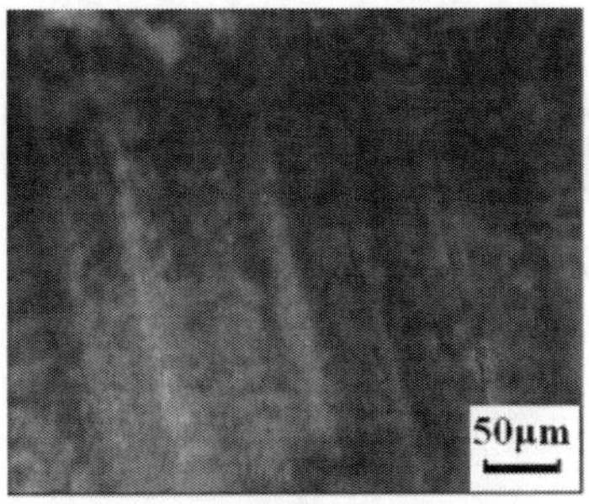

(a)

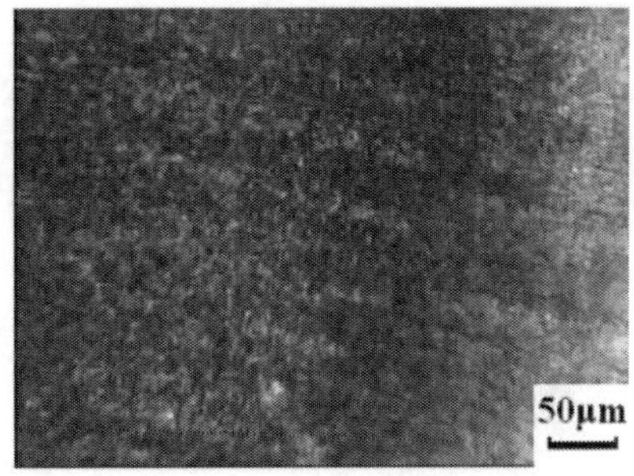

(b)

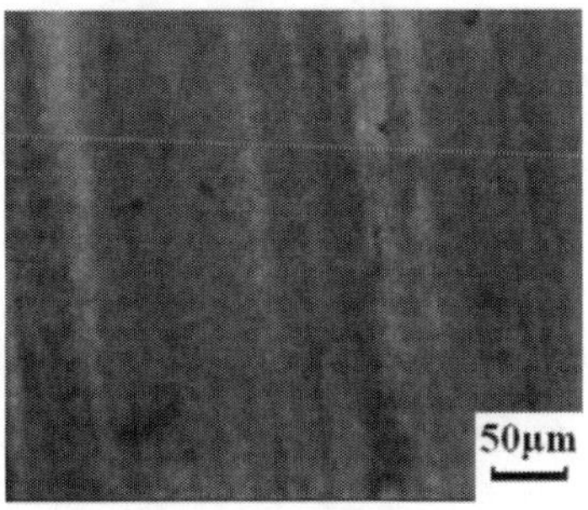

(c)

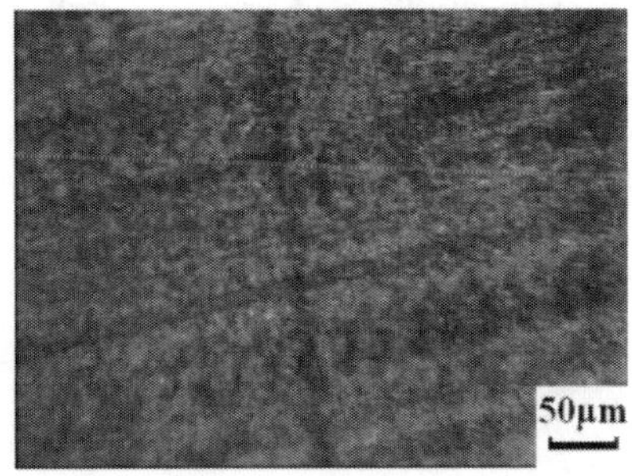

(d)

Figure 23. PLM pictures of skins and cores of injection molded specimens nucleated by NT-C. (a) 0.05wt% NT-C (skin) (b) 0.05wt% NT-C (core) (c) 0.1wt% NT-C (skin) (d) 0.1wt% NT-C (core).

SEM Observation

Figure 24 shows SEM images of injection molded specimens nucleated by NT-A and NT-C. We can see that the skin is orientated structure without any spherulites, while spherulites exist in the core. The spherulitic size in the core of nucleated iPP is smaller than that of pure iPP. This is in accord with the results of PLM observation.

Mechanical test

Figure 25 (a) shows the dependence of tensile strength and tensile strain at break of injection molded specimens on the content of NT-A. Tensile strength of nucleated iPP is lower than that of pure iPP, while tensile strain at break of nucleated iPP is higher than that of pure iPP. When the content of NT-A is 0.2wt%, tensile strain at break is the highest.

Figure 25 (b) shows the dependence of flexural modulus and Izod notched impact strength on the content of NT-A. Flexural modulus decreases from 1435 MPa to 1368 MPa when the content of NT-A is 0.025wt%; after that, flexural modulus increases with the content of NT-A. Izod notched impact strength increases with the content of NT-A; when the content of NT-A is 0.075wt%, impact strength reaches the maximum, which is 2.6 times more than that of pure iPP; after that, impact strength deceases somewhat. Varga [24] summarized that the superior toughness of β-iPP to α-iPP can be attributed to a combined effect of (1) the α to β phase transformation induced by mechanical load, (2) enhanced mechanical damping of β-iPP, and (3) the peculiar lamellar morphology of β-iPP.

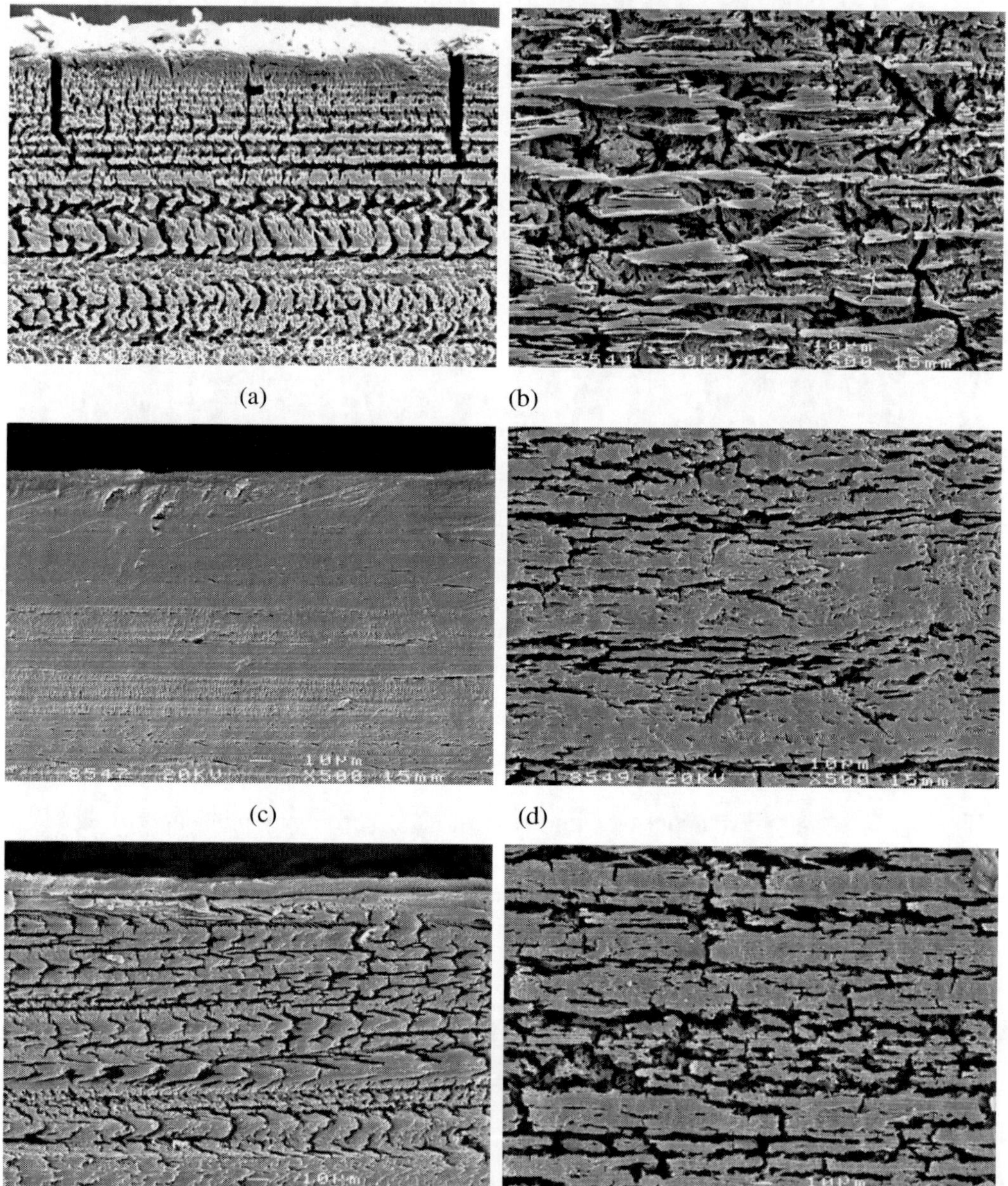

Figure 24. SEM images of injection molded specimens. (a) pure iPP (skin) (b) pure iPP (core) (c) 0.05wt% NT-A (skin) (d) 0.05wt% NT-A (core) (e) 0.05wt% NT-C (skin) (f) 0.05wt% NT-C (core).

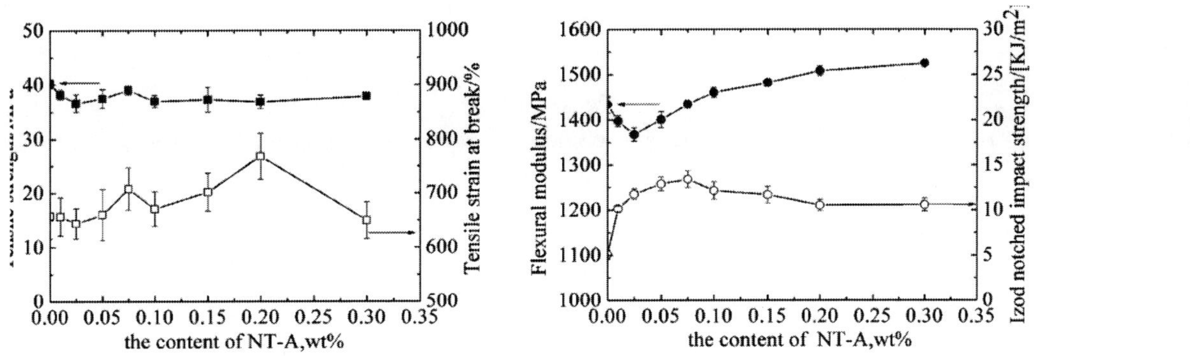

(a) (b)

Figure 25. Mechanical properties of injection molded specimens nucleated by NT-A. (a) tensile strength and tensile strain at beak (b) flexural modulus and Izod notched impact strength.

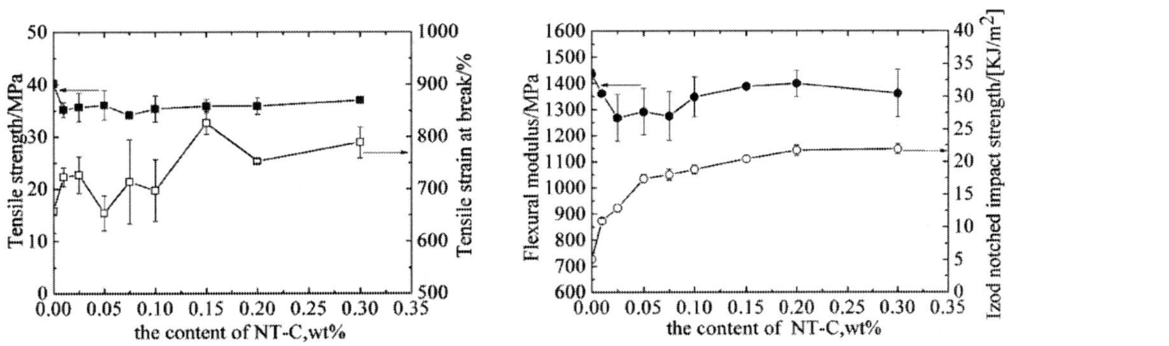

(a) (b)

Figure 26. Mechanical properties of injection molded specimens nucleated by NT-C. (a) tensile strength and tensile strain at beak (b) flexural modulus and Izod notched impact strength.

Figure 26 (a) shows the dependence of tensile strength and tensile strain at break on the content of NT-C. Tensile strength of β nucleated iPP is lower than that of pure iPP. With the addition of 0.01 wt% NT-C, tensile strength deceases from 40.2 MPa to 35.2 MPa. Tensile strain at break of β nucleated iPP is higher than that of pure iPP. With the addition of 0.15wt% NT-C, tensile strain at break is the highest, increases by 26%.

Figure 26 (b) shows the dependence of flexural modulus and Izod notched impact strength on the content of NT-C. Flexural modulus decreases from 1435 MPa to 1269 MPa when the content of NT-C is 0.025wt%; between 0.025wt% and 0.075wt%, flexural modulus changes little; after 0.075wt%, flexural modulus increases somewhat. Notched impact strength increases with the content of NT-C, and when the content of NT-C is 0.3wt%, the impact strength is 4.3 times more than that of pure iPP.

It is shown that β-iPP has lower tensile strength and flexural modulus, and higher tensile strain at break and impact strength than those of α-iPP.

HDT Test

The effect of NT-A on HDT of injection molded specimens is illustrated in Figure 27. It is shown that HDT increases with the content of NT-A. Fujiyama [4,20,25] and Varga [24] summarized that heat distortion temperature of β-iPP exceeds that of α-iPP. In Jarus and his coworkers' research [26], the HDT of injection-molded bars is directly related to the anisotropic morphology gradient produced by the injection process. The crystallization temperature determines the extent of orientation. The higher crystallization temperature of nucleated polypropylene results in increased retention of melt orientation. This increase in orientation correlates with an increase in the high temperature modulus and therefore with an increase in the HDT.

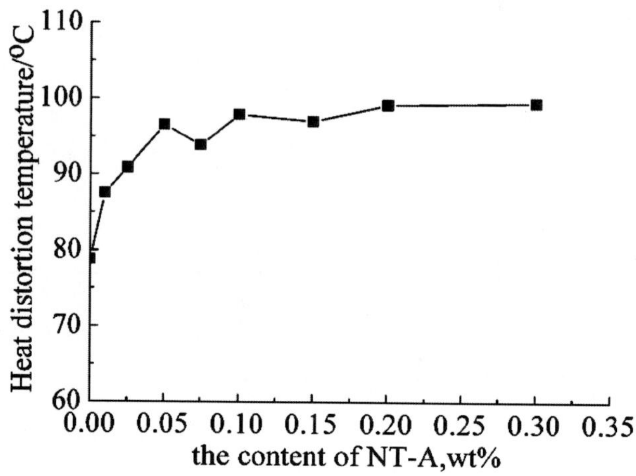

Figure 27. HDT of injection molded specimens nucleated by NT-A.

Figure 28 shows the dependence of heat distortion temperature (HDT) on the content of NT-C. It is also found that HDT increases with the content of NT-C.

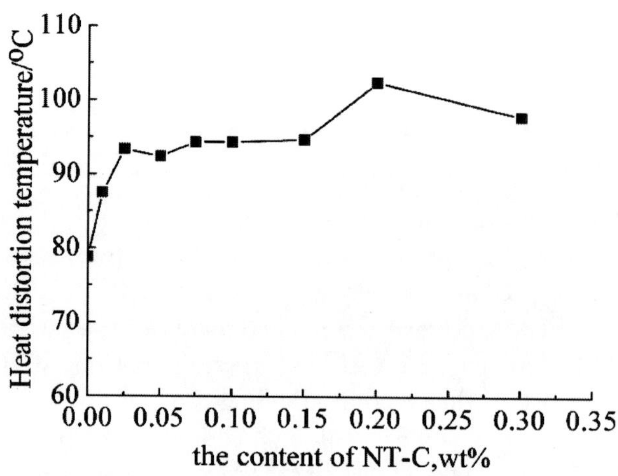

Figure 28. HDT of injection molded specimens nucleated by NT-C.

CONCLUSION

In this chapter, two commercial β nucleating agents, NT-A and NT-C, were used to prepare β-iPP, and its effect on melting and crystallization behavior, crystal structure, morphology, mechanical properties and heat distortion temperatures of iPP were investigated. Skin-core structures are observed in injection molded specimens. The skin is orientated structure without any spherulites, while spherulites exist in the core, and the content of β crystal of the core is higher than that of the skin. Formation of skin-core structure will influence properties of iPP. β nucleating agents induce abundant β crystals and make spherulites small, which improve Izod notched impact strength. When the contents of NT-A and NT-C are 0.075 wt% and 0.3wt%, Izod notched impact strengths reach the maximums, which are 2.6 times and 3.3 times more than that of pure iPP, respectively. Compared to pure iPP, nucleated iPP has lower tensile strength and higher tensile strain at break. Heat distortion temperature is also improved by the addition of β nucleating agents.

REFERENCES

[1] D. R. Norton and A. Keller, *Polymer*, 26, 704 (1985).
[2] J. X. Li and W. L. Cheung, *Polymer*, 40, 2085 (1999).
[3] J. Varga and J. Karger-Kocsis, *J. Polym. Sci. Part B: Polym. Phys.* 34, 657 (1996).
[4] M. Fujiyama, *Int. Polym. Proc.* XI, 271 (1996).
[5] [5*Polym. Degrad. Stab.* 91, 459 (2006).
[6] J. Varga and A. Menyhard, *Macromolecules*, 40, 2422 (2007).
[7] Q. L. Lu and Q. Dou, *e-Polymer*, no.76 (2008).
[8] Q. Dou, *J. Appl. Polym. Sci.* 111, 1738 (2009).
[9] Q. L. Lu and Q. Dou, *J. Polym. Res.* 16, 555 (2009).
[10] S. C. Tjong, J. S. Shen and R. K. Y. Li, *Polym. Eng. Sci.* 36, 100 (1996).
[11] S. C. Tjong, J. S. Shen and R. K. Y. Li, *Polymer*, 37, 2309 (1996).
[12] J. X. Li, W. L. Cheung and J. Demin, *Polymer*, 40, 1219 (1999).
[13] P. Juhász, J. Varga and K. J. Belina, *J. Therm. Anal. Calorim.* 69, 561 (2002).
[14] M. R. Meng and Q. Dou, *Mater. Sci. Eng. A*, 492, 177 (2008).
[15] . C. Feng, M. C. Chen, Z. T. Huang, Y. Q. Guo and H. Q. Hu, *J. Appl. Polym. Sci.* 85, 1742 (2002).
[16] M. R. Kantz, H. D. Newman and F. H. Stigale, *J. Appl. Polym. Sci.* 16, 1249 (1972).
[17] G. Menges, G. Wubken and B. Horn, *Colloid Polym. Sci.* 254, 267 (1976).
[18] D. R. Fitchmum and Z. Mencik, *J. Polym. Sci. Polym. Phys. Ed.* 11, 951 (1973).
[19] J. Karger-Kocsis and K. Friedrich, *Int. J. Fatigue*, 11, 161 (1989).
[20] M. Fujiyama, *Int. Polym. Proc.* X, 172 (1995).
[21] R. Cermak, M. Obadal, P. Ponizil, P. Martina, S. Karel and L. Anezka, *Eur. Polym. J.* 41, 1838 (2005).
[22] A. Turner-Jones, J. M. Aizlewood and D. R. Beckett, *Makromol. Chem.* 75, 134 (1964).
[23] R. H. Olley, A. M. Hodge, D. C. Bassett and J. J. Thomson, *J. Polym. Sci. Polym. Phys. Ed.* 17, 627 (1979).
[24] J. Varga, *J. Macromol. Sci. Phys.* 41, 1121 (2002).

[25] M. Fujiyama, *Int. Polym. Proc.* X, 251 (1995).
[26] D. Jarus, A. Scheibelhoffer, A. Hiltner and E. Baer, *J. Appl. Polym. Sci.* 60, 209 (1996).

In: Polypropylene
Editors: L. P. Silva and E. F. Barbosa

ISBN: 978-1-62417-142-0

Chapter 5

POLYPROPYLENE BINARY AND TERNARY BLENDS WITH POLY(LACTIC ACID), AND ACRYLONITRILE–BUTADIENE–STYRENE: PREPARATION AND CHARACTERIZATION

Kotiba Hamad* and Fawaz Deri
Faculty of Science, Department of Chemistry, Laboratory of Materials Rheology, Damascus University, Damascus, Syria

ABSTRACT

Polypropylene (PP) is one of the most important polymers; it has a number of advantages over other polymers in plastic and textile applications. PP has a high work of rupture, indicating that it is a tough plastic, and a high degree of crystallinity, resulting in a fiber that is strong and resilient. However, PP has poor degradability and is high hydrophobic. This makes it difficult to dispose the PP articles after use. Therefore, PP was blended with poly(lactic acid) (PLA) in efforts to enhance the degradability of PP. It was found that the blend (PP/PLA) has good processability and dyeability. Nevertheless, drawbacks were noted in this blend (PP/PLA), such as lower mechanical properties (tensile strength and modulus) comparing with those of pure PLA. From another side, acrylonitrile-butadiene-styrene (ABS) was used for improving the impact strength of PLA. In this chapter binary and ternary blends with polypropylene, poly(lactic acid), and ABS were prepared and characterized in efforts to obtain new materials with improved properties, such works are very important in the development of composites and blends from biodegradable polyesters.

INTRODUCTION

Increasing concerns over the environmental impact and sustainability of conventional polymer materials have motivated academia and industry to devote considerable efforts to the

Email: kotibahamad@yahoo.com

development of polymers from renewable resources. Among a few commercially available bio-based or partially bio-based thermoplastic polymers, poly(lactic acid) (PLA) has undergone the most investigation. PLA is one of the most important biodegradable polyesters which could be processed using the traditional processing techniques (extrusion, injection…) and it has better thermal processability compared to other biopolymers such as poly(hydroxyl alkanoates) (PHAs), poly(ethyele glycol) (PEG), polycaprolactone (PCL), etc [1]. Due to its initial production costs, the starting applications of PLA have been focused on high value products such as medical devices [2]. Advances in the polymerization technology have significantly reduced the production cost and have contributed to make PLA economically competitive with petroleum-based polymers. PLA's potential for food packaging is very high due to its transparency, mechanical properties and acceptable moisture processability for dry food stuff [1]. Nevertheless, there are some drawbacks, such as its high brittleness and poor crystallization behavior that limit its current use in food packaging [3]. In order to overcome these limitations, the copolymerization of lactic acid with other monomers has been a widely used approach, but without any success for commercial application. A less expensive and more practical strategy to overcome these drawbacks is the blending of PLA with other polymers. Hence, several synthetic polymers and copolymers such as, polyethylene (PE) [4-7], polypropylene (PP) [8-11], polystyrene (PS) [12-18], poly(ethylene terephtalate) (PET) [19,20], polycarbonate (PC) [21], polyamide (PA) [22], and acrylonitrile–butadiene–styrene (ABS) [23] or biodegradable polymers such as polycaprolactone [24-29], poly(butylene succinate) (PBS) [30-34], poly(butylenes adipate-co-terephtalate) (PBAT) [35-41], thermoplastic starch (TPS) [3, 42-47] and poly(butylene succinate adipate) (PBSA) [48,49] have been used.

Preparation and characterization of PLA/PP polymer blends were reported in many works [8-11], they found that blending PLA with PP improves the resistance of PLA to hydrolysis and enhance its dyeability, also they found that the degradability of PP can be improved by blending with PLA. Nevertheless, drawbacks were noted in this blend (PLA/PP), such as lower mechanical properties (tensile strength and modulus) comparing with those of pure PLA. Yoo et al. [10] studied the effect of different compatibilizers on the mechanical and interfacial properties of PLA/PP (20/80) polymer blend, PP-g-MA and styrene-ethylene-butylene-styrene-g-maleic anhydride (SEBS-g-MA) were used as compatibilizers in this system. The results showed that PP-g-MA does not affect on the tensile strength of the blend while tensile strength of the blend decreases with increasing SEBS-g-MA content in the blend. Also it was found that PP-g-MA does not affect on the impact strength of the blend, but the impact strength of the blend increases with increasing SEBS-g-MAH content. From another side ABS was used for improving the impact strength of PLA, Li and Shimizu [23] used ABS for toughening PLA. Blends of PLA and ABS in different ratios were prepared and as the in-situ compatibilizer for PLA/ABS blends to improve the compatibility between PLA and ABS, the reactive styrene-acrylonitrile-glycidyl methacrylate copolymer (SAN-GMA) by incorporating with ethyltriphenyl phosphonium bromide (ETPB) as the catalyst was used. The mechanical results showed that impact strength of the compatibilized blend is higher than that of uncompatibilized and pure PLA, but the modulus of the compatibilized blend was less than that of pure PLA. The using of compatiblezers in improving mechanical properties of the polymer blends still faces some difficult related to the high cost of these materials. In this work ternary blends of PLA, PP, ABS were prepared in efforts to obtain new materials with improved mechanical properties, such works are very important in the development of composites and blends from biodegradable polyesters. The aim of this chapter is to prepare and characterize binary (PLA/PP) and ternary blends (PLA/PP/ABS).

MATERIAL AND METHODS

Materials

Poly(lactic acid) (PLA) (ESUN™ A-1001) [density =1.25 g/cm^3 (21.5°C), *MFI* =12.5 g/10 min (190°C/2.18 Kg)] was supplied by Bright China Industrial Company. Ltd (Shenzhen, China), the selected grade is an extrusion material; it was dried at 70°C for 6 hours before using. Polypropylene (PP) (PETOPLEN EH 251) [*MFI* = 24 g/10 min (230°C/2.16 Kg)] was supplied by PETKIM Petrokimya Holding A.S. (Turkey). Acrylonitrile – Butadiene – Styrene (ABS) (Kumho ABS 750SW) was supplied by Korea Kumho Petrochemical Co., Ltd. (Korea), [density = 1.04 g/cm^3, *MFR* = 50 g/10 min (200°C /21.6 Kg)].

Blends Preparation

Binary and ternary blends of PLA, PP, and ABS in different ratios were prepared using a laboratory scale single screw extruder (SSE) (*D*=20mm, *L/D*=25) [SHAM EXTRUDER 25D Performance: *Kreem Industrial Establishment*, Damascus - Syria]. The screw has a fluted type mixing section located before the metering zone [50, 51], in this type of mixers the material is forced to pass at a high shear stress. This brings in some level of dispersing action besides reorienting the interfacial area, and increasing the imposed total strain. The screw speed was set at 20 rpm in the blends preparation, and the extruder temperature profile along the barrel was 145, 155, 165, 175°C (from feed zone to die). The blends were extruded through a multi holes die (3mm), the extrudates were then fed into a granulator, which converted them into granules. The obtained granules were dried at 70°C for 6 h before studying. The compositions of the blends are shown in Table 1.

Table 1. Compositions of the binary and ternary blends

Sample	PP30/ ABS0	PP50/ ABS0	PP70/ ABS0	PP50/A BS5	PP50/A BS10	PP50/A BS15	PP50/A BS20	PP70/A BS20	PP30/A BS20
PLA (wt %)	70	50	30	50	50	50	50	30	70
PS (wt %)	30	50	70	50	50	50	50	70	30
ABS (wt%)*	0	0	0	5	10	15	20	20	20

*wt %: part per hundred of the binary blend (PLA/PP)

Tensile Samples Preparation

The prepared granules were injection molded at 200-230°C using NEGRI BOSSI (NB 25) injection machine (Lessona Corporation, Italy). The tensile samples were prepared according to the following injection conditions, cooling time in the mold was 15 sec, the mold temperature was room temperature with water-cooling (25°C) and injection pressure was 9 MPa. The molded samples were dog bone-shaped samples with a thickness and width of 4 mm and 10 mm respectively. The gauge length of the sample was 80 mm (Figure 1). The obtained products were immediately packed in plastics bags and stored in a dark cool surrounding.

Rheological Properties

Rheological experiments were carried out using a capillary rheometer (Davenport 3/80), the rheological experiments were carried out at 170, 175, 180 and 185°C, and by using *L/R* = 8, 15, 25 and 36 capillaries. Bagley's correction [52] was performed by using the data from the four capillary dies. The apparent shear rate (γ_a) is given by:

$$\gamma_a = \frac{4Q}{\pi . R^3} \quad (1)$$

where R is the capillary radius, and Q is the volumetric flow rate. The true shear rate (γ_r) is given by:

$$\gamma_r = (\frac{3n+1}{4n})\gamma_a \quad (2)$$

where n is the non-Newtonian index depending on temperature, the term ($\frac{3n+1}{4n}$) was the Rabinowitsch correction factor. The apparent shear stress (τ_a) is given by:

$$\tau_a = \frac{\Delta P}{2(\frac{L}{R})} \quad (3)$$

where ΔP is the pressure at the capillary entrance, and L is the capillary length. The true shear stress (τ_r) is given by:

$$\tau_r = \frac{\Delta P}{2(\frac{L}{R} + e)} \quad (4)$$

where e is the Bagley's correction factor. The true viscosity (η_r) is given by:

$$\eta_r = \frac{\tau_r}{\gamma_r} \quad (5)$$

The values of flow activation energy at a constant shear rate (E_γ) were determined using Arrhenius equation form:

$$\eta_r = A.e^{\frac{E_\gamma}{R.T}} \quad (6)$$

where A is the consistency related to structure and formulation and R is the gas constant (8.314 J/mol. K).

Mechanical Properties

Tensile testing to study stress at break (N/mm^2), strain at break (%) and Young's modulus (N/mm^2) were performed using Testometric M350-10KN (The Testometric Company Ltd, Rochdale, UK) at room temperature, all samples were strained at 50 mm/min. Samples were conditioned at room temperature for a period of 48 h prior to testing. Results from 6-8 specimens were averaged.

Figure 1. Injection molded sample of PP50/ABS20 blend.

Results and Discussion

Rheological Properties

Flow Curves

Polymer rheology in the molten state is very important for determining the processing conditions of the materials. Figure 2 shows the relationship between apparent shear rate and apparent shear stress (flow curves) for PP50/ABS20 at different temperatures.

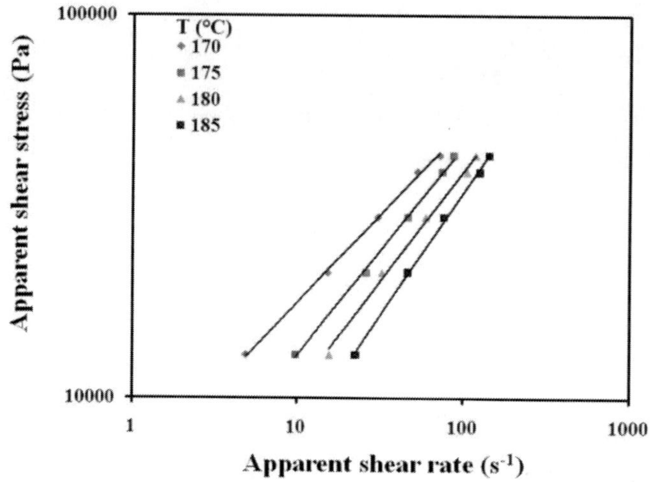

Figure 2. Flow curves of PP50/ABS20 blend.

It is clearly seen from Figure 2 that the relationship between shear stress and shear rate obeys the power law in the range of the studied shear rates and temperatures.

$$\tau = K\gamma^n \tag{7}$$

where K is the consistency index and n is the non-Newtonian index (power law index), which can be calculated from the slope of the lines in Figure 2. The same relationship was obtained for both polymers and binary polymer blends. Figure 3 shows the effect of ABS on the n of the blend at 170°C. It could be noted from Figure 3 that the values of n were less than 1, implying that PLA/PP/ABS blend melts were pseudo plastic; similar to most polymer and binary polymer blends in the molten state. No attention has been given in the past to flow behavior of the ternary polymer blends during the capillary rheometer.

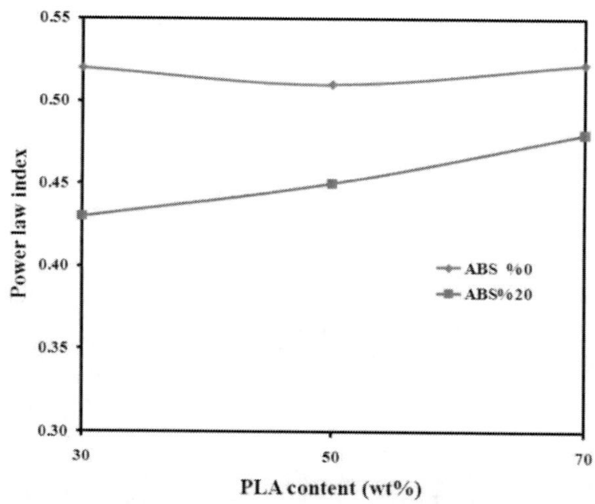

Figure 3. The effect of ABS content on n of the blends at 170°C.

Also It could be noted from Figure 3 that the values of n for the ternary blend (PLA/PP/ABS) is less than that of the binary blend (PLA/PP). The value of n describes the deviation from the Newtonian fluids about flow behavior, so it is also called the flow behavior index. A higher value of n reveals less influence of shear rate on flow behavior. In other words, the changes in viscosity upon shear rate are not obvious, so it could be said that the flow behavior of ternary blend melts is more sensitive to shear rate as compared with the binary blend (PLA/PP).

Viscosity Curves

Figure 4 shows the relationship between true shear viscosity and true shear rate (viscosity curves) of PP50/ABS20 blend at different temperatures. It could be noted from Figure 4 that the true shear viscosity of the blends decreases with increasing true shear rate in the studied range of the shear rates and temperatures. The decreasing trend of viscosity with increasing shear rate, commonly referred to as shear-thinning behavior, is believed to arise from the stretching of an entangled state of polymer chains to an oriented state when the applied shear rate is higher than a certain critical value. The high shear viscosity at a low shear rate provide the integrity of the extrudate during extrusion, and the low shear viscosity at a high shear rate enables low injection pressure, high injection speed and less time of the injection cycle.

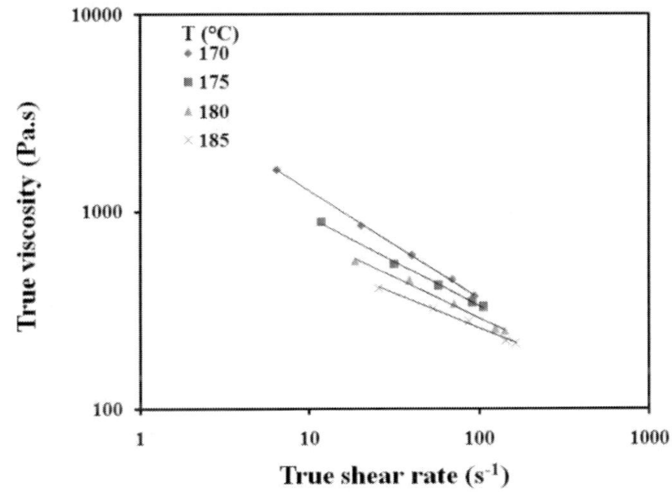

Figure 4. True shear viscosity versus true shear rate of PP50/ABS20 blend.

Figure 5 shows the plots of true viscosity versus blending ratio at 170°C and at shear rate = 10 s^{-1}. It could be seen from Figure 5 that the true shear viscosity of the binary blend (PLA/PP) decreases with increasing PLA content in the blend and this behavior can be attributed to the low melt viscosity of PLA comparing with that of pure PP. Also it could be noted from Figure 5 that the viscosity of the binary blends is less than that of the ternary blends, where the addition of 10 wt% of ABS to PLA/PP (50/50) increases the true shear viscosity by a factor of 0.39 and the addition of 20 wt% of ABS to the binary blend PLA/PP (30/70) increases the viscosity by a factor of 0.1.

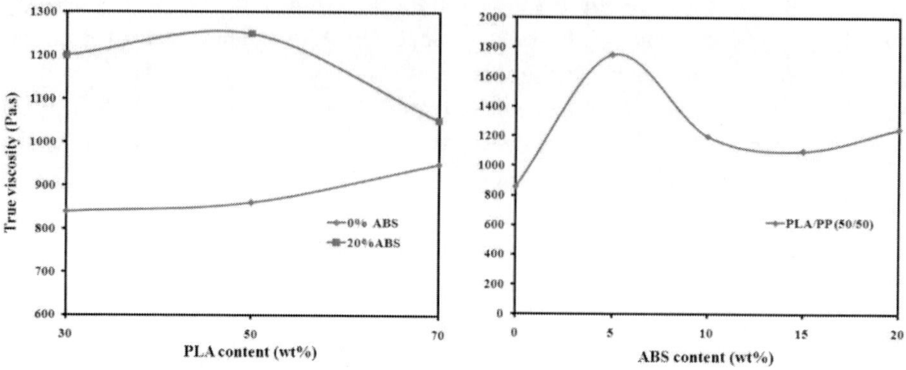

Figure 5. True shear viscosity of the blends at 170°C and $\gamma = 10\ s^{-1}$.

Flow Activation Energy

Figure 6 shows the relationship between true shear viscosity and *1/T* of the ternary blends at shear rate = 10 s^{-1}, it could be noted from Figure 6 that the true shear viscosity decreases with increasing temperature. With a rise of temperature the motion ability of polymer chains enhances, and the resistance between the melt layers decreases relevantly, leading to reduction of the melt viscosity

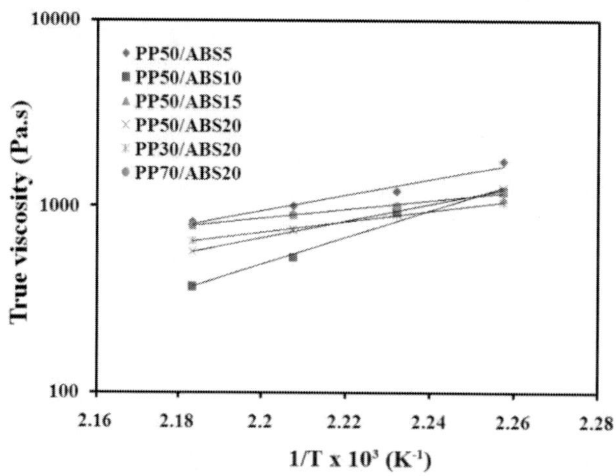

Figure 6. True viscosity versus *1/T* of the blends at $\gamma = 10\ s^{-1}$.

The plots of true shear viscosity versus *1/T* shown in Figure 6 give a group of straight lines, which indicates that the relationship between the true viscosity and temperature follows the form of Arrhenius equation (equation (6)) so the flow activation energy at a constant shear rate (E_γ) could be calculated from the slopes of these lines.

$$E_\gamma = R\left(\frac{d\log\eta_r}{d(1/T)}\right)_\gamma \quad (8)$$

Figure 7 shows the effect of ABS on the flow activation energy of the binary blends at a constant shear rate ($\gamma = 10\ s^{-1}$).

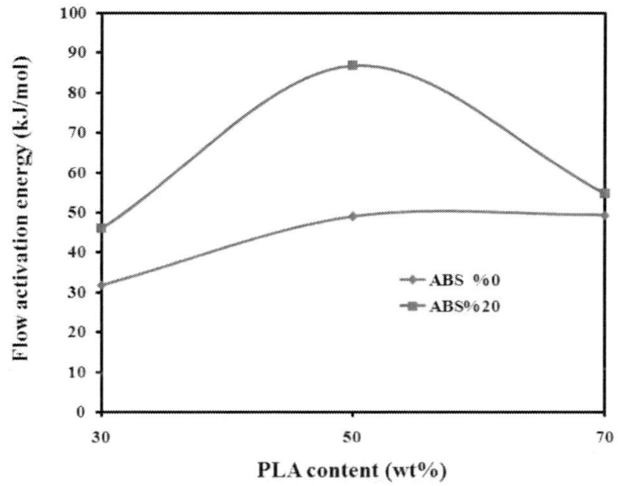

Figure 7. The effect of ABS on the flow activation energy at a constant shear rate ($\gamma = 10\ s^{-1}$).

It could be noted from Figure 7 that the flow activation energy of the binary blend (PLA/PP) is less than that that of the ternary blends (PLA/PP/ABS), so it could be said that the addition of ABS caused in increasing flow activation energy. It is well known that the value of flow activation energy reflects the temperature – sensitivity of viscosity, so the more E_γ was the more sensitive the behavior of blends were to the temperature so it could be said that ternary blends are more sensitive to the temperature than the binary blend.

Mechanical Properties

Mechanical properties of the blends were determined in term of tensile properties. Stress at break, strain at break, and Young's modulus were determined. Figure 8 shows the stress at break of the blends.

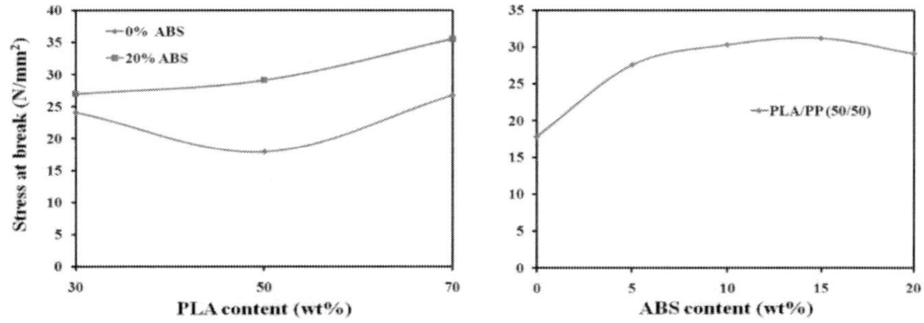

Figure 8. Stress at break of the blends.

It could be noted from Figure 8 that the addition of ABS improves the stress at break of binary blend (PLA/PP), where adding 15 wt% ABS to PLA/PP (50/50) enhanced the stress at break nearly by a factor of 0.77, also the addition of 20 wt% ABS to the binary (PLA/PP) blend caused in improving the stress at the break of the blend.

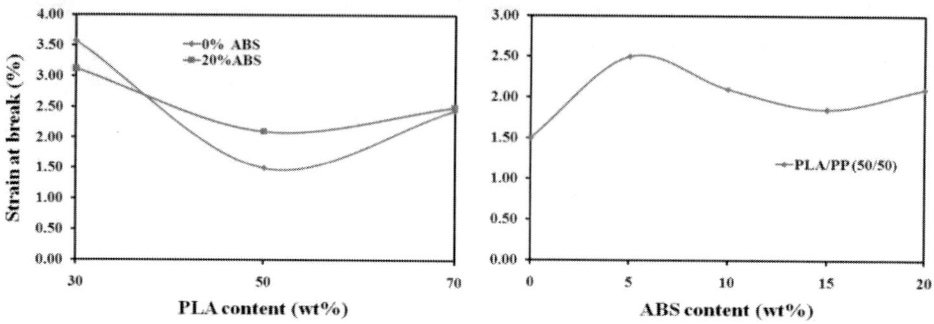

Figure 9. Strain at break of the blends.

Figure 9 shows the strain at break of the blends, it could be noted from Figure 9 that the strain at break of the ternary blends is more than that of the binary blend (PLA/PP (50/50)). Figure 10 shows the Young's modulus of the blends, it could be noted from Figure 10 that the Young's modulus of the ternary blends also is more than that of the binary blends. It could be noted from Figures 8 and 10 that the ternary blend (PP30/ABS5) has good mechanical properties (stress at break and Young's modulus), which are similar to those of pure PLA (stress at break = 56 N/mm^2 and Young's modulus = 2220 N/mm^2) [11], and better than that of the binary blend (PLA/PP) [11]. Also it has low melt viscosity comparing with the other compositions of the ternary blend which means good processibility and low cost comparing with pure PLA so it could be said that this blend (PP30/ABS5) is suitable to be used in a lot of applications as alternative to pure poly (lactic acid) such as automotive fibers applications.

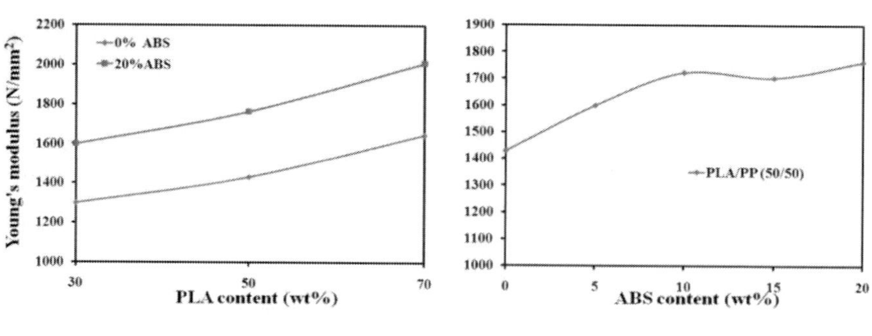

Figure 10. Young's modulus of the blends.

CONCLUSION

In this chapter, rheological and mechanical properties of binary and ternary blends with poly(lactic acid) (PLA), polypropylene (PP) and acrylonitrile-butadiene-styrene (ABS) were studied. The blends were prepared using a single screw extruder. Rheological results showed that the ternary blend (PLA/PP/ABS) is pseudo plastic in nature; its viscosity decreases with increasing shear rate similar to most polymer and polymer blend melts. Also it was found that the true shear viscosity of the binary blend (PLA/PP) decreases with increasing PLA content which was attributed to PLA low viscosity, and the true shear viscosity of the ternary blend is more than that of the binary blend. The mechanical results showed that the stress at break and Young's modulus of the ternary blend are better than those of the binary blend. PP30/ABS20 blend showed good mechanical properties near to the mechanical properties of pure PLA.

REFERENCES

[1] R. Auras, B. Harte, and S. Selke, *Macromol. Biosci.* 4, 835 (2004).

[2] J. Lunt, Polym. *Degrad. Stab*. 59, 145 (1998).

[3] M. A. Huneault and H. Li, *Polymer*. 48. 270 (2007).

[4] K. S. Anderson and M. A. Hillmyer, *Polymer*. 45, 8809 (2004).

[5] H. Balakrishnan, A. Hassan and M. U. Wahit, *Elast. Plast.* 42, 223 (2010).

[6] G. Singh, H. Bhunia, A. Rajor, R. N. Jana and V. Choudhary, *J. Appl. Polym. Sci.* 118, 496 (2010).

[7] K. Hamad, M. Kaseem and F. Deri, *Adva. Chem. Eng. Sci.* 4. 208 (2011).

[8] A. Bourmaud and S. Pimbert, *Composites: Part A*. 39, 1444 (2008).

[9] N. Reddy, D. Nama and Y. Yang, *Polym. Degrad. Stab.* 39, 233 (2008).

[10] T. W. Yoo, H. Yoon, S. J. Choi, M. S. Kim, Y. H. Kim and W. N. Kim, *Macromo. Res.* 18, 583 (2010).

[11] K. Hamad, M. Kaseem and F. Deri, *J. Polym. Res.* DOI 10.1007/s10965-011-9586-6 (2011).

[12] G. Biresaw and C. J. Carriere, *J. Polym. Sci. Part B*. 40, 2248 (2002).

[13] G. Biresaw and C. J. Carriere, *J. Appl. Polym. Sci.* 83, 3145 (2002).

[14] G. Biresaw and C. J. Carriere, *Polym Prepr* (Am Chem Soc, Div Polym Chem). 43, 367 (2002).

[15] G. Biresaw and C. J. Carriere, *Composites: Part A*. 35, 313 (2004).

[16] A. Mohamed, S. H. Gordon, and G. Biresaw, *J. Appl. Polym. Sci.* 106, 1689 (2007).

[17] K. Hamad, M. Kaseem and F. Deri, *Polym. Bull.* 65, 509 (2010).

[18] K. Hamad, M. Kaseem and F. Deri, *J. Mater. Sci.* 46, 3013 (2011).

[19] B.G. Girija, R. R. N. Sailaja and G. Madras, *Polym. Degrad. Stab.* 90, 147 (2005).

[20] H. Chen, M. Pyda and P. Cebe, Thermochimica. Acta. 492, 61 (2009).

[21] J. B. Lee, Y. K. Lee, G. D. Choi, S. W. Na , T. S. Park and W. N. Kim, *Polym. Degrad. Stab*. 96, 553 (2011).

[22] G. Stoclet, R. Seguela and J. M. Lefebvre, *Polymer*. 52, 1417 (2011).

[23] Y. Li, H. Shimizu, Eur. Polym. J. 45, 738 (2009).

[24] C. L. Simoes, J. C. Viana and A. M. Cunha, *J. Appl. Polym. Sci.* 112, 345 (2009).

[25] Y. Zhang, D. Wu, M. Zhang, W. Zhou and C. Xu, *Polym. Eng. Sci.* 49, 2293 (2009).

[26] D. Wu, Y. Zhang, L. Yuan, M. Zhang, and W. Zhou, *J. Polym. Sci. Part A*. 48, 756 (2010).

[27] T. Takayama, M. Todo, H. Tsuji and K. Arakawa, *J. Mater. Sci.* 41, 4989 (2006).

[28] T. Takayama and M. Todo, *J. Mater. Sci.* 41, 6501 (2006).

[29] T. Takayama, M. Todo and H. Tsuji, *J. Mecha. Bihav. Biomed. Mater.* 4, 255 (2011).
[30] J. W. Park and S. S. Im, *J. Appl. Polym. Sci.* 86, 647 (2002).
[31] A. Bhatia, R. K. Gupta, S. N. Bhattacharya and H. J. Choi, *Korea-Australia. Rheo. J.* 19, 125 (2007).
[32] T. Yokohara and M. Yamaguchi, *Eur. Polym.* J. 44, 677 (2008).
[33] R. Wang, S. Wang, Y. Zhang, C. Wan and P. Ma, *Polym. Eng. Sci.* 49, 26 (2009).
[34] P. Bae, H. Yeon, M. Whan, I. Soon and Y. Sang, *Macromo. Res.* 18, 463 (2010).
[35] L. Jiang, M. P. Wolcott and J. Zhang, *Biomacromolecules*. 7, 199 (2006).
[36] S. Y. Gu, K. Zhang, J. Ren and H. Zhan, *Carbohydr. Polym.* 74, 79 (2008).
[37] F. Signori, M. B. Coltelli and S. Bronco, *Polym. Degrad. Stab.* 94, 74 (2009).
[38] H. Xiao, W. Lu and J. Yeh, *J. Appl. Polym. Sci.* 112, 3754 (2009).
[39] H. Yuan, Z. Liu and J. Ren, *Polym. Eng. Sci.* 49, 1004 (2009).
[40] N. Zhang, Q. Wang, J. Ren and L. Wang, *J. Mater. Sci.* 44, 250 (2009).
[41] J. T. Yeh, C. H. Tsou, C. Y. Huang, K. N. Chen, C. S. Wu and W. Chai, *J. Appl. Polym. Sci.* 116, 680 (2010).
[42] N. Wang, J. Yu, and X. Ma, *Polym Inter.* 56, 1440 (2007).
[43] N. Wang, J. Yu, P. R. Chang and X. Ma, *Starch*. 59, 409 (2007).
[44] N. Wang, J. Yu, and X. Ma, *Polym Compos.* 29, 551 (2008).
[45] N. Wang, J. Yu, P. R. Chang and X. Ma, *Carbohydr. Polym.* 71, 109 (2008).
[46] E. Schwach, J. Six and L. Averous, *J. Polym. Environ.* 16, 286 (2008).
[47] M. A. Huneault and H. Li, *J. Appl. Polym. Sci.* 119, 2439 (2011).
[48] S. Lee and J. W. Lee, *Korea-Australia. Rheo.* J. 17, 71 (2005).
[49] Y. Wang and J. F. Mano, *J. Appl. Polym. Sci.* 105, 3204 (2007).
[50] Z. Tadmor and C. G. Gogos, Principles of polymer processing. John Wiley & Sons Publishing, New Jersey(2006).
[51] C. D. Han, Rheology and processing of polymeric materials (Polymer Rheology). Oxford University Press, New York (2007).
[52] E. B. Bagley, *J. Appl. Phys.* 28, 624 (1957).

In: Polypropylene
Editors: L. P. Silva and E. F. Barbosa

ISBN: 978-1-62417-142-0

Chapter 6

MELT RHEOLOGY OF POLYPROPYLENE/ THERMOPLASTIC STARCH BLENDS

Mosab Kaseem* and Fawaz Deri
Department of Chemistry, Laboratory of Materials Rheology,
Faculty of Science, Damascus University, Syria

ABSTRACT

Starch as an inexpensive and renewable source has been used as a filler for environmentally friendly plastics for about 2 decades. In order to improve the compatibility between hydrophilic starch granules and hydrophobic polypropylene (PP), glycerol used as a plasticizer for starch to enhance the dispersion and the interfacial affinity in thermoplastic starch (TPS)/PP blend. Rheological properties were studied using a capillary rheometer and the Bagley's correction was performed. The rheological properties showed that the viscosity of TPS decreases with increasing shear stress. Also it was found that PP/TPS blends are pseudo plastic in nature and the flow activation energy of the blends is greater than that of PP and the flow activation energy of the blends at a constant shear stress is more than that at a constant shear rate. The results of this work indicated to good processibility of this system.

INTRODUCTION

Conventional polymers are usually made from petroleum resources such as polyolefins, and they are ideal for many applications such as packaging, building resources, commodities and consumer goods. Polyolefin-based plastics have become a foundation of modern civilization and are low cost, durable, resistant to solvents, waterproof and resistant to physical ageing. The resistance of polyolefin materials to degradation by micro-organisms is both an advantage and in the long term, a serious environmental problem.

* E-mail address: mosabkaseem@yahoo.com

Biopolymers serve two key purposes as a base for thermoplastics; they are environmentally friendly, degrading in soil first into sugars and other organic compounds, and they are derived from naturally abundant regenerable resources.

Starch is an important productive (it has the best properties between all polysaccharides) polysaccharide in plants. Because of its low cost, large availability as a renewable resource, biodegradable nature and the innocuous degradation products, it has already been widely researched as an important raw material for environmental and biomedical applications. There are many hydrogen bonds between the starch macromolecules, which reduce the movement of these molecules, so the native starch has poor processing ability. Starch was used at first as a filler in the traditional polymers which helps in cost reduction and rising biodegradability of resultant, in this case starch acts as filler increasing the rigidity of the material, and so, the starch concentration at which useful material can be obtained is therefore limited to a low range after which the materials properties suffer dramatically [1].

Thermoplastic starch (TPS) can be obtained by destruction of the granules in the presence of plasticizers under specific conditions. Several plasticizers have been used for plasticization process to convert starch into thermoplastic starch (TPS) to be used in polymer blends such as glycerol [2-11], water [10-12], urea [13-15], formamide [16] ethylenebisformamide [17-19], sorbitol [20], citric acid [21], N-(2-Hydroxyethyl)formamide [22] and amino acids [23]. Water is more effective as a plasticizer than glycerol, but the most used plasticizer in TPS preparation is glycerol due to its high boiling point, availability, and low cost. TPS has two main disadvantages compared to most plastics currently in use, i.e. it is highly water-soluble and has poor mechanical properties. These features can be improved by mixing it with certain synthetic polymers such as PE [2], PP, [4, 24], PB-1[25], ABS [26], PS [5, 27-29]. Salcido et al. [2] studied the biodegradation of LDPE/TPS blends and they found that mechanical properties of the prepared blend decreases after the biodegradation tests which was attributed to the formation of holes on the surface which behave as fracture propagation sites. Kaseem et al. [24] studied the rheological and mechanical properties of PB-1/TPS blends and the found that the blend is pseudo plastic in manner, where its non-Newtonian index was less than 1, and the melt viscosity of the blend is less than that of PB-1 which was attributed to the lubrication role of glycerol. Also Kaseem et al. [25] studied the rheological and die swell properties of ABS /TPS blends. It was found that the ABS/TPS blends are pseudo plastic in nature, and their viscosity decreases with increasing shear rate, similarly to most polymer and polymer blend melts. Also it was found that the melt viscosity of the blend decreases with increasing TPS content, which was attributed to the lubrication effect of the used plasticizer (glycerol).

In other works Schlemmer et al. [27-29] studied the biodegradation of PS/TPS blends and they found that the addition of TPS to PS is an effective technique to achieve biodegradability. Rosa et al [4] studied the influence of the plasticizer type on the thermal and mechanical properties of PP/TPS blends, where they blended TPS, which has 20% of plasticizer, with PP. They found that the incorporation of TPS to PP has generally reduced the mechanical properties in PP. The aim of this chapter is to present rheological properties of PP/TPS blends, such works are very important of new materials with improved properties.

MATERIAL AND METHODS

Materials

Polypropylene (PP) (PETOPLEN EH 251) was supplied by PETKIM Petrokimya Holding A.S (Turkey) [*MFR*= 24 g/10min (230 °C /21.6 Kg)]. Native corn starch is a commercial material; it was brought from local supply and used as received. Glycerol 99.5% is a commercial grade used without any treatment.

Thermoplastic Starch (TPS) and TPS/PP Blends Preparation

Corn starch sample was mixed manually with (20% glycerol), the obtained mixtures (sludge) were then fed into a laboratory scale single screw extruder (SSE) (L/D = 25, D = 20) [SHAM EXTRUDER 25D, Performance: *Kreem Industrial Establishment*, Damascus – Syria], which could be operated at different speeds, varied from 0 to100 RPM. The temperatures profile along the barrel of extruder were set at 90,100,110,100 °C (from feed zone to die) and the screw speed was 30 RPM in TPS preparation. TPS were then extruded through a multi holes die (3 mm) and the extrudates were left to cool in air and then fed into a granulator which converted them into granules Figure 1. The obtained TPS granules were then blended with PP by using the single screw extruder in different ratios Table 1. The temperatures profile along the barrel of the extruder in the PP/TPS blends were set at 125,160,170,180 °C (from feed zone to die) and the speed was 15 RPM. Also the blends were extruded through the multi holes die, left to cool in air, and the extrudates were fed into the granulator, the obtained granules were then dried in 85 °C for a 6 h before using. The compositions of TPS and the blends are shown in Table 1.

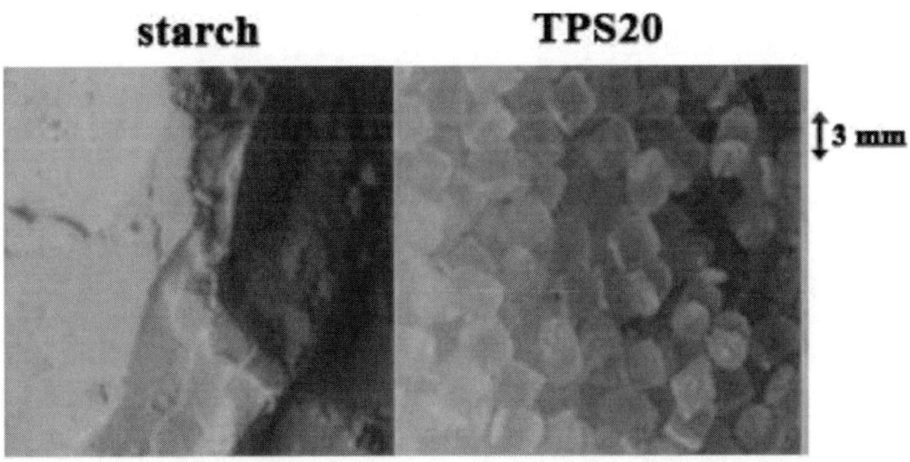

Figure 1. Native starch and thermoplastic starch samples.

Table 1. The compositions of PP/TPS blends

Sample	TPS composition		TPS (wt%)	PP (wt%)
	Starch (wt%)	Glycerol (wt%)		
PP1/TPS20	80	20	10	90
PP2/TPS20			20	80
PP3/TPS20			30	70

Rheology

Rheological properties of the blends were studied using a capillary rheometer (Davenport 3\80), it consists of a barrel into which material was loaded before begin pushed by a plunger through a capillary, the load in the plunger provide the total pressure drop in the barrel and capillary, and the volume flow rate. The rheological experiments were carried out at 185, 190, 195, 200 °C, and by using *L/R* =8, 15, 25, 36 capillaries. Bagley's correction was performed using the data from the four capillary dies. The true shear rate (γ_r) is given by:

$$\gamma_r = (\frac{3n+1}{4n}) \cdot \frac{4Q}{\pi R^3} \quad (1)$$

where R is the capillary radius, n is the non-Newtonian index depending on temperature and Q is the volumetric flow rate. The term $(\frac{3n+1}{4n})$ was the Rabinowitsch correction factor [30]. The true shear stress (τ_r) is given by:

$$\tau_r = \frac{\Delta P}{2\left(\frac{L}{R} + e\right)} \quad (2)$$

where ΔP is the pressure at capillary entrance, L is the capillary length and e is the Bagley's correction factor [31]. True viscosity is given by:

$$\eta_r = \frac{\tau_r}{\gamma_r} \quad (3)$$

Flow activation energy at a constant shear rate ($E_{\tau,\gamma}$) was determined by using Arrhenius equation:

$$\eta_r = A.e^{\frac{E\tau,\gamma}{RT}} \quad (4)$$

where A is the consistency related to structure and formulation and R is the gas constant (8.314 J/mol.K).

RESULTS AND DISCUSSION

Rheological Properties

Flow Curves

Figure 2 shows the flow curves of pure PP and PP/TPS blends at 185, 190, 195 and 200 °C and *L/R*=15. It could be noted from Figure 2 that the linearity of these lines is excellent and they obey the power law in a certain range of shear rate:

$$\tau = K\gamma^{n} \tag{5}$$

where K is the consistency index and n is the non-Newtonian index. The non-Newtonian index (n) was calculated from the slope of the fitted lines in Figure 2. Figure 3 shows the effect of TPS20 content on *n*. It could be noted that the values of *n* of PP/TPS blends were less than 1, implying that PP/TPS blends were pseudo plastic. The same behavior was noted in other TPS systems such as PB-1/TPS [25], so it could be said that the TPS polymer blends are non-Newtonian materials similar to all polymers and polymer blends [25, 26].

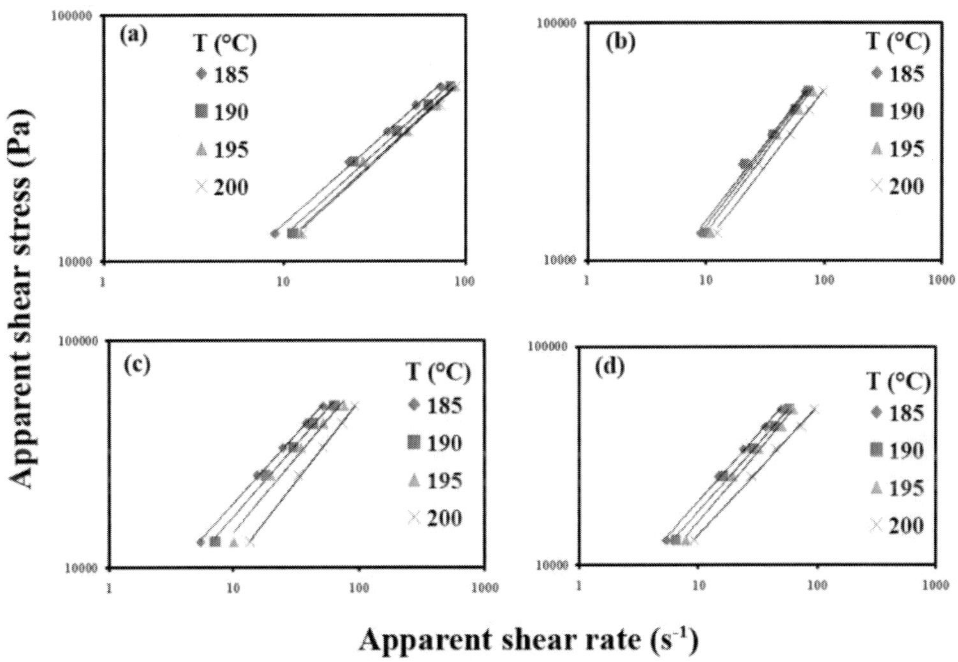

Figure 2. Flow curves of PP/TPS20 (a) PP, (b) PP1/TPS20, (c) PP2/TPS20, (d) PP3/TPS20.

Also it could be noted from Figure 3 that the *n* value increases with increasing TPS content in the blend in a range of 0-10% where the maximum value was observed. Increasing TPS higher than 10% caused a decreasing trend of the non-Newtonian index this indicates the greater pseudoplasticity in the flow for the melts of PP/TPS blends at a loading level of TPS higher than 10%. The value of *n* describes the deviation from the Newtonian fluids about flow behavior, so it is also called the flow behavior index. A higher value of *n* reveals less influence of shear rate on flow behavior, so it could be said that, at a high TPS content, the blend will be more sensitive to shear rate which means good process ability [32].

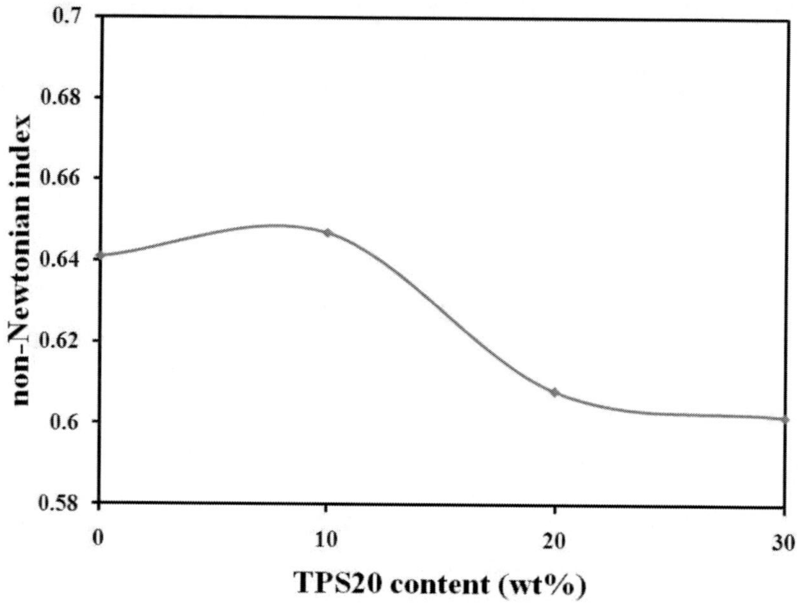

Figure 3. Non-Newtonian index versus TPS20 content (wt %) at 185°C.

Viscosity Curves

The relationship between true viscosity and true shear stress (i.e., viscosity curves) of pure PP and PP/TPS blends is shown in Figure 4.

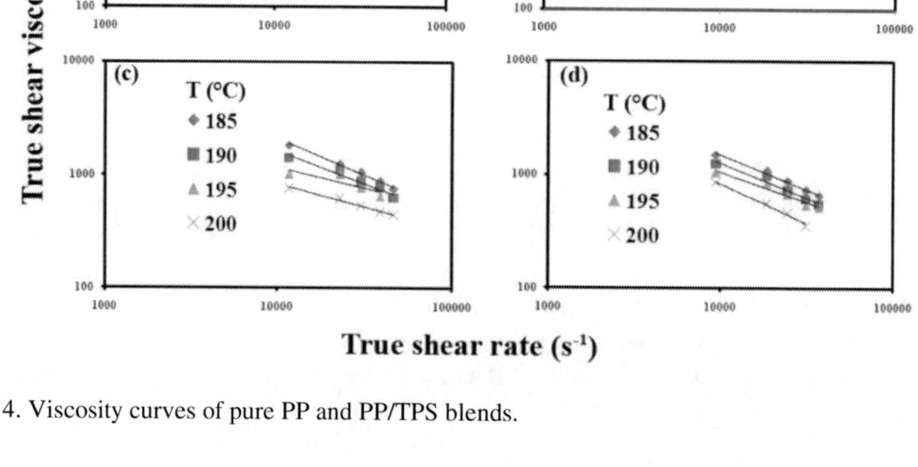

Figure 4. Viscosity curves of pure PP and PP/TPS blends.

It could be noted from Figure 4 that the true viscosity of the blend decreases with increasing shear stress in a certain range of shear stress, this behavior was attributed to the alignment of chain segments of polymers in the direction of applied shear stress. The high viscosity at a low shear rate provide the integrity of the extrudate during extrusion and the low viscosity at a high shear rate enables low injection pressure and less time of the injection cycle [33].

Figure 5 shows the effect of TPS20 content on the true viscosity of the blend at different shear rates and shear stresses and a constant temperature (185°C).

It could be noted from Figure 5 that at a given true shear rates and shear stress the true viscosity of the blend decreased with loading levels of TPS20 until 10%, where the minimum value was observed that is, increasing TPS20 higher than 10% caused an increasing trend of the true viscosity. Also, it could be noted from Figure 5 that the true viscosity of the blend at 10% of TPS20 indicates negative deviation blends (NDBs), while above 10% it indicates a positive deviation blends (PDBs) according to the following log additives rule:

$$\log\eta_B = \sum_i W_i \log\eta_i \qquad (6)$$

where η_i and η_B are the viscosity of the *ith* component and that of the blend and *wi* is the weight fraction of the *ith* component.

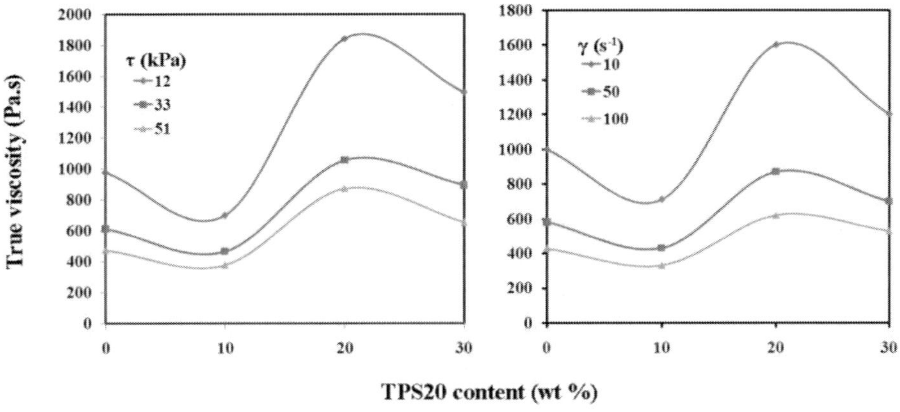

Figure 5. True viscosity versus TPS20 content (wt%) at 185°C.

Flow Activation Energy

Figure 6 shows the relationship between true viscosity and *1/T* for sample melts at a constant shear rates and shear stresses. It is observed that the linearity for all curves in Figure 6, is good, which indicates that the relationship between true viscosity and *1/T* obeyed by Arrhenius equation form, so the flow activation energy at a constant shear stress ($E_{\tau,\gamma}$) could be determined from the slope of these lines.

Figure 7 shows the effect of TPS20 content on the flow activation energy at a constant shear rate and shear stress of the blends. It could be noted from Figure 7 that the flow activation energy of the blends is greater than that of PP. These results indicate that the flow behavior of the blends is more sensitive to temperature compared with PP. Also it could be noted that the flow activation energy at a constant shear stress is more than that at a constant shear rate which can be attributed to the non- Newtonian behavior of the blend [34].

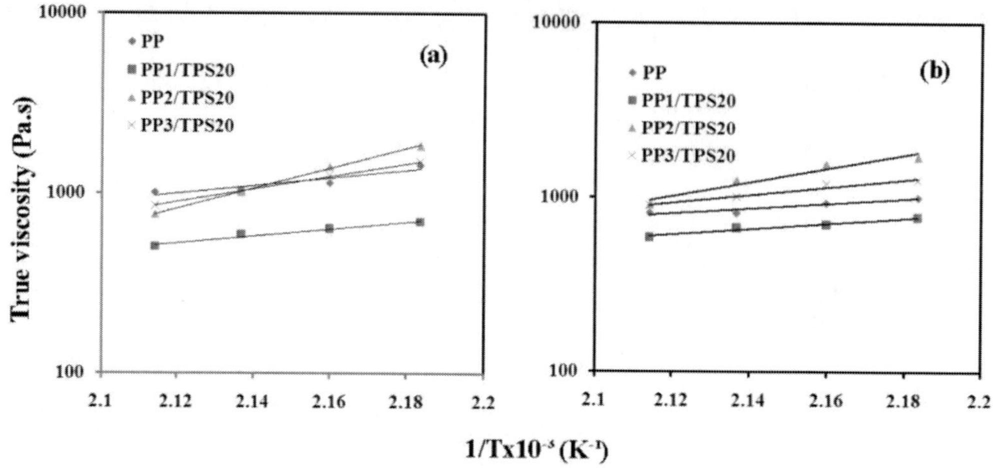

Figure 6. True viscosity versus *1/T* for sample melts (a) τ=12 kPa, (b) γ =10 s^{-1}.

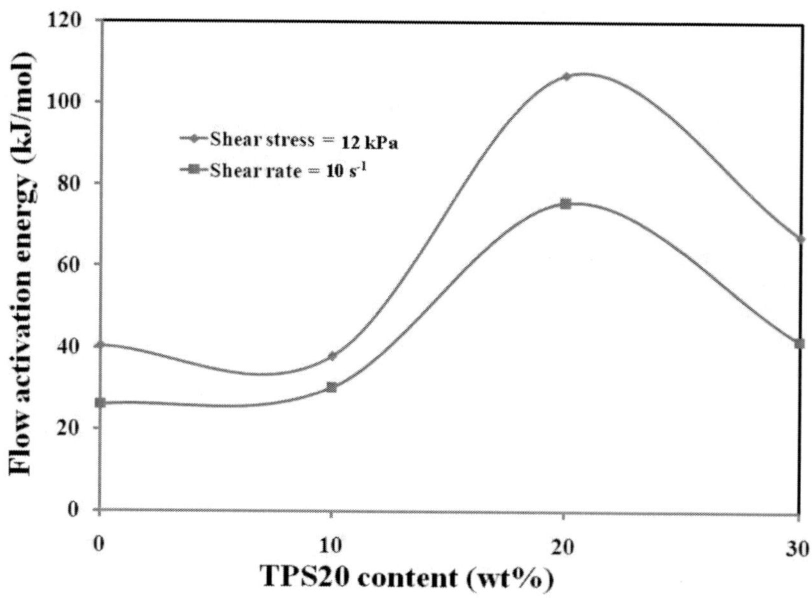

Figure 7. Flow activation energy versus TPS20 content (wt%).

Conclusion

PP/TPS blends were prepared using a single screw extruder, rheological tests showed that PP/TPS blends were pseudo plastic and exhibited shear-thinning behavior where the viscosity decreased with increasing shear stress. Also the flow behavior of the blends is more sensitive to temperature and shear rate compared with PP, where the values of *n* and *E* are higher than that of pure PP. Also it was found that the true viscosity of the blend at 10% of TPS20 indicates negative deviation blends (NDBs), while above 10% it indicates a positive deviation blends (PDBs). The blend has good processability and low cost comparing with pure PP, and it can be used as alternative to pure PP in some of its applications.

REFERENCES

[1] M. Kaseem, D. Hamad and F. Deri, *Inter. J. Polym. Mater.* Accepted (2011).

[2] C.S. Salcido, F.J. Gonzalez, M. L. Hernandez and J. C. Esquivel, *Polym. Bull.* 60, 677 (2008).

[3] D. S. Rosa, C. G. F. Guedes and C. L. Carvalho, *J. Mater. Sci.* 42,551(2007).

[4] D.S. Rosa, M. A. G. Bardi, L. D. B. Machado, D. B. Dias, L. G. A. Silva and Y. Kodama, *J. Them. Anal. Calorim.* 97, 565 (2009).

[5] M. Mihai, M. Huneault and B. D. Favis, *J. Cell Plast.* 43, 215 (2007).

[6] F. J. Gonzalez, B. A. Ramasy and B. D. Favis, *Carbohydr. Polym.* 58, 139 (2004).

[7] J. Yu, N. Wang and X. Ma, *Starch.* 57, 494 (2005).

[8] J. Yu, S. Chen, J. Gao, H. Zheng, J. Zhang and T. Lin, *Starch.* 50, 246 (1998).

[9] E. Corradini, E. Medeiros, A. Carvalho, A. Curvelo and L. Mattoso, *J. Appl. Polym. Sci.* 101, 4133 (2006).

[10] W. Orts, G. Nobes, G. Glenn, G. Gray, S. Imam and B. Chiou, *Polym. Advc. Techn.* 18, 629 (2007).

[11] E. Teixeira, A. Roz, A. Carvalho and A. Curvelo, *Carbohydr. Polym.* 69, 619 (2007).

[12] N. Wang, J. Yu, P. Chang and X. Ma, *Carbohydr. Polym.* 71,109 (2008).

[13] X. Ma, J. Yu and J. J. Wan, *Carbohydr. Polym.* 64, 267 (2006).

[14] X. Ma and J. Yu, *Carbohydr. Polym.* 57,197 (2004).

[15] X. Ma and J. Yu, *Starch*, 56,545 (2004).

[16] X. Ma and J. Yu, *J. Appl. Polym. Sci.* 93, 1769 (2004).

[17] J. H. Yang, J. Yu and X. Ma, *Chine. Chem. Lett.* 17,133 (2006).

[18] J. H. Yang, J. Yu and X. Ma, *Carbohydr. Polym.* 63,218 (2006).

[19] J. H Yang, J. Yu and X. Ma, *Carbohydr. Polym.* 66, 110 (2006).

[20] T. Bourtoom, Songk. lankarain. *J. Sci. Technol.* 30, 149 (2008).

[21] N. Wang, J. Yu, X. Ma and W. Ying, *Carbohydr. Polym.* 67, 446 (2007).

[22] H. Dai, P. Chang, F. Peng, J. Yu and X. Ma, *J. Polym. Res.* 16, 529 (2008).

[23] T. Stein and R. Greene, *Starch.* 49, 245 (1997).

[24] M. Kaseem, H. Hamad and F. Deri, *Polym. Bull.* DOI 10.1007/s00289-011-0611-z (2011).

[25] M. Kaseem, K. Hamad and F. Deri, *J. Appl. Polym. Sci.* DOI: 10.1002/app.35350 (2011).

[26] M. Kaseem, K. Hamad and F. Deri, *Malays. Polym. J.* Accepted (2010).

[27] D. Schlemmer, E. R. Oliveira and M. Sales, *J. Them. Anal. Calorim.* 87, 635 (2007).

[28] T. Pimentel, J. Duraes, A. Drummond, D. Schlemmer, R. Falcao and M. Salas, *J. Mater. Sci.* 42, 7530 (2007).

[29] D. Schlemmer, M. Sales and I. S. Resck, *Polym.* 75, 58 (2009).

[30] C. D. Han, *Rheology and processing of polymeric materials* (Polymer Rheology). Oxford University Press, New York (2007).
[31] E. B. Bagley, *J. Appl. Phys.* 28,624 (1957).
[32] G. S. Hu, B. B. Wang and F. Z. Gao, *Mater. Sci. Eng. A.* 426, 263 (2006).
[33] W. Sinthavathavorn, M. Nithitanakul, B. Grady and R. Mangaraphan, *Polym. Bull.* 63, 23 (2009).
[34] R. S. Lenk, *Polymer rheology*. Applied Science Publishers Ltd, London. (1978).

In: Polypropylene
Editors: L. P. Silva and E. F. Barbosa

ISBN: 978-1-62417-142-0

Chapter 7

EVOLUTION OF INJECTION-MOLDED MORPHOLOGY ON POLYPROPYLENE AND ITS APPLICATIONS

Yutaka Kobayashi*
Advanced Composites, Inc., Sidney, OH, US

ABSTRACT

According to the fountain-flow concept, the surface morphology of injection-molded plaques is created at the advancing flow front because moving, melted polypropylene (PP) is immediately frozen at the cold cavity wall. The rubber dispersion and the orientation of a PP/EBR blend were the same where the flow front speed was constant, proving that the surface formation mechanism follows the fountain flow concept. However, increasing the mold temperature led to the thinning of the skin layer and the formation of a trans-crystalline β–form at the surface of the injection-molded plaques. The thickness gap caused the difference in the β–ratio and the crystallinity between the flow-in and the flow-out positions. The extensional deformation at the flow-out section enhanced the β–form crystals and the skin thickness. The PP characteristics also change the morphology induced by the injection molding. The molecular weight dependency of the crystallization was investigated from a rheological perspective. Under a constant melt flow rate, a storage modulus (G') at low frequency was directly related to the flexural modulus of injection-molded specimens, in which the MWDs were controlled by mixing PPs of low and high molecular weights.

INTRODUCTION

Polypropylene (PP) is widely used in the automotive industry because it is lightweight, inexpensive, recyclable and easily processed by injection molding. Since the inception in the 1970s, research on the use of PP for automotive applications has mainly focused on improving the polymer's stiffness and impact resistance [1]. To increase the material's crystallinity, the isotacticity of PP can be improved using a Ziegler-Natta catalyst. Highly

* E-mail address: Yutaka.Kobayashi@advcmp.com.

crystalline, commercially produced PP currently has a meso-pentad ratio of 97% as measured by ^{13}C-NMR. Control of the injection molding-induced morphology is another important area of research. These developments enabled material engineers to control the stiffness, as captured by the flexural modulus.

Recently, the surface properties of injection-molded PP have become more important for automotive parts for environmental reasons. Volatile organic solvents are being removed from paints; thus, painting systems require PP with better paintability. The painting itself is sometimes foregone, in these cases, excellent scratch resistance is essential for plastic automotive parts. To enhance these surface properties, it is important to understand the relationship between the surface morphology and the properties near the surface of an injection-molded plaque. This study provides an overview of the surface phenomena related to injection-molded PP, although many researchers are attempting to determine the formation mechanism of the surface morphology under specific conditions.

Some Preliminary Information

The morphology of injection-molded PP is a skin-core structure [2] in which several layers have been observed under polarizing optical microscopy (POM). The layers are characterized by three to six structures, from the surface to the core [3]. The three layers related to the structure formation mechanism are typically discussed in the literature. The melted polymer in contact with the mold surface freezes immediately and a few spherulites appear in this frozen layer. The shear flow inside the frozen layer gradually cools, and an oriented crystal structure with high birefringence is observed in this shear-oriented layer. The core layer contains spherulites because the relaxation of the flow history occurs during the slow cooling in this region [1].

X-ray scattering has been used to analyze the crystal orientation of the skin-core structure. The melted polymer is subjected to a large stress and a high deformation rate at the skin layer, causing lamellae oriented toward the c- and a*-axis to occur at higher frequencies than the isotropically distributed lamellae. When the specimen wall thickness is thin (1 mm), the c-axis orientation ratio is higher than that of the a*-axis at the skin layer [4]. The lamellae thickness, the long period and the crystallinity of the injection molded PP specimens are measured from the surface inward by synchrotron small angle X-ray scattering (SAXS). For 1-mm-thick specimens these three values are minimized at the core layer and maximized at the shear-oriented layer. The depth profile of the lamellae changes as the thickness increases. A 5-mm-thick specimen has the maximum (minimum) lamellae thickness, long period and crystallinity at the core (frozen layer) [5].

The crystal structure difference for injection molding is interesting and has been studied from the perspective of a formation mechanism for the surface morphology. The skin-core structure is typically discussed using the fountain-flow mechanism [6]. During the surface formation process, hot melted polymer moves into the mold cavity through the fountain flow. At the advancing flow front, the melt flows from the middle of the cavity toward the wall and is subjected to extensional deformation. The oriented melt immediately freezes when it comes into contact with the cold wall. As a result of the subsequent flow during the filling process, a shear-oriented layer forms inside the frozen layer. Although the fountain-flow mechanism is commonly used as the above explanation of morphology creation, many numerical simulation of the depth profile of the injection-molded parts utilize an approximation, which leads to a Hele–Shaw flow that neglects the effect of the fountain flow at the flow front [7]. Birefringence quantifies the presence of a highly oriented molecular structure and relates the injection mold-induced structure and the resulting properties. The depth profile of the birefringence from the surface to the shear-oriented layer is generally v-shaped [8], and has been estimated using a viscoelastic flow analysis [9-11].

The created morphology affects the PP's performance, which is typically captured by its mechanical properties, such as the yield strength, tensile modulus and Izod impact strength. The typical factors of injection-molded morphology that influence these mechanical properties are characterized by the skin-core ratio, molecular orientation to the flow direction and crystallinity. The flexural modulus and the flexural strength increase when the skin ratio and the molecular orientation increase [2]. The increased crystallinity accompanying the increased molecular orientation increases the high-temperature modulus and therefore the heat-deflection temperature (HDT) [12]. Although the desired properties can be achieved using suitable morphology of the injection-molded specimens, the skin ratio, molecular orientation and crystallinity change simultaneously with the injection-molding conditions. The skin ratio increases with the stress level and decreases with the mold temperature. The orientation level at the skin increases with the wall shear stress and decreases with the mold temperature. The skin crystallinity increases with both the wall shear stress and the mold temperature. Thus, the PP stiffness is directly related to the molding conditions [13, 14].

However, the impact behavior of the injection-molded specimens is not clearly related to the surface morphology of homo-PP. Although a falling weight impact test (FWI) with biaxial out-of-plane loading is not significantly influenced by the skin-core morphology, a thick skin decreases the tensile impact strength at 3.7 m/s [15]. The skin orientation also reduces the tensile impact strength at 3.0 m/s, and as the tensile velocity decreases, the contribution of the core crystallinity is enhanced with a gradual decreasing of the influence of the skin orientation. [16].

The molecular orientation in the injection-molded specimens on PP is the greatest at the surface. As shown in Figure 1.1, the surface area of an injection-molded specimen is reduced by the relaxation of the stretched molecules after the specimen heated near the melting temperature. This shrinkage suggests that the distribution of the molecular orientation in the depth influences many properties of injection-molded products, especially the surface characteristics. This chapter studied the mechanism of the morphology formation near the surface and the effects of the resulting morphology on the physical properties of injection-molded products.

Previous studies have numerically analyzed the surface structure using the fountain-flow concept [17], although a large amount of research has been conducted using the lubrication approximation, which ignores the fountain flow at the advancing flow front. The original 2-D flow analysis program for viscoelastic fluids simulated the advancing flow front of a 3-mm thick plaque. The morphology near the surface was formed just behind the advancing flow front of the injection-molded plaque. During the fountain flow, the melt was subjected to planar extensional deformation and was oriented parallel to the surface, as shown in Figure 1.2. At the flow front, the PP chains stretched extensionally, and a strain of approximately 500% was estimated for the moment when the PP made contact with the wall. In the simulation of the actual molded specimens, a frozen layer occurred from the surface to a depth of 0.06 mm during the filling process of the injection molding. An intermediate layer between the surface and the shear-oriented layers was divided into two parts: the surface side retained a flow history in the flow front and the inner side was strongly affected by the shear flow. Figure 1.3 illustrates the simulation result above, which is different from the formation mechanism of the oriented morphology based on the shear deformation [2, 9].

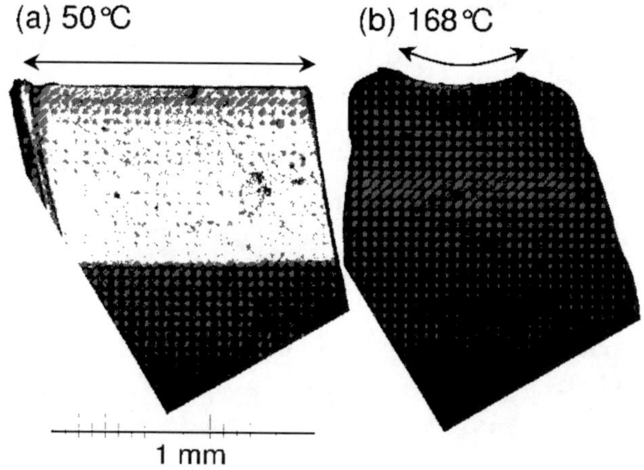

Figure 1. POM photograph taken at the cross section of an injection-molded plaque with hot stage temperatures of a) 50°C and b) 168°C.

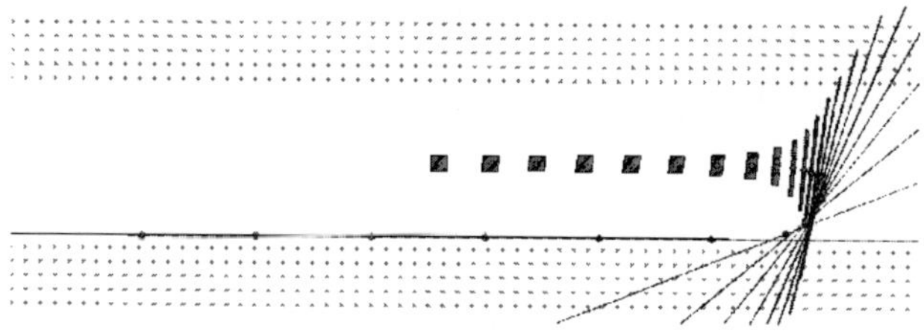

Figure 1.2. Conceptual diagram of the particle motion in the flow front. A particle starting at the initial position is subject to large deformations as it moves to the final position.

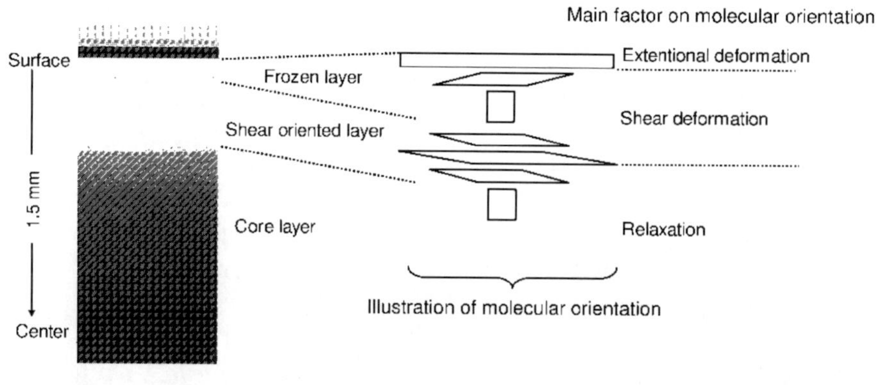

Figure 1.3. Conceptual diagram of the molecular orientation to the flow direction near the surface of the injection-molded plaques. The polymer chains that are subjected to large extensional deformations in the advancing flow front freeze at the surface. Beneath the surface, the polymer chains crystallize with a shear deformation. In the core layer, the stretched chains are relaxed before the crystallization.

The crystal morphology of PP is determined by the crystallization in addition to the molecular orientation. Highly stretched molecular chains at the flow front can crystallize when cooled. However, there is typically a low crystallity at the frozen layer. The skin morphology formation of PP in the injection-molded plaques was investigated using X-ray diffraction with a 0.273 μm × 0.389 μm X-ray beam [18]. The crystallinity decreased constantly from the core outward when the skin morphology was observed from the surface to a 100-μm depth. Nucleating agents (NAs) did not affect PP crystallization at a 1-μm depth, and the mesomorphic phase was observed. However, the NAs did promote crystallization from a 5-μm depth inward. The observed skin morphology was compared to the factors calculated by a numerical viscoelastic analysis of the fountain flow. The crystallinity was affected by the cooling rate and the temperature after quenching as well as the deformation rate of the melt. It is likely that the oriented chains required time to move the crystal lattice during the cooling process at the surface.

This chapter considers three experiments related to the fountain flow concept. In section 2, the consistency of the morphology and the mechanical properties at the surface is confirmed using a large plaque molded at a constant injection speed. Sections 3 and 4 investigated the effect of the flow path and the molecular weight distribution on crystallization, respectively.

MICROHARDNESS AT THE INJECTION-MOLDED SURFACE OF PP

Introduction

This section describes the relationship between the crystal morphology and the microhardness in the injection-molded PP. Because the motivation of this study is to develop enhanced PP with superior surface properties such as paintability and scratch resistance, this investigation focused on the area from the surface to a depth of 20 μm. However, few methods test the mechanical properties in such small area besides the microhardness test. For example, the popular indentation hardness measured by a durometer is not suitable for this experiment because the steel rod of the durometer has a diameter of approximately 1 mm.

The differences in the polymer crystal structures affect the microhardness, which is measured by the instrumented microindentation method [19]. The microhardness on the surface of a compression-molded sheet is proportional to the tensile modulus of several different PPs and related to the crystallinity as measured by differential scanning calorimetry (DSC) [20]. For the injection-molded PP, the microhardness on the cross-cut surface indicates the local maximum at the shear-oriented layer and the local minimum at the skin layer. As a result, the crystallinity near the surface is small [21]. For linear low-density polyethylene, increasing the amount of comonomer in the polyethylene decreases both the thickness of the crystalline lamellae and the crystallinity. In this case, both factors are proportional to the microhardness [22]. The high deformation rate and the quenching in the injection molding affect the crystal size and the polymorphism of the polymers in addition to the crystallinity. Although an α-form crystal of PP (with a density of 936 kg/m^3) converts into the mesomorphic form (with a density of 880 kg/m^3) due to the rapid cooling in the quiescent crystallization, the microhardness is only related to the density of the PP [23]. The effect of the lamellar size on the microhardness remains unclear. A lower microhardness has been observed for the stacking lamellar structure, in which there is a large amorphous region between the lamellae [24].

The microhardness is used to evaluate the microscopic deformation of polymer blends [25-28]. Polymer blends consisting of soft and hard components frequently show drastic deviations from the overall microhardness estimated by the law of additivity. Many morphology types are observed in these blends, and the microhardness must be calculated according to the volume fraction and the glass transition temperature of each component [25]. The deviation was also measured for the reactor blends of PP and EPR, which have spherical EPR domains within a PP matrix. The overall microhardness is affected by the reduced partial microhardness of the PP crystals because the EPR increases the interfacial energy of the PP crystals [26]. The precise polymerization of the styrene-butadiene block copolymers results in the formation of unique matrix-domain structures. The microhardness values of the samples containing the same amounts of styrene depend on the morphologies such as the cylinders and the alternatively stacking lamellae [27]. For the lamellar morphology, the microhardness is proportional to the thickness of the polystyrene lamellae [28].

In previous studies [29], the crystal structure was determined along the depth direction of the cross-cut surface of the specimen using a 0.35 μm-wide X-ray beam. Although many earlier studies have characterized the frozen layer, the crystal structures obtained in those studies were averaged over the specimen thickness or limited by the aperture size of the analytical instruments. Two distinct structures were observed in the frozen layer. The molecular and the crystal orientations changed sharply with depth in the surface region, while the structures remained constant in the inner region. The morphology did not affect the microhardness along the depth direction on the cross-cut surface, but the crystallinity was proportional to the hardness. When comparing the microhardnesses at the gate and center positions in the molded plaque, the microhardness was higher near the gate despite the lower crystallinity at this position. For the case of multi-extensional flow near the gate, the microhardness is influenced by biaxially extended orientations, which easily form superstructures such as cross-hatched lamellae in addition to the crystallinity. For the PP and the ethylene-butene rubber (EBR) blend [30], the breakup and coalescence of the EBR particles rarely occurred during the injection process. The EBR particles near the surface were continuous fibers gradually changing their elliptical shape in the depth direction. The morphology of the injection-molded specimen was related to the depth profiles of the mechanical factors, including the microhardness and the shear stress measured by Vickers and Knoop micro-indenters and a "Surface and Interface Cutting Analysis System", respectively. The crystal structure of the PP matrix had a greater effect on the microdeformation than the EBR phase. The large oriented EBR domains disrupted the continuity of the PP matrix and acted as a weak layer in the specimens.

This section investigates the relationship between the morphology and the microhardess at the surface of the injection-molded PP along the flow direction. The gate part had a more visible skin-core structure and a thicker skin layer than the distant part and had a higher impact strength and a higher tensile yield strength [31]. However, the surface properties are approximately the same at the gate and the flow end when the velocity of the advancing flow front is constant in the flow direction. In this section, the experimental methods for evaluating the morphology were the polarized optical microscopy (POM) and transmission electron microscopy (TEM). The microhardness was measured by an instrumented micro-indenter with a Vickers pyramid diamond.

Experimental

Materials

This study used a PP/EBR polymer blend. The PP was a Ziegler–Natta isotactic homopolymer containing a nucleating agent (MFR 13 g/10 min for 2.16 kg at 230°C, pentad fraction 97%) with a molecular weight of 306 kg/mol and an Mw/Mn ratio of 8.3. The EBR was A4050S, a commercial product of Mitsui Chemical (Tokyo Japan). The PP/EBR blend (77.2 wt%/22.8 wt%) was mixed by a twin-screw extruder (Toshiba TEM-30) with a barrel temperature of 200°C.

Specimens Preparation

Rectangular plaques ($^{w}100$ mm × $^{l}420$ mm × $^{t}3$ mm) with a side gate ($^{w}3$ mm ×$^{t}3$ mm) were molded by a 350-ton injection-molding machine, UBE MDS-III. The PP/EBR pellets were melted during the plasticizing process at a barrel temperature of 200°C. The melted resin was injected for the 5-s filling time into the tool's cavity at the 40°C, and then the holding pressure was applied for 15 s to avoid deformation caused by the volumetric shrinking during cooling. Finally, the molded plaque was cooled for 20 s after the holding pressure was turned off. The injection pressure was automatically controlled to maintain the constant injection velocity. Each set of conditions was indicated when the settings were changed.

Microhardness Measurements

The microhardness was measured with an Elionix ENT-1100a tester at room temperature. A Vickers tip was used to obtain the experimental microhardness data, which were denoted as H_V (Figure 2.1). The contact area was used to calculate the Vickers hardness. The studied positions in the injection-molded plaques were located 80 mm, 180 mm, 280 mm and 380 mm from the gate, close to the center of the part (Figure 2.2). These distances were measured from the gate on the surface of the molded plaques.

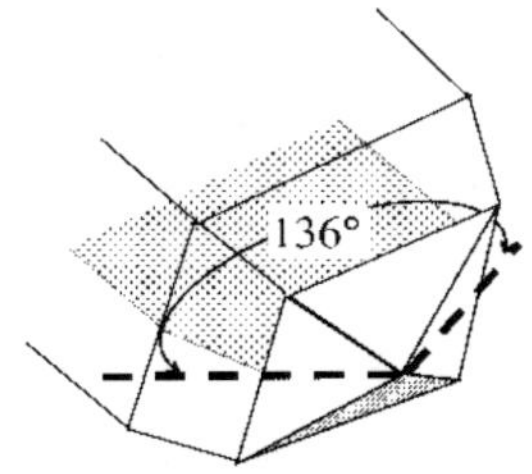

$$H_V = \frac{F}{A_{surface}} = \frac{F}{26.43h^2}$$

Figure 2.1. Geometry of an indenter, where H is the microhardness for a Vickers indenter, F is the applied force, A is the contact area and h is the indentation depth.

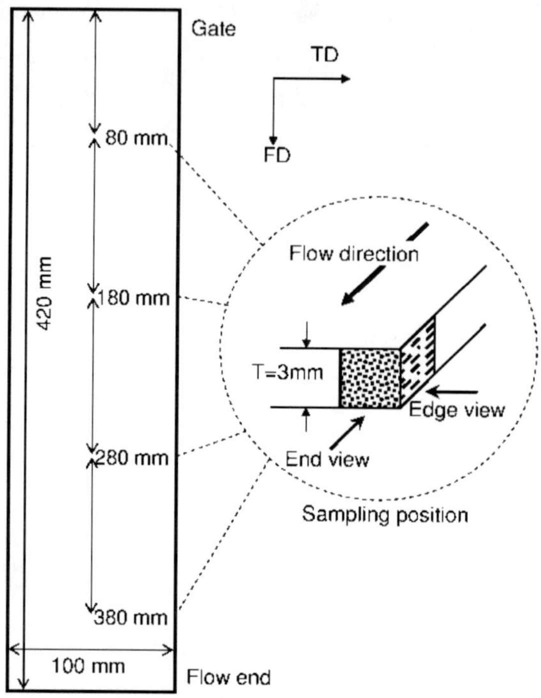

Figure 2.2. Schematic diagram of an injection molded plaque and the specimens cut 80 mm, 180 mm, 280 mm and 380 mm from the gate.

Morphological Characterization

The molded specimens were characterized by two experimental techniques: POM and TEM. The measurements were taken 80 mm, 180 mm, 280 mm and 380 mm from the gate (Figure 2.2). The POM observations were carried out using an Olympus BH-2 microscope with a Berek compensator at room temperature. The morphology of the microtomed section was viewed from the direction parallel to that of the flow (edge view).

The morphology of the EBR in the PP matrix was observed by a TEM (JEOL JEM-1010). The specimens were stained with ruthenium tetroxide (RuO_4) vapor at 25°C for 15 hrs. The EBR domains dispersed in the PP matrix were mainly stained, and there was little change in the PP matrix. Thus, there was a sharp contrast between the matrix and the domains. The stained dark phases represent rubber portions. The ultrathin sections were microtomed from the stained specimens by a Leica EM UC6. The observations were carried out in the direction parallel to the flow (edge view).

Results and Discussion

Rubber Morphology along the Flow Direction

The elastomer morphology in an injection-molded specimen is related to the microhardness of the PP/EBR blend [30]. First, the morphologies near the surface at different positions in a plaque were precisely characterized using POM and TEM. The thin observed sections were cut parallel to the flow direction at the center of the injection-molded plaque. The molecular and crystal orientations in the specimen were observed at the edge thin film

(0.03 mm) of the PP/EBR blend by POM, as shown in Figure 2.3. The bright section is a skin layer, which includes the frozen and the shear-oriented layers. This typical skin-core microstructure occurs when the molecular chains in the skin layer are highly oriented to the flow direction due to the high strain rates and the rapid cooling during the processing [32]. The skin thickness decreased as the flow distance from the gate increased. The frozen layer formed at the constant speed of the advancing flow front is homogeneous along the flow direction [17]. Thus, the different skin thickness is caused by the flowing quantity within the frozen layer. At the flow end, the shear flow immediately stops at the filling and the holding processes so that the depth of shear-oriented layer does not expand.

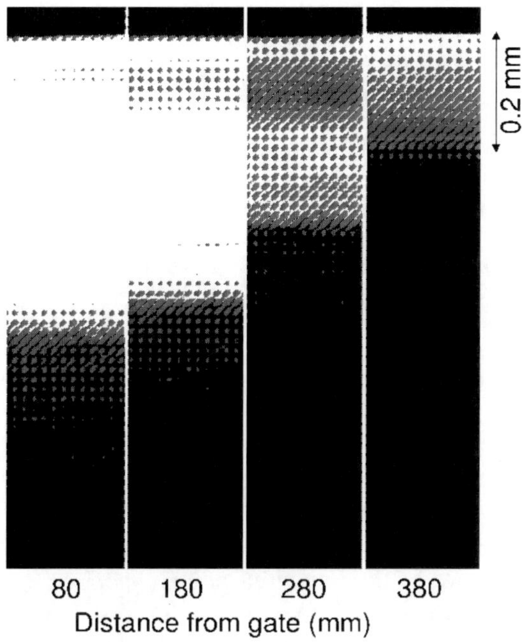

Figure 2.3. POM photograph taken at the cross sections of an injection-molded plaque, where the measured points are 80 mm, 180 mm, 280 mm and 380 mm from the gate.

The morphology of the EBR phase at the surface of the injection-molded specimen was observed by TEM. The black layers in Figure 2.4 indicate the ruthenium tetroxide residues on the surface. The EBR domain was fibrous from the edge view. The rubber particles deform to rod shapes during the injection molding, and the deformation gradually decreases from the surface to the core [30]. The morphology along the flow direction was expected to be the same at each position because the constant-speed condition of the injection molding utilized in the plaque preparation. The visual impression provided by the TEM micrograph of the rubber dispersion was rougher at the 80-mm position than at the 180-, 280- and 380-mm positions, although the EBR fiber sizes were indistinguishable because the EBR particles near the surface were highly stretched and observed as continuous fibers. The biaxial stretching of the rubber particle by the radial flow reduces the deformation of the flow direction due to the side gate of the plaque. After the radial flow is finished, the surface morphology does not relate to the skin-core structure. Therefore, this result supports the fountain flow concept of the morphology formation mechanism.

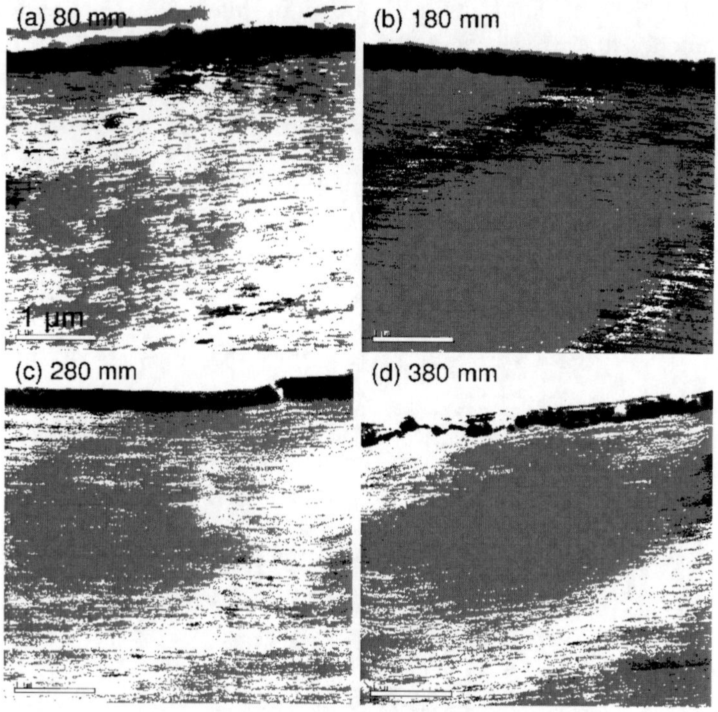

Figure 2.4 TEM photographs at the surface of the injection-molded PP/EBR from the gate to the flow end, where the measured points are a) 80 mm, b) 180 mm, c) 280 mm and d) 380 mm from the gate.

Influence of the Morphology on the Microhardness

Figure 2.5 presents the changes in the microhardness measured along the flow direction with a Vickers tip at the surface of the injection-molded specimen. The microhardness at the 80-mm position was greater than those at the 180-mm, 280-mm and 380-mm positions, coinciding with the morphological observations. The indentation depth of the 10-mN applied load was approximately 3 μm based on the equation in Figure 2.1.

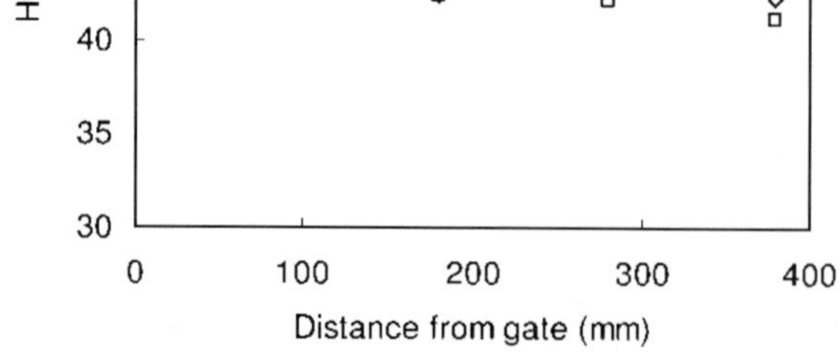

Figure 2.5. Microhardness (H_V) at the surface of an injection-molded plaque, where 1-mN, 10-mN and 100-mN indentation loads were applied at room temperature.

Previous studies have suggested that the increase in the interfacial energy resulting from the extension of the EBR particles decreases the microhardness [26], and the small H_V observed at the injection-molded PP surface is the low crystallinity of the PP resulting from quenching on the mold wall [21]. However, the difference between the positions is due to the continuity of the PP matrix, which is the major factor affecting the microdeformation in the PP/EBR blend. In other words, the fibrous EBR domains in the injection-molded specimens act as a weak phase for the microhardness [30].

Conclusion

This section investigated the rubber morphology of a PP/EBR blend at the surface along the flow direction and the relationship between the resulting morphology and the resulting microhardness. A highly stretched rubber domain was observed by TEM and changed at the position near the gate and the other positions in a plaque. Although the radial flow decreased the rubber orientation in the flow direction, the constant speed of the advancing flow front resulted in the same morphology at the surface supporting the fountain flow concept of the morphology formation mechanism.

The observed morphology coincided with the microhardness at the surface. The fibrous EBR domains in the injection-molded specimens act as a weak phase for the microhardness.

Injection Molding-Induced Structures

Introduction

The previous study investigated the effects of nucleating agents (NAs) under quenching conditions to observe whether crystallinity at the surface is increased and the flow-induced crystallization dominates the NA-induced nucleation in the PP near the surface during injection molding [18]. Using a 0.273 μm × 0.389 μm X-ray beam, the skin morphology was observed from the surface to a 100-μm depth. The orientation degree measured on the mesomorphic and the α-form crystals showed a v-shaped pattern with typical birefringence behavior, although the crystallinity decreased constantly from the interior to the surface. In the injection molding, the NAs did not work on the crystallization of the PP at the 1-μm depth, and the mesomorphic phase was observed. The NAs promoted the crystallization from 5-μm deep inwards. The observed skin morphology was compared to the factors calculated by a numerical viscoelastic analysis of the fountain flow. The crystallinity was affected by the cooling rate and the temperature after the quenching instead of the deformation rate of the melt.

Increasing the mold temperature is a potential industrial method to achieve higher crystallinity at the surface because the melted PP is extensionally stretched by the fountain flow mechanism and immediately freezes after coming into contact with the cold wall in the injection molding. Thus, the morphology of the non-nucleated PP near the surface in the 3-mm thick plaques molded at three temperatures was studied [33]. At a mold temperature of 20°C, the skin layer was 0.2-mm thick and β–form spherulites were observed in the core region. When the mold temperature increased, the skin thickness decreased and trans-crystalline β–form appeared near the surface at 100°C. At 20°C and 65°C, few spherulites appeared in this frozen layer because the melted polymer that came into contact with the mold surface froze immediately. At 100°C, the frozen layer was not observed.

The purpose of this section is to clarify the effect of the mold temperature on the crystal morphology at the local position of an injection-molded product using WAXD analysis. Because the shapes of injection-molded products are typically complex, the actual mold temperature varies at each local section and the flow pattern is unsteady. The delayed cooling of the cavity wall and the shear-rate jump are caused by a thickness gap in a flat plate and analyzed with POM and WAXD.

Experimental

Materials and Molding Conditions

The PP used in this study was a Ziegler-Natta isotactic homopolymer (MFR 30 g/10min for 2.16 kg at 230°C, pentad fraction 97%) with an Mw of 200 kg/mol and an Mw/Mn ratio of 4.0. The nucleating agents were not doped in the PP. The plaques (^{w}80 mm × ^{l}80 mm × ^{t}3mm) were molded under the following conditions: a barrel temperature of 200°C, a filling time of 1 s, a holding time of 20 s and a cooling time of 360 s. The mold temperatures were set to 20°C, 65°C and 100°C with a high-pressure water-circulation temperature controller.

Morphological Characterization

The molded specimens were characterized by two experimental techniques: POM and WAXD. The POM observation was performed using an Olympus BH-2 at room temperature. As shown in Figure 3.1, the morphology of the local position of the plaques was viewed in a direction parallel to the flow (edge view). The specimens were microtomed in a sledge microtome at room temperature in the flow direction to a thickness of 30 μm.

The WAXD experiment was performed using a Rigaku RINT 2550 with an imaging plate. The circle-shaped beam had a radius of 0.15 mm and a wavelength of 1.54059 Å (CuKα, 40 kV, 370 mA). The measurements were performed at room temperature with the beam along the flow direction (edge view), and the X-ray beam was irradiated at the nine positions denoted in the specimens as G1 to G3, C1 to C3 and E1 to E3. All of the X-ray data were corrected for background scattering (air and instrument) before the analysis. The crystallinity was calculated as the ratio between the integral of the intensities diffracted by the crystalline part and the total integral of the intensities [34], i.e., a relative value related to the maximum crystallinity measured in the crystalline part. Although the PP exhibited polymorphism, the crystallinity included all of the crystal forms: α, β and γ. The β-ratio is defined as (300) / ((300) + (110) + (040) + (130)), where (110), (040) and (130) are the intensities of the lattice planes in the α-form crystal on the 1-D WAXD profile, and (300) is the intensity of a characteristic lattice plane in the β-form crystal shown in Figure 3.2.

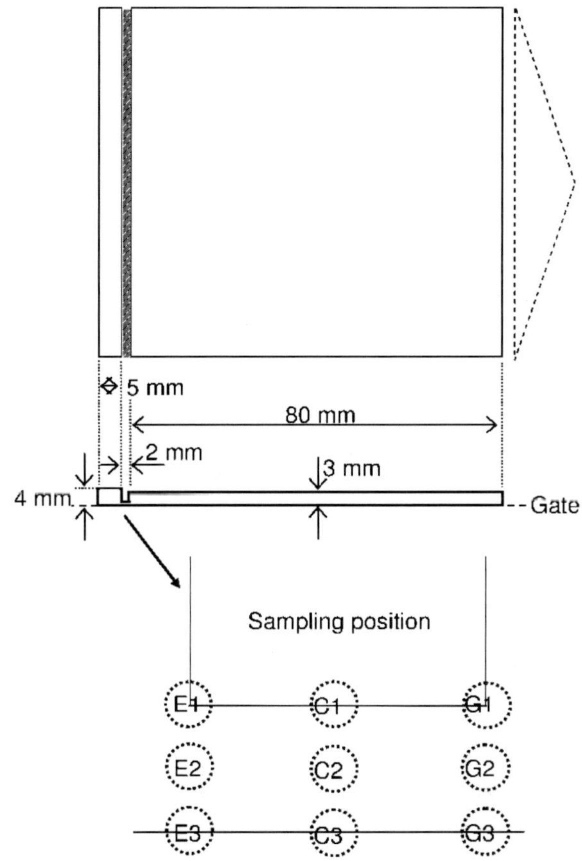

Figure 3.1. Schematic diagram of the injection molded plaques and the observed positions.

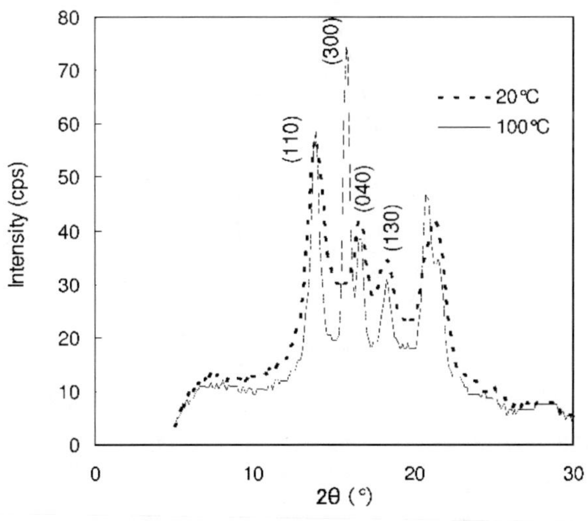

Figure 3.2. Changes in the 1-D WAXD patterns due to the mold temperature at the E3 position. The WAXD patterns were taken at room temperature from an edge view with a 0.3-mm beam diameter.

Results and Discussion

Crystal Morphology Observed by POM

Figure 3.3 shows POM micrographs of the local position in the plaques molded at 20°C, 60°C and 100°C. The typical skin-core structure was observed, but the skin thickness was thinner at position 1 (G1 to E1) than position 3 (G3 to E3). The skin thickness was negligible at the G1 corner and increased from G1 to E1. After the E1 corner, the wall thickness increased from 1 mm to 4 mm and a highly oriented layer appeared. In contrast, there was no oriented structure before the G1 corner when the wall thickness decreased from 3 mm to 1 mm. When the mold temperature increased, the spherulites became more common and the optical orientation and skin layer thickness decreased.

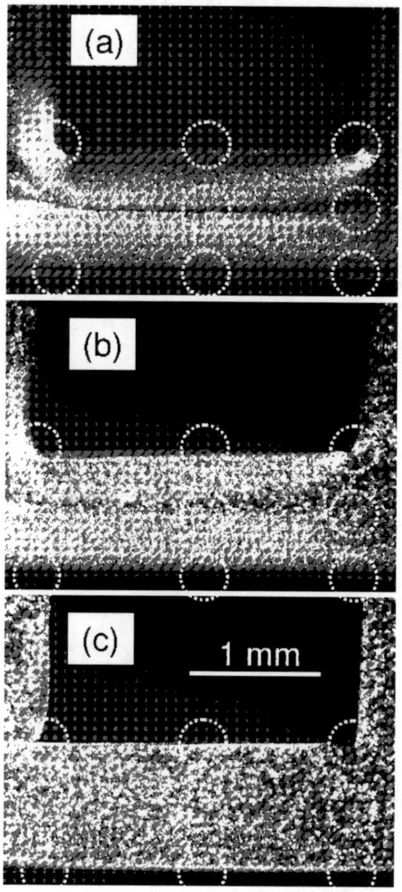

Figure 3.3. POM photographs taken at the cross sections molded at the mold temperatures of a) 20°C, b) 60°C and c) 100°C.

Crystal Structure Analyzed by WAXD

Table 3.1 lists the crystallinities and a β-ratio at the nine positions of the plaques molded at 20°C, 60°C and 100°C. The crystallinity at position 1 was greater than that at position 3 at the 20°C mold temperature, but the crystallinity at position 1 was similar to those at position

3 at 60°C and 100°C. Thus, the actual mold temperature at position 1 was higher than 20°C because the cooling at the obstacle on the cavity wall was delayed compared that on the flat wall. Figure 3.4 shows changes in the crystallinity along the flow direction. The crystallinity of position 3 at 20°C gradually increased from the gate side to the flow end and coincided with the increasing skin thickness, which had a higher crystallinity, as stated in the next section. At position 1, however, G1 had a high crystallinity without the skin layer. Because the hot melt keeps flowing at the G1 corner until the filling process finished, and the square obstacle delays cooling of the melt. The crystallinity at position 2 did not depend on G, C or E due to the delayed cooling at the core area. When the mold temperature was 100°C, the crystallinities at the different positions were very similar.

Table 3.1. Crystallinity and β-ratio at the local positions in an asymmetrical flow path with mold temperatures of 20°C, 60°C and 100°C

		Crystallinity (%)			β-ratio (-)		
Mold temperature	(°C)	**20**	**60**	**100**	**20**	**60**	**100**
Position	G1	55	56	60	nd	nd	0.27
	G2	62	63	65	nd	nd	0.20
	G3	34	63	61	nd	nd	0.34
	C1	49	57	59	nd	0.24	0.37
	C2	62	60	63	nd	0.07	0.18
	C3	38	58	59	nd	sd	0.32
	E1	60	57	61	sd	0.25	0.40
	E2	63	62	64	sd	0.11	0.20
	E3	41	57	60	nd	0.16	0.42

Previous studies have analyzed the near-surface morphology of a simple rectangular plaque molded at 20°C, 65°C and 100°C [33]. At a 20°C mold temperature, β–form spherulites were present in the core region but not the skin layer. When the mold temperature increased, the skin thickness decreased and a trans-crystalline β–form was present near the surface at 100°C. The β–form ratio was 0.4 at the 100°C mold temperature from the surface to a depth of 0.15 mm. At the mold temperatures of 20°C and 65°C, no β–form crystals were detected near the surface by WAXD. Figure 3.5 shows the formation of the β-form crystals at the local positions in the plaque molded at 100°C. The formations of the β-form crystals at positions 1 and 2 were similar to those of the normal plaques. However, the local flow influenced on the β–form ratio: a) the extensional flows at E1 and E3 increased the ratio, b) the ratio at G1 was less than that of G3 and c) the ratio at position 2 was less than those at positions 1 and 3. The crystal growth rate of the β-form crystals at the quiescent state is greater than that of the α-form crystals when the temperature ranges from 90°C to 133°C [35]. Thus, the difference at the cavity wall with the 100°C mold temperature is induced by the shear deformation of the melt [36, 37].

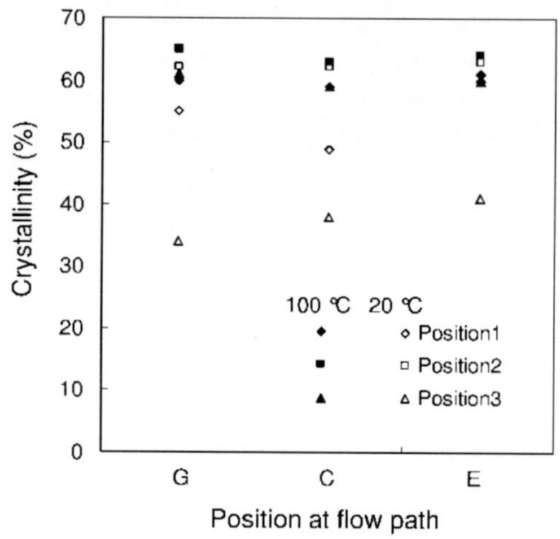

Figure 3.4. Changes in the crystallinity from the gate side to the flow end. The measured plaques were molded at 20°C and 100°C.

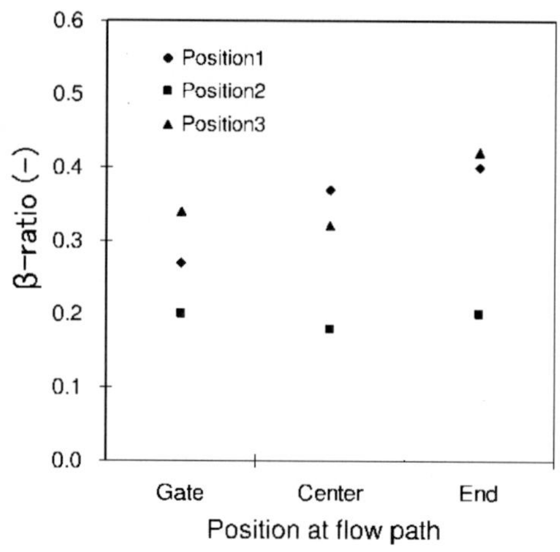

Figure 3.5. Changes in the β-ratio from the gate side to the flow end. The measured plaques were molded at 100°C.

Crystal Morphology Formation in the Complex Shape

For simple rectangular plaques, the overall crystallization mechanism of the PP near the surface in the injection molding is as follows: (a) The melted PP is extensionally stretched at the flow front; (b) the melt is quenched after coming into contact with the cold wall, but crystallization is delayed during the transient temperature; and (c) the crystallinity near the surface depends on the temperature at the depth. This study considered the special case in which the melt was not quenched.

Although the mold temperature was typically changed by thermolators, the setting temperature is not always equal to the actual temperature. The surface temperatures of actual products with complex shapes vary based on the local position. When the mold temperature of the flat smooth surface was 20°C, the mold temperature on the top of a 2 mm × 2 mm obstacle was overestimated, and the resulting crystallinity exceeded that at 20°C, shown in Figure 3.1. When the mold temperature rose from 20°C to 60°C, the frozen layer remained. At the 100°C mold temperature, the trans-crystalline β–form formed by the rearrangement of the stretched polymer chains instead of by quenching.

The melt that flowed into a thin section had a thinner skin layer and a lower β–ratio than that melt that flowed out of a thin section. Although the mass flow rate was the same, the compression or the expansion of the melt influenced on the morphology formation.

Conclusion

This study intended to clarify the effect of the mold temperature on the crystal morphology at the local position of an injection-molded product using WAXD analysis. In conclusion, the top of an obstacle in a cavity exhibited a cooling delay for the melt relative to a flat smooth wall. The resulting crystallinity of the molded product was higher on the obstacle side than on the normal side, despite identical mold temperature settings. The thickness gap caused the difference in the β–ratios and crystallinities between the flow-in and flow-out positions. The extensional deformation at the flow-out section enhanced the β–form crystals and the skin thickness.

Influence of the Molecular Weight Distribution on the Surface Property

Introduction

To increase the surface hardness, PP crystallization at the surface is required. Although the high mold temperature eliminated a frozen layer, the cycle time of the injection molding was too long. This section studied the influence of the molecular weight distribution on the injection molding-induced morphology formation.

Crystallization in the Quiescent State

Nascent nuclei are observed in the early stage of PP crystallization [38] indicating that PP nucleation begins by a statistical process explained by classical nucleation theory [39]. For a one-component system, the formation free energy of a nucleus is

$$\Delta G = \Delta\mu N + \sigma A \,, \tag{4.1}$$

where $\Delta\mu$ is the change in the chemical potential; N is the number of molecules in the nucleus; σ is the surface energy; and A is the surface area. When the nucleus is spherical with diameter r, as shown in Figure 4.1, there is an energy barrier to increase the size of the sphere because the increase in the surface energy is greater than the decrease in the chemical potential. The reduction of the energy barrier is important for industrial applications. Thus, nucleating agents (NAs) are often used to enhance the PP crystallinity. Heterogeneous nucleation occurs on the surface of the NA [40], as shown in Figure 4.2. A simplified polymer nucleus with a dimension of $l \cdot a_0 \times m \cdot b_0 \times n \cdot c_0$ is considered. The free energy of the formation in homogeneous nucleation is

$$\Delta G = -lmna_0b_0c_0\Delta g + 2lma_0b_0\sigma_e + 2\sigma(nmc_0b_0 + lna_0c_0) \tag{4.2}$$

where Δg is the free energy of fusion; σ_e is the surface energy of the end surface; and σ is the surface energy of the side surface. When heterogeneous nucleation occurs at the surface of a substrate, the free energy of formation is

$$\Delta G = -lmna_0b_0c_0\Delta g + 2lma_0b_0\sigma_e + 2nmc_0b_0\sigma + lna_0c_0\Delta\sigma$$

$$\Delta\sigma \equiv \sigma + \sigma_{0s} - \sigma_s \tag{4.3}$$

where σ_{0s} is the interfacial energy between the substrate and the nucleus and σ_s is the surface energy of the substrate. At coherency, $\Delta\sigma$ is zero because the substrate is the same material, with σ_{0s} = 0 and σ_s = σ. The mechanism of the NA's acceleration in the PP crystallization is the reduction of the energy barrier to form a critical nucleus by epitaxy on the NA [41, 42].

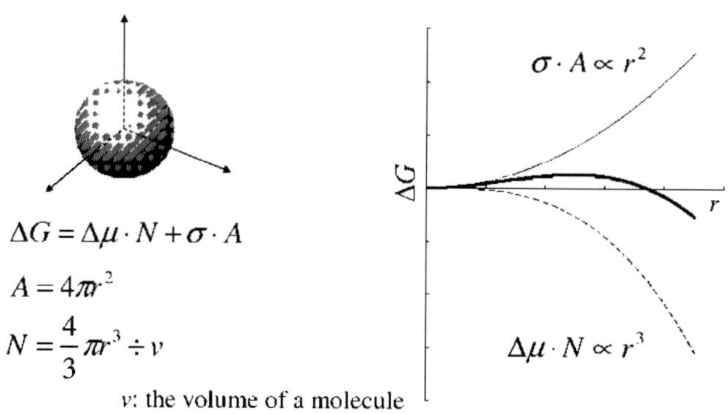

Figure 4.1. Schematic diagram of the free energy during the growth of a spherical nucleus.

Crystallization typically occurs under supercooling conditions, which increases the change in the free energy [43]. However, the maximum crystallization rate occurs between the melting temperature and the glass transition temperature. A bell-shaped pattern in the crystallization rate is expressed by the following equations for the nucleation rate and growth rate, respectively [44]:

$$I = I_0 \exp\left[-\frac{\Delta E_i}{RT} - \frac{K_i T_m^{o2}}{RT\Delta T^2}\right] \tag{4.4}$$

$$G = G_0 \exp\left[-\frac{\Delta E_g}{RT} - \frac{K_g T_m^o}{RT\Delta T}\right] \tag{4.5}$$

where I_0, K_i, G_0 *and* K_g are constants, ΔE_i and ΔE_g are the activation energies of diffusion, and ΔT (= T_m - T) is the supercooling temperature. The overall crystallization rate consists of the nucleation rate and the growth rate, which both decrease at low temperatures due to the transport terms ($\Delta E_i/RT$ and $\Delta E_g/RT$). These terms show that at low temperatures, the polymer chains are nearly stationary and are located in the crystal lattice.

The molecular weight dependence of the maximum nucleation rate (I_{max}) and the maximum growth rate (G_{max}) can be expressed as

$$I_{max} \propto M^{1.5} \tag{4.6}$$

and

$$G_{max} \propto M^{-0.5} \tag{4.7}$$

respectively, where M is the molecular weight [44]. Large molecules decrease G_{max} according to the transport term of Eq.4.5. However, I_{max} increases with M. Further investigation is required because the opposite molecular weight dependency of the nucleation rate has also been reported [45]. PP of a lower molecular weight typically has a higher crystallinity in industrial polymer processing.

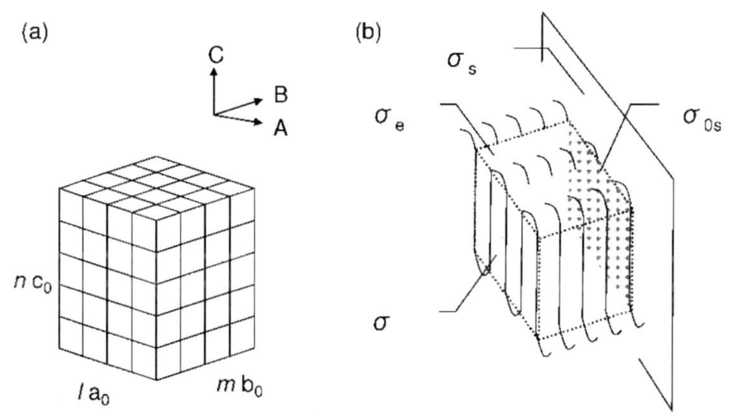

Figure 4.2. Schematic diagrams of the (a) homogeneous and (b) heterogeneous nucleations of simplified polymer nuclei with dimensions of $l{\cdot}a_0 \times m{\cdot}b_0 \times n{\cdot}c_0$ where σ_e is the surface energy of the end surface, σ is the surface energy of the side surface, σ_{0s} is the interface energy between the substrate and σ_s is the surface energy of the substrate.

Flow-Induced Crystallization

The nucleation rate of PP is greater under shear and extensional flows than under quiescent conditions. The molecular orientation is a cause of the increasing nucleation. Because the low entropic state of the oriented melt leads to a high equilibrium melting temperature, ΔG substantially decreases under supercooling. However, the nucleation rate increases even when the estimated entropy change under shear is low according to the measured birefringence [46]. The polymer chains are locally stretched and roughly oriented with the shear flow direction. The organized chains then form nascent nuclei [47]. Although long helical sequences of PP chains immediately appear in the melt after the chains are subjected to shear flow, spatial rearrangement of the stretched chains is required for nucleation [48]. As shown in Figure 4.3, the stretched chains relax under lower-magnitude supercooling conditions and freeze at the high cooling rate. Thus, flow-induced crystallization requires a high degree of orientation in the chains in the fluid and adequate mobility of the chains to form a crystal lattice.

Numerous studies have been conducted on the molecular weight dependency of flow-induced crystallization on PP. The critical orientation molecular weight depends on a shear rate and is approximately 300 kg/mol at high shear rates [49]. When the z-averaged molecular weight (Mz) increases for a given number-averaged molecular weight (Mn), shear flow-induced crystallization exhibited a higher degree of crystal orientation, a higher amount of the oriented crystal fraction and faster crystallization kinetics [50]. Two PPs of high and low molecular weights are blended to determine the relationship between the characteristics of specific molecules and their role in the crystallization process and to specify the length of "long chains" and the concentration of these chains in the melt [51]. However, adding long chains in PP increases the injection pressure at the practical injection molding because of the high viscosity. This section studied the changes in the morphology and the mechanical properties of several PPs with a constant melt flow rate (MFR).

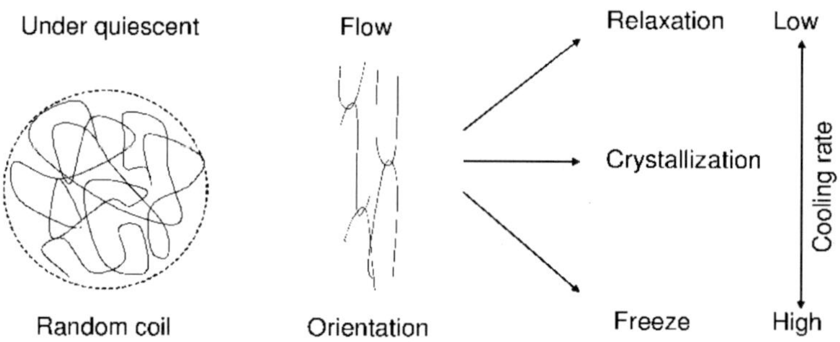

Figure 4.3. Schematic illustration of the flow-induced crystallization.

Experimental

The study used three Ziegler–Natta isotactic homopolymers with a pentad fraction of 97 %. The MFRs of L-PP, M-PP and H-PP were 0.5, 30 and 200 g/min, respectively, for 2.16 kg at 230 °C. As shown in Table 4.1, the five samples named MWD0 to MWD4 were prepared by blending the L-PP, M-PP and H-PP polymers with suitable ratios to maintain a constant MFR in the blended samples. The PPs were mixed with a twin-screw extruder (Toshiba TEM-30) at 200°C.

Table 4.1. Blend ratio and molecular weight of the samples

	unit	MWD0	MWD1	MWD2	MWD3	MWD4
L-PP	wt%	0	5	10	21	43
M-PP	wt%	100	90	80	59	17
H-PP	wt%	0	5	10	20	40
Mn	kg/mol	42.7	43.7	36.0	34.3	30.3
Mw	kg/mol	157	160	171	200	249
Mz	kg/mol	480	580	661	1190	1960
Mw/Mn	-	3.7	3.7	4.8	5.8	8.2
Mz/Mw	-	3.1	3.6	3.9	6.0	7.9

As shown in Figure 4.4, a plaque (^{w}70 mm × ^{l}270 mm × ^{t}3 mm) was molded with an injection-molding machine (Toshiba, IS100F-III) to study the crystallinity morphology. The molding conditions were as follows: a mold temperature of 40°C, a barrel temperature of 200°C, a filling time of 3 s, a holding time of 8 s and a cooling time of 20 s. The injection pressure was controlled automatically to maintain a constant injection velocity. The holding pressure was manually adjusted to 70% of the filling pressure.

The molded specimens were characterized by POM (Olympus BH-2). The sampling positions are shown in Figure 4.4. The morphologies of the microtome sections were viewed parallel to the flow direction (edge view).

The molecular weight distributions of the samples were determined by gel permeation chromatography (GPC). GPC was performed using two Tosoh GMHHR-H(S)HT columns, an RI detector (Waters Alliance GPC2000) and a 1.0-mL/min flow rate. The temperature was set at 145°C, and the solvent was 1,2,4-trichlorobenzene (TCB) with the addition of the antioxidant Irganox 1010 to prevent oxidative degradation.

The mechanical properties were tested based on the procedure of the International Organization for Standardization (ISO). The rheological (flow curve) measurements were performed on a cone-plate mode rheometer (REOLOGICA Instruments AB).

Results and Discussion

Molecular Weight Distribution and Morphology

Figure 4.5 shows the MWDs of the five samples. MWD0 was M-PP blended with L-PP and H-PP to broaden the MWD. The fractions, which between molecular weights of 10^6 and $10^{4.5}$ increased when L-PP and H-PP were both added. According to previous studies reporting that long chains enhanced the shear-induced crystallization [49-51], a high molecular weight fraction thickened the skin layer, as shown in Figure 4.6.

Changes in the Mechanical Properties

Table 4.2 shows typical mechanical properties of the samples measured by the ISO testing. The flexural modulus increased when the MWD widened, although MWD2 and MWD3 were very similar. While the amount of long chain in MWD3 was clearly greater than that of MDW2, as shown in Figure 4.5, the shear-induced crystallization was not effectively accelerated. At low shear rates, the rheological measurement showed that G', which was sensitive to the amount of long chains, showed similar patterns in MWD2 and MDW3, as shown in Figure 4.7. As shown in Figure 4.8, the flexural modulus was proportional to G' at 0.01 rad/s, which represented the actual flow characteristics of the melt. The mixing of MD3 is not sufficient to homogenize long chains, despite MD3 containing more long chains than MD2.

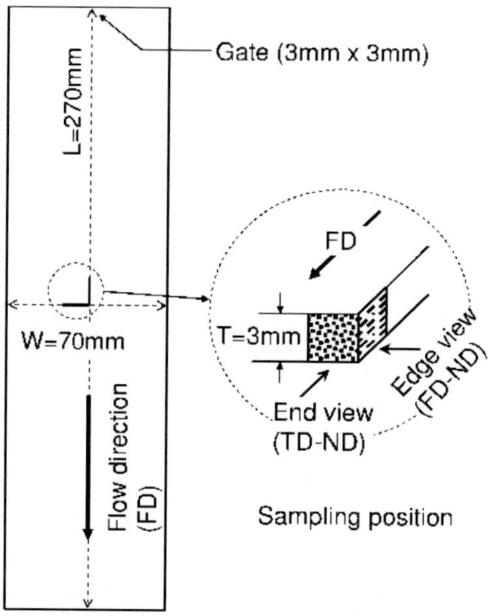

Figure 4.4. Schematic diagram of an injection-molded plaque and the specimens cut at the central position.

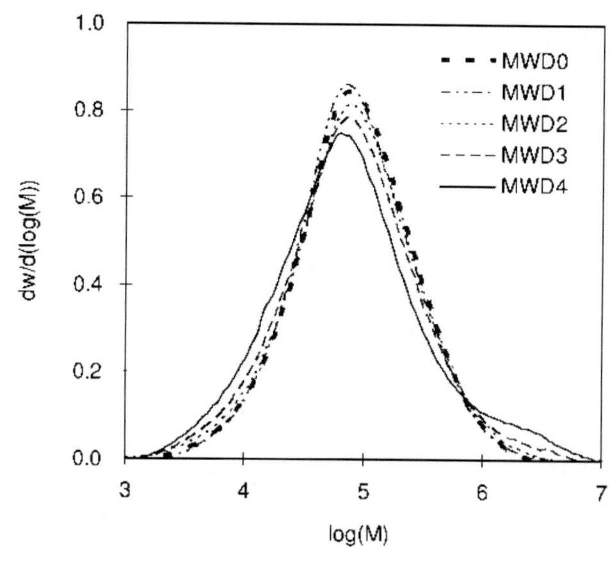

Figure 4.5. GPC curves of the samples with different molecular weight distributions.

Table 4.2. Typical mechanical properties of the samples

	Condition	Unit	MWD0	MWD1	MWD2	MWD3	MWD4
MFR	ISO 1133	g/10min	36	31	29	29	25
Density	ISO 1183	kg/m^3	908	910	910	910	911
Tensile yeild strength	ISO 527	MPa	37.7	36.5	36.6	36.1	40.1
Elongation at break		%	10.3	12.3	8.5	7.6	6.2
Flexural strength	ISO 178	MPa	50.0	52.6	54.7	54.7	62.3
Flexural modulus		MPa	1770	1990	2160	2170	2470
Charpy impact strength	ISO 179	kJ/m^2	2.5	3.1	3.2	3.2	2.6
Rockwell hardness (R)	ISO 2039-2	-	107	107	107	107	107
HDT	ISO 75	°C	110	121	131	130	136
Gloss	ASTM D 523	%	83	83	82	82	83

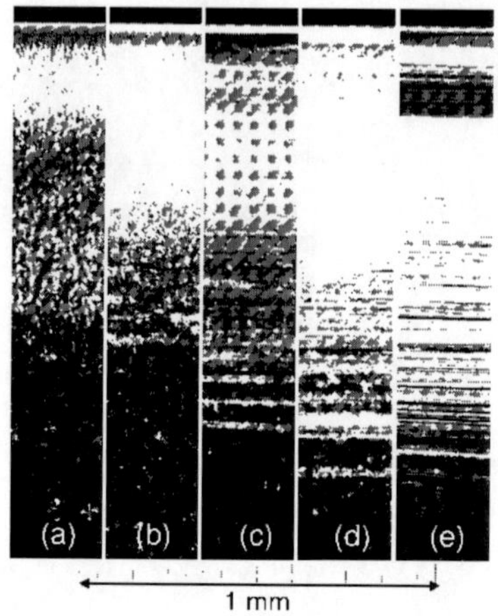

Figure 4.6. POM photographs (edge view) of a) MWD0, b) MDW1 c) MWD2, d) MWD3 and e) MWD4.

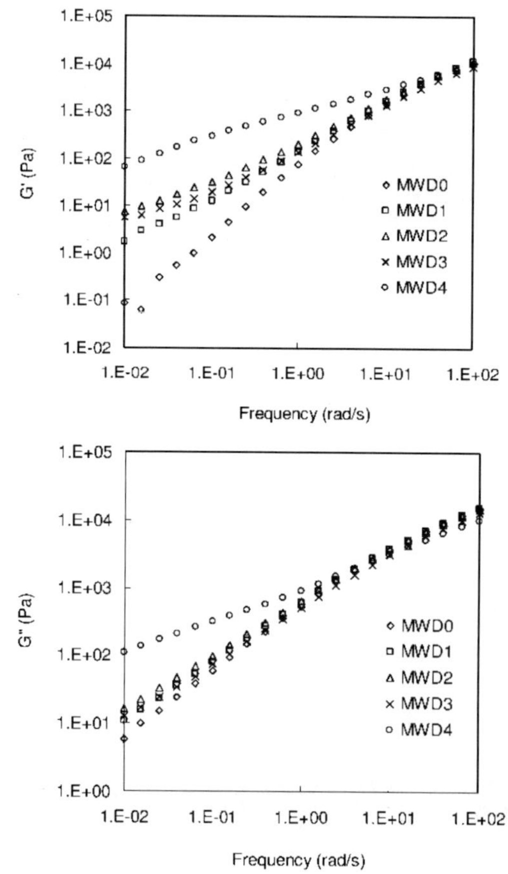

Figure 4.7. Storage and loss moduli given as function of the angular frequency at 210°C.

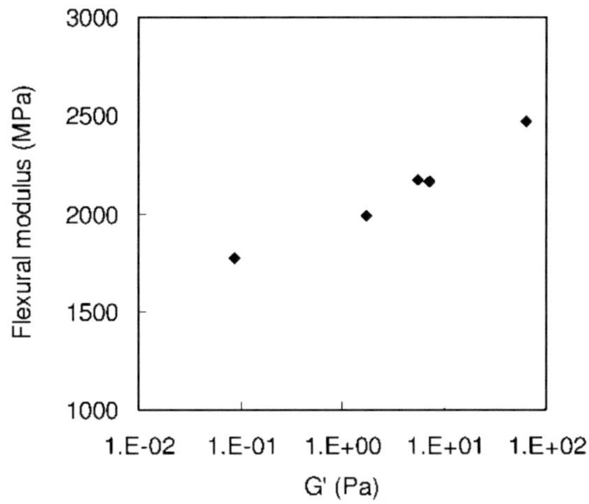

Figure 4.8. Relationship between the storage modulus at the 0.01-rad/s angular frequency and the flexural modulus of the samples.

Conclusion

The molecular weight dependency of the morphology and the mechanical properties were investigated in injection molding. A thicker skin layer corresponded to an increased stiffness of the injection-molded specimens. For a given MFR, G' was directly related to the flexural modulus samples, while the MWDs were controlled by mixing PPs of low and high molecular weights.

CONCLUSION

This chapter considers with three topics related to the formation of the skin morphology and its properties on injection-molded PP. The surface morphology is different from that at a given depth because the moving melted PP is immediately frozen at the cold cavity wall. In contrast, the constant speed of the advancing flow front yielded the same morphology at the surface in the flow direction. TEM revealed a highly stretched rubber domain at the surface along the flow direction on a PP/EBR blend. The observed morphology coincided to the surface microhardness. The fibrous EBR domains in the injection-molded specimens act as a weak phase for the microhardness.

The injection molding-induced morphology of an asymmetrical cavity was analyzed using a WAXD. The top of the obstacle in a cavity created a cooling delay of the melt compared to a flat smooth wall, and the resulting crystallinity of the molded product was higher at the obstacle side than at the normal side, despite equal mold temperature settings. The thickness gap caused the difference in the β–ratio and the crystallinity between the flow-in and flow-out positions. The extensional deformation at the flow-out section enhanced the β–form crystals and the skin thickness.

The molecular weight dependency of the skin morphology and the mechanical properties was investigated. For a given MFR, G' was directly related to the flexural modulus of the samples, in which the MWDs were controlled by mixing PPs with low and high molecular weights.

ACKNOWLEDGMENT

The author is grateful to Prof. Toshitaka Kanai for all of the discussions and Dr. Yasuhiko Otsuki for providing me with the viscoelastic simulation of the advancing flow front. Mr. Makoto Ando, Mr. Satoshi Ikeuch and Ms. Emi Otsuka assisted me in the analysis of microdeformation and TEM. Mr. Yasuhiro Hanamoto helped me with WAXD. Mr. Toshifumi Shimazaki and Mr. Takayoshi Tanaka prepared the injection-molded plaques. My experiments were performed at Idemitsu Kosan Co., Ltd. and Prime Polymer Co., Ltd. I would like to thank both companies for allowing me to release the findings of this study.

REFERENCES

[1] N. Pasquini, Ed., *Polypropylene Handbook 2nd*, Carl Hanser Verlag, Munich (2005).
[2] M. Fujiyama, *Polypropylene*, Karger-Kocsis, Chapman & Hall: London (1995).
[3] M. R. Kantz, H. D. Jr. Newman, F. H.; Stigale, *Appl. Polym. Sci.* 16, 1249 (1972).
[4] R. Mendoza, G. Régnier, W. Seiler, J. L. Lebrun, *Polymer* 44, 3363 (2003).
[5] P. Zhu, G. Edward, *Polymer* 45, 2603 (2004).
[6] Z. Tadmor, *J. Appl. Polym. Sci.* 18, 1753 (1974).
[7] P. K. Kennedy, *Thesis of doctorate*, Technische Universiteit Eindhoven, (2008).
[8] D. Choi, J. L. White, *Polym. Eng. Sci.* 42, 1642 (2002).
[9] T. N. Chung, G. Mennig,. *Rheol. ACTA* 40, 67 (2001).
[10] A. A. M. Flaman, *Polym. Eng. Sci.* 33, 193 (1993).
[11] K. H. Kim, A. I. Isayev, K. Kwon, C. van Sweden, *Polymer* 46, 4183 (2005).
[12] D. Jarus, A. Scheibelhoffer, A. Hiltner, E. Baer, *J. Appl. Polym. Sci.* 60, 209 (1996).
[13] A. M. Cunha, S. Fakirov, *Structure Development during Polymer Processing*, Kluwer Academic Publishers, 255 (2000).
[14] J. C. Viana, N. Billon, A. M. Cunha, *Polym. Eng. Sci.* 44, 1522 (2004).
[15] J. Karger-kocsis, D. E. Mouzakis, G. W. Ehrenstein, J. Varga, *J. Appl. Polym. Sci.* 73, 1205 (1999).
[16] J. C.Viana, A. M. Cunha, N. Billon, *Polym. Eng. Sci.* 39, 1463 (1999).
[17] Y. Kobayashi, Y. Otsuki, T. Kanai, *Polym. Eng. Sci.* 50, 2182 (2010).
[18] Y. Kobayashi, Y. Otsuki, H. Kanno, T. Sasakawa, Y. Hanamoto, T. Kanai, *Polym. Eng. Sci.* 51, 1236 (2011).
[19] F. J. Baltá Calleja, J. *TRIP* 2, 419 (1994).
[20] J. Arranz-André, B. Penã, R. Benavente, E. Pérez, M. L. Cerrada, *Eur. Polym. J.* 43, 2357 (2007).
[21] J. Aurrekoetxea, M. A. Sarrionandia, I. Urrutibeascoa, M. L. Maspoch, *Polymer* 44, 6959 (2003).

[22] S. Fakirov, M. Krumova, D. R. Rueda, *Polymer* 41, 3047 (2000).

[23] D. Tranchida, S. Piccarolo, *Polymer* 46, 4032 (2005).

[24] T. Koch, S. Seidler, E. Halwax, S. Bernstorff, *J. Mater. Sci.* 42, 5318 (2007).

[25] S. Fakirov, *J. Mater. Sci.* 42, 1131(2007).

[26] A. Flores, J. Aurrekoetxea, R. Gensler, H. H. Kausch, F. J. Baltá Calleja, *Colloid. Polym. Sci.* 276, 786 (1998).

[27] G. H. Michler, F. J. Baltá-Calleja, I. Puente, M. E. Cagiao, K. Knoll, S. Henning, R. Adhikari, *J. Appl. Polym. Sci.* 90, 1670 (2003).

[28] F. J. Baltá-Calleja, M. E. Cagiaoa, R. Adhikarib, G. H. Michler, *Polymer* 45, 247 (2004).

[29] Y. Kobayashi, H. Kanno, Y. Hanamoto, M. Ando, T. Kanai, *J. Appl. Polym. Sci.* 116, 1823 (2010).

[30] Y. Kobayashi, M. Ando, T. Kanai, *J. Appl. Polym. Sci.* 116, 2590 (2010).

[31] X. Yu, H. Wu, J. Li, S. Guo, J. Qiu, *Polym. Eng. Sci.* 49, 703 (2009).

[32] J.C. Viana, *Polymer* 46, 11773 (2005).

[33] Y. Kobayashi, Y. Otsuki, H. Kanno, Y. Hanamoto, Y. Kanai, *J. Appl. Polym. Sci.* 120, 141 (2011).

[34] G. Natta, P. Corradini, M. Cesari, *Atti della Accademia nazionale dei Lincei, Rendiconti* 22, 11(1957).

[35] K. Nakamura, S. Shimizu, S. Umemoto, A. Thierry, B. Lotz, N. Okui, *Polym. J.* 40, 915 (2008).

[36] C. M. Wu, M. Chen, J. Karger-Kocsis, *Polymer* 40, 4195 (1999).

[37] J. Varga, *J. Macromol. Sci.* B41, 1121(2002).

[38] P. Panine, E. D. Cola, M. Sztucki, T. Narayanan, *Polymer* 49, 676 (2008).

[39] H. Vehkamäki, *Classical Nucleation Theory in Multicomponent System*, Springer-Verlag, Berlin, (2006).

[40] F. P. Price, *Nucleation*, A.C Zettlemoyer Ed., Marcel Dekker, Inc., New York, (1969).

[41] K. Okada, K. Watanabe, Y. Urushihara, A. Toda, M. Hikosaka, *Polymer* 48, 401 (2007).

[42] T. Urushihara, K. Okada, K. Watanabe, A. Toda, N. Kawamoto, M. Hikosaka, *Polym. J.* 41, 228 (2009).

[43] S. Shanbhag, R. G. Alamo, *Polymer* 49, 2515 (2008).

[44] N. Okui, S. Umemoto, R. Kawano, A. Mamun, *Progress in Understanding of Polymer Crystallization*, G. Reiter, editor, Springer-Verlag, Berlin (2007).

[45] M. Hikosaka, K. Watanabe, K. Okada, S. Yamazaki, S. *Adv. Polym. Sci.* 191, 137 (2005).

[46] H. Janeschitz-Kriegl, *Crystallization modalities in Polymer Melt Processing*, Springer-Verlag, Wien, (2010).

[47] Y. Ogino, H. Fukushima, N. Takahashi, G. Matsuba, K. Nishida, T. Kanaya, *Macromolecules* 39, 7617 (2006).

[48] Y. Geng, G. Wang, Y. Cong, L. Bai, L. Li, C. Yang, *Macromolecules* 42, 4751 (2009).

[49] R. H. Somani, B. S. Hsiao, A. Nogales, S. Srinivas, A. H. Tsou, I. Sics, F. J. Balta-Calleja, T. A. Ezquerra, *Macromolecules* 33, 9385 (2000).

[50] R. H. Somani, L. Yang, B. S. Hsiao, *Polymer* 47, 5657 (2006).

[51] M. Seki, D. W. Thurman, J. P. Oberhauser, J. A. Kornfield, *Macromolecules* 35, 2583 (2002).

In: Polypropylene
Editors: L. P. Silva and E. F. Barbosa

ISBN: 978-1-62417-142-0

Chapter 8

MECHANICAL PROPERTIES OF POLYPROPYLENE/TALC COMPOSITES: ROLE OF THE MINERAL MORPHOLOGY AND PARTICLE SURFACE MODIFICATION

Luciana Castillo, Numa Capiati and Silvia Barbosa*
PLAPIQUI (UNS-CONICET)
Bahía Blanca, Argentina

ABSTRACT

Talc filled polypropylene is a very used versatile composite as it combines a relative good performance with easy processability and low cost. Talc addition strongly affects polypropylene (PP) properties, mainly by their effect on crystallization. It is an efficient nucleating agent; changing the morphology, chain orientation, crystal type and crystallinity degree of PP. Such properties can indeed be tailored by changing the volume fraction, shape, and size of talc particles. According to the genesis, talc has different morphologies and different associated minerals. These aspects conferred it specific whiteness, particle shape, oil absorption, and amphiphilic behavior. On the other hand, the geomorphologic and geochemical characteristics, as well as the mineral processing (milling, treatments, among others), also affect thermal, mechanical and optical properties of PP/talc composites. In order to obtain a higher mechanical performance of PP/talc composites it is necessary to assure an homogeneous dispersion and distribution of particles, having very large aspect ratio, and a good particle/matrix interfacial adhesion. Generally, to improve the filler/polymer interaction it is a common practice to modify the filler surface. In this work, a systematic analysis of the influence of talc geomorphology and different surface treatments on the PP/talc composites structure and their mechanical properties was performed. Two well characterized talc samples, macro and microcrystalline, were used. Thus, a low cost surface treatment is proposed in order to improve the mechanical performance, by the way of adhesion enhancement and particle size reduction, but without increasing the material price. An accurate particle characterization, before and after the treatment, was carried out in order to determine

* E-mail: sbarbosa@plapiqui.edu.ar

their morphological and surface aspects. A methodology to improve the dispersive and distributive talc mixing in PP, using a twin screw extruder, is also proposed. Composite structure was characterized by Differential Scanning Calorimetry (DSC), Scanning Electronic Microscopy (SEM), Fourier Transform Infrared Spectroscopy (FTIR) and X-ray Diffraction (XRD). Tensile properties were assessed and their relationship with the structure, carefully analyzed. Very strong adhesion improvements were found either when macrocrystalline talc or treated one is used. The results confirm that composites processed with the higher aspect ratio particle, show a better mechanical behavior than the other. In the case of PP/treated talc composites, an improvement on dispersion, distribution and interfacial adhesion are observed, as a consequence of talc functionalization. However, the talc surface treatment has more influence on mechanical properties in only one of the composites, due to its efficiency depends on the intrinsic talc characteristic.

INTRODUCTION

In the last decades, many studies have been carried out in order to tailor the final properties of commodity polymers to specific requirements in the final part. The driving force to develop new materials based on polyolefins, the main commodity plastics in the market; is their good cost/performance relationship. In this sense, the challenge is to obtain a new material with desired specific properties by using an economical process, focusing to conserve this advantageous relationship.

In this sense, the use of cheap minerals as fillers is an attractive idea completely aligned with the above goal, as they reduce the final composite costs. Moreover, filler addition increments the polymer heat resistance (heat deflection temperature), stiffness and electrical conductivity, as well as, reduces its shrinkage and flammability. Consequently, the dimensional stability and thermal conductivity result improved, giving higher process productivity [1].

Particularly, talc filled polypropylene (PP) is one of the most used composites in a wide range of applications, as household appliances, packaging, automotive and construction industries. The main reason resides in the synergetic combination of properties between PP and talc. PP is a semicrystalline polymer with good mechanical properties. These properties are so improved with the presence of talc, mainly because their particles act as a good crystal nucleating, increasing the PP crystallinity [2-3].

Talc is a natural layered silicate with platy morphology. However, the specific chemistry and morphology depend on the geological formation, indeed each talc deposit is unique [4]. Pressure, either at the time of the metamorphic transformation or later, determines the mineral degree of lamellarity (low pressure produces low lamellarity, high pressure produces high lamellarity) [5]. This condition also determines the size of an individual talc platelet (a few thousand elementary sheets). It can vary from approximately 1 μm to over 100 μm, depending on the genesis conditions [6-8]. Different talc morphologies are observed, ranging from long and well-defined and stacked up flakes (so-called 'macrocrystalline' type, high lamellarity) to an heterogeneous stack of small and irregular plates ('microcrystalline' type, low lamellarity). These two terms are related with particle having high and low basal/lateral surface ratios, respectively. Also, during geological talc formation, new mineral phases are created from preexisting ones. As a consequence, the ores may consist of mixtures of associated minerals, including mainly chlorites (hydrous magnesium and aluminum silicates), magnesite (magnesium carbonate), calcite (calcium carbonate) and dolomite (calcium and magnesium carbonate) [4]. In this sense, chemical and structural parameters might be indicative of the origin and crystallization conditions of talc. For this reason, talc ores differ in chemical composition, morphology, associated minerals content, whiteness and other properties. In the plastic industry, pure and high aspect ratio talc grades are used. However, the performance of industrial talc is not only dependent on these inherent characteristics, but also the ones imparted by the mining and manufacturing processes, e.g. beneficiation, grinding, classification, calcination, slurring, surface treatment on talc [9].

The final PP-talc composite properties depend on the talc intrinsic characteristics, their dimensions, dispersion, distribution, matrix crystallinity and degree of adhesion between this mineral and PP. The last two aspects occur in a three dimensional region, surrounding the filler, known as the interphase. The degree of adhesion depends on the surface properties of the filler and of the matrix [10, 11]. In general, stronger filler-matrix interactions result in improved processability, impact strength, and surface quality; while too weak interactions lead to decrease strength and to increase deformability of the composite [12]. Inorganic particles usually have higher surface energy than polymers because of their hydrophilic ionic nature, and the surface energy also depends on morphology and chemistry. Hydrophobic polymers, like PP, have low surface energy, then PP-mineral fillers interaction is weak and the adhesion is low, reducing composite final properties [13-15]. One of the most common ways for improving the adhesion in PP/talc composites is by using surface modifiers on talc particles [16-17].

The role of surface modification is to enhance the interaction between the components, affecting both the macroscopic and microscopic behavior. This has been deduced from previous investigations on the tensile composite behavior. Different authors studied the influence of surface treatment on the mechanical properties [18-24]. They concluded that the surface treatment of filler gives rise to higher mechanical properties than unmodified one, showing an improved stiffness and strength.

Concerning to the thermal behavior, studies revealed that surface treated filler shows an improved nucleation activity, modifying the crystalline morphology, and as a consequence, the composite performance [25-31]. The traditional method of treating filler surface by low-molecular coupling agents or surfactants has been found to be reasonably effective due to reducing interfacial tension between filler and polymer matrix [32-34].

The aim of this work is to analyze the influence of talc geomorphology and different surface treatments on structure and mechanical properties of PP/talc composites. The study is performed preparing composites with two well characterized talc samples, micro and macrocrystalline. Thus, a low cost surface treatment is proposed in order to improve the mechanical performance, by the way of adhesion enhancement and particle size reduction, but without increasing the material price. The results are analyzed in terms of the structure-tensile properties relationship.

Talc Charactrerization

Talc is a naturally occurring magnesium silicate, with a chemical formula of $Mg_3Si_4O_{10}(OH)_2$. Its layered structure consists of an octahedral brucite sheet sandwiched between two tetrahedral silica sheets. Van der Waals' gaps between the layers are formed due to stacking of layers [35], which may assist in the delamination behavior of talc particles during composite processing. From the geological point of view, talc can be formed by the alteration of a dolomite or a serpentinite host rock. The talc originated from a dolomite host is typically found in China, India, United States, Australia, France, among others. These ores are massive, compact and dense and talc shows a microcrystalline structure. Particularly, the Australian talc (A10) used in this study has this characteristic morphology. Figure 1a shows micrographs that evidence the microcrystalline morphology, where small platelets stack up heterogeneously, giving particles with a low aspect ratio (length/thickness). In addition, rounded and anisotropic agglomerates, where particles are organized in laminar concentric domains (onion-like), are observed. These agglomerates tend to disaggregate upon milling, resulting in smaller laminar particles.

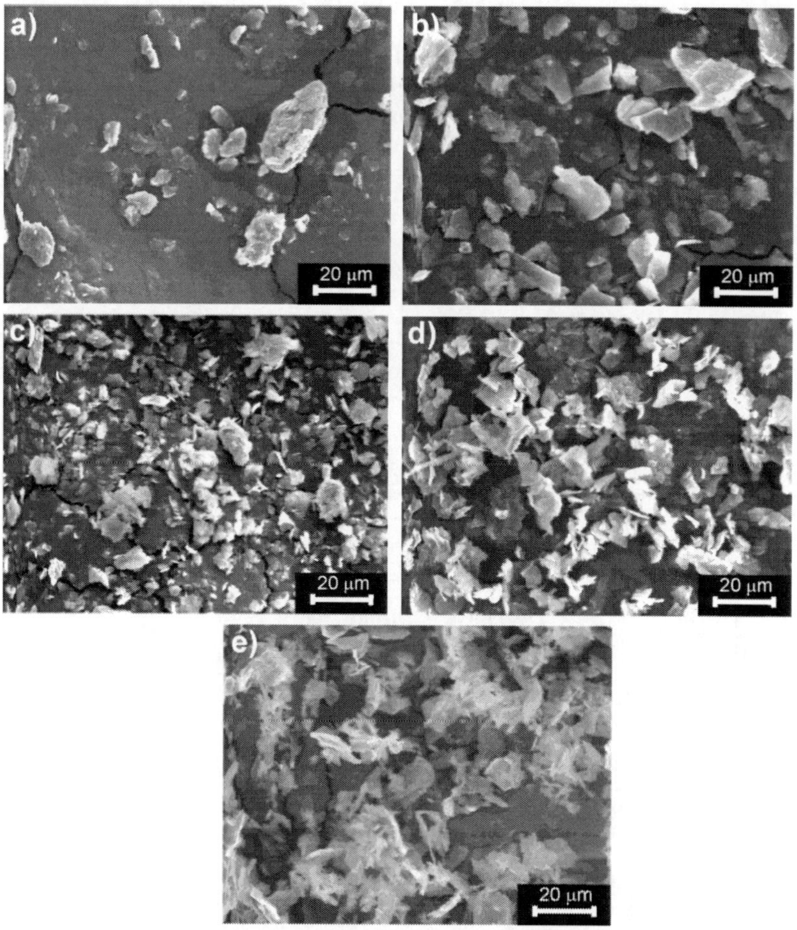

Figure 1. SEM micrographs of talc samples (1000x): a) A10, b) A10A, c) SJ10, d) SJ10C and e) SJ10CA.

On the other side, talc/magnesite ores are derived from serpentinite hosts and are commonly found in Canada, France, India, Argentine, Italy, among others. These types of ores are macrocrystalline. In Figure 1b, an Argentinean talc (SJ10), showing long and well stacked up platelets with high aspect ratio, is presented. These particles appear as blocks, with abrupt and well-defined borders.

Talc samples from different ores can be differentiated by the crystallinity degree, which is determined by the pressure and deformations during the metamorphism, both concurrent with and subsequent to this transformation [36]. As it can see from the X-ray Diffraction (XRD) spectra in Figure 2, the presence of long and well defined peaks from Argentinean talc (SJ10) spectrum, with higher intensity maximum, allowed inferring the higher crystallinity degree of this talc, comparing to Australian one (A10). So, talc having a macrocrystalline character matches with a higher crystallinity degree.

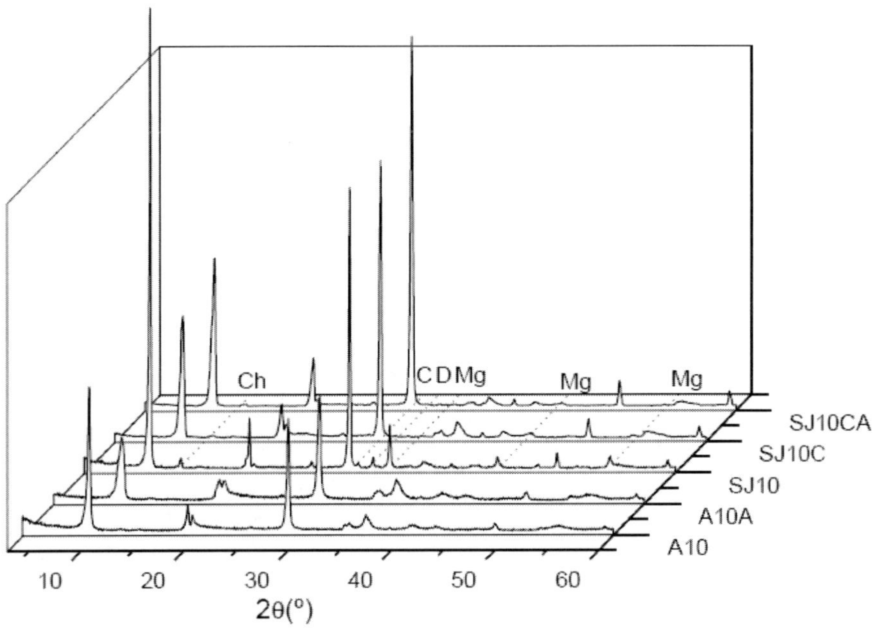

Figure 2. XRD spectra of talc samples. C: calcite, Ch: chlorite, D: dolomite, Mg: magnesite.

As a result of the origin, besides the morphology and degree of crystallinity; talc can also vary in their purity degree. The chemical composition of talc and its associated minerals result from the original rock type, the nature of the hydrothermal alteration and metamorphic history [37]. Generally, the microcrystalline talc has a high purity degree ranging from 93 to 99%, while macrocrystalline one is composed approximately 45-65% talc by weight, containing carbonates as characteristics associated minerals. The differences in purity degree between macro- and microcrystalline talc samples are reflected in XRD spectra. As it can be observed, the microcrystalline sample (A10) is almost pure talc, having a low amount of chlorite as impurity. In the case of macrocrystalline talc (SJ10), peaks corresponding to carbonates as calcite ($CaCO_3$), magnesite ($MgCO_3$) and dolomite ($CaMg(CO_3)_2$), and chlorite ($Mg_5(Al,Fe)(Al,Si)_4O_{10}(OH)_8$) are detected. The mineralogical composition of both talc samples can be corroborated by Fourier Transform Infrared (FTIR) in Figure 3.

Talc Surface Modification

The surface modification of talc affects its intrinsic characteristics, as particle size, associated minerals amount, and surface energy, among others. The traditional method of treating talc surface by low-molecular coupling agents or surfactants has been found to be reasonably effective and to have a relatively low cost [32-34]. The modification of talc surface with acids seems to be an interesting alternative considering the obtained results [38].

The effect of treatment with inorganic and organic acid on particle size in micro- and macrocrystalline talc is observed in Figure 1c, 1d, and 1e. A reduction in particle length is produced, and delamination is favored, both for the organic treatment (A10A and SJ10CA) and the inorganic one (SJ10C). A more uniform particle distribution is obtained for all the samples, independent of the crystalline character. Concerning to the surface modification with carboxylic acid, organic groups are grafted on microcrystalline talc surface. The bands at 2926 and 2854 cm^{-1} were assigned to asymmetric vibrations of C–H bonds (A10T in Figure 3) due to the reaction between the OH groups of talc and the carboxyl group of the acid.

When the treatment is directly applied on macrocrystalline talc, containing great amount of associated minerals, acid is consumed in order to solubilize the carbonates. So, during the organic acid attack at least two competitive reactions took place: dissolution of associated minerals and esterification of talc OH groups.

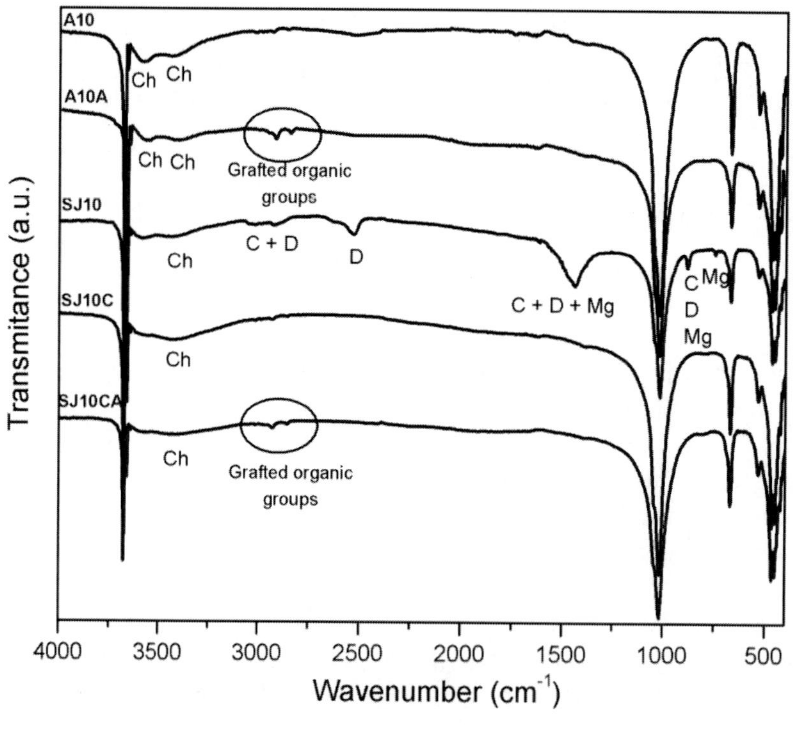

Figure 3. FTIR spectra of talc samples. C: calcite, Ch: chlorite, D: dolomite, Mg: magnesite.

In SJ10C spectrum, the bands corresponding to the associated minerals are absent, confirming that the dissolution of associated minerals prevails over the esterification between talc and acid. Once again, if the organic acid treatment is applied on the macrocrystalline talc free of impurities, it can be observed the presence of organic bands (SJ10CA spectrum in Figure 3).

The acid treatment effects analyzed by XRD also shows the removal of carbonates and the reduction of crystallinity degree, by decreasing the maximum intensity of the spectra in the treated talc. However, it is not possible to detect any change on talc structure related to the grafted organic groups.

Depending on the acid used in the treatment, talc surface energy is affected. In the case of organic acid, surface becomes more hydrophobic due to the grafting of the organic groups. By the inorganic acid treatment, hydrophilicity was developed on talc surface, due to the acid breaks siloxane bonds at the surfaces, increasing the –OH groups content.

Composites Characterization

In order to obtain good and homogeneous final properties, PP/talc compounding is an important step. Effective mixing is crucial for achieving optimal dispersion and distribution of talc and for optimizing the composite properties. The method used in this work includes compounding in a twin screw extruder, feeding it with a physical mixture of PP powder and talc. In order to analyze the influence of morphology and surface modification of talc on mechanical properties of PP/talc composites, initially the dispersion and distribution of talc particles in the composites were assessed. In this sense, the morphological analysis of injected tests specimens was performed using a Scanning Electronic Microscope, exploring the whole cross section of composites. As it can see in Figure 4, for PP/talc composites containing 10 wt% of talc, the particles are well distributed in all of them, but the dispersion degree is even better in the composites with talc treated with organic acid (PP10A10A and PPSJ10CA). This claim is based on the reduced particle-particle interaction due to the organic groups grafted onto talc surface by the acid treatment. The same behavior was observed for composites with 1, 3 and 5 wt% of talc.

The final composite properties depend on the type of compounding process, concerning to the filler orientation. Since the mechanical properties tests were carried out on injected samples, the effect of injection molding on the orientation talc particles in PP was studied. This kind of orientation results from the laminar morphology and the filler movement on a viscous fluid.

Due to the talc particles align following the flow lines during the mold filling, the injected specimens show a gradient of particle orientation along the specimen thickness, as it can be appreciated in Figure 5. At the outer side of the specimen, the particles tend to lie parallel to the mould surface, although they organize in ellipsoidal curves at the inner part of the injected specimen.

This study about the state of particle dispersion in PP also allows inferring that the injection molding has an effect on break up of talc particles. It seems that high levels of shear stress, involved in the injection process, cause a dramatic decrease in talc particle size. The previous observation is confirmed when the length of particle itself with its analogous in the composite, are compared.

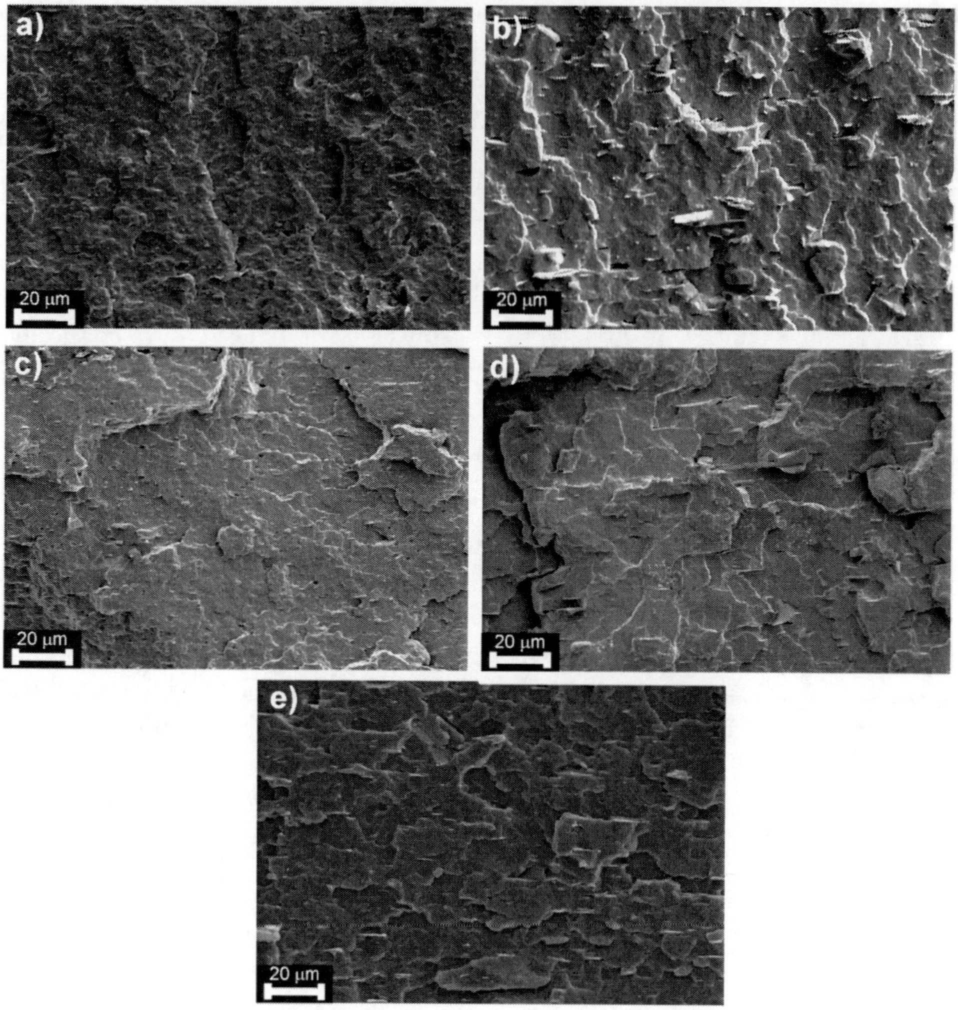

Figure 4. SEM micrographs of composites (2000x): a) PP10A10, b) PP10SJ10, c) PP10A10A, d) PP10SJ10C and e) PP10SJ10CA.

It is well known that talc acts as a nucleating agent that strongly influences the crystallization process, degree of crystallinity, and nucleation of crystallites of semicrystalline polymers. As a result, this mineral favors the nucleation of PP crystallization, which indirectly induces significant changes on thermal, optical, barrier and mechanical large scale properties in the composite. In this sense, the crystallinity degree is considered. The values of crystallinity degree were calculated on the basis of Dynamic Scanning Calorimetry (DSC) curves. In Figure 6, the crystallinity degree as a function of talc volumetric fraction is shown. The values for composites with microcrystalline talc (PPA10 and PPA10A) are low compared with the PP/macrocrystalline talc (PPSJ10, PPSJ10C and PPSJ10CA). This fact can be related to the talc crystalline character as well as the treatment effect. The untreated macrocrystalline talc (SJ10) is much more crystalline than microcrystalline one (A10) and, as a consequence, the crystallinity induction in the composite could be more significant with the addition of the first talc. In addition, treated talc induces a high crystallinity degree than untreated ones. The reason of this difference resides on the particle size reduction favoured by the treatment, and consequently, the increased number of PP nuclei. Although there are differences among the crystallinity values of the composites, these are about in a range of 5%.

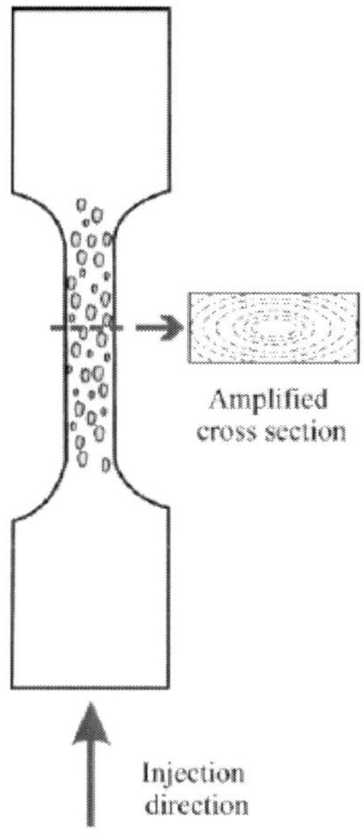

Figure 5. Particle orientation along the specimen thickness.

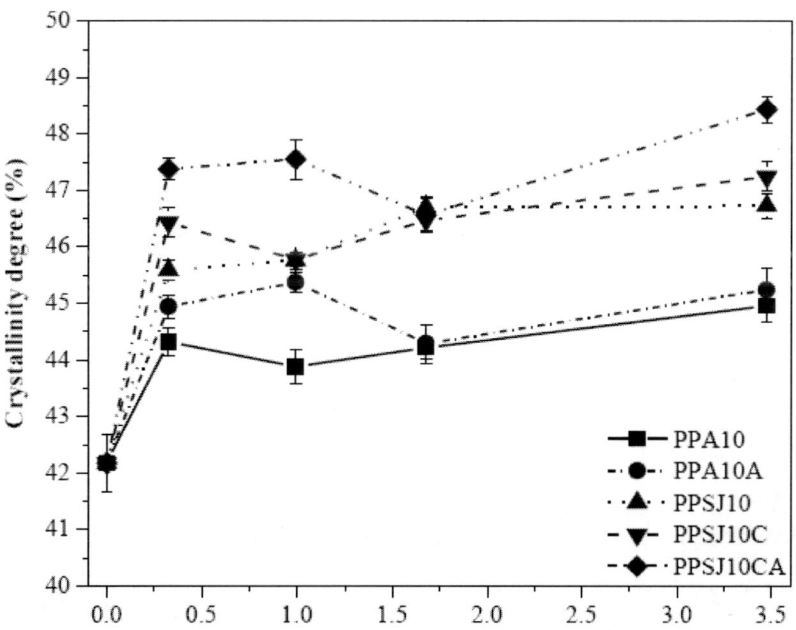

Figure 6. Crystallinity degree of composites as a function of talc volumetric fraction.

The introduction of fillers or reinforcements, besides modifying the crystallinity degree, also influences on the crystal type of PP. In fact, except for glass fiber, almost all the widely used reinforcing fillers, such as talc, are strong promoters for α modification and they may suppress the occurrence of β form even in β nucleated PP composites. Generally, fillers with high aspect ratio induce β crystal. The XRD spectra of all composites, at a same talc concentration, are shown in Figure 7. PP has a little peak at 2θ=16° corresponding to β form, mainly due to the presence of stabilizant agent in the commercial PP, which induces this kind of crystallization. The addition of microcrystalline talc (A10 and A10A) in PP suppresses completely β crystal. When a macrocrystalline talc, as SJ10, SJ10C and SJ10CA, is added to PP, some modifications are observed: an increment of *β*-form induced by the stabilizing agent, and a variation in (110) and (040) intensities of *α*-form reflections. In Figure 7, a shoulder assigned to *β*-form can be observed in PPSJ10, PPSJ10C and PPSJ10CA, taking into account that the presence of this phase improves the elongation at break and impact strength in comparison with the other crystal forms.

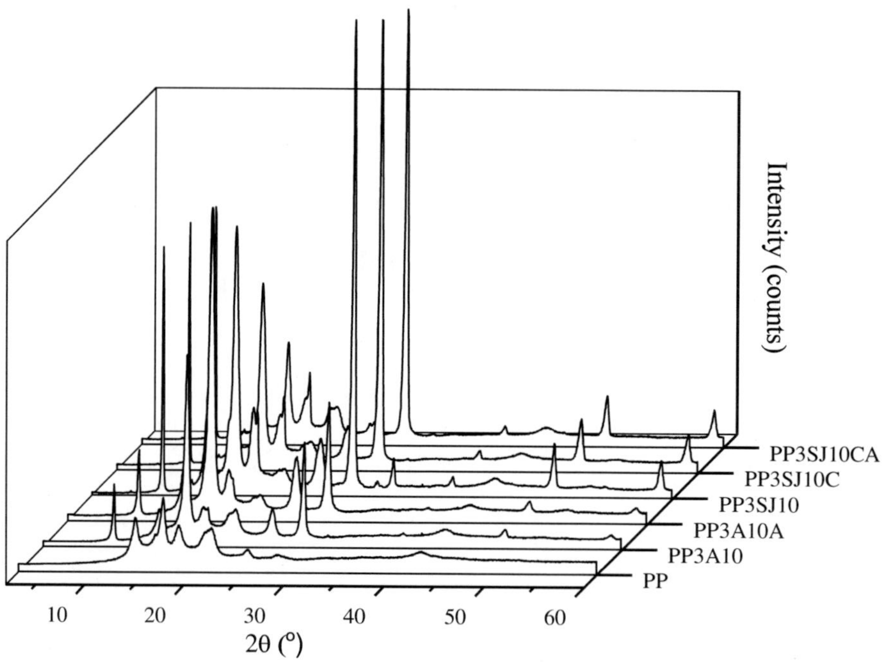

Figure 7. XRD spectra of all composites at a same talc concentration (3 wt%).

Table 1. XRD intensity ratio of specific reflection planes for PP and PP/talc composites

	PP	PP3A10	PP3A10A	PP3SJ10	PP3SJ10C	PP3SJ10CT
(040)/(110)	1.54	6.81	6.26	2.44	3.13	1.91
(002)/(006)	-	0.58	0.58	0.52	0.54	0.54
(020)/(002)	-	0.61	0.59	0.34	0.29	0.29
(020)/(006)	-	0.36	0.34	0.18	0.16	0.16

The platy talc particles lie parallel to the surface of the mold, inducing a preferential orientation in the morphology of PP crystals placed near the filler (*b*-axis). In order to evaluate the degree of *b*-axis orientation of the PP unit cell, the ratio of the intensity between (040) to that of the (110) reflection was obtained [39-44]. Other ratios were also measured to study the orientation of talc particles, as it shown in Table 1. As deduced from XRD data, the plain polymer showed a non-isotropic crystalline structure: the ratio of the reflection intensities was about 1.54, whereas the reported value for the isotropic case was about 0.54 [45]. The analysis of (110) and (040) PP diffraction peaks of the composites with microcrystalline talc (PP3A10 and PP3A10A) show that the intensity of (040) reflection is strongly favoured in relation to (110) reflection. Besides the different orientations of PP crystals in the composites with macro- and microcrystalline talc, there is a difference in the talc crystal orientation between them (Table 1). The analysis of (002), (020) and (006) talc diffraction peaks shows that the crystalline character has a different effect on the talc crystals orientation.

Composite Mechanical Properties

The stress-strain curves of PP and PP/talc composites, for each talc concentration, are shown in Figure 8. For composites with untreated microcrystalline talc (PPA10), the curves have a different form from PP one, in the range of the studied concentrations. These composites form neck, but it can propagate and they fail at lower elongations than the others. This mechanical behavior could be attributed to a low interfacial adhesion between talc particles and PP, reducing the yield strength and showing a low elongation at break. In the case of treated microcrystalline composites (PPA10A), the mechanical performance is improved, showing the treatment effect on the increment of PP/talc adhesion. The form of stress-strain curve at low talc concentrations seems to be similar to the PP one. However, when the concentration increases, PPA10A curve is similar to PPA10 one. This behavior can be attributed to two main reasons: talc-PP adhesion would be higher than the cohesion among particles or posterior delamination of treated particles ocurred during the mixing. In this case, the attrition could delaminate them, exposing the surface that was not treated to PP. So, the interfacial adhesion particle-matrix is reduced. The influence of these possible reasons is increased with the talc concentration, as it is observed in Figure 8. During the mixing of PP with low talc concentrations, the interaction among particles and the degree of delamination is low, allowing that the adhesion between PP and talc would be more effective. However, the mechanical properties of composites with treated talc were improved over the concentration range. Young Modulus for these composites increases slightly, toughness increases about 20%, while elongation at break is doubled at low concentrations, but increases 40% at 10wt% of talc. The photographs of tensile specimens of all composites, after mechanical testing, are shown in Figure 9. For both composites containing untreated and treated microcrystalline talc (PP/A10 and PP/A10A), a neck formation is observed; but the difference between them is the propagation. For PP/A10A composites, the necking region propagates at higher proportions, compared to PP/A10, due to the treatment effects.

Stress-strain curves of PP with untreated macrocrystalline talc (PP/SJ10) show a mechanical performance analogous to PP, but they have higher Young modulus and yield strength. In all the composites, the neck formation occurs and it propagates, reaching higher elongations than untreated microcrystalline filled PP. The photographs in Figure 9 confirm the neck propagation and the higher elongation at break values of PP/SJ10. If the properties of composites with untreated micro- and macrocrystalline talc are compared, it is observed that PP/SJ10 has better properties than PP/A10. Young modulus values are around 4-14% higher, yield strength increases 25% and elongation at break is doubled, along the concentration range. This fact shows the influence of talc crystalline character on mechanical behavior. There are differences in the mechanical performance of composites with micro and macrocrystalline talc, mainly in yield strength and elongation at break. These properties depend on the talc surface characteristics, what it means that A10 and SJ10 interact in a different way with PP. This claim is based on that the associated mineral presence and the particle size of SJ10 talc are not the responsible of these differences because they produce the inverse effect on the composites mechanical properties. On the other hand, crystallinity degree induced by each talc, is not so different. The composites containing treated macrocrystalline talc, as PPSJ10C and PPSJ10CA, show mechanical behavior slightly improved respect to PPSJ10 one. Yield strength and elongation at break are increased, being more noticeable the change in the last one (14%), at low concentrations. However, there are no significant differences between the mechanical behavior of PPSJ10C and PPSJ10CA.

From a comparative analysis of the mechanical properties improvement, induced by the treatment applied on micro and macrocrystalline talc, is evident that this depends on mainly the talc crystalline character. The relative improvement of the properties, using untreated and treated talc, is higher in PP/A10 composites than PP/SJ10 ones.

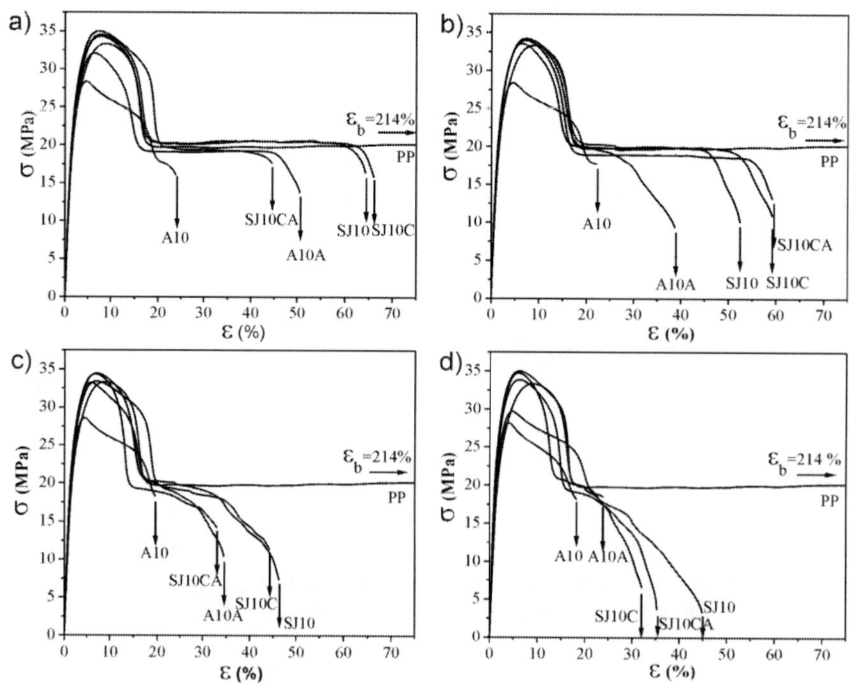

Figure 8. Stress-strain curves for PP and PP/Talc composites containing: a) 1, b) 3, c) 5 and d) 10 wt%. In each figure, the filler name ller was used for identifying each composite.

Concerning to the elastic modulus, an increase of 6% is reached for PP/A10A respect to PP/A10, while the increment in PP/SJ10C and PP/SJ10CA is lower than 3%, compared to PP/SJ10. In the case of yield strength, it is obtained an improvement up to 21% for PP/A10A and 3% for PP/SJ10C and PP/SJ10CA, respect to the composites containing the corresponding untreated talc. However, the major difference is observed in the deformation at break. For PP/A10A this property increases up to 120%, but only 4% for PP/SJ10C and PP/SJ10CA, comparing to the composites containing the corresponding untreated talc. If composites with talc having the same purity degree and analogous treatment are compared, i.e. PP/A10 with PP/SJ10C and PP/A10A with PP/SJ10CA, it is observed that the mechanical properties of composites containing treated talc from SJ10 are superior. From this fact, it is evident that the incorporation of macrocrystalline talc (SJ10) in PP increases the mechanical properties further that the microcrystalline ones.

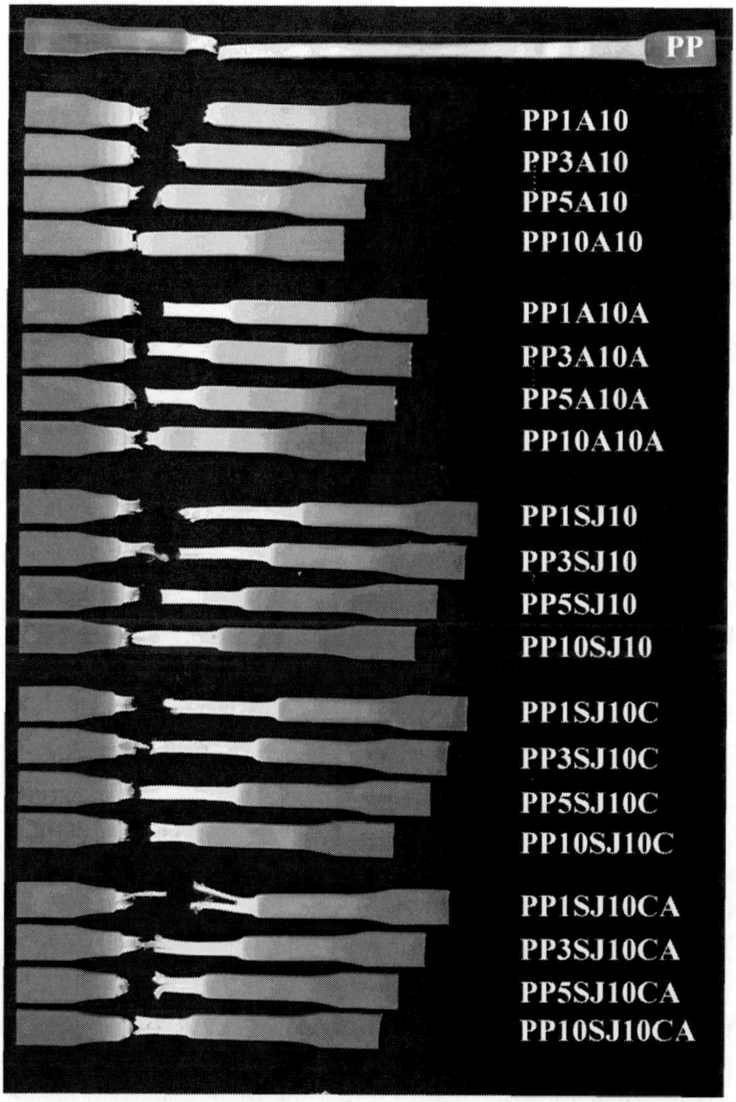

Figure 9. Photographs of tensile specimens of PP and PP/talc composites, after mechanical testing.

CONCLUSION

Much work has been done on the mechanical properties of particle filled composites, and a lot of experimental data have been collected on the effect of different parameters, such as the volume fraction and the surface treatment of particles. However, the influence of the filler geomorphology has not been considered in these studies. In this work, a systematic analysis of the influence of geomorphology and different surface treatments of talc on structure and mechanical properties of PP/talc composites was performed. The main concluding remarks are:

Depending on the geological origin, talc samples differ in their chemical composition, morphology, associated minerals content, crystalline character and other properties. Basically, talc can be micro or macrocrystalline, depending on the individual platelet size. These morphologies range from long and well-defined and stacked up flakes (so-called 'macrocrystalline' type) to a heterogeneous stack of small and irregular plates ('microcrystalline' type).

The modification of talc surface with acids seems to be an interesting alternative taking into account the obtained results [38]. A more uniform particle distribution is achieved for all the samples, independent of the crystalline character. Concerning to the surface modification with carboxylic acid, organic groups are grafted on microcrystalline talc surface, due to the reaction between the OH groups of talc and the carboxyl group of the acid. However, if this treatment is directly applied on macrocrystalline talc, which contains a great amount of associated minerals, acid is consumed in order to solubilize the carbonates. So, during the organic acid attack at least two competitive reactions take place: dissolution of associated minerals and esterification of talc OH groups.

Depending on the acid used in the treatment, the surface energy is affected. In the case of organic acid, talc surface becomes more hydrophobic due to the grafting of organic groups. By the inorganic acid treatment, hydrophilicity was developed on talc surface due to the acid breaks siloxane bonds at the surfaces, increasing OH groups content.

For composites containing micro and macrocrystalline talc, particles are well distributed in all of them, but the dispersion degree is even better in the composites with talc treated with organic acid (PP10A10A and PPSJ10CA). This claim is based on the reduced particle-particle interaction due to the organic groups grafted onto talc surface by the acid treatment. Also, a good filler modification improves the mechanical behavior of PP/talc composites.

The values of crystallinity degree for composites with microcrystalline talc (PPA10 and PPA10A) are low compared to PP/macrocrystalline talc (PPSJ10, PPSJ10C and PPSJ10CA). This fact can be related with the talc crystalline character as well as the treatment effects. The macrocrystalline talc (SJ10) is much more crystalline than microcrystalline one (A10) and, as a consequence, the crystallinity induction in the composite could be more significant with the incorporation of the first talc. In addition, treated talcs induce a high crystallinity degree than the initial ones. The reason of this difference resides on the particle size reduction favoured by the treatment, and consequently, the increased number of PP nuclei. Although there are differences among the crystallinity values of the composites, these are about in a range of 5%.

The incorporation of talc induces PP crystals orientation. The composites containing microcrystalline (PP/A10 and PP/A101A) talc suppress β phase, detected for PP, while in composites with macrocrystalline talc (PP/SJ10, PP/SJ10C and PP/SJ10CA) this reflection is hold and increased.

Talc microcrystalline (A10 and A10A) induces a higher PP crystal orientation than macrocrystalline talc (SJ10, SJ10C and SJ10CA) because they strongly favored the intensity of (040) reflection in relation to (110) one.

Besides the different orientations of PP crystals in composites with macro- and microcrystalline talc, there is a difference in talc crystal orientation between them. The analysis of (002), (020) and (006) talc diffraction peaks shows that crystalline character has a different effect on talc crystals orientation.

For composites with untreated microcrystalline talc (PPA10), stress-strain curves have a different form from PP one, in the range of the studied concentrations. These composites form neck, but it can propagate and they fail at lower elongations than the others. This mechanical behavior could be attributed to a low interfacial adhesion between talc particles and PP. In the case of composites containing treated microcrystalline talc (PPA10A), mechanical performance is improved, showing the treatment effect on the increment of PP/talc adhesion.

Stress-strain curves of PP with untreated macrocrystalline talc (PP/SJ10) show a mechanical performance analogous to PP, but they have higher Young modulus and yield strength. In all the composites, neck formation occurs and it propagates, reaching higher elongations than untreated microcrystalline filled PP.

REFERENCES

[1] M. Xanthos, *Functional Fillers for Plastics*; Wiley-VCH Verlag GmbHandCo KGaA, Weinheim, 207 (2005).

[2] L. Castillo, S. Barbosa, N. Capiati, *J. Appl. Polym. Sci.*, accepted for publication (2012).

[3] E. Ferrage, F. Martin, A. Boudet, S. Petit, G. Fourty, F. Jouffret, P. Micoud, P. de Parseval, S. Salvi, C. Bourgerette, J. Ferret, Y. Saint-Gerard, S. Buratto, J. P. Fortune, *J. Mater. Sci.*, 37, 1561 (2002).

[4] R. J. Piniazkiewicz, E. F. Mc Carty, N. A. Genco, N. *Industrials Minerals and Rocks*; Carr, D. D.; Ed.; 6 th Edition. SME Inc.: Littleton, Colorado, (1994).

[5] Eurotalc (2002). What is talc. http:/www.ima-eu.org/en/ talcwhat.html.

[6] J. P. Fortuné, B. Gavoille, J. Thiebault, 26th International Geological Conference, Paris, France, 10, (1980).

[7] B. Moine, J. P. Fortuné, P. Moreau, F. Viguier, *Econ. Geol.*, 84, 1398 (1989).

[8] P. de Parseval, B. Moine, J. P. Fortuné, J. Ferret, *Current Research in Geology Applied to Ore Deposits* (P. Fenoll Hach-Ali, J. Torres-Ruiz y F. Gervilla, Editores), Universidad de Granada, España (1993).

[9] J. M. Adams, *Clay Miner.*, 28, 509 (1993).

[10] L. Lavielle, *Ann. Phys. Fr.*, 14, 1 (1989).

[11] G. Fourche, Polymer Eng. Sci., 35, 957 (1995).

[12] B. Pukansky, *Polypropylene: Structure, Blends and Composites*, Karger-Kocsis, J.; Ed.; Chapman and Hall: London, 3, (1995).

[13] M. Y. Boluk, H. P. Schreiber, *Polym. Compos.*, 7, 295 (1986).

[14] F. Candau, F. *Polymerization in Organized Media*; Paleos, C.; Ed.; Gordon and Breach Science Publishers: Philadelphia, PA (1992).

[15] E. Ruckenstein, J. S. Park, *Polymer*, 33, 405 (1992).

[16] H. P. Schlumpf, *Synthetis,* 12, 31 (1981).

[17] T. Nakatsuka, *Polym. Sci. Technol.*, 27, 51 (1985).

[18] J. Taranco, J. M. García-Martínez, O. Laguna, E. P. Collar, *J. Polym. Eng.*, 13, 287 (1994).

[19] J. Taranco, O. Laguna, E. P. Collar, *J. Polym. Eng.*, 11, 359 (1992).

[20] J. Taranco, O. Laguna, E. P. Collar, *J. Polym. Eng.*, 11, 325 (1992).

[21] J. Taranco, O. Laguna, E. P. Collar, *J. Polym. Eng.*, 11, 335 (1992).

[22] E. P. Collar, J. M. García-Martínez, O. Laguna, J. Taranco, *J. Polym. Mater.*, 13, 111 (1996).

[23] J. M. García-Martínez, O. Laguna, E. P. Collar, *J. Polym. Eng.*, 17, 269 (1998).

[24] J. M. García-Martínez, O. Laguna, E. P. Collar, *J. Polym. Mater.*, 15, 127 (1998).

[25] E. P. Collar, S. Areso, O. Laguna, J. M. García-Martínez, *J. Polym. Mater.*, 15, 237 (1998).

[26] E. P. Collar, S. Areso, O. Laguna, J. M. García-Martínez, *Polym. Mater.*, 15, 355 (1998).

[27] E. P. Collar, S. Areso, O. Laguna, J. M. García-Martínez, *J. Polym. Mater.*, 15, 363 (1998).

[28] J. M. García-Martínez, O. Laguna, S. Areso, E. P. Collar, *J. Polym. Sci. Polym. Phys.*, 38, 1554 (2000).

[29] J. M. García-Martínez, O. Laguna, S. Areso, E. P. Collar, *J. Appl. Polym. Sci.*, 81, 625 (2001).

[30] J. M. García-Martínez, O. Laguna, S. Areso, E. P. Collar, *Eur. Polym. J.*, 38, 1583 (2002).

[31] J. M. García-Martínez, O. Laguna, S. Areso, E. P. Collar, *J. Polym. Sci. Polym. Phys.*, 40, 1371 (2002).

[32] S. N. Lu, L. Yan, X. G. Zhu, Z. N. Qi, *J. Mater. Sci.*, 27, 4633 (1992).

[33] Y. C. Qu, Z. Z. Yu, *Polym. Int.*, 37, 113 (1995).

[34] D. Sekutowaski, *Plast. Eng.*, 43 (1987).

[35] Helmy, E. Ferreiro, S. de Bussetti, *J. Colloid Interface Sci.*, 285, 314 (2005).

[36] Luzenac Talc For The World - Talc geology. At http://www.luzenac.com/geology.htlm. (2002)

[37] P. W. Harben, M. Kužvart, *Industrial minerals: a global geology*. Industrial Minerals Information Ltd, London (1996).

[38] L. Castillo, S. Barbosa, N. Capiati, *J. Mater. Sci.*, 46, 2578 (2011).

[39] M. Fujiyama, T. Wakino, Y. Kawasaki, *J. Appl. Polym. Sci.*, 35, 29 (1988).

[40] Z. Chen, C. Finet, K. Liddell, D. P. Thompson, J. R. White, *J. Appl. Polym. Sci.*, 46, 1429 (1992).

[41] Z. Menzik, D. R. Fitchmun, *J. Polym. Sci. Polym. Phys. Ed.*, 11, 973 (1973).

[42] M. Fujiyama, T Wakino, *J. Appl. Polym. Sci.*, 42, 2739 (1991).

[43] M. Fujiyama, T. Wakino, *J. Appl. Polym. Sci.*, 43, 57 (1991).

[44] M. Fujiyama, T. Wakino, *J. Appl. Polym. Sci.*, 43, 97 (1991).

[45] E. J. Addink, J. Beintema, *Polymer*, 2, 185 (1961).

In: Polypropylene
Editors: L. P. Silva and E. F. Barbosa

ISBN: 978-1-62417-142-0

Chapter 9

CRYSTALLIZATION KINETICS OF POLYPROPYLENE IN COMPOSITE WITH FLY ASH

Dilip C. D. Nath*[*], *Rita Khanna and Veena Sahajwall
Centre for Sustainable Materials Research & Technology (SMaRT@UNSW),
School of Materials Science and Engineering,
University of New South Wales, Australia

ABSTRACT

We have prepared polypropylene (PP) composite materials reinforced with 20, 45 and 60 wt% fly ash (FA) by injection moulding at 210°C and measured tensile strengths of the composite specimens at 25, 50 and 70°C using a controlled oven in Instron. At 25°C, the composites suffered significant loss in strength, as much as 47%, whereas, at 50 and 70°C, there was up to 15% gain in strength. Dynamic mechanical analysis shows that the addition of FA leads to the significant enhancement of both storage (E′) and loss modulus (E″) of the composites compared to those of neat PP at a given temperature. It supports the formation of interfacial interaction between the surfaces of PP and FA. The tangent δ curves of neat PP and composites showed a broader pattern of peak. The glass transition temperature (T_g) of PP was assigned at 70°C in neat PP and 73°C in composites. Crystallization kinetics of PP in composites has been studied non-isothermally and isothermally using differential scanning calorimetry at cooling/heating rates 10°C, 15°C and 20°C per min between 200°C to -30°C. Whilst neat PP showed a mono modal α crystalline phase structure only, and the presence of FA led to bimodal thermographs revealing partial transcrystallisation of α into β, to maximum 14 %. The onset and peak crystallization temperatures of all samples decreased by approx 3°C with each 5°C /min increase in cooling rate. Parameters such as crystal growth rate, dimensions and activation energy were determined using a series of established models. The Avrami graphs showed that contrary to the published data, there are two sets of straight lines a) with a lower slope at low cooling rate and b) with a distinctly higher slope for high cooling rate. Activation energy of the materials reached a maximum at 45 wt% FA. In the isothermal crystallization kinetics condition, the lowest points of the

[*] Email: dilip.nath@unsw.edu.au

exotherm peaks shifted to higher crystallization times in the ranges of 0.75 - 1.50 min with the increasing of crystallization temperature (T_c) of PP in neat PP and composites regardless of FA percentage addition.

INTRODUCTION

The semicrystalline polymer, polypropylene (PP) and its composites are widely using in many engineering and industrial applications. The mechanical and thermal properties of PP composites depend on the degree of crystallinity and the nature of the crystalline morphology, which may constitute crystallographic phases of α (monoclinic) and ß (pseudohexagonal). The ß phase melts at a relatively lower temperature, whereas the more stable α phase melts at a higher temperature [1-2]. In the literature, PP reinforced with inorganic natural clay shows low gas permeability, low thermal expansion coefficient, and excellent solvent and chemical resistance [3-7]. The nature of filler plays a significant role in transcrystallization of α to ß form and the dimensional distribution of spherulites of PP in composites [8].

Synthesis of Polypropylene

Polypropylene is a plastic synthesized by Zeigler and Natta in 1950s using transition metal chloride (e.g. $TiCl_3$) as catalyst in combination with organo-compounds of group III metal, like triethylaluminium [$(C_2H_5)_3Al$] as co-catalyst. The raw material oil-derived gas, propylene that is naturally abundant in inexpensive petroleum source uses as monomer in the synthesis of PP [9, 10].

Zeigler and Natta had won the novel prize for their wonderful research work in stereo-regular polymerization of propylene in 1963. Stereo-regularity of polypropylene is a key factor in determining the qualitative properties of polymer. Stereo-regularity described the term of tacticity of the polymer formed. For example, if all the stereo-centres in the polymer are the same, then it is "*isotactic*" as shown in Figure 1. If the stereo-centres alternatively arrange in back and forth mode in the polymer backbone, then polymer is "*syndiotactic*" as shown in diagram, Figure 2. If there is no control of the placement of stereo-centres, then the polymer is "*atactic*" as shown in diagram, Figure 3 [11].

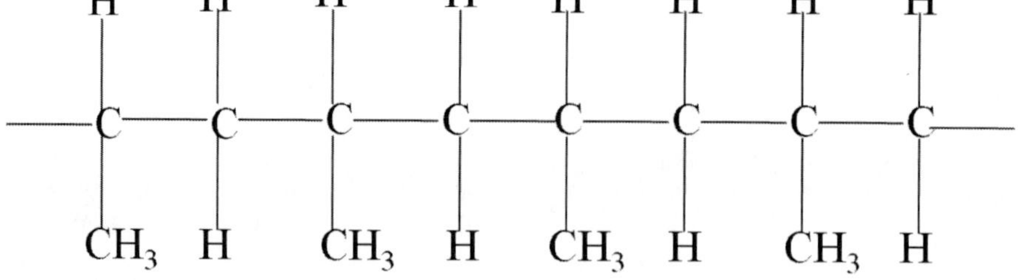

Figure 1. *Isotactic* Polyproyplene.

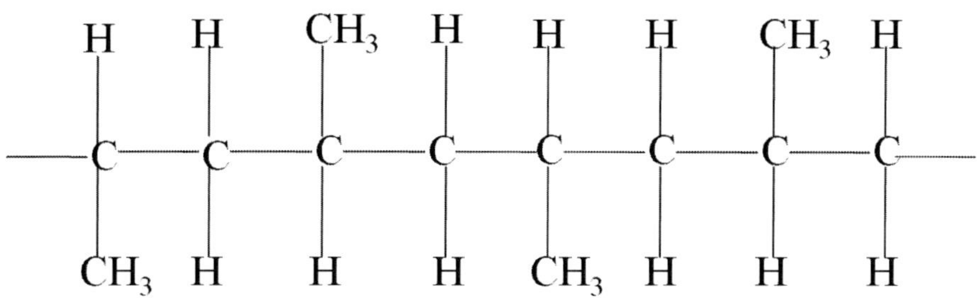

Figure 2. *Syndiotactic* Polypropylene.

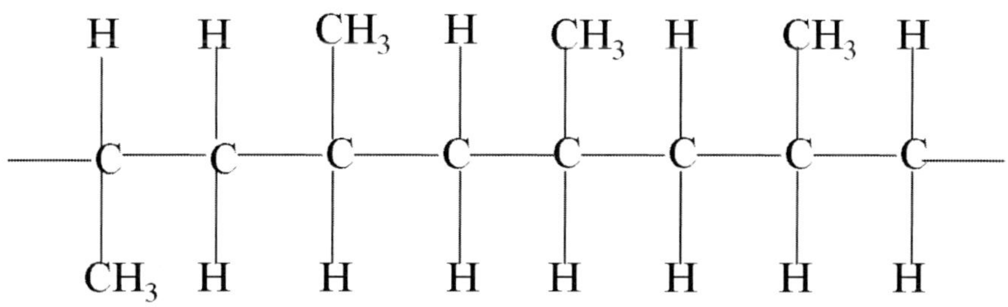

Figure 3. *Atactic* Polypropylene.

Since discovery of Zeigler-Natta catalyst, tremendous research works have found different catalytic systems for polymerization of PP. Metallocence complexes of Ti, Zr and Hf use in polymerization of propylene in combination with methylaluminoxane (MAO) as cocatalyst. The discovery of cocatalyst, MAO by Kaminsky and Sinn significantly changed the phenomena of olefin polymerization in Zeigler-Natta catalytic systems [12]. Another effective cocatalyst, Lewis acidic tris(pentafluorophenyl)borane $[B(C_6F_6)_3]$ strongly promotes olefin polymerization in combination with group IV alkylmetallocenes discovered by Marks and Ewen independently [13,14].

Metallocene compounds, which have a sandwich structure carrying two cyclopentadienyl (Cp) ligands, have been the most famous and sophisticated homogeneous catalyst. Homogeneous metallocene catalysts have highly changed the stereo-specificity of polypropylene. In addition, heterogeneous catalysts supported on silica also offer the industrial gaseous or slurry polymerization of polypropylene.

Polypropylene Composite Materials

The industrial sectors use the polypropylene composite materials due to its controlled physical and mechanical properties in relatively low processing cost. Different types of reinforcing filler materials such as, talc, mica, clay, whiskers, glass fiber, carbon fiber, silica nanoparticles, carbon nanotubes and calcium carbonate reinforce polypropylene in the fabrication of composites [5]. There is increasing pressure to reuse industrial by-products discarded in the past. Coal-burning power stations generate amount of fine powder byproduct, known as fly ash (FA). The storage and handling of FA are challenging tasks that affect the cost effectiveness of the relevant industry. The landfill method mainly disposes FA in ash dams and lagoons. Research on recycling and reuse of FA has lately become an interesting field of study in the emerging markets for green and eco-friendly manufacturing processes and products [1].

There are several interrelated factors, which influence FA properties, including the nature of the coal, coal preparation before firing, the firing system, the method of collection and any subsequent processing. Knowledge on surface morphology and elemental composition of FA are of importance for incorporating them into metal, ceramic or polymer. Optical microscopy (OM), scanning electron microscopy (SEM), transmission microscopy (TEM), atomic force microscopy (AFM) as well as wide-angle X-ray diffraction (WAXRD) have extensively visualized the macro- and micro-structural features of the FA surfaces [2-4]. Chemically, FA itself is principally a mixture of alkali and transition metal oxides mainly of silicon, aluminium and iron, and small percentage of oxides of calcium, magnesium, potassium, sodium, titanium depending on the processing and coal composition [5].

FA is currently using in the production of cement, concrete, liquid waste stabilization, hydraulic mine backfill, mining application, road constructions and as fertilizer in paddy field [15]. Recent research has used FA as filler/reinforcing material in metal [8] and polymer matrix such as polyester [16], epoxy [17], PP [18-20] and PVA [21]. Whilst PP is a very attractive thermoplastic with moderately high strength and elevated temperature capability however, the tensile strength of PP–FA composite materials decreased with the addition of FA [18]. In contrast, the use of naturally abundant clay particle as reinforcement material in PP has increased its strength because of the smaller size particles of clays causing a much better dispersion in PP matrix [5].

The resulting properties of composites generally depend on the filler's nature, size and distribution, aspect ratio, volume fraction, and the intrinsic adhesion between the surfaces of filler and polymer [6]. High aspect ratio (fibre type) fillers generally increase the yield strength because the filler is capable of attaining high local stress transferred from the polymer matrix [22], which does not happen with lower aspect ratio fillers such as FA. The reduced tensile strength in particulate fillers such as fly ash may arise from, a) poor adhesion at the interface of the filler and the polymer, b) inability of the particulate filler to hold the local stress from the polymer matrix and c) generation and growth of cracks in the particle and subsequent splitting of the composites [22, 23].

Despite the above-mentioned drawbacks, fly ash reinforced PP composites can still find numerous engineering applications particularly at elevated temperatures, where the fly ash becomes a source of strength whilst the polymer provides enhanced ductility. One major environmental value of using the already available huge quantities of fly ash is that it can lead to conservation of other regular filler materials routinely used in polymers in significant proportions [24].

Crystallization Kinetics of Polypropylene in Composite Materials

A number of studies have reported that talc [25], kaolin [26], carbon black [27], calcium carbonate [28], clay [29], ZnO [30], glass fibre [31], silicon oxide [32,33] and carbon nanotube [34] have the potential to act as enhanced nucleating agents for crystalline PP chain.

Such information leads to the understanding of crystallisation kinetics of semi-crystalline polymer, which in turn control their mechanical and other engineering properties.

The study with different models on non-isothermal and isothermal parameters in PP composites has reported in the literature. The Equations 1 and 2 calculate the degree of crystallinity of polymer [27, 35].

$$X_{c\,(\%)} = (\Delta H_m / \Delta H) \times 100 \tag{1}$$

where, ΔH_m is the enthalpy of melting of samples and ΔH is the enthalpy of melting of 100 % α- or ß-crystalline phase taken as 178 J/mg for α, and 168 J/mg for ß forms [36].

$$X_c (\%) = \Delta H_c / (1\text{-}Ø)\, \Delta H_m \times 100 \tag{2}$$

where ΔH_c is apparent enthalpy of crystallization, ΔH_m enthalpy of melting of 100 % crystalline polymer and Ø is the weight fraction of filler in the composite. The difference of melting and crystallization temperatures indicates the supercooling temperature (ΔT), which also characterizes the crystallization behaviour of the melt [37]. Dynamic (non-isothermal) crystallization, the common industrial route, is much more complicated than isothermal situation due to alteration of thermal environment with time. Equation 3 leads to the Avrami non-isothermal crystallization kinetic parameters [37].

$$1 - X_t = \exp(-Z_t\, t^n) \tag{3}$$

where X_t is the relative degree of crystallinity at time t, Z_t is crystallization constant involving both nucleation and growth rate of crystallization, n is Avrami exponent related to the crystalline dimension. The crystallization time obtained from the crystallization temperatures during cooling using Equation 4 [38].

$$t = (T_0 - T)/ \Phi \tag{4}$$

where t is crystallization time, T_o the onset crystallization temperature, T the temperature at crystallization time t, and Φ the cooling rate from melt state. The double-logarithmic form of the above Avrami equation 5 yields:

$$\ln [-\ln(1 - X_t)] = \ln Z_t + n \ln t \tag{5}$$

According to Jeziorny model 6 [39], the constant $\ln Z_t$ can be expressed as

$$\ln Z_c = \ln Z_t / \Phi \tag{6}$$

where, Z_c is the corrected Jeziorny crystallization constant.

Ozawa developed a new extended form of Avrami Equation 3 to determine the crystallization parameters and dimensional growth. Equation 7 proposed the investigation of non-isothermal crystallization kinetics by integration of infinitesimally small isothermal crystallization steps [37].

$$\ln [-\ln(1-X_t)] = \ln K(T) - m \ln\Phi \quad (7)$$

where K(T) is crystallization constant, and m is Ozawa parameter related to the crystal dimensional growth. On the other hand, Mo [40] combined the Avrami n and Ozawa m exponents and derived another form of kinetic equation 8 as given below:

$$\ln \Phi = \ln F(T) - a \ln t \quad (8)$$

where, F(T) is the crystallization constant, *a* the ratio of n and m exponents related to the crystallization dimension. Kissinger model 9 provides the following expression for the determination of activation energy in non-isothermal condition [41].

$$d[\ln(\Phi/T_p^2)]/d(1/T_p) = -\Delta E/R \quad (9)$$

where, T_p is the crystallization peak temperature, R universal gas constant, and ΔE the activation energy. Most of the kinetics research as of date conducted the systems other than FA-polymer composites. Recently, available high quality ceramic waste, FA has marked the reinforcement of polymers such as polyester, epoxy, PVA and PP. Whilst one of these papers has undertaken limited thermal studies, the bulk of the publications have not investigated these important crystallization kinetics and/or the activation energy parameter of the matrix under transient cooling in the presence of FA. The composites of 20 wt% FA with PP showed 14% improvement in tensile strength at elevated test temperatures, 50 and 70°C with reduced brittle nature [42]. The aims of this chapter are to study the effect of FA on crystallization behaviours of PP in composites and to evaluate the characteristic parameters of non-isothermal crystallization kinetics composite systems using the Avrami, Jeziorny, Ozawa and Mo models including activation energy. Another aim is to report the isothermal crystallization kinetics of composites using Avrami equation and calculation of the initial crystal thickness using supercooling temperature.

MATERIAL AND METHODS

Materials

Gladstone Coal Fire Plant, Qld, Australia and Martogg Company Ltd. Sydney, Australia supplied FA sample and white powdery *isotactic* PP, respectively. The ASTM D1505 -03, 10 minute melt flow index (MFI) of the PP was 60, confirming that it was a low viscosity grade resin suitable for injection moulding.

Sample Preparations

Injection moulding in a Boy 15 S equipment at 210°C from pre-mixed mixture of PP with FA (ball mill for 72 h) fabricated PP-FA composite samples having concentrations of a) 20,

b) 45, and c) 60 wt% FA as well as pure PP samples as control under similar conditions compared to the composites.

Sample Analysis

Coulter Light Scattering Particle Size Analyser (CLSPSA) determined the FA particle size and size distribution using water as a fluid in 60 seconds run. Inductively Coupled Plasma Atomic Emission spectroscopy (ICP-AES) and X-ray Fluorescence Spectroscopy (XRF) determine the chemical composition of FA. Tensile properties were determined using an Instron 1122 5kN testing machine with crosshead movement 5 mm/min at 25, 50 and 70°C using temperature-controlled furnace chamber. We set the specimens inside the furnace and allowed time to stabilize to the desired temperature, and the load-displacement data collected using a Picolog ADC100 Datalogger instrument. Perkin Elmer dynamical mechanical analyser (DMA) determined dynamic thermo-mechanical measurements on the 35 mm x 12 mm x 5.0 mm strip samples in tension-torsion mode in the ranges 25 to 150°C at a constant frequency of 1 Hz. The heating rate was 2°C/min with constant strain amplitude of 0.5%.

Perkin Elmer Differential Scanning Calorimeter (DSC) TAC-7/DX recorded the thermal phenomena. About 5-7 mg samples were sealed in 30 μl aluminium pan with an empty aluminium pan used as reference, under nitrogen atmosphere. Furnace in the non-isothermal crystallization condition heated the sample at 10, 15 and 20°C heating rate in the range of -30 to 200 °C, and hold 2 min at 200 °C to remove all the thermal history properly. We also set the isothermal condition to quench the samples from 200°C with a rapid cooling rate 200°C /min to a fixed isothermal temperature of crystallization (T_c) 130°C, and held for 10 min then cool down to 25°C by 10°C /min. The cycles collect the crystallization exotherms at 132 and 134°C under an identical condition. The T_c has chosen in the middle of the melting (163°C) and glass transition temperatures (<100°C) of neat PP. The thermal conditions, molecular weight and stereospecficity of PP strongly influence the glass transition temperature, for example, *isotactic*-PP 100°C, *atactic*-PP -20°C and *syn*-PP -2°C.

Wide-angle X-ray diffraction (WAXRD) investigated the structure of FA, PP, and the composites. This conducted in a multipurpose diffraction (MPD-Scherrer) with diffracted beam graphite monochromators using Ni filtered CuKα radiation. Diffractometer conditions for data collection were 40 kV and 20 mA, 2θ scanning range from 5~90° with step size 0.02, time per step 0.5s. Hitachi 4500-II SEM images give the morphology of FA and composites samples studied in this chapter. The chromium coating makes the non-conductive FA particles and fractured surfaces of the composites on carbon tape to conductive surface for better images.

Results and Discussion

Mechanical Properties of Composite Materials

CLSPSA (Coulter Light Scattering Particle Size Analyser) determined the particle size and distribution profiles of FA as a function of FA shown in Figure 4. One peak in the range

10-35 μm and another broad peak in the ranges 30-70 μm appeared in all cases. SEM shows the morphology of particles in in Figure 4, which shows poly-dispersed spherical particles. XRD spectra of powdery FA showed mainly 70 % orthorhombic shape of mullite with 2:1 stoichiometric compositions of Al_2O_3 and SiO_2 along with 30 % α-quartz. The ratio of SiO_2, Al_2O_3, Fe_2O_3 and CaO are slightly different in the two methods: 47, 31, 15 and 2 for ICP-AES, whereas it is 47, 29, 12 and 5 by XRF [21]. The percentage of the minor compounds in FA is almost the same using the two methods. Our previous works reported the results of the constituent phases of FA and composites with PP obtained by WAXRD [43]. Comparison with standard reference by search and match method using X-Pert software indicates the contents of FA to be 68.7% orthorhombic mullite (aluminium silicate, $3Al_2O_3.2SiO_2$) (essentially a type of glass), 23.2% hexagonal quartz (silicon oxide, SiO_2), and 8.1% of hematite (Fe_2O_3) [43].

Table 1 displays the calculated values of tensile strength, strain to failure, and percentage change in modulus of the materials from stress–strain diagrams. The modulus values as obtained from the data file- in the absence of an extensometer- appear to be somewhat lower than published values, perhaps showing a systematic error. Pukanszky model at different test temperatures calculates the values of interfacial parameter B_y shown in Table 1 [44]. The increasing of FA decreases the tensile strength of the composites than neat resins, as reported in the literature [43].

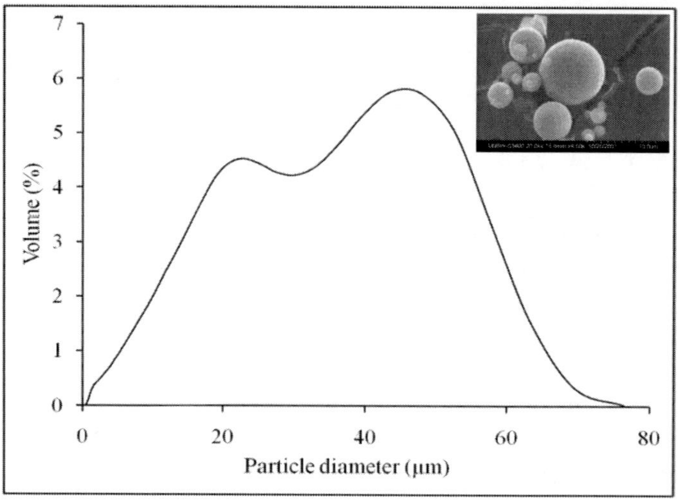

Figure 4. Particle size distribution patterns of FA, as analysed by CLSPASA technique and SEM images in inset.

At 50 and 70°C, the similar linear drop in strength is observed between 20 to 60 wt% FA. However, between 0 and 20 wt% FA, there is a small yet noticeable increase in the strength, which could be attributed to the enhanced free volume of the thermoplastic polymer at the elevated temperature, which is being filled up by the fly ash apparently causing preferable wetting between PP and FA, thereby increasing the strength. The system of talc and PP composite showed the similar behaviour. The embedment of talc in void sites in PP enhanced the adhesion exceptionally with PP and gave the high performance of mechanical strength [44]. In the reduction of free volume environment, the segmental chain of PP gets restriction strongly, reflecting higher T_g values.

Table 1. Properties of Polypropylene-Fly ash Composites

FA (%)	Volume fraction, φ_f	Test temp (°C)	Tensile stress (MPa)	Strain at break (%)	*Y* (%) Increase in modulus	B_y Interfacial parameter
0	--	25	32.2±0.8	11.5±0.2	--	--
20	0.14	25	25.4±0.5	11.0±0.4	9±0.6	1.5
45	0.34	25	21.2±0.7	7.6±0.2	23±0.5	1.7
60	0.49	25	17.0±0.6	6.5±0.3	25±0.4	1.7
0	--	50	17.3±0.5	12.2±0.5	--	--
20	0.14	50	19.5±0.5	22.5±0.5	12±0.7	4.1
45	0.34	50	15.0±0.5	18.0±0.5	24±0.4	2.6
60	0.49	50	12.0±0.5	9.6±0.5	29±0.5	2.3
0	--	70	13.0±0.5	44.0±0.5	--	--
20	0.14	70	14.7±0.5	51.5±0.5	9±0.6	4.1
45	0.34	70	11.0±0.5	23.5±0.5	53±0.6	2.5
60	0.49	70	9.1±0.5	12.8±0.5	59±0.7	2.3

The secondary binding forces -OH groups on the surface of FA may have a role in the formation of weak interfacial interactions in PP fibril [45]. However, 20 wt% FA seems to be the limiting amount that can be absorbed in the free volume expansion, and any further increase in FA content results in increased linear elastic fracture. Exceeding 20 wt% addition rate may reduce intimate contact between the surface of FA and PP lessening the interfacial interaction [46].

Figure 5. SEM images of fractured surface of neat PP and composites at 25°C, a) neat PP, b) 20, c) 45 and d) 60 wt% FA.

The SEM micrographs show the fracture surfaces of PP and composites tested at 25, 50 and 70°C in Figures 5-7, respectively. Figure 5a at 25°C shows the presence of α-spherulite fragments, voids, and bundles of drawn fibrils in neat PP. In Figure 5b with 20 wt% FA, it appears that a portion of the fly ash particles are occupying heterogeneous nucleus nucleated by the presence of fly ash particles sites consequently giving PP-FA bonding interaction, which led to a significant increase in strain to failure.

For 45 and 60 wt% FA composite the excess FA are obviously occupying sites within the spherulites and interspherulitic boundaries. In the absence of any chemical bonding agent as in this case, FA particles take away cleanly from the PP surface, as evident in Figures 5c and 5d. The polymer fibrillation happens at a considerably lower applied stress, because the spherical FA particles can raise the localised stress by a factor as high as 3, thereby reducing the overall strength of the composites [43].

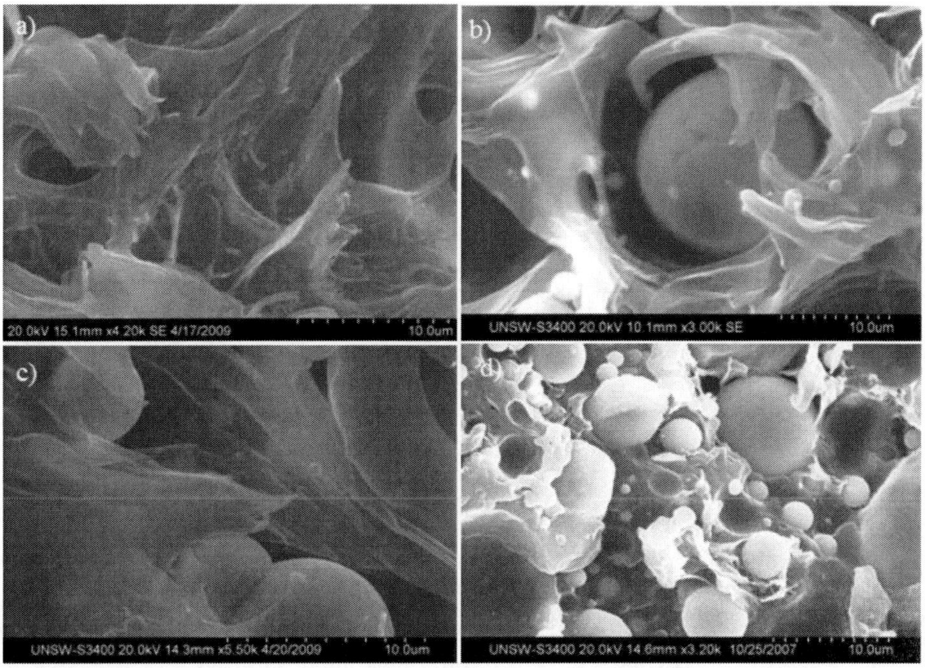

Figure 6. SEM images of fractured surface of neat PP and composites at 50°C, a) neat PP b) 20, c) 45 and d) 60 wt% FA.

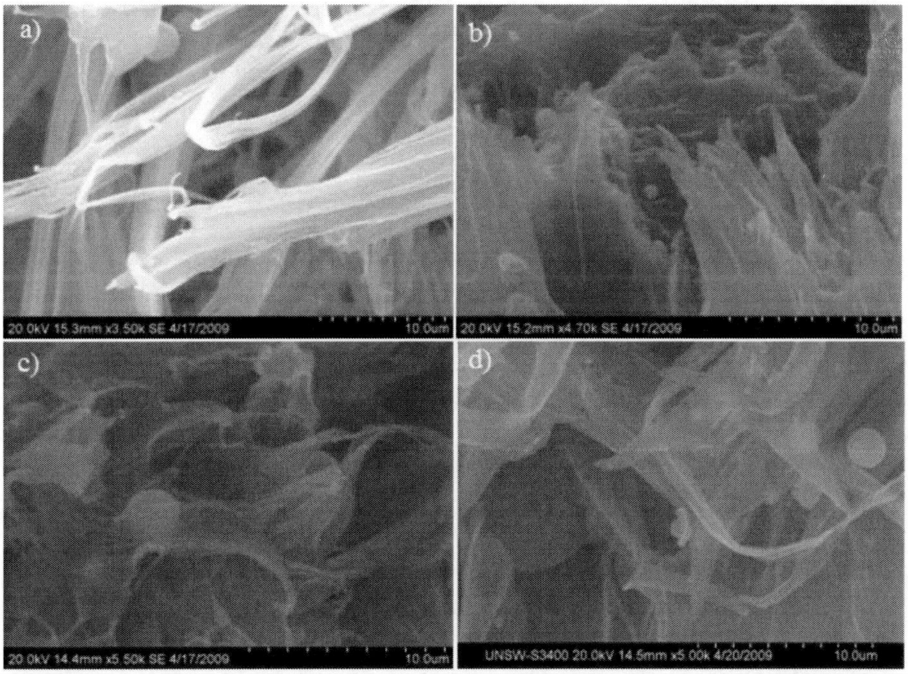

Figure 7. SEM images of fractured surface of neat PP and composites at 70°C, a) neat PP b) 20 c) 45 and d) 60 wt% FA.

At 50°C there is evidence of fragmentation, voids and more fibrillar flow of the polymer chains (Figure 6a). In 20 wt% FA composites (Figure 6b), the PP fibrils appear to fully cover the FA particles that are within the spherulites. The phenomenon supports the presence of physical interaction between FA and PP chain. The increase of FA content from 45 to 60 wt% (Figure 6c and Figure 6d) shows some evidence of the polymer fibril wetting the FA surface.

Figure 7(a-d) at 70°C supports the enhancement of the phenomena as observed at 50°C as shown in Figure 6. PP covers the less density FA particles in the 20 wt% composite, whereas FA particles in the higher concentration PP-FA composites are in closer proximity with PP. Once in close proximity the ash particles can act on each other mechanically as levers transferring mechanical load.

Dynamic Mechanical Properties of Composite Materials

Table 2 shows the dynamic mechanical properties of neat PP and PP-FA composites against temperatures 40, 60, 80, 100 and 120°C. Although the decreasing trends of E′ values of all materials with temperature are prominent, the magnitudes of E′ values of composite systems are higher than neat PP at a particular temperature. The differences of storage modulus (E′) between composites and neat PP are increasing with the FA concentration. The calculated percentages shows the E′ is increased by 9.7% with 20 wt% FA, 54.9% with 45 wt% FA and 110.1% by 60 wt% FA in PP-FA composites as compared to E′ values of neat PP as standard. The degree of interfacial strength plays an important role in the compensation of viscoelastic deformation of neat PP. The presence of higher intimate contact of FA and PP in 60 wt% FA composite enhances E′ values than 20 wt% FA [47].

Table 2. Dynamic Mechanical Properties of PP-FA composites

FA (%)	Temp. (°C)	Storage modulus (MPa)	Loss tangent δ
20		1291.1	0.058
45	40	1811.9	0.063
60		2318.8	0.061
20		879.7	0.081
45	60	1238.1	0.088
60		1641.7	0.086
20		546.5	0.106
45	80	780.7	0.113
60		1176.8	0.112
20		380.8	0.109
45	100	545.0	0.120
60		749.9	0.125
20		264.7	0.107
45	120	368.5	0.126
60		480.9	0.139

The increasing of E′ values in composites systems supports the mechanism that the highly populated and well-dispersed spherical particles are able to make a bridge by connection one by one and consequently the bridge allows stress transfer from the neat PP matrix to FA. In the case of lower concentration of 20 wt% FA, formation of the bridge is incomplete therefore, non-transferable stress makes a localized strain and the localized strain deforms the materials quicker [48].

The loss tan δ values increase with temperature and reach a maximum at the level of glass transition temperature of the amorphous fraction of neat PP. The broad peak refers to the α-transition of rigid amorphous molecular segments present in PP crystal [49].

The peak point of neat PP curve marks at approximately 70°C, indicating the T_g temperatures of α- phase in PP [50]. Due to broadness of T_g peak, it is really difficult to differentiate the effect of FA on T_g values in PP-FA composites, apparently it seems that FA has some effect in increase of T_g values.

The increase of T_g values in PP-FA composites with FA addition reflects the decreasing trend of the mobility of amorphous domain of molecular segments in PP due to the possible formation of interfacial interaction. The effect of filler on T_g values completely depends on the nature of filler. The increasing or decreasing phenomenon of T_g with the addition of filler in PP are found by several authors and reported elsewhere. [51]

At any given temperature, The loss modulus (E″) of PP-FA composites shows higher than rather than that of PP. E″ vales of all the materials increase slowly, and reach the maximum level at 70°C and form a broaden inflection point [47]. The sharpness and intensity of the loss peak in general were greater of amorphous polymer rather than semicrystalline polymer. The broaden and inflated peaks of E″ in neat PP and PP-FA composites were referred T_g temperature of amorphous domain PP due to the semicrystallinity of PP chain.

The loss modulus of polymer in general shows the decreasing propensity with the function of temperature due the sliding of polymer chain. In the presence of FA, the smooth sliding of PP segments gets interruptions due to dual functions of FA. The spherical shape FA particulates may create a possible steric hindrance between inter-segments of PP and the physical bonding between the existence of –OH groups on FA surfaces and PP [44].

As a result, FA increases the loss modulus of PP-FA composites, although the propensity remains in decrease with increasing temperatures. The corresponding E″ values of neat PP and PP-FA composites are shown in Table 3 at several given temperatures, 40, 60, 80, 100 and 120°C. The possible causes for the increasing propensity support the consideration of increased energy losses due to interfacial friction between FA and PP in the environment of external oscillation mode [49].

Table 3. Loss Modulus of neat PP and PP-FA Composites

FA (%)	E″(MPa)				
	40°C	**60°C**	**80°C**	**100°C**	**120°C**
0	66.5	64.4	52.3	37.4	25.0
20	74.9	71.4	58.0	41.8	28.6
45	114.9	109.6	88.5	65.8	46.5
60	142.7	142.3	121.7	94.1	67.3

Thermal Properties of PP and Composites

The DSC scans of the materials in Figure 8, show mainly mono modal peak of monoclinic α-crystalline phase [20] for neat PP. With increasing FA addition, the appearance of pseudo hexagonal ß-phase becomes increasingly pronounced. Whilst in these composites FA is seen as a heterogeneous ß nucleator, a number of papers have reported similar phenomena but using different fillers e.g. calcium carbonate, quinacridon dye and wollastonite [40], talc [25], kaolin [26], carbon black [27], calcium carbonate [28], clay [29], ZnO [30], glass fiber [31], silica [33] and carbon nanotube [34].

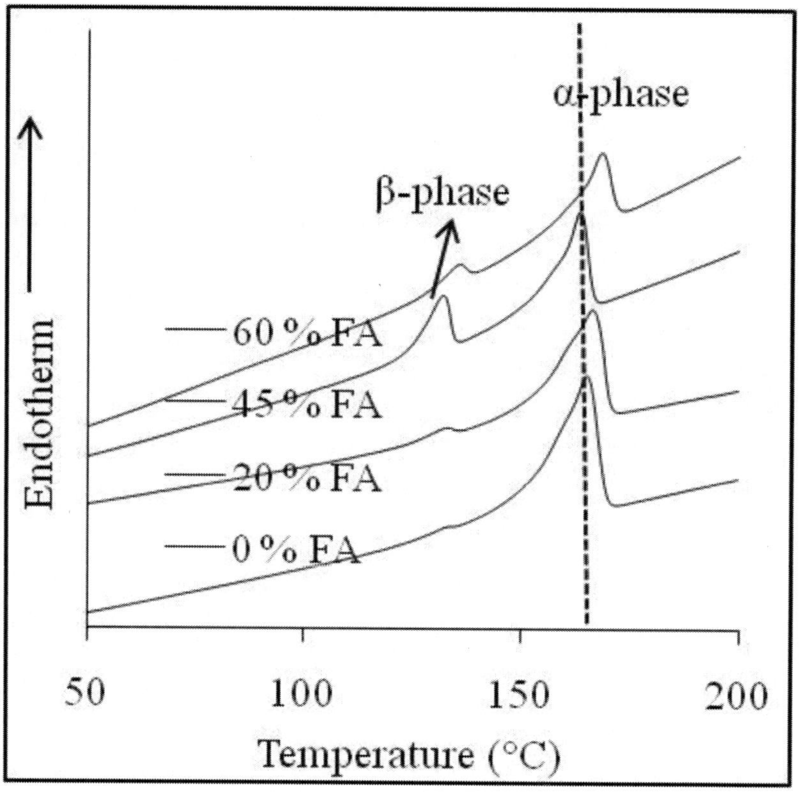

Figure 8. DSC chart of melting behaviour of PP and composites with the variation of FA at heating rate 10°C/min.

Table 4 shows the melting point, enthalpy, and percentage crystallinity calculated using equation 1 of PP-FA composites. FA in 45 wt% composite generates the β form crystallinity of PP with maximum 11%. The addition of more FA, the molar fraction of the neat PP goes down, which results in a decrease in melting enthalpy values for the composites. The melting point, enthalpy, crystallinity and the concentrations of α- and ß-crystalline phases steadily go down as DSC heating rate goes up, indicating that high heating rate inhibits equilibrium development in that. It does not allow enough time for the molecular movements of the polymer chains to happen.

Table 4. Thermal Behaviour of Neat PP and Composites with Different Heating Rates

FA (%)	Heating rate	Melting point α-phase	ΔH_m (J/g) α-phase	α -Phase (% X_α)	Melting point β-phase	ΔH_m (J/g) β -phase	ß -Phase (% X_β)
	10	165.0	77.84	44	136.6	0.51	0.3
0	15	164.6	70.25	39	136.6	0.25	0.2
	20	163.1	45.21	25	136.4	0.13	0.2
	10	166.4	61.52	35	132.4	1.66	1.0
20	15	163.6	58.73	32	131.6	0.52	0.8
	20	163.4	42.59	24	130.4	1.07	0.7
	10	163.4	41.75	24	132.7	17.13	11
45	15	162.1	32.59	19	131.1	12.09	8
	20	162.1	31.56	17	131.1	12.19	7
	10	168.4	28.15	16	135.4	4.56	3
60	15	162.6	21.55	12	133.1	3.15	2
	20	162.7	20.36	11	132.1	3.44	2

Crystallization Behaviour of PP and Composites

Figures 9 (a-d) present the normalized non-isothermal crystallization thermographs of PP and composites at different cooling rates. Table 5 shows the onset, peak crystallisation, melting and supercooling temperatures, along with calculated crystallinity using equation 1.

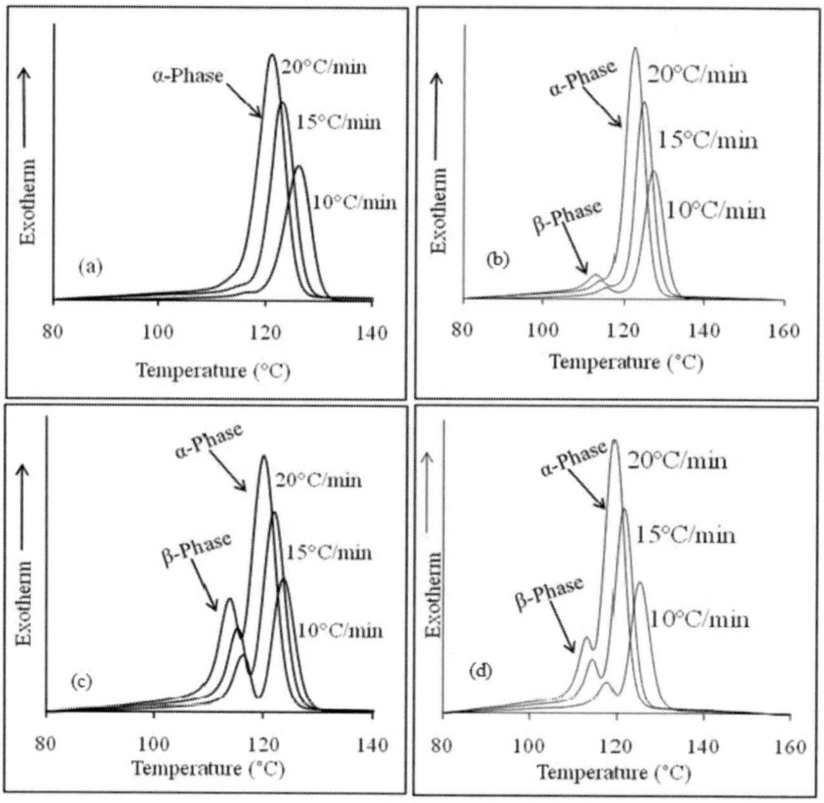

Figure 9. DSC chart at different cooling rate in non-isothermal crystallization of a) neat PP, b) 20, c) 45 and d) 60 wt% FA composites.

The mono modal thermogram of neat PP in Figure 9a shows only α crystalline peak at 126.3°C for 10°C /min cooling rate, 123°C for 15°C /min, and 121°C at 20°C /min. In the presence of FA these transformed into bimodal graphs, Figures 9 (b-d), reflecting the partial transformation of α to β. The amount of β phase increased with increasing FA content, and maximum 14 % ß was achieved with 45 wt% FA addition. β content decreased to 4 % with further addition of FA to 60 wt% FA. This effect may arise from diffusion and movement of molecular segments of PP suffering from steric hindrance at higher level of FA and increased viscosity [18].

The onset and peak crystallisation temperatures of the α → β peaks shift to lower positions by approximately 2 - 3°C with increasing cooling rate from 10°C /min to 15°C /min and from 15°C/min to 20°C / min irrespective of FA concentration. The supercooling, ΔT, increased approximately by 2 – 4°C in neat PP and composites based on cooling rates, indicating the higher cooling rates decrease the crystallization rate.

Table 5. Crystallization Behavior of Neat PP and Composites at Different Cooling Rates

FA (%)	Cooling rate (°C)	T_{cry} α-phase	T_m α-phase	T_{cry} β-phase	ΔT (T_m-T_{cry}) α-phase	X_α(%) α-crystall.	X_β (%) β-crystall.
	10	126.3	165.0	--	38.7	46	--
0	15	123.0	164.6	--	40.6	43	--
	20	121.3	163.1	--	42.8	42	--
	10	127.3	166.4	115.6	39.1	47	1.29
20	15	125.0	163.6	114.5	38.6	44	0.75
	20	122.6	163.4	113.3	40.8	42	0.65
	10	123.6	163.4	116.3	39.7	45	14
45	15	122.1	162.1	115.0	40.1	43	12
	20	120.0	162.1	113.3	42.1	40	10
	10	123.3	168.4	117.6	43.1	46	4
60	15	121.5	162.6	115.0	41.1	42	3
	20	119.3	162.7	113.3	43.5	38	1

Zhang *et al.* [28] have reported a similar observation and attributed these to a higher rate of heterogeneous nucleation induced by FA. The formation of β-crystalline PP phase peak may come from the partial transformation of α-crystalline PP segments or from the amorphous phase of PP, therefore, the new β-crystalline PP peak is smaller than α-crystalline peak. The crystallization starts almost at the same time irrespective of neat PP and composites but the crystallization was completed earlier in the composites with higher percentage of FA particles. We can conclude that FA particles have enhanced the crystallization rate.

The degree of β-crystallinity was increasing with the addition of FA, although the higher cooling rate slightly decreases the degree of β-crystallinity. The maximum 14 % of β-crystalline phase were appeared by 45 wt% FA addition, and the value decreased further with addition of FA. This may result the diffusion and movement of molecular segments of PP suffered hindrance effect from higher concentration of FA and viscosity. Different kinetic models of Avrami, Jeziorny, Ozawa and Mo calculate the crystallinity and crystallization kinetics of PP in composites.

Avrami Equation

The relative crystallinity X_t as a function of temperature was calculated by integrating the heat evolved during cooling of the melt (dH/dt) in Figures 9 (a–b) [38]. The results shown in Figures 10(a-b) are the plots of ln[-ln(1- X_t)] as a function of ln t using equation 4. In 11 out of 12 cases (with the exception of neat PP at 10°C/ min), the graphs show two straight lines with distinctly different slopes; i) slow region with $1<n_1>2$ indicating crystals with 1 to 2 dimension lamellar growth and ii) fast region $2<n_2>3$, indicating crystals shapes approaching three dimensions.

Table 6. Avrami crystallization kinetics calculated by equation 4 in neat PP and composites

FA (%)	Cooling rate (°C)	n_1 slope of slow regions	n_2 slope of fast regions	Z_{t1} Intercept of slow regions	Z_{t2} Deviation regions
	10	1.27	--	0.558	--
0	15	1.09	1.82	0.757	0.944
	20	1.05	1.61	0.869	1.261
	10	1.07	2.48	0.684	0.685
20	15	1.09	2.60	1.237	2.577
	20	1.12	2.44	1.676	4.369
	10	1.05	1.59	0.438	0.453
45	15	1.08	1.95	0.944	1.383
	20	1.05	1.53	1.037	1.625
	10	1.03	1.86	0.505	0.599
60	15	1.07	1.96	0.985	1.673
	20	1.04	2.08	1.324	3.250

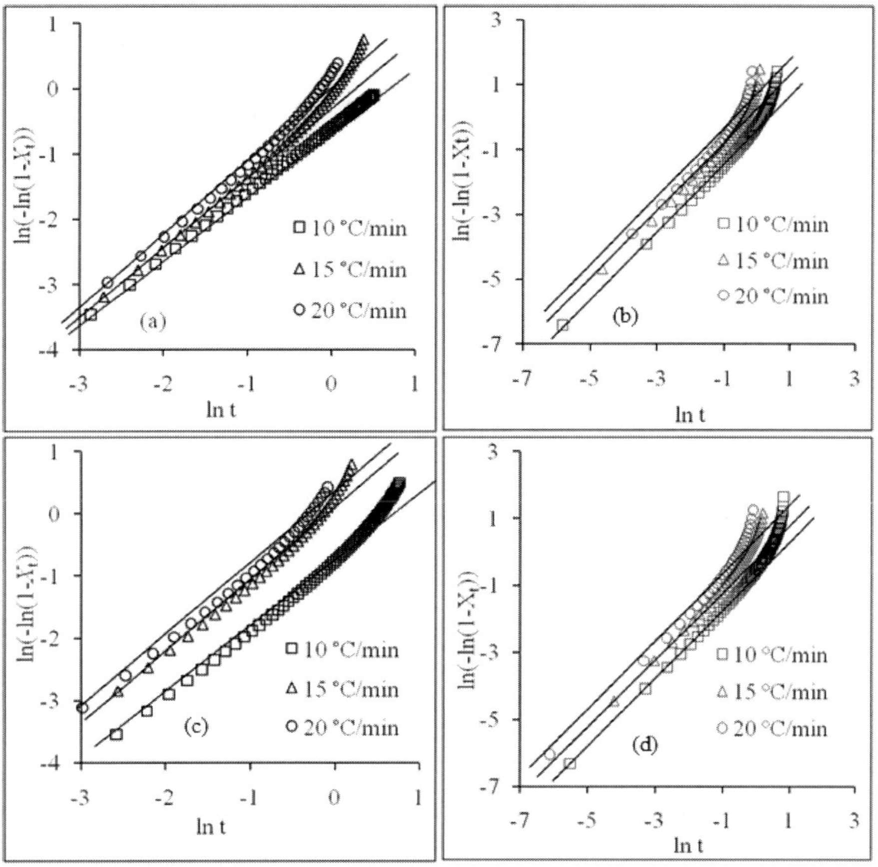

Figure 10. Avrami plot at different cooling rate in non-isothermal crystallization of, a) neat PP, b) 20, c) 45 and d) 60 wt% FA composites.

The lower slopes happened in the first half of the cooling time, whilst the higher slopes occurred in the second half of cooling. Table 6 displays the slopes and the intercept valuesThe higher slope may come from the disorder diffusion levels of PP chains in the presence of poly-disperse FA particles along with quicker cooling system. The quicker cooling rates in general give less time for rearrangements of molecular chains in the polymer. Z_t values also increased steadily with increasing cooling rates. Apparently, the values of n and Z_t were influenced by the nature of nucleation and unsteady thermal condition. It is worth mentioning that work on glass fibre and kenaf fibre reinforced PP [52] reported the presence of a second straight line with increased cooling time, albeit with reduced slope.

Jeziorny Equation

The calculated values of Z_{c1} in the primary straight line and Z_{c2} in the secondary regions using equation 5 are displayed in Table 7. The corrected crystallization constant Z_c increased with increasing cooling rate, in agreement with the results obtained by other researchers [53].

Table 7. Jeziorny non-isothermal crystallization kinetics calculated by equation 5

FA (%)	Cooling rate (°C)	Z_{t1}	Z_{c1}	Z_{t2}	Z_{c2}
		Straight line regions		Deviation regions	
	10	0.558	0.943	--	--
0	15	0.757	0.981	0.944	0.996
	20	0.869	0.993	1.261	1.011
	10	0.684	0.962	0.685	0.962
20	15	1.237	1.014	2.577	1.065
	20	1.676	1.026	4.369	1.076
	10	0.438	0.920	0.453	0.923
45	15	0.944	0.996	1.383	1.021
	20	1.037	1.001	1.625	1.024
	10	0.505	0.934	0.599	0.950
60	15	0.985	0.999	1.673	1.034
	20	1.324	1.014	3.250	1.060

Ozawa Model

Figures 11 (a-d) show the plots of $\ln[-\ln(1-X_t)]$ against $\ln\Phi$ using equation 6 at different temperatures for the four materials. All the plots are straight lines, confirming the applicability of the Ozawa model. The occasional changes in slope m in the plots indicate that the crystal growth dimension is altering with temperature. Values of m and K(T) are displayed in Table 8.

Table 8. Ozawa non-isothermal crystallization kinetics calculated by equation 6

FA (%)	Temperature (°C)	m	K(T)
	118	1.48	50.65
0	122	0.90	6.73
	126	0.81	2.86
	118	1.06	33.75
20	122	0.83	12.34
	126	0.87	7.21
	118	0.17	1.13
45	122	0.35	1.15
	126	0.83	1.64
	118	0.84	8.76
60	122	1.29	15.21
	126	2.44	21.50

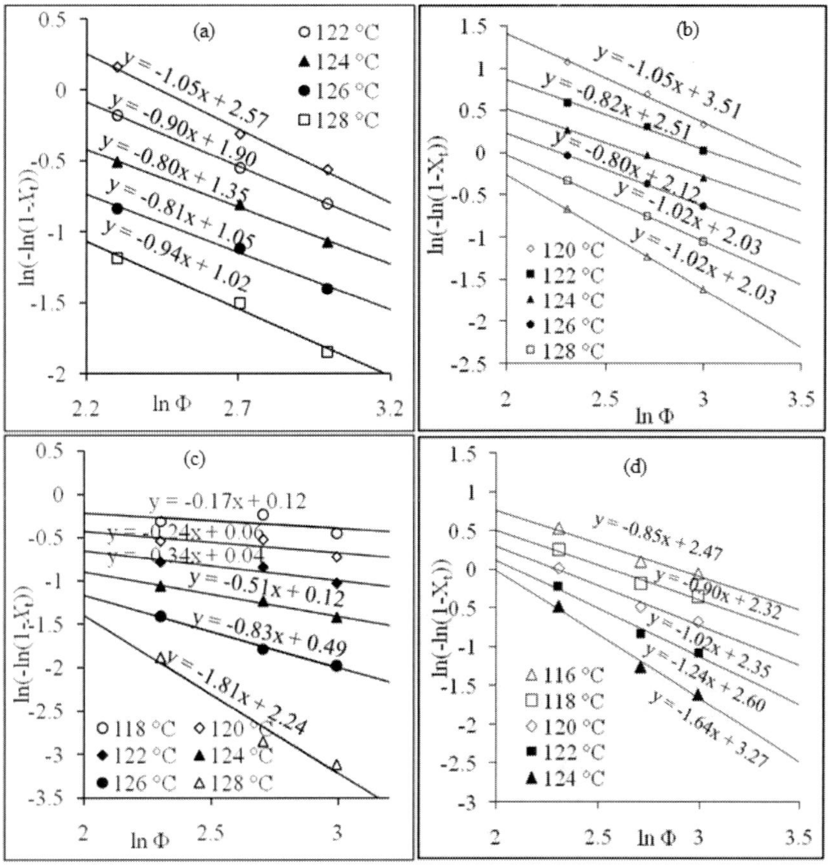

Figure 11. Ozawa plot at different cooling rate in non-isothermal crystallization of, a) neat PP, b) 20, c) 45 and d) 60 wt% FA composites.

A decrease of m values in the range of $0<m<2$ and K(T) with increasing temperature is observed in neat PP and 20 wt% FA composite, whereas the increase in K(T) and m in the range $0<m<3$ is observed in 45 and 60 wt% FA composites. It is possible that neat PP and PP in 20 wt% FA composites form two-dimensional growths of crystal in a lamellar fashion, whilst in 45 and 60 wt% FA composites the presence of large amount of FA particles causes hindrance to such planar flow, thereby resulting in localised 3 – dimensional crystal growth [38].

Mo Equation

The plots of lnΦ vs ln t using equation 7 at a given degree of crystallinity for neat PP and the composites show straight lines in Figures 12 (a-d). Table 9 shows the values of a from the slope and F(T) from the intercept values. It can be seen that a is between 3 and 4 in neat PP for all crystallinities, but a is below 1 in PP-FA composites with slightly decreasing trend as the FA concentration increases. F(T) on the other hand increases with the degree in crystallinity in PP as well as that of the composites. There was one exception for 80%

crystallinity, the F(T) values of composites are almost half of those of neat PP, indicating that the crystallization rates are faster in the presence of FA [52].

Table 9. Mo non-isothermal crystallization kinetics calculated by equation 7

FA (%)	X_t (%)	*a*	F(T), from equations in graphs
	20	3.72	0.156
0	40	2.75	6.410
	60	3.40	11.712
	80	3.23	25.503
	20	0.80	4.463
20	40	0.84	7.682
	60	0.84	10.730
	80	0.90	13.921
	20	0.64	6.721
45	40	0.70	10.350
	60	0.72	13.861
	80	0.78	17.372
	20	0.67	5.983
60	40	0.77	9.181
	60	0.79	12.422
	80	0.76	15.550

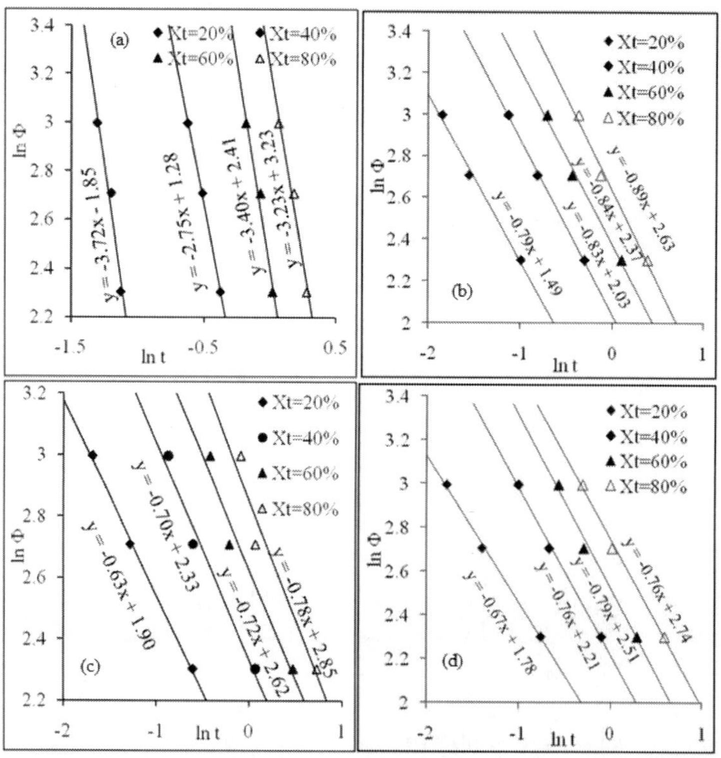

Figure 12. Mo plot at different cooling rate in non-isothermal crystallization of, a) neat PP, b) 20, c) 45 and d) 60 wt% FA composites.

Effect of Isothermal Conditions on Crystallization Kinetics of PP

Figure 13 (a-d) shows the DSC exotherms at 130, 132 and 134°C of neat PP and composite systems, respectively. At higher crystallization time, a broader mono-modal exotherms of α-crystal phase of PP chains with increasing T_c appeared. At higher T_c, the translational mobility of PP chain in the crystal phase is slower due to the presence of higher level of free energy, representing the broaden peaks. A certain minimum temperature needs to overcome the kinetic barriers to complete a crystallization process. In the presence of spherical particles of FA, PP chains start nucleation to form a new crystalline β-phase and crystallize at lower temperature than the α-phase [54].

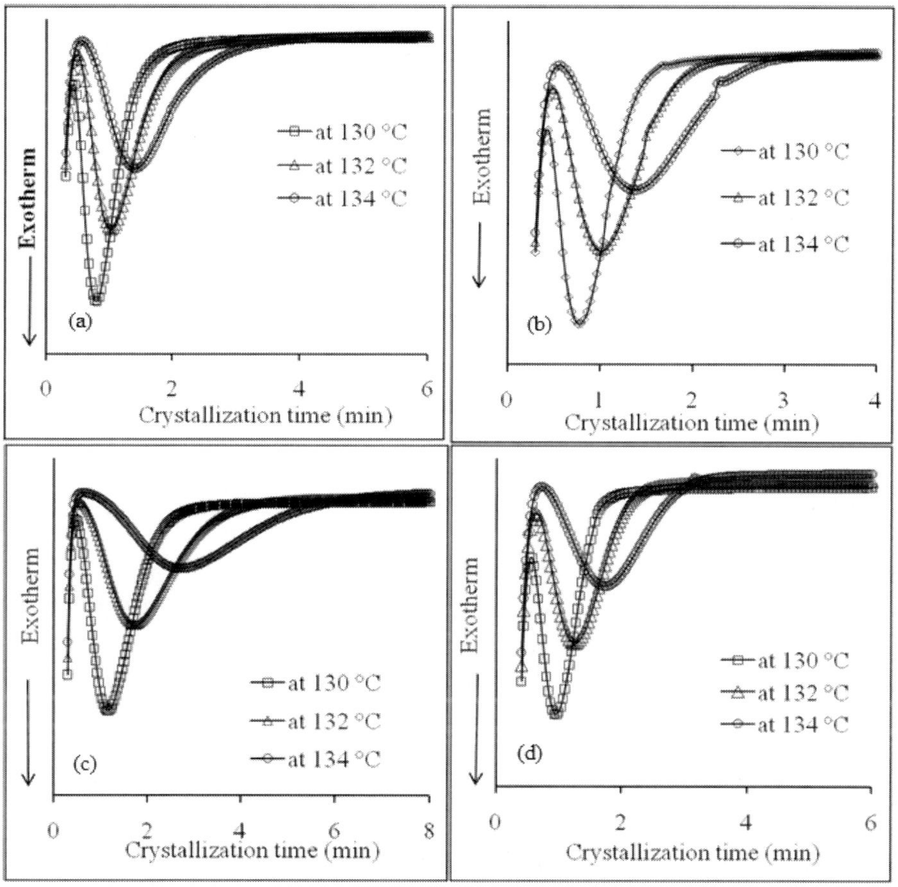

Figure 13. (a-d) DSC of isothermal crystallization at different temperatures of a) neat PP, b) 20, c) 45 and d) 60 wt% FA.

Figure 14 shows the influence of FA on the crystallization process at 130°C as model. The crystallization peaks of composites shifted to higher crystallization time, although the onset crystallization times were almost identical. The slower crystallization rate in composites may occur due to the interruption of segmental motion of PP chains. Table 10 displays the onset and end times of crystallization of neat PP and composites at different T_c along with half time crystallization ($t_{½}$).

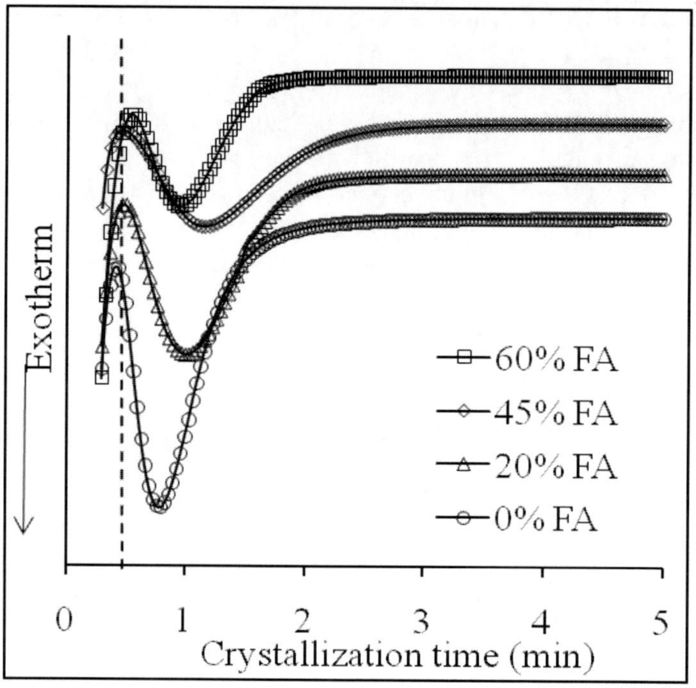

Figure 14. DSC of crystallization of neat PP and composites at 130°C.

Table 10. *Iso*-thermal crystallization time at different crystallization temperatures

FA(%)	Crystallization temp.(°C)	Onset time t (min)	End time t (min)	$t_{1/2}$ (min)
	130	0.43	3.1	0.83
0	132	0.53	3.8	1.06
	134	0.56	5.0	1.43
	130	0.43	2.8	1.06
20	132	0.50	3.4	1.06
	134	0.56	4.8	1.43
	130	0.46	3.6	1.20
45	132	0.53	5.4	1.76
	134	0.63	6.9	2.90
	130	0.56	2.8	0.93
60	132	0.60	3.1	1.20
	134	0.76	4.3	1.73

Figures 15 (a-b) show the relationship of $t_{½}$ with T_c and the percentage of FA, respectively. The $t_{½}$ in neat PP and composites increased with increasing T_c, and the slope of 45 wt% FA is higher than that of the other composites. On the other hand, $t_{½}$ values show valley-shaped curves as a function of FA at a given T_c, and the higher vertex comes from 45 wt% FA composite sample.

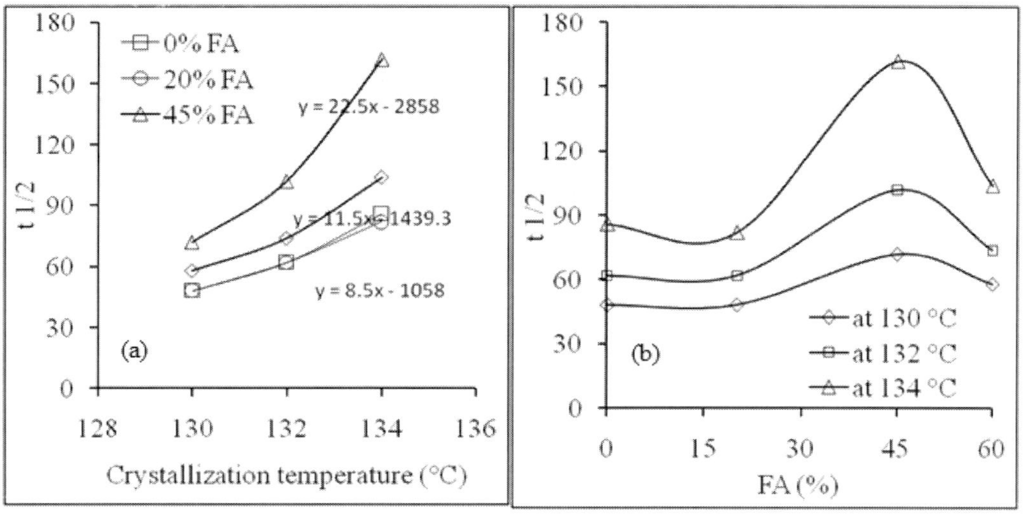

Figure 15. a) Relationship of half time crystallization with crystallization temperatures and b) Relationship of half time crystallization with volume fraction of the filler in Composites.

The crystalline phenomena in neat PP, 20 and 60 wt% FA composites showed marginal differences in crystallization time and $t_{½}$ at any given T_c except in the case of 45 wt% FA system. The case 45 wt% FA showed an unusual broader peak and relatively higher $t_{½}$ values with increasing T_c.

The author by considering the molecular structural arrangement of PP chains offers a plausible explanation of the phenomena. The crystallization of the polymer comes from the aligned mobility of chain. In the case of composites, the surface of filler linked to the polymer chains, and the level and strength of linking depends on the nature and concentration of the filler. Several reports have contributed to the correlation of mechanical and thermal behaviours of composites with the morphological information provided schematic models by SEM images on FA and filler particles in polymer chains [55, 56].

The segmental motion of PP chains in neat PP can diffuse smoothly from the melt state even in the less populated composite i.e. 20 wt% FA concentration. On the other hand, in the highly populated 60 wt% FA composite system, the crystallization of PP chains may occur in a localized manner at a given T_c due to entanglement of PP chains in interstitial voids between spherical filler particles. The inter-particle space arises from the agglomeration of the spherical particles [57]. In the case of 45 wt% FA, the diffusible PP chains may face strong interruption in the inter-layer of two particles.

The crystallization therefore may occur in sharing of free and inter layer of FA particles. As a result, PP segments in 45 wt% FA system form a transitional crystalline phase, and reveal a broaden scan [42]. The SEM images in Figure 16 of neat PP and composites support the plausible interpretation of crystallization. The inter-particle space may come from the agglomeration of spherical particle. In the case of 45 wt% FA, the diffusible PP chains may face strong interruption in the inter-layer of two particles.

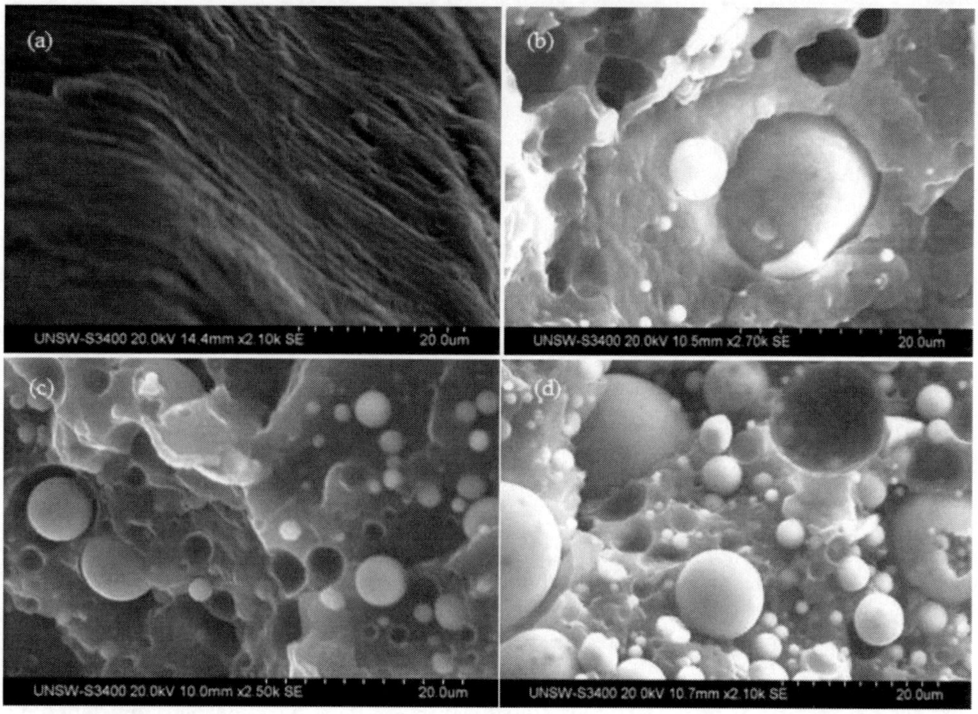

Figure 16. SEM photographs of a) neat PP, b) 20, c) 45 and d) 60 wt% FA composites.

Isothermal Crystallization Kinetics of PP in Composites

The differential heat (dH/dt) evolved during rapid cooling from melt was recorded as a function of crystallization time at a given crystallization temperature. The relative crystallinity X_t of crystallized material can be calculated using the integral equation reported in references and using Figures 17(a-d) [58]. The Avrami equation 4 calculates the isothermal crystallization kinetics of PP in neat PP and in composites [36]. The integrated DSC exotherms of neat PP and composites give straight lines and fit well with the Avrami model at different T_c. The values of n and Z_t were calculated from the slope and the intercept of the straight-line curves, respectively, and are displayed in Table 11.

The n values of PP and composites in the ranges of 2-4 were unaffected by the addition of FA and different T_c. However, the Z_t values decreased slightly with respect to T_c. The values of n and Z_t in general are influenced by the nature of nucleation and molecular weight of polymer as well as the thermal condition. In non-isothermal condition, n values of composites show lower values than in isothermal condition [42].

The Avrami exponent n values of composite systems are similar to those of glass fibre/PP composites [59] and kenaf fibre-PP system gave higher values. The higher n values reflect the existence of a specific number of three-dimensional spherulite growths in PP and composites [60]. Figures 18 (a-b) shows the relationship of n as a function of T_c and the percentage of FA in composites.

Table 11. Avrami crystallization parameters calculated using equation 4

FA(%)	Crystallization temperature (°C)	n	Zt
0	130	2.03	0.179
	132	2.14	0.133
	134	2.28	0.139
20	130	2.69	0.321
	132	2.83	0.256
	134	2.61	0.163
45	130	2.33	0.140
	132	2.17	0.105
	134	2.05	0.071
60	130	3.05	0.203
	132	3.09	0.176
	134	2.84	0.11

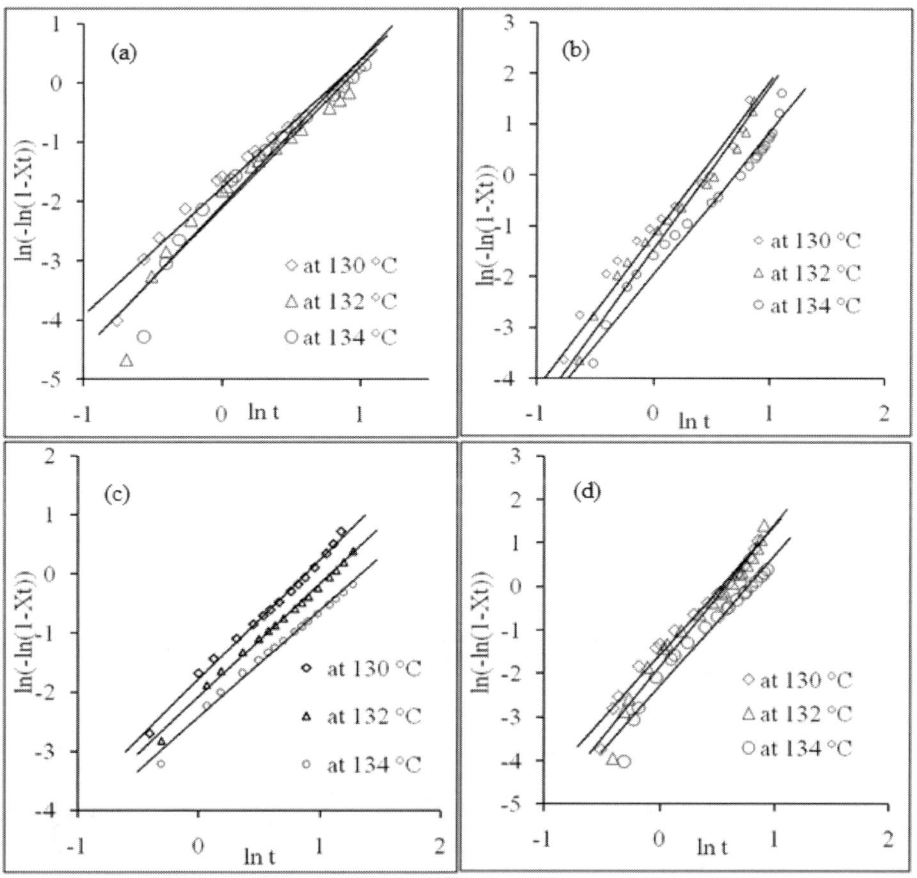

Figure 17. Avrami plot of isothermal crystallization for: a) neat PP, b) 20, c) 45 and 60 wt% FA composite materials.

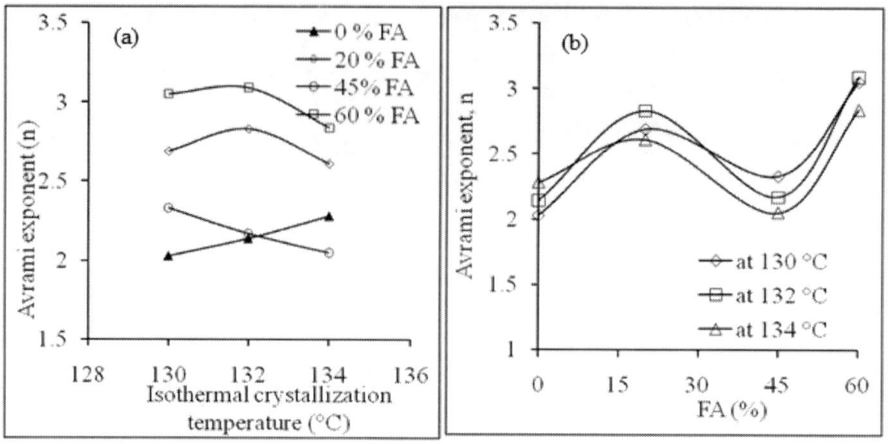

Figure 18. a) Relationship of Avrami exponent to crystallization temperatures of PP and composites and b) Relationship of Avrami exponent to the percentage of FA in composites.

The slope of n vs T_c (Figure 18a) curves decreases in PP while it shows slight increase and then decrease in the case of composites. The n vs FA content curves also show a similar trend in Figure 18b. At 20 and 60 wt% FA, the higher n values were observed compared to neat PP and 45 wt% FA containing composite.

Effect of FA on Crystallization Temperature and Crystallinity of PP

FA in composite generates the β-crystalline phase of PP chains considered as a β-nucleating agent. The intensity of β -crystalline phase increased with increasing concentration of FA in composite, and the maximum degree of β -crystallinity was calculated as 14% with 45 wt% FA addition [42]. The effect of FA content on crystallization temperature and degree of crystallinity are shown in Figures 19 (a-b) with cooling rates, 10, 15 and 20°C per min, respectively.

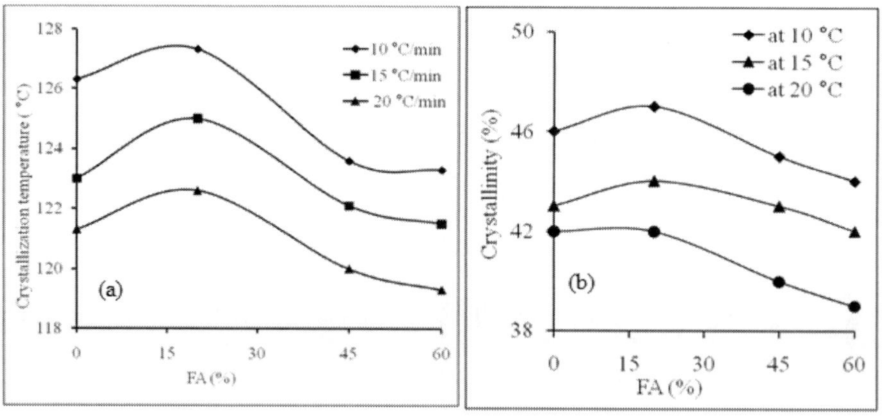

Figure 19. a) The effect of FA on crystallization temperature on PP in composites and b) The effect of FA on the degree of crystallinity of PP in composites.

Figures 19 (a-b) show the cooling rates have significant level of influence on the crystallization temperatures and the degree of crystallinity. A slight increase in crystallization temperature in 20 wt% FA system results in an early crystallization due to the possible enhanced crystallization effect of FA. The accelerated nucleation effect of 20 wt% FA helps to convert some part of amorphous chains to crystalline chains, and initiates the formation of a new β-crystalline phase. With increasing FA content in the composites, the β-phase increases with decrease of α-phase (not shown the β-phase in Figure 19a).

Calculation of Initial Crystal Thickness

Crystallization of polymer primarily is reflecting the generation and growth of lamella in solution and/or in melt state. The thickness of lamella implies the corresponding fold length. Equation 10 provides a relationship between initial lamella thickness and the super-cooling temperature [61].

$$Lc^* \text{ (in nm)} = 276.4/\Delta T + 4.16 \tag{10}$$

where Lc* is initial crystal thickness and ΔT is degree of supercooling. The initial crystal thickness is inversely proportional to the supercooling temperature and this dependence has led to the development of the kinetic theories of chain folding. The kinetic theory permits prediction and calculation of the initial thickness of crystal and growth rates quantitatively. Equation 10 calculates the initial crystal thickness of PP for neat PP and composite systems from melt states. Table 12 shows the values of Lc*, and Figure 20 plots Lc* as a function of ΔT.

Table 12. Calculation of thickness of initial crystal of PP in neat and composites

FA (%)	Cooling rate (°C)	ΔT (T_m-T_{cry}) α-phase	Lc* (in nm) (Thickness of initial crystal)
	10	38.7	11.3
0	15	40.6	10.9
	20	42.8	10.6
	10	39.1	11.2
20	15	38.6	11.3
	20	40.8	10.9
	10	39.7	11.1
45	15	40.1	11.0
	20	42.1	10.7
	10	43.1	10.5
60	15	41.1	10.8
	20	43.5	10.5

All the data points Lc* in fact are in one line and show inverse proportionality to ΔT. The research paper in the case of crystallization of polyethylene [54] reports the similar trend. The initial crystal thickness of PP is not influenced significantly in the presence of spherical particles FA and different concentrations; although it is believed that the growth of crystal will face some interruption due to the presence of poly-dispersed spherical particles FA as immobilized solid phase in the systems as shown in SEM images in Figure 16.

In the absence of FA, the lamella appears unidirectional-rodlike structure initially, and then develops as bundle growth along with branching, fanning twisting, and finally forms spherulites by diffusion of molecular segments [61-62]. In the presence of FA particles, the diffusible molecular segments face hindrance effect as a result the trend of unidirectional lamella growth turns to multidirectional and occurs in scattered fashion. The scattered behaviour of lamella growth is more severe in the highly populated 60 wt% FA system, and completely turns in only localized crystallization in the entangled area around the FA particles.

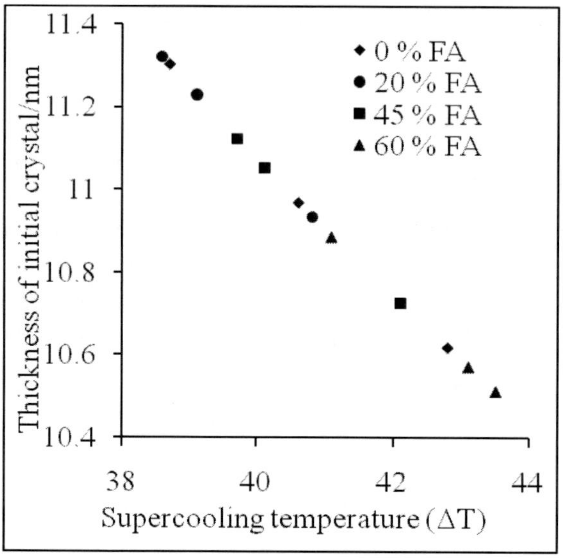

Figure 20. The relationship of initial thickness of crystal to supercooling temperatures in neat PP and composites.

Activation Energy in Composites

Kissinger's model established the differential thermal patterns of inorganic minerals with constant heating rates. There are several researcher groups in the field of polymer and polymer composites who applied Kissinger's equation in cooling system for calculation of activation energy in crystallization process and heating system too [63]. There are several reports, which concern the using of Kissinger model in cooling system. Augis-Bennett [64] Takhor and others [64-68] Friedman [67] and Vyazovkin [68] have modified Kissinger model. The calculated activation energy using former three methods did not show any big difference [66]. We therefore calculated the activation of PP and composites in heating and cooling systems using Kissinger method only.

Figure 21 shows the plots of $\ln(\Phi/T_p^2)$ as function of $1/T_p$ using equation 9 for PP and composites. Table 13 displays the values of ΔE, slope and r from the straight lines. The use of integral iso-conversional kinetic equation of Vyazovkin in PP degradation case [68] showed the activation energy nearly 180 kJ/mol^{-1} at 50% conversion, whereas our calculation shows 185 kJ/mol-1 using Kissinger method. The activation energy of composites are higher than that of neat PP, reaching the highest value at 45 wt% FA then showing a 10 % drop when FA is 60 wt%.

The results suggest that in the presence of FA, the viscosity of the system goes up, the rearrangement of molecular segments of PP get a hindrance effect in diffusion or formation of interfacial interaction between FA and PP and it takes a higher energy to be crystallized. The addition of filler in polymer composites changes the mechanical and thermal properties of the materials due to the formation interfacial linking between two surfaces. The linking may be physical adsorption or chemical bonding and this linking restricts the mobility of polymer chains. There are several reports where they have modelled the dispersion of FA or particles in polymer composite based on the experimental results of mechanical and thermal in correlation of SEM images [55].

The crystallization of the polymer happens from the alignment of mobile polymer chain. In the case of polymer composite materials, the polymer chains may adhere on the surface of filler, and the level and strength of adherence solely depends on the nature and concentration of filler. The nature of filler defines the level of interaction to polymer chains. Three distinguishable polymer chains in composite materials characterize the immobilized, intermediate and mobile regions [69]. The immobilized regions are the surface area of the filler adherence to polymer chains. The structural motions were strictly restricted and no role in crystallization.

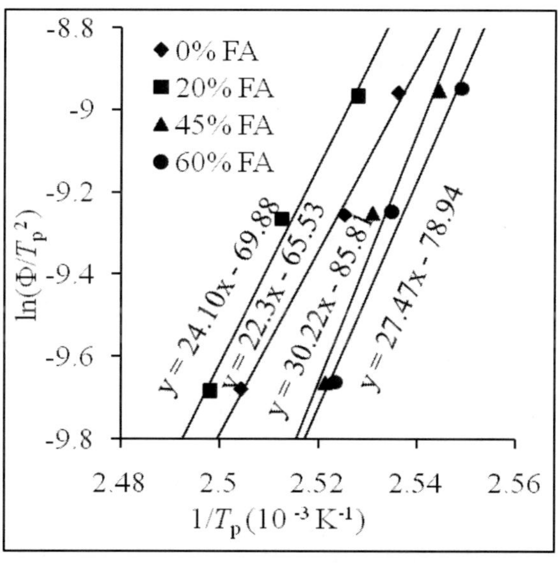

Figure 21. Kissinger plot for determination of activation energy.

The strength of the filler and polymer interaction force influences the intermediate region of the polymer located near the surface of the filler particles. The mobile region is further from the filler particles such that the motions are not restricted and it has the characteristics major role in crystallization [55, 69]. We also observed the higher mechanical strength of the composites at elevated temperatures, indicating the presence of interfacial interaction between

FA and PP. The regions of the PP chain attached onto the surface of FA restrict the mobility of PP chains during crystallization process, indicating higher activation energy.

Table 13. Activation energy of PP and composites calculated by equation 8

FA (%)	Cooling rate(°C)	Slope	r	ΔE kJ/mol^{-1}
	10			
0	15	22.3	0.99	185.4
	20			
	10			
20	15	24.1	0.98	200.4
	20			
	10			
45	15	30.2	0.96	251.1
	20			
	10			
60	15	27.5	0.98	228.6
	20			

CONCLUSION

The tensile strength at test temperature, 25 °C decreases with FA content, whereas at 50 and 70°C there is initial increase in strength up to 20 wt% FA, but after that strength goes down with more FA addition: linear drop in strength is observed between 20 to 60 wt% FA. The drop in strength with addition of FA results from the weak interface formation with PP. The average percentage increases of storage modulus were, 9.7% for 20 wt% FA, 54.9% for 45 wt% FA and 110.1% for 60 wt% FA composites. The addition of FA also leads to the enhancement of loss modulus of the composites compared to those of PP at a given temperature, and the average percentages of loss modulus were 12.4% for 20%, 74.8% for 45 wt% and 137.7% for 60 wt% FA composites compared to PP. The tangent δ curves of PP and composites showed broad peaks and T_g was approximately 70°C regardless of FA addition.

Addition of FA to PP acts as a heterogeneous β -nucleating agent and induces a secondary β-crystalline phase in the composites, a maximum amount of 14 % β being formed with 45 wt% FA. Increasing cooling rate by 5°C/min resulted in 2-3°C drop in onset crystallization temperature of PP in composites. The overall Avrami exponent n shows non-integral number $1<n>3$ suggesting the existence of a size- range of three-dimensional spherulites of PP in neat PP. The intercept constants Zt, Zc, K(T) and F(T) of the four models increase with increasing of cooling rate in all four categories materials. The constant values for the composites were consistently higher than neat PP at a given temperature. Finally, the initial crystal thickness was calculated using supercooling relationship equation, and the values are almost identical irrespective of FA added to the systems. The calculated values have a good agreement with reported values elsewhere.

REFERENCES

[1] B. Lotz, J. C. Wittmann and J. A. Lovinger, *Polymer,* 37, 4979 (1996).
[2] J. X. Li and W. L. Cheung, *J. Therm. Anal. Calorim.* 61, 757 (2000).
[3] P. B. Messersmith and E. P. Giannelis, *J. Polym. Sci. Part A: Polym. Chem.* 33, 1047 (1995).
[4] A. Usuki, T. Kawasumi, Y. Fukushima and A. Okada, *J. Mater. Res.* 8, 1179 (1993).
[5] S. S. Ray and M. Okamoto, *Prog. Polym. Sci.* 28, 1539 (2003).
[6] E. P. Giammelis, *Adv. Mater.* 8, 29 (1996).
[7] P. H. Nam, P. Dubois, T. Sun, J. M. Garces and R. Jerom, *Polymer,* 42, 2123 (2002).
[8] J. L. Acosta, M. C. Ojeda, E. Morales and A. Linares, *J. Appl. Polym. Sci.* 32, 4119 (1986).
[9] K. Zeigler, *Angew. Chem.* 67, 541 (1955).
[10] G. Natta, *J. Am. Chem. Soc.* 77, 1708 (1955).
[11] K. B. Wagener, *Science,* 267, 191 (1995).
[12] H. Sinn, W. Kaminsky, H. J. Vollmer and R. Woldt, *Angew. Chem. Int. Ed. Engl.* 19, 369 (1980).
[13] X. Yang, Cl. L. Stern and T. J. Marks, *J. Am. Chem. Soc.* 113, 3623 (1991).
[14] J. A. Ewen, M. J. Elder, Eur. Pat. Appl. 0427697, US Pat. 5561092, 1996 (1991).
[15] R. W. McDowell, *Aust. J. Soil Res.* 43, 853 (2005).
[16] S. Guhanathan and M. Sarojadevi, *Compos. Interf.* 11, 43 (2004).
[17] N. Gupta, B. S. Brar and E. Woldesenbet, *Bull. Mater. Sci.* 24(2), 219 (2001).
[18] K.W.Y. Wong and R. W. Truss, *Comp. Sci. Tech.* 52, 361 (1994).
[19] P. A. Jarvela and P. K. Jarvela, *J. Mat. Sci.* 31, 3853 (1996).
[20] X. Huang, J. Y. Hwang, J. M. Gillis, *J. Min. & Mat. Charact. & Eng.* 2(1), 11 (2003).
[21] D. C. D. Nath, S. Bandyopadhyay, P. Boughton, A. Yu, D. Blackburn and C. White, *J. Appl. Polym. Sci.* 117, 114 (2010).
[22] D. M. Bigg, *Polym. Compos.* 8, 115 (1987).
[23] J. X. Li, M. Silverstein, A. Hiltner and E. Baer, *J. Appl. Polym. Sci.* 52, 255 (1994).
[24] J. A. Brydson, *Plastics materials,* 2001, Elsiver Science.
[25] E. Ferrage, F. Martin, A. Boudet, G. Petit, F. Jouffet, P. Micoud, P.D. Parseval, S. Salvi, S. Courgerette, J. Ferret, Y. Saint-Gerard, S. Buratto and J. P. Fortune, *J. Mater. Sci.* 37, 1561 (2002).
[26] M. Mucha and Z. Krolikowski, *J. Therm. Anal. Calorim.* 74, 549 (2003).
[27] J. Chen, X. Li and C. Wu, *Polymer, J.* 39(7), 722 (2007).
[28] Q. X. Zhang, Z. Z. Yu, X. L. Xie and Y. W. Mai, *Polymer,* 45, 5985 (2004).
[29] B. Kim, S. H. Lee, D. Lee, B. Ha, J. Park, K. Char, *Ind. Eng. Chem. Res.* 43, 6082 (2004).
[30] J. G.Tang, Y. Wang, H. Y. Liu, L. A. Belfiore, *Polymer,* 45, 2081 (2004).
[31] G. B. Gaceva, A. Janevsky and E. Mader, E. *Polymer,* 42, 4409 (2001).
[32] G. Z. Papageorgiou, D. S. Achilias, D. N. Bikiaris, G. P. Karayanndis, *Thermochim Acta,* 427, 117 (2005).
[33] M. Z. Rong, M. Q. Zhang, S. L. Pan, B. Lehmann and K. Friedrich, *Polym. Intl.* 53, 176 (2004).

[34] L. Valentini, J. Biagiotti, J. M. Kenny and M. Santucci, *J. Appl. Polym. Sci.* 87, 708 (2003).

[35] J. X. Li, W. L. Cheung and D. Jia, *Polymer,* 40, 1219 (1999).

[36] M. J. Avrami, *Chem. Phys.* 9, 177 (1941).

[37] W. Xu, M. Ge and P. He, *J. Polym. Sci. Polym. Phys.* 40, 408 (2002).

[38] T. Ozawa, *Polymer,* 12, 150 (1971).

[39] T. Liu, Z. Mo, S. Wang and H. Zhang, *Polym. Eng. and Sci.* 37, 568 (1997).

[40] J. Liu, X. Wei and Q. Guo, *J. Appl. Polym. Sci.* 41, 2829 (1990).

[41] H. E. Kissinger, *Anal. Chem.*, 29(11), 1702 (1957).

[42] D. C. D. Nath, S. Bandyopadhyay, A. Yu, D. Blackburn and C. White, *J. Appl. Polym. Sci.* 115, 1510 (2010).

[43] D. C. D. Nath, S. Bandyopadhyay, A. Yu, D. Blackburn and C. White, *J. Mater. Sci.* 45, 1354 ((2010).

[44] Y. W. Leong, M. B. A. Bakar, Z. A. M. Ishak, A. Ariffin and B. Pukanszky, *J. Appl. Polym. Sci.* 91, 3315 (2004).

[45] R. Mueller, H. K. Kammler, K. Wegner and S. E. Pratsinis, *Langmuir,* 19, 160 (2003).

[46] L. E. Nielsen, *J. Appl. Polym. Sci.* 17, 97 (1979).

[47] D. C. D. Nath, S. Bandyopadhyay, J. Rider, A. Yu, D. Blackburn and C. White, *Macromol. Res.,* 19(4), 338 (2011).

[48] P. V. Joseph, G. Mathew, K. Joseph, G. Groeninckx, and S. Thomas, *Composites Part A: Appl. Sci. Manu.* 34, 275 (2003).

[49] T. T. L. Doan, H. Brodowsky and E. Mader, E. *Compos. Sci. Tech.* 67, 2707 (2007).

[50] X. L. Xie, R. K. Y. Li, S. C. Jong, Y. W. Mai, *Polym. Comp.* 23(3), 319 (2002).

[51] T. Murayama, *Dynamical Mechanical Analysis of Polymeric Material,* 1978, ISBN 0-444-41648-6.

[52] T. Liu, Z. Mo, S. Wang and H. Zhang, *Polym. Eng. Sci.* 37(3), 568 (1997).

[53] A. Jeziorny, *Polymer,* 19, 1142 (1978).

[54] P. J. Barham, R. A. Chivers, A. Keller and J. Martinez-Salazar, *J. Mater. Sci.* 20, 1625 (1985).

[55] V. Sridhar, Z. Z. Xiu, D. Xu, S.H. Lee, J. K. Kim, D. J. Kang and D. S. Bang, *Waste Manag.* 29, 1058 (2009).

[56] D. C. D. Nath, S. Bandyopadhyay, J. Rider, A. Yu, D. Blackburn, C. White and S.Varughese, *J. Therm. Anal. Calorim.* 99, 423 ((2010).

[57] L. S. Tan and A. J. Mchugh, *J. Mater. Sci.* 31, 3701 (1996).

[58] W. Xu, H. Zhai, H. Guo, N. Whitely and W. P. Pan, *J. Therm. Anal. and Calorim.* 78, 6457 (2004).

[59] M. Aella, E. Martuscelli, C. Sellitti and Garagnani, *J. Mater.. Sci.* 22, 3185 (1987).

[60] A. Grozdanov, A. Burzarovska, G. Bogoeva-Gaceva, M. Avella, M. E. Errico and G. Gentile, *Polym. Eng. and Sci.* 5, 47 (2007).

[61] KTH *Fibre and Polymr Technol "Mophology of semicrystalline polymers"* Chapter 6, P186, accessed July 1, 2009: http://www.polymer.kth.se/forskarutbildning/doktorandkurser/Chapter%206-1.pdf, http://www.polymer.kth.se/forskarutbildning/doktorandkurser/Chapter%206-3.pdf

[62] S. C. Tjong, R. K. Li and T. Cheung, *Polym. Eng. and Sci.* 37, 166 (2007).

In: Polypropylene
Editors: L. P. Silva and E. F. Barbosa

ISBN: 978-1-62417-142-0

Chapter 10

A REVIEW OF MICROENCAPSULATION OF FLAME RETARDANT FORMULATIONS SUITABLE FOR APPLICATION IN POLYPROPYLENE TEXTILE SUBSTRATES

Fabien Salaün[1,2*], Stéphane Giraud[1,2], Isabelle Vroman[3] and François Rault[1,2]

[1] Univ Lille Nord de France, Lille, France
[2] ENSAIT, GEMTEX, Roubaix, France
[3] Groupe de Recherche En Sciences Pour l'Ingénieur (GRESPI), Laboratoire d'Etudes des Matériaux Polymères d'Emballage (LEMPE), Ecole Supérieure d'Ingénieurs en Emballage et Conditionnement (ESIEC), Esplanade Roland Garros - Pôle Henri Farman, France

ABSTRACT

The thermal performance of protective clothing is primarily based on the thermophysical properties of the materials and the type of textile construction used to make the clothing. The improvement of this thermal performance can be obtained through different ways, and one of the novel approaches presented in this review uses flame-retardant microparticles. Microencapsulation of flame retardant formulations has attracted considerable attention in recent years, since they are more environmentally friendly than the traditional halogen-containing flame retardant. Among the principal types of flame retardant systems indentified for inclusion or padding in PP fibers or textile structures, phosphate or polyphosphate may cause deterioration in the mechanical properties of the polymer and in some cases, it has a tendency to exude through migration to the surface, and therefore leads to a decrease in the properties of the polymer composites. Moreover, it not only decreases flame/fire retardancy, but also some other properties. These problems can be solved by the technique of microencapsulation, which consists in enclosing a substance inside a miniature capsule. Thus, microencapsulation of

* E-mail address: fabien.salaun@ensait.fr

flame retardant for polypropylene (PP) fibers and/or fabrics as well as the influence of the microcapsules on the polypropylene structure is reviewed.

INTRODUCTION

The world consumption of polyolefins, and more specially polypropylene (PP) has represented more than 50 million tons in 2009, and it is in expansion at an average annual growth rate estimated to be 4% in 2010-2020 [1]. In contrast to other polymers like polystyrene or polyvinyl chloride, it can be denoted that PP is widely used to produce engineering plastics, since its cost is not a function of the increase in price for crude oil and gas. Due to its universal properties, it is allowed in various applications in industrial fields such as electronic and electrical, automobile, pipeline, among others, and since its introduction from the 1950s, the application of PP in textile has increased exponentially [2]. Furthermore, the declining price and the quality improvements of polypropylene resins have contributed to the development of new products with an added value. Polypropylene has replaced other polymers in production of a wide range of fibrous materials as well as in household textiles (carpet backing and yarns, upholstery fabrics, rugs, and others), in health and medicine for care materials or disposable diapers, in protective clothing, in geo and/or agrotextile, nonwovens, and others. Thus, the production of PP fibres represented 10% of the volume of fibers manufactured in 2007 in the world. Even if the consumption of polypropylene textile materials in Europe has decreased about 4.8% from 2008 to 2009, polyolefin fibres represented 44.3% of the consumption of man-made fibre in 2010 mainly in spunbond and meltblown (31.2%), tapes and slit films (21.8%), multifilaments (19.1%) and staple fibres (20.7%).

Among the three polypropylene forms, i.e atactic, syndiotactic and isotactic, the first was synthesized under an unvulcanizable elastomer. In the early 50's, Giulio Natta discovered a catalyst for the family of catalysts known as "Ziegler-Natta" to provide a new polymer class, isotactic polypropylene, characterized by a high mechanical strength, inertia to chemical attack and operating temperatures above 100°C. Nevertheless, the fibres based on this kind of PP had presented some limitations, i.e. a lack of dyeability, poor heat stability with low melting temperature, poor sunlight stability, but with their lower prices and greater versatility, they have been introduced to the U.S. textile industry from 1960s. Furthermore, PP fibres have been obtained from Ziegler Natta catalyst for over 30 years; and in the 1970s the improvement of extrusion and manufactured technology led to an increase of fibrous product uses. Most of these drawbacks, except the low temperature and the dyeability, are now overcome. Since the 1990s, the improvement of control over the molecular architecture with the introduction of metallocene catalyst led to the improvement of the manufacturing of polypropylene fibers and the properties of the fibers, and also commercially isotactic polypropylene with various levels of tacticity are now becoming available. Even if, syndiotactic polypropylene, was synthesized in the 1960s by Natta and coworkers, the development of metallocene catalyst has allowed producing highly stereoregular compounds with a wide range of chain structure, morphologies and new properties, i.e. toughness and elastic properties, compared to isotactic polypropylene in recent years.

Enhancing flame retardancy of materials, taking in account the new regulations and environmental concerns, increases the interest in research for new solution in flame-retardant treatments and in particular in intumescent systems [3]. In fact, one of the main drawbacks or disadvantages of the use of PP fibres is that it is flammable because of its wholly aliphatic hydrocarbon structure, and most of the fire retardant treatments have relatively limited durability and need high add-on. Thus, this combustible material ignites spontaneously at about 360°C and can be ignited at around 345°C. The burning produces droplets that can spread the fire, thus one of the essential primary safety considerations is to reduce the PP flammability. Furthermore, polypropylene has a high heat of combustion, about 46.4 kJ/g [4] and this polymer is a low char-forming material with an oxygen index about 18%. Its ignition time is relatively lower than for some other thermoplastics, and even if it can be classified in the category of polymers with low smoke emission, but it should be taken into account that it is accompanied by a high peak of heat release during the rapid production of smoke, representing a hazard in real situation [5].

Most of the main methods used to decrease the combustibility or to control the flammability of PP are the addition of flame or fire retardant materials (FR) to the PP matrix by blending or by finishing treatment on the surface. These compounds can act at different stages of the combustion process and to enhance their activities and are mainly employed in combination either to decrease the cost or to optimize the properties [6]. Therefore, flame retardant compounds can act either in the condensed phase to interfere with the combustion process by an endothermic degradation, e.g. hydrates of aluminum and magnesium; by dilution, e.g. inert fillers as talc, $CaCO_3$, clays; by creating a thermal shield with an intumescent system; or in the gas phase by radical inhibition with halogenated or phosphorus compounds, or by dilution to decrease the flame temperature. Irrespective of the species used for fire retardant application in polypropylene, it should be thermally stable during the manufacturing process, and compatible with the PP matrix. Therefore, it should have no migratory or leaching properties and along with this, it can keep intact its flame retardant property. Moreover, a low loading content (less than 10%-wt) is needed to keep good mechanical properties of the textile, and to reduce to the smoke toxicity.

Among the FR formulations presented above the intumescent flame retardant system has aroused a great attention in the last years as an alternative solution to improve polypropylene's fire resistance. Indeed, this class of formulation is not only more environmentally friendly than the other traditional halogen-containing flame retardant but it shows also higher efficiency. However, since a part of component used in these formulations is water-soluble, an encapsulation step is necessary to overcome this problem [7]. In the present review, the authors concentrate this chapter on the presently used and developing microencapsulation technologies for intumescent flame retardant.

Classification of Flame Retardant Compound For Potential Uses in Polypropylene

In order to be used in textile application, flame retardant compounds must have specific requirements. Indeed, they should be easily processable and would act as safeguard for the physical, mechanical and aesthetic properties of the textile substrate. Thus, if they are incorporated into the fibers, they should be thermally stable up to the temperature of the spinning process, compatible with the matrix to avoid any leaching or migratory phenomena, and also their loading content should be less than 10 wt-% to minimize their effect on the thermo-mechanical properties of the fibers. They should act at a lower temperature than the degradation temperature of the PP and should reduce the toxicity of gas and smoke produced during burning to an acceptable level, and the coloration aspect should also be taken into account. It can be noticed that the research efforts to flame retardant additives introducing by melt spinning in polypropylene have been described in details by Zhang and Horrocks [8], and their commercial use by Weil [9].

Although PP is broadly used in many fields, this polymer presents several difficulties in flame retardancy because of its aliphatic hydrocarbon structure and high crystallinity. Furthermore flame retardancy properties of the PP fabrics can be achieved either by incorporating flame retardant compounds during the spinning process, or by finishing and back-coating formulation. According to Horrocks, some functional treatments are as durable as the fiber structure itself with varying and limited success [10]. In most of the cases, flame retardant additives may have a mode of action in gas phase to extinguish the flame by radical inhibition or dilution, that leading to the decrease of the flame temperature or to reduce the availability of flammable gases and oxygen; and/or in condensed phase, in the vicinity of the flame to perturb the thermal degradation of the substrate by dehydration and char formation. Therefore, according to the chemical nature and the stage of the combustion process acting, the action modes can be classified into five categories, i.e. gas dilution involving the use of additives which decompose into non-flammable gases and lead to reduction of both fuel and oxygen levels; thermal quenching in which flame retardants reduce the rate of burning by endothermic decomposition; protective coating when the additives induce the formation of a barrier or a thermal shield; physical dilution with the incorporation of an inert and non-flammable component; and chemical interaction, when the flame retardant additive decomposes into radical species in competition with the burning process. The flame retardant compounds can be classified either according to their chemical structure or their mode of action to provide fire resistance.

Halogen-Containing Flame Retardants

The halogen-containing flame retardants, used in most of cases in combination with metal compounds, show a chemical action in gas phase, even if the condensed phase reactions with the polymer could be involved. These compounds are based on chlorine and bromine to react with flammable gases to significantly reduce the combustion rate by radical transfer. Thus, during the first step of the reaction, the halogen compounds interrupt the chain reaction of combustion by the release of halogen hydracids which transform the highly reactive radicals, $OH^•$ and $H^•$ by less reactive radicals. Thus, the effectiveness of halogen compounds depends mainly on the ease of release the halogen and on the nature of the group containing the halogen. To achieve good levels of flame retardancy, high concentrations, 40%-wt for chloride and 20 %-wt for bromide, are required [11]. Therefore to reduce the halogen concentration and to enhance significantly flame retardant properties, several metal compounds such as metal oxide, could be used as synergists. Amongst these synergistic agents, antimony trioxide plays the role of an inhibitor in the gas phase, others common synergists are phosphorus, nitrogen, zinc and tin-containing compounds. Nevertheless, these compounds are less and less used since they release during their degradation toxic gases and corrosives as dioxins, benzofurans, HCl and HBr and they also present some environmental health hazards.

Inorganic Compounds

Inorganic compounds, as hydrates of aluminium and magnesium based on endothermic additives are used to slow down the rate of pyrolysis. They have rather low prices, low toxicities but high loading content is required to provide the fire performances of interest [8-9; 12-13], and therefore it results in the decrease of the mechanical properties of the polymer. This class of compounds acts simultaneously in gas and condensed phase, when their endothermic decomposition induces the water release to cool down the system and to dilute fuels in the flame zone. Aluminium and magnesium hydroxide, which decompose thermally in the temperature range of 180-240°C and 330-360°C, respectively [14], are widely used as halogen free alternative to brominated flame retardants, as well as inorganic phosphorus. This inorganic group, containing also boron salts, inorganic antimony, tin, zinc and red phosphorus, represent around 50% by volume of the global flame retardant production. Rault et al. investigated the introduction of manganese oxide and manganese oxalate in PP fibers with a loading content about 10%-wt [15]. Even if, they observed an improvement of thermal stability of their knitted fabrics, the fire properties were not improved.

Phosphorus-Containing Flame Retardants

Phosphorus-containing flame retardants, inorganic or organic, mainly react in the condensed phase to influence pyrolysis and char formation, and can be used as the acid-generating part of intumescent system. They are sometimes associated to halogen or nitrogen-containing compounds, or contain bromide and chloride to enhance the flame retardant effect. In the condensed phase, they are converted by thermal decomposition to phosphoric acid which leads to the dehydration of organic compound to promote carbonaceous layer formation. The most used phosphorus compounds as flame retardants in polypropylene are halogen phosphate, phosphine oxide and ammonium phosphate or polyphosphate, phosphate, phosphonate derivatives, organic phosphinate. Among the commercially available flame retardant additives for melt spinning polypropylene, Tris(tribromoneopentyl) phosphate (ICL's FR-370®) is one of the most promising compounds and can be used in combination of a free radical generator.

Nanocomposites

Nanocomposite fibers have attracted considerable attention, since it is suggested that the presence of clay or poly(vinylsilsesquioxane) particles in polymer, and more specially in polypropylene can enhance the formation of a protective layer and slow down the combustion of the polypropylene matrix. Thus, Zhang et al. reported that the flammability of PP can be reduced by the introduction of clay with phosphorus compounds and a hindered amine for a loading content of 6 %-wt [16]. Nevertheless, the burning behaviour of PP fabrics containing only montmorillonite clays is modified and encouraged some char formation without conferring flame retardancy properties [17]. Similar conclusions were drawn by Rault et al. [18]. Indeed the authors showed that the thermal stability of PP fibers can be enhanced by addition of graphite, montmorillonite clays or combination of both without conferring the fire properties. However, the use of clay for developing effective flame retardant system for polypropylene continues actually to draw attention of the scientific community. For example, it has been reported a reduction of 24% on the maximum Heat Release Rate (HRR) of knitted fabric made with polypropylene multifilament yarn filled with a very low amount of fiber-like clay (sepiolite) [19]. About the use of polysilsesquioxanes in PP, the latter appears not sufficient to obtain performance required in several fields of application, even if, the authors have observed some improvements in fire behaviour [20-21]. Apart from clays and POSS other nanoparticles have been studied from both academic and industrial point of view. This is particularly the case of carbon nanotubes. Thus, Bellayer et al. showed that the introduction of multiwalled carbon nanotubes allows to increase the thermal stability and to decrease of 50% the HRR peak of PP knitted fabrics [22]. The same phenomenon has been highlighted by Vargas et al. with polypropylene knitted fabrics containing carbon nanotubes, without providing nevertheless a decrease of HRR [19].

INTUMESCENT SYSTEMS

Among the flame retardants for polypropylene, intumescent flame retardant (IFR) has been very carefully studied in recent years, since they represent an environmental friendly alternative to the traditional halogen-containing compounds. Even if intumescent systems have existed since the 1940s in paint industry, they have only appeared in the textile field market for about 30 years. Furthermore, IFR presents many advantages compared to halogen-containing flame retardants, such as less toxicity, less acidic fumes during the burning process [23]. Flame retarding textile substrate by intumescence process has been known for several years and is essentially a special case of a condensed phase mechanism [24-30]. Intumescent systems interrupt the self-sustained combustion of the polypropylene at the thermal degradation with evolution of gaseous fuels. And thus, the intumescence process results from a combination of charring and foaming of the surface, limiting heat and mass transfer between the gas and the condensed phases and the flame and the substrate [26]. Furthermore, the layer inhibits the evolution of volatile fuels via an entrapment process, which limits the oxygen diffusion to the polymer bulk [24]. Intumescent formulations contain three kind of active additives [31-32], i.e. an acid source or precursor for catalytic acidic species which can be an acid (phosphoric, sulphuric, boric), ammonium salts (phosphates or polyphosphate, sulfates, halides), a phosphate of amine or amide (urea, melamine phosphate,...) or an organophosphorus compound as tricresyl phosphate, alkyl phosphate or haloalkyl phosphate; a carbonic or polyhydric source such as starch, dextrin, sorbitol, pentaerythritol or methylol melamine; and a blowing or spumific agent like urea, urea-formaldehyde resin, dicyandiamide, melamine or polyamide. Contrary to others polymers having intrinsic charring behaviour, the presence of charring agent in IFR system is required for PP. The most widely reported IFR for replacing halogen-containing agent is phosphorus-nitrogen containing compounds. Thus, the IFR formulations for PP include generally these kinds of system, i.e. ammonium polyphosphate – melamine- pentaerythritol [33]; ammonium polyphosphate – triazine compound, mixture of melamine phosphate with pentaerythritol [34] or their reaction products and these derivative compounds [35-36] in addition with a charring polymer [37-38]. However, these conventional intumescent flame retardant systems need to be improved and especially by the addition of synergistic agent such as zeolithe [39-41], montmorillonite [42-43] or sepiolite [40], silicotungistic acid [44], metal oxide [45-48], silica or alumina [49], or in working condition to adjust the relative ratio of each component [50]. Furthermore, the high loading content of IFR additive leads to a decrease in the mechanical properties of the polymer, and most of the time these compounds being highly polar induce the problem of water leaching. Several authors reported the possibility to decrease the polarity and therefore to improve the compatibility with polypropylene by using reactive extrusion to synthesize melamine salt of pentaerythritol phosphate [51] [52].

Unfortunately, conventional IFR systems have presented some drawbacks. Acid sources and charring agents are globally hydrophilic chemicals. They have some problems such as moisture sensitivity and poor compatibility in polymer matrix in particularly in non-polar polymer like PP. Migration phenomenon could occur during the processing, the application or the ageing of the material. Thus, exudation and high water solubility are encountered problems which lead to the loss of FR properties [26]. Moreover conventional IFR systems show efficiency only from 20 %-wt loading. The minimum loading of IFR systems and the poor compatibility of their components induce homogeneity problems; create damage of the surface quality and decrease of mechanical properties for the final material. Finally, the necessity to have the three components of IFR system is not convenient for the industrial implement of final PP materials. These drawbacks make impossible the introduction of conventional IFR systems in the core of PP fiber by melt spinning process. Indeed, it is well-known that moreover the high level of fillers is incompatible and induces some problems during the drawing step. Thereby, the coating process still needs a possible way to apply IFR formulation on PP fabrics. However the lost of handle is one of disadvantage of this process without to solve problems of IFR systems.

The concept of microencapsulation is a way to solve the different drawbacks of conventional IFR systems. Microencapsulation is a process of enveloping microscopic amounts of matter (solid particles, droplets of liquids or gas bubbles) in a thin film of polymer which forms a solid wall (microcapsule) or inside a polymeric matrix (microsphere). These structures allow the isolation of the encapsulated substance from the immediate surroundings and thus protect it from any degrading factors such as water. Also the microencapsulation processes allow encapsulating acid source of IFR system and protects it from any degrading factors such as moisture. The polymeric membrane of microcapsule permits also to improve the compatibility of acid source in the materials and thus to keep FR and mechanical properties. Moreover, the polymeric matrix or membrane of microcapsules should be appropriately chosen in order to have the chemical characteristics of charring agent and/or blowing agent (polyurethane, aminoplast resin, isocyanurate). Also, the combination of the encapsulated acid source and the microcapsule membrane with charring and/or blowing properties form an intrinsic IFR formulation. The concept to have "all in one system" makes easier the IFR incorporation system in polymeric material such as PP.

Microencapsulation of Intumescent Flame Retardant

For the last ten years, the market and the needs for technical or smart textiles have been constantly growing, that leading to more technical developments and innovations of advanced textiles for new functionalities. In the same way, the formulation and understanding of polymeric surfaces have progressed tremendously to obtain systems with well-defined functionality. Microencapsulation represents not only one possible way to incorporate new functionalities to textile, but also protect the active substance from the environment [53-56]. So far, these microcapsules have been mostly applied to achieve the functional finishes, which have relatively short lifetime. For a textile use, FR compounds should be surrounded by a protective shell to avoid any liquid migration during the manufacture process and therefore to preserve FR in the polymer shell as long as possible through the ageing or the textile use. Besides, the encapsulation step allows increasing the heat transfer area, to decrease the reactivity of the core materials, to limit the interference with other materials parameters, to enhance the low heat conductivity and to make the manipulation of IFR easier. Microencapsulated FRs have attracted more and more attention since the end of the 1990s, and at the initial phase the materials were used in order to overcome the modification of the textile properties such as softness and drape [57], to avoid undesirable properties, i.e. their chemical activity, volatility or migration to the polymer surface [58], and afterwards the application was mostly focused to improve their compatibility with the polymer matrix [59-61]. The choice of the polymer for the membrane synthesis of the microcapsules is linked to the considered application and the required material processes. In the textile field, the polymers used should have good thermo-mechanical properties to resist to the thermal and mechanical requests during the implementation processes, and also, during the daily use by the wearer. Thus, the selection of an appropriate shell not only prevents contact of the IFR with the surrounding medium, but can also increase its efficiency. Moreover, this encapsulation step reduces the moisture sensitivity of the active substance by decreasing the water absorption and therefore should increase its water resistance in a polymeric matrix [62-65]. The use of microencapsulated FR or IFR in textile requires a high thermal stability of these particles that is influenced by the shell and the core materials. A review on the various techniques available for the microencapsulation of flame retardant compounds is presented. The methods can be classified as being either physical or chemical based.

Principle and Generalities

Microencapsulation refers to the formation of polymeric particles with a range of diameters from nanometers to millimeters, and which exhibits several kinds of morphologies, in which the mains are so called microcapsules and microspheres [66]. Microspheres consist of polymeric network structure in which an active substance is enclosed, whereas in microcapsules or core/shell structures exhibit a reservoir structure, i.e. the core substance is surrounded by a polymeric layer. Even if, microencapsulation principle was firstly introduced by Green and Schleicher in the 1950s to create microcapsules for carbonless copying paper (Green & Schleicher, 1956 a & b, as cited in [67]), textile applications were mainly developed in laboratories since the early 1990's and actually we can see much more applications in the textile industry for developing new properties and added values [54].

Although, there are more than 200 microencapsulation methods described in scientific literature and patent, most of them include three basic steps, i.e. enclosure of core component, formation of the microparticles, and hardening of it. These methods are generally divided or classified in three main groups, which are based on the mechanisms of microparticle formation, such as mechanical, chemical and physical-chemical processes. The choice of a method rather than another is based on the cost of processing, the use of organic solvents for health and environment considerations. Nevertheless, the microencapsulation processes used in textile industry are generally focused on chemical and physical-chemical methods, because mechanical processes lead to the formation of particles with a mean diameter higher than 100 μm. Even if microparticles with mean size distribution in the range of 20 to 40 μm should be useful for textile application [68], very small microparticles (mean diameter ranging from 1 to 10 μm) are preferably incorporated within textile fibers [69-70] but according to Colvin [71] particles with a mean diameter less than 1 μm are suitable for fiber fabrication; and larger microcapsules (mean diameter up to 100 μm) should be used into foams or coated formulation [72-73].

Therefore, the most suitable methods used to encapsulate FR include interfacial, *in situ* and suspension or dispersion polymerization methods for chemical processes; and phase coacervation and solvent evaporation for physical-chemical processes, and supercritical fluids for mechanical processes. In all cases, the microencapsulation process includes two main steps, e.g. emulsification step which determines the size and the size distribution of the microcapsules; and the formation of the capsules. The emulsification step may be influenced at once by physical parameters such as apparatus configuration, stirring rate and volume ratio of the two phases, and by physicochemical properties such as interfacial tension, viscosities, densities and the chemical compositions of the two phases. The formation of microcapsules is greatly affected by the surfactant, which influences not only the mean diameter but also the stability of the dispersion. The surfactants used in the system have two roles, one to reduce the interfacial tension between oil and aqueous phases allowing formation of smaller microcapsules and one to prevent coalescence by its adsorption on the oil-water interface and therefore by forming a layer around the oil droplets. The second step, or the synthesis of a core/shell particle or other possible morphologies, is mainly governed by the kinetic factors, i.e. the ability of the monomers, pre-polymer or polymer to react or to cross-link, and thermodynamic factors, i.e. the minimum total free energy exchange in the system.

STATE OF THE ART IN MICROENCAPSULATED IFR'S

The microencapsulation of IFR has been few investigated until the late 1990s to provide significantly enhanced thermoprotection properties and flammability of polyolefins for various application fields. Nevertheless, preliminary works on microencapsulation of FRs and more mainly for paints applications started from the 1970s. The aim was to improve the flame retardancy of materials by the use of flame retardant additives, but the addition of theses additives has a detrimental effect upon the properties of the substrate. Thus, the microencapsulation has been suggested as a mean for overcoming most of the problems, i.e.

encapsulating flame retardant compound within a substantially impermeable shell should prevent the migration and/or volatilization of the active substance, and reduce the potential toxicity. The incorporation of microcapsules containing flame retardant and having diameters of 20 microns into various polymeric materials such as polystyrene, polyurethane and polyepoxides, has been cited in the US Patent n° 3,660,321 [74]. Eichel also described the use of microcapsule containing gas-producing agents [75]. Polyurethane microcapsules having an average diameter below 5 microns and containing an halogenated phosphate had been incorporated into polyurethane foam [76-77].

The use of melamine formaldehyde resin and styrene derivatives to entrap flame retardant compound like hexabromocyclododecane, was patented under US Patent n° 5,043,218 [78]. European patent n° 180795 disclosed flame retardant agents comprising ammonium polyphosphate microencapsulated within a melamine formaldehyde resin [79]. The invention patent n° EP0542359 related to ammonium polyphosphate microencapsulated with aminoplastic resins was obtained by polymerizing with aldehydes, preferably formaldehyde, derivatives of 2,4,6-triamino-1,3,5-triazine [80-81]. The stability of flame retardant to hydrolysis is an another aspect described in the literature patent and more specially with the EP 0131097 and US Patent n° 4,639,331 and n° 5,599,626 which disclosed the formation of microencapsulated ammonium polyphosphate and their use in polyurethane foams [82-84]. In 1996, Bolid GmbH (Germany) commercialized microencapsulated in a polyurethane sheating APP-FR CROS-484 EC for fire protection of various polymers, including polyolefins. During the same period, Zubkova *et al.* published a series of papers about the interest and the use of microencapsulated FR [59-61]. The various processes used in these studies were based on chemical or physico-chemical methods, that indicating the coating or phase polymer separation to coat the active substance, a binary mixture of diamide of methylphosphonic acid and ammonium chloride. Furthermore, the authors recommended the use of polyvinyltriethoxysilane and poly-p-xylene to hydrophobize the flame retardant core.

From the beginning of the 21st century, one of the strategies developed to synthesize new intumescent flame retardant for textile application is based on the formulation of intumescent microcapsules. The approach developed by Giraud *et al.* is to microencapsulated phosphate salt with polyurea-urethane membrane [85-87]. Thus, two microencapsulation methods were established, i.e. interfacial polymerization and solvent evaporation, to succeed in. The first process provides microcapsules with a low mean diameter (<5 μm) and very low encapsulation yield, whereas in the second process the yield was improved to 50% and the microcapsules diameter reaches 100 μm even if small particles about 10 μm should be obtained. And, in 2009, a weight to weight ratio of ammonium polyphosphate particles microencapsulated within a melamine or melamine-based resin from 3:1 to 1:3 was patented [88].

The microencapsulation of IFRs provides some beneficial effect on the physical-chemical properties of the compounds, since the improvement of the coating durability reduces the solubility; modifies polarity. Nevertheless the first researches denote that the preparation of the microencapsulated agents is not an easy process and different techniques are being developed in order to simply the synthesis and to improve yields. The product is always in a dried state making easier its incorporation into polymer matrix. The microcapsules embedment is relatively simple, and the various IFRs formulation with different acid sources or charring compounds should be used to obtain the formation of the thermal shield at the desired temperature when degradation occurs. Furthermore, the use of microcapsules increases the specific exchange area, which is provided for effective heat transfer. Therefore, to maintain the core compounds inert, they are encapsulated within the impermeable microcapsules walls for the whole product life. The microcapsules need to be resistant to mechanical and thermal stresses.

Microencapsulation of flame retardants is now at the research stage. The significant development of research works described in the next part shows that they will lead to performance apparel offering enhanced protection against heat and flame. At the present, the microencapsulation technologies in the industrial applications of inorganic flame retardant are mainly focused on the use of red phosphorus.

MICROENCAPSULATION METHODS USED FOR FRS

One of the most typical methods which may be used to microencapsulate an IFR is to disperse droplets of this material in a non miscible solution and to form walls around the droplets. A literature survey on microencapsulated IFRs indicates that the most suitable methods include interfacial, *in-situ* and suspension polymerization methods for chemical processes and phase coacervation or and solvent evaporation for physical-chemical processes and supercritical fluids for the mechanical processes.

The choice of the shell materials has been widely investigated by several researchers (Table 1) to enhance the structural anti-permeability or to bring up a solid protection and to improve the thermal stability of these particles. The selection of the microencapsulation process is mainly governed by the choice of the chemical nature of shell materials and by the desired mean diameter. Thus, the *in situ* polymerization is suitable for the melamine derivatives, the interfacial polymerizations are used to obtain polyurea and polyurethane walls, whereas acrylates are obtained from dispersion polymerization reactions, respectively. Besides, microencapsulation of IFR is essentially restricted to phosphate derivatives, which are moisture or water sensitive materials, whereas the process technology allows obtaining a polymeric shell sufficiently tightened to prevent the small water molecules diffusion.

Table 1. Shell composition and some properties of microcapsules suitable for thermal retardancy

Microencapsulation process	Microcapsule shell	core material	mean size µm	loading core content wt-%	reference
Interfacial polymerization	poly(ether-urethane-urea)	DAHP	~1	3	[85]
	poly(urethane-urea)	DAHP	13.3	62	[89]
	polyurethane	APP	---	---	[90]
	polyurethane	APP	---	---	[91]
	cellulose acetate butyrate - TDI	APP	<10	n.a	[92]

Table 1. (Continued)

Microencapsulation process	Microcapsule shell	core material	mean size µm	loading core content wt-%	reference
	polyurea	Chlorocyclo-phosphazene	---	---	[35]
Coacervation	gelatin-polyurea	DAHP	5.5	74	[93]
	poly(vinyl alcohol) -polyurethane	DAHP	3.8	63.9	[93]
	poly(vinyl alcohol) -glutaraldehyde	DMMP	5.3-12.5	n.a	[94]
In situ polymerization	melamine formaldehyde	DAHP	3-5	26.4	[95]
	melamine formaldehyde	DAHP-PHMA	3-5	21.2-6.8	[95]
	melamine-formaldehyde	APP-DPER	n.a	n.a	[96]
	melamine-formaldehyde	APP	2.8	90	[65]
	melamine formaldehyde	APP	n.a	89.1	[97]
	melamine-formaldehyde	DBDPO	4.5	7.5(**)	[98]
	melamine-formaldehyde	RP	n.a	n.a	[99]
	melamine-formaldehyde	RP-MMT	n.a	n.a	[99]
	melamine-cyanurate	RP	~1	35-80	[100]
	melamine-urea-formaldehyde	PEPA	2-5	n.a	[101]
	urea-melamine-formaldehyde	APP	15	91.6	[63]
	melamine-formaldehyde & urea-formaldehyde	APP	24.6	62.9	[64]
	PEG-melamine-formaldehyde	APP	10.7	68.4	[102]
	Starch-melamine-formaldehyde	APP	13.5	75.2	[103]
	PVA-melamine-formaldehyde	APP	20.0	65.2	[62]
	Silica	APP	n.a	1.84(*)	[104]
	Silica	APP	n.a	n.a	[105]
Solvent evaporation	poly(ester-urethane-urea)	DAHP	20-100	1	[87]
Dispersion polymerization	poly(pentabromobenzyl acrylate)	PBBA	0.06-1.7	70(**)	[106]
Supercritical fluids	paraffin	RP	50-65	n.a	[107]

(*): phophorus atom; (**): bromine atom

APP: ammonium polyphosphate; DAHP: di-ammonium hydrogen phosphate; DBDPO:dedabromodiphenyl ether; DMMP: dimethyl methyl phosphorate; DPER: dipentaerythritol; MMT: montmorillonite; PBBA: pentabromobenzyl acrylate; PEPA:1-Oxo-4-methoxy-2,6,7-trioxa-1-phosphabiccyclo[2,2,2]octane; PHMA: poly(1,6-hexamethylene adipate glycol); RP: red phosphorus; TDI : toluene-2,4-diisocyanate

INTERFACIAL POLYMERIZATION

The microencapsulation of flame retardant compounds by interfacial polycondensation or polymerization involves the dispersion of the liposoluble phase in an aqueous continuous phase, or an aqueous phase in an organic phase depending on the solubility medium of the FR, and also to induce the precipitation of the polymeric materials at the droplet interface. Each phase contains a dissolved specific monomer suitable to react with the other present in the other phase. The dispersed phase acts as a good solvent for the monomers but also serves as a non-solvent for the produced polymer. Therefore during polymerization, the system is composed of three mutually immiscible phases. Once, the various monomers were added to the system, the reactions occur at the interface resulting in the formation of insoluble oligomers in the droplet with the tendency to precipitate at the interface to form primary shell around the droplets. Further crosslinking reactions lead to the formation of growing shell by diffusion of monomers. When multifunctional monomers are used, a three-dimensional cross-linked system is obtained. Nevertheless, the thickness of the shell increases slowly due to the restricted diffusion of the starting monomers inducing the changes in layer morphology. The porosity of the capsules increases with the precipitation of the polymer at the interface and the solvent molecule diffusion through the wall. However, the main drawback of this method is the use of organic solvent, e.g. toluene or ethyl acetate, to solubilize both core material and reactive liposoluble monomer. Based on this method, polyurea, polyurethane microcapsules shells should be prepared from isocyanate and diols or diamine, the reaction of isocyanate groups with water molecules was promoted during the shell formation process. Furthermore, Saihi et al. [89] have observed that the formation of a polyurethane shell should be also provided from the reaction of the isocyanate with the hydroxyl groupments of the surfactant.

The formation of the membrane around the droplets is governed by the swelling power of the both phase, a porous structure may be obtained when the solvency of the medium for the polymer is low [108]. Furthermore, the morphology of the particles is influenced by the rate of precipitation of the polymer, and a higher rate promotes the formation of less uniform and more porous membrane. Therefore, the knowledge of the mechanism of the shell formation and polycondensation reaction which occur during the synthesis is necessary to control the shape and the morphology of the particles [109-110]. Various studies have shown that the polycondensation reactions mainly occur in the organic phase [111]. The mechanism of the polycondensation of the diisocyanate at the droplet interface can be illustrated through the following reaction schemes (Figure 1). In the first reaction, diisocyanate or isocyanate end groups react with hydroxyl groups of polyol to form a urethane linkage and therefore the polyurethane chains. In the second reaction, isocyanate monomers can be slowly hydrolyzed at the interface to form an instable amino acid group which dissociates in amine end group and carbon dioxide. This reaction occurring in the oil side of the interface depends mainly on the water molecules diffusion in the medium or through the polymer shell network. The formation and release of CO_2 during the membrane formation contribute to increase the porosity of the particles. The amine end group reacts with the isocyanate end group to form a urea linkage in the third reaction. Other reactions can occur during the shell growth, as the reaction between an isocyanate end group and urethane or urea unit results in the formation of an allophanate or a biuret, respectively, creating interconnections between the polymer chains.

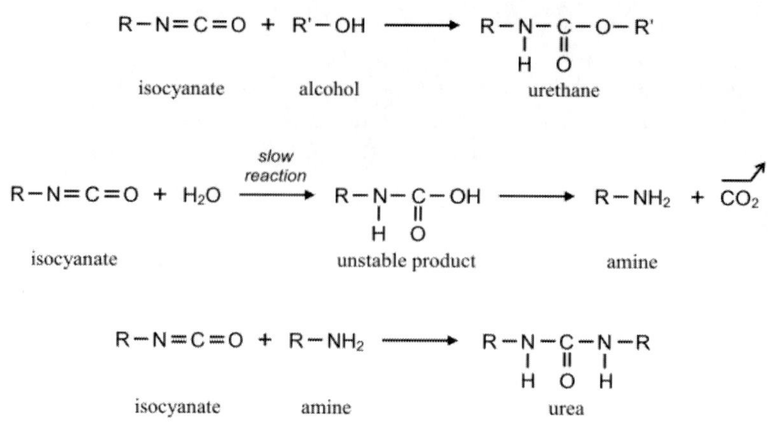

Figure 1. Reactions of isocyanate functions.

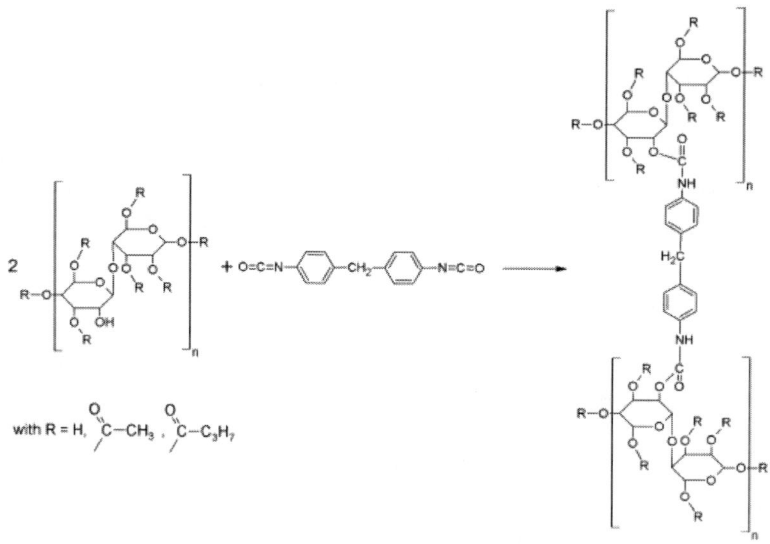

Figure 2. Synthetic scheme for the preparation of a cross-linked CAB/isocyanate microcapsule shell.

In the same way, the polyurethane shell can also be formed by reaction of the hydroxyl groups of cellulose acetate butyrate (CAB) with isocyanate groups at the interface [92] (Figure 2). To enhance the reactivity of the synthesis, isocyanate monomers can be replaced by polyurethane prepolymer, which reduces the porosity of the thicker shell since the isocyanate hydrolysis is limited, allows obtaining a thermoplastic rather than a thermosetting with no covalent bond between the shell and the active substance.

Furthermore, the formation of microcapsules with polyurea or polyurethane shell, suitable to be used as a carbon source, prevents not only the active substance degradation [112], but also enhance the thermal stability and even low loading content due to interactions between the core compounds and the polymeric shell during the degradation steps [7].

Nevertheless, the encapsulation yield is relatively poor, even if the shell layer can effectively protect the phosphate derivative from water.

PHASE COACERVATION

Phase coacervation is one of the oldest and most widely used techniques of microencapsulation, and can be divided into two groups, i.e. simple coacervation which implies the use of one colloidal solute such as gelatine, and complex coacervation, in which the aqueous polymeric solution is prepared from the interaction of two oppositely charged colloids, such as gelatine-arabic gum. Coacervation can be defined as the separation of a macromolecular solution into two immiscible liquid phases, i.e. a dense coacervate phase and a dilute equilibrium phase. The general outline of this method consists in three consecutive steps carried out under agitation Figure 3.

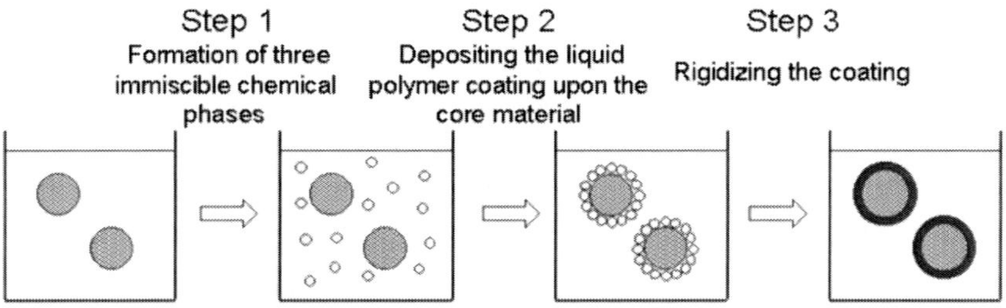

Figure 3. Schematic representation of the various stages of the microencapsulation process by the coacervation method.

Firstly, the water soluble core material is emulsified into organic solvent, e.g. toluene or cyclohexane. Secondly, microcapsules are obtained by dispersing the previous emulsion into the hydrophilic polymer solution, e.g. gelatine or PVA. The deposition of the coating material is promoted by a reduction in the total free interfacial energy of the system, followed by the decrease of surface area of the coating material during coalescence of the liquid polymer droplets. And thirdly, solid particles are obtained by increasing the rigidity of the coating material by thermal, cross linking or desolvation techniques. The size of particle depends on various parameters, such as the concentration of hydrophilic polymer in the aqueous medium, the content of surfactant, and the content of crosslinking agent. Saihi et al. have observed that the use of PVA rather than gelatin allowed to decrease the mean diameter, but the loading content in DAHP and the encapsulation yield obtained were significantly lower [93]. Furthermore, they used methylene diphenyl diisocyanate (MDI) to crosslink the polymeric shell, which improves thermal stabilization and also the microcapsules exhibit an intumescent behaviour. Lin et al. have studied the microencapsulation of water soluble flame retardant by PVA, but in their study they used glutaraldehyde as crosslinker [94]. They found that the size distribution of microcapsules was mainly affected by the PVA, and surfactant concentrations and by the phase volume ratio. Furthermore, the capsules presented good flame retarding efficiency.

In Situ Polymerization

Until now, melamine-formaldehyde and urea-formaldehyde resins were usually used or selected as microcapsule wall materials for the FR protection and also to improve the thermal stability of these particles. The use of this pre-polymer offers some advantages to succeed in the synthesis of microcapsules due to their high reactivity, and therefore a short reaction time and a controlled amount of polymer are required to form the particles shell with a high loading content.

The *in situ* polymerization microencapsulation process, described in Figure 4, consisted in the following steps: preparation of an oil in water emulsion from the dispersion of an IFR solution in an aqueous continuous phase containing surfactant under high stirring rate; addition of melamine-formaldehyde pre-polymer for shell formation; induction of polycondensation by rising temperature to 60-80°C and/or by adjusting the pH of the emulsion to 4; after attaining desired size distribution, the stirring speed was decreased and polymerization occurred during 1 to 3 hours with continuous agitation according to the temperature chosen; then the solution was cooled to room temperature, and neutralized, free formaldehyde was removed by adding ammonia. The suspension was then filtered, and the microcapsules were washed twice with methanol and distilled water.

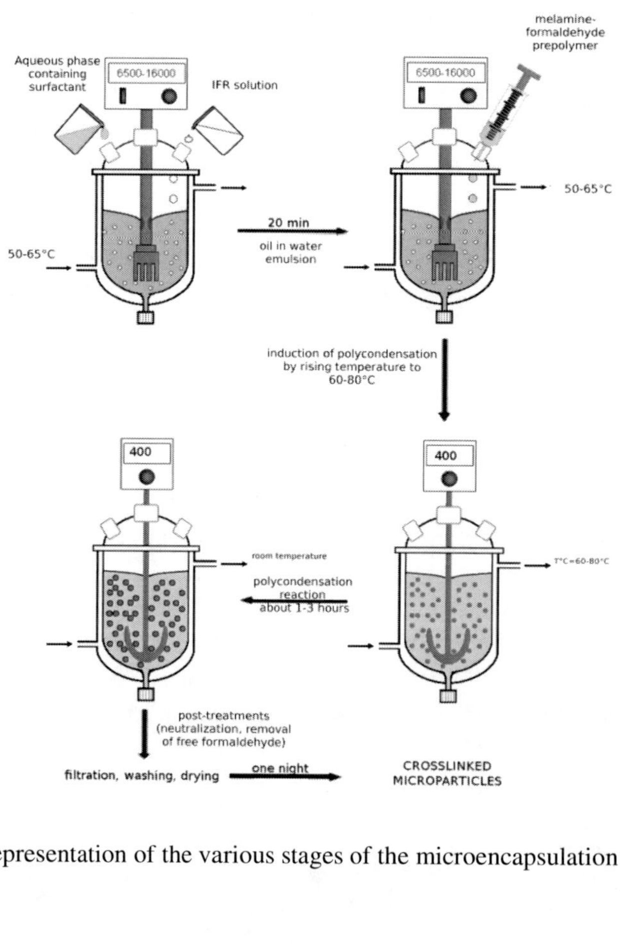

Figure 4. Schematic representation of the various stages of the microencapsulation process by *in situ* polymerization.

This microencapsulation method can start either with monomers, (i.e. melamine and formaldehyde, urea and formaldehyde or urea, melamine and formaldehyde), or with commercial prepolymers such as partially methylated trimethylomelamine, hexamethoxy-methylolmelamine, etc [113]. In all of these cases, the ability of amino resin for self condensation around the core material droplet is linked to its surface activity and this process is governed by an enrichment of resin molecules within the interface. The reaction scheme for the synthesis was given in Figure 5.

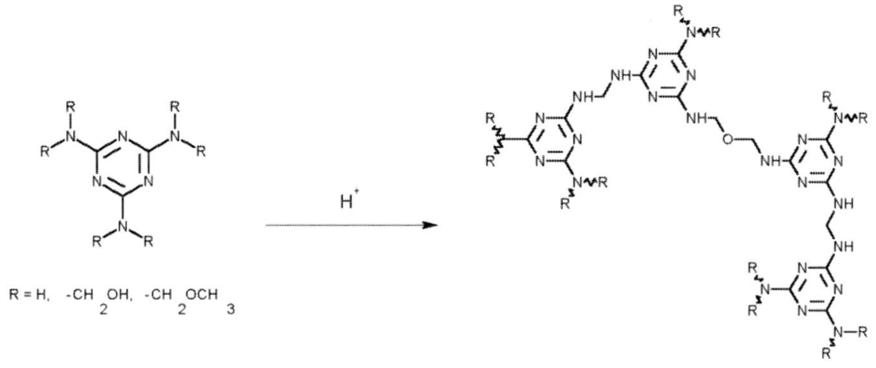

Figure 5. Formation of cross-linked amino resin shell.

The formation of the microcapsule shell described in the literature shows that it occurs in three consecutive stages [114]. In the first step, the melamine-formaldehyde pre-polymers obtain H^+ by changing into active pre-polymers under acidic condition, which results in an increase of the etherification degree due to a reduction in hydroxyl group concentration. Then, the water solubility of the pre-polymer decreases and the material separates from the continuous phase as a liquid solution relatively concentrated in active pre-polymer. During the second step, the amino resins wet the IFRs droplets to create primary shell around them. The concentration of resin in the boundary layer is enhanced by the hydrophilic/hydrophobic interactions of the partial methylolated melamine, and therefore it occurs by an enrichment of resin molecules within the interface. Thus, the resin condensation proceeds much faster in the boundary layer than in the volume phase, allowing the formation of tougher capsule walls. From this step, it is possible to recover particle under powder form. Nevertheless the capsule walls are not yet thermo-mechanically stable and not completely hardened, and the particles might have collapsed during drying. In the last step the reaction rate of the polycondensation at the interface is increased with the result of microcapsule shell formation. During the polycondensation process, the particle surface becomes more and more hydrophobic, and according to the pH level, the shell is crosslinked either by ether linkage or methylene bridges as shown in Figure 6. Furthermore, the surface morphology depends mainly on the pH value, thus, a low pH value promotes the formation of methylene bridges affecting the amino resin solubility and therefore to obtain a rough surface. The formation of higher oligomers of melamine-formaldehyde polymer with the formation of methylene bridges can be described as the reaction of bound CH_2OH group with H. However, the presence of free-formaldehyde after forming the shell may cause environmental and health problem; and it is relatively difficult to eliminate all formaldehyde which can be continuously released from the resultant microcapsules. To solve this problem, Li et al.[115] proposed to incorporate formaldehyde once and urea or melamine for three times; on the other hand, Šumiga et al. [113] observed that the use of ammonia during the post-treatment process allowed to decrease formaldehyde content level in the suspension.

a) $-NCH_2OCH_3 + ROH \xrightleftharpoons{H^+} -NCH_2OR + CH_3OH$

b) $-NCH_2OR + HOCH_2N- \xrightleftharpoons{H^+} -NCH_2OCH_2N- + ROH$

c) $-NCH_2OR + HN- \xrightarrow{H^+} -NCH_2N- + ROH$

d) $-NCH_2OR + HOCH_2N- \xrightarrow{H^+} -NCH_2N- + ROH + CH_2O$

e) $-NCH_2OR + ROCH_2N- \xrightarrow{H^+} -NCH_2N- + CH_2(OR)_2$

f) $-NCH_2OH + HN- \xrightarrow{H^+} -NCH_2N- + H_2O$

Figure 6. Crosslinking reaction of melamine resin.

The morphology of melamine-formaldehyde microcapsules is deeply influenced by the pH, the temperature and the core content. The surface of the particles is sometimes found with increased roughness due to the presence of fragments, tiny particles or unreacted amino resin removal during the recovering. Salaün & Vroman in their study observed that the presence of charring polymer during the synthesis limited the interpenetration between the phosphate salt and the amino resin, and therefore the surface appeared smoother [95]. Furthermore, thermal conductivity depends on the chemical structure of the polymeric shell whereas heat capacity is related to the core content of particle.

Ammonium polyphosphate (APP) is a typical acid source used in the intumescent flame retardant formulation [26; 37-38], but unfortunately it is water and moisture sensitive, migrates to the surface, and leads to a decrease of the flame retardancy and mechanical properties of the polymer [116]. Therefore, microencapsulation of APP with water insoluble polymers has been widely studied to deal with these problems and to lead to an increase in its water resistance and limiting oxygen index (LOI) value, for several years. Thus, Wu et al. proposed several amino polymer, i.e. melamine-formaldehyde [65], melamine-formaldehyde and dipentaerythritol [96-97], urea-melamine-formaldehyde [63], PVA-melamine-formaldehyde [62], starch-melamine-formaldehyde [103], urea-formaldehyde and melamine formaldehyde to succeed in the formation of a double shell [64], and melamine-poly(ethylene glycol)-formaldehyde [102]. APP was firstly dispersed in ethanol under stirring, before adding suitable amount of amino pre-polymer. Then, the pH was adjusted to 4-5 with sulphuric acid, and the temperature was kept constant to 80°C for 2h. The particles were recovered by filtration, washing and drier at 105°C. This encapsulation step allows also to decrease the size distribution and therefore it can be expected that microencapsulated APP would have better dispersion in PP than raw APP. This modification in size distribution can be attributed to the fact that since APP is moisture sensitive, the particles tend to agglomerate.

The shell polymer prevents the core from the attack of moisture, and moreover, the lowering of particles size results from the increase in the hydrophobic nature of shell polymer, and thus, the lowest size distribution was established for amino resin modified with PEG and starch which is in correlation with the water solubility. Furthermore, recovered particles have rougher surface than raw APP due to the deposition of melamine-formaldehyde macromolecular chains onto the surface, then the shell appears similar to a multilayer structure.

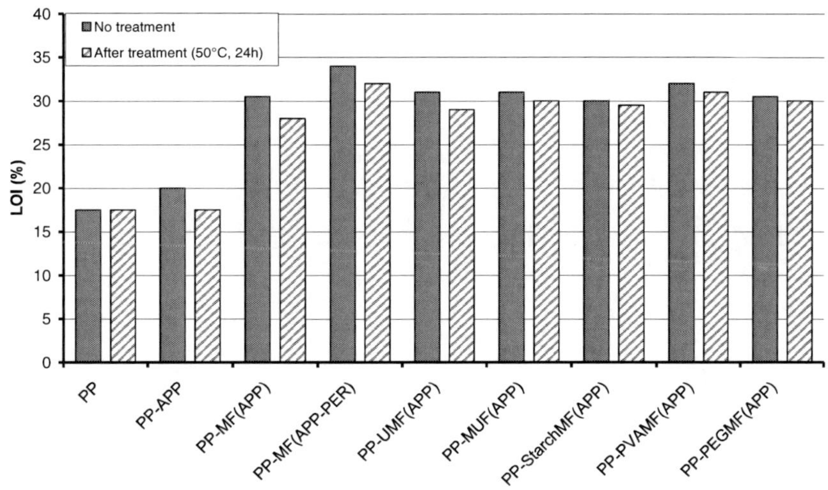

Figure 7. LOI values of PP composites containing 30% of IFR or microencapsulated IFR formulations (PP: polypropylene; APP : ammonium polyphosphate; MF(APP): microencapsulated APP in melamine-formaldehyde resin; UMF(APP) : microencapsulated APP in urea-melamine-formaldehyde resin; MUF(APP) : microencapsulated APP in double shell of urea-formaldehyde and melamine-formaldehyde; StarchMF(APP) : microencapsulated APP in starch-melamine-formaldehyde resin; PVAMF(APP) : microencapsulated APP in PVA-melamine-formaldehyde resin; PEGMF(APP): microencapsulated APP in poly(ethylene glycol)-melamine-formaldehyde resin) [62-65; 96; 102-103].

The influence of the polymeric shell used in the microencapsulation on the LOI values of PP microcapsules composites (loading content about 30%) is shown in Figure 7 From this figure, it can be seen that the encapsulation step allows to increase the LOI value before and after water treatment, since the LOI values of all the samples containing microcapsules are more than that of the sample labelled PP-APP. Moreover, the incorporation of PER in the IFR formulation increases this value, as well as the formation of a double shell, sample labelled PP-MUF(APP). The modification of the amino resin with a carbonization agent, i.e. polysaccharide, poly(vinyl alcohol) or poly(ethylene glycol) decreases the water solubility of these composites and therefore, after water treatment the LOI values are closed to 30 %. Then, it can be expected that the formation of an intumescent char during the LOI test protect the underlying materials, and this protection is more important for PVA compounds due to the formation of ammonia which induces a swelling behaviour of the sample.

Amino resin is also used to coat red phosphorus in presence of surfactant or layered silicate, or with cyanurate to lead the formation of a self assembly aromatic structure [99-100; 117].

Other Microencapsulation Methods

For several years, research on microencapsulation process has been focussed on new processes, either to obtain environmentally friendly products or to develop an easier, cheaper and robuster method, such as the use of supercritical fluids to encapsulate red phosphorus [107]. Goldstein and Margel suggested a new microencapsulation method based on a dispersion polymerization of their flame retardant compound [106]. Furthermore, it was found that hydroxyl silicon can be used in the IFR formulation and improve APP flame retardant properties. Thus, Chen et al. reported the encapsulation of APP with silica shell, which leads to an increase of flame retardance and water resistance [105].

Polypropylene Containing Microencapsulated IFRs

Bulk Polymeric Components

The entrapment of IFR or FR with an amino resin results in a decrease of the polarity of the IFR system, which leads to enhance the dispersion of the active substance in PP matrix. The microencapsulation of IFR decrease their water solubility and water leaching rate, but also enhance their dispersion and compatibility with the PP matrix. Thus, the microencapsulation step has a positive effect on water resistance of the PP composite. The effect of microparticles on the thermal stability of the iPP/microparticles composites can be studied by means of thermogravimetric experiments carried out in oxidative or inert conditions. In oxidative conditions, the thermal behaviour of raw PP displayed in single step degradation, and the decomposition of this matrix characterized by the onset temperature at 5% weight loss, started at 295°C. The main decomposition peak corresponding to the PP backbone decomposition took place at 400°C, and the maximum rate of decomposition was fast; i.e. 10.9 %.min^{-1}; and there is no char residue at over 420°C [118]. The thermal stability of PP samples with microcapsules containing acid source, such as DAHP or APP, is not declined, since the onset degradation. However, the DTG curves presented in the literature show clearly that the composites displayed double step or triple degradation process. The first stage was in the range of 225-420°C with a weight loss depending on the loading content in microcapsules and in which the iPP/microparticles composites present a sharper peak than that of raw PP with a highest decomposition rate. The maximal weight loss points are found lower than that of raw PP. From 420°C, the thermal degradation of melamine takes place with the break-down of the three-dimensional polymer network and in presence of DAHP or APP in the microparticles, a stable residue was observed above 600°C. These results suggest that the PP/microparticles composites decompose early in comparison with PP. WU et al. [62; 102; 119] have reported the combustion behaviour of FR polypropylene. It appears that the presence of intumescent system in the bulk polymer decreases significantly the HRR values compared with the raw PP. The HRR curve exhibits two peaks, typical of IFR systems, i.e. the first is assigned to the ignition and the formation of an expanded protective shield; whereas, the second one corresponds to the destruction of this intumescent structure and the residue formation. Furthermore, the encapsulation step of APP allows delaying the active substance decomposition.

The incorporation of rough microcapsules in polypropylene matrix at 5 wt-% modifies the crystallinity index and the polymorphic form, i.e. α and β forms, which lead to a significant enhancement of the materials' Young's modulus, as well as to a small improvement in the elongation at break. Furthermore, the flame retardant compounds can act as nucleating agent and should promote the formation of β-crystals [119]. The thermal stability is also substantially improved due to the presence of intumescent particles [118].

Fibers and Textiles

To achieve good overall flame retardant properties, IFR should be incorporated into the PP fibers rather than applied by finishing or back-coating [10]. They should therefore be compatible with the polymer matrix to avoid any migratory problems, and thermally stable to be incorporated during the spinning process. Furthermore, the loading content should not exceed 10 %-wt to minimize its effect on the thermomechanical properties of the fibers as well as cost. The use of IFR in microencapsulated form is interesting because the microcapsule can be formulated so that the whole system, both microcapsule shell and core materials, act as an efficient intumescent system. Microcapsules also present the advantage of being a permanent incorporation, since migration problems are drastically reduced compared to direct incorporation of the flame retardant agent which tends to exudate more easily from the polymer matrix.

Up to now, the main way envisaged to incorporate microcapsules directly into fibres was solvent spinning rather than melt-spinning because of their tendency to aggregate themselves under dried powder form, and also sometimes the high temperature requirement of melt-spinning could cause degradation of the microcapsules. The proper care should therefore be taken to formulate thermally stable microcapsules intended to be used in melt-spinning. Very few works are reported in literature on this subject. Recently, Leskovšek et al. [120] shown the possibility of introducing microcapsules into polypropylene fibres in the presence of a lubricant. In their study, the added amount of microcapsules was relatively small (1%, 2% and 5%) and the PP was in the form of a single monofilament yarn with a large diameter. They denoted that the microcapsules did not impair the mechanical properties of the fibres.

The incorporation of IFR microcapsules in PP fibers had been realized by Salaün et al [121]. The microcapsules were melt-compounded at 5%-wt with an isotactic polypropylene matrix using a twin-screw extruder, and multi-filaments had been spun afterwards from the various extrudates. The manufactured fibres were thermo-mechanically characterized by measuring their tensile properties, and their thermal properties. The differential scanning analysis showed that the introduction of microparticles does not affect much the relative crystallinity of the fibers. Nevertheless, they denoted an increase of the crystallisation temperature, when the fiber had been loaded with microparticles having a rough surface, and then some particles acted as a nucleating agent. It was also stated that the influence of the microparticles on the mechanical properties of the PP composite fibers, particularly their reinforcing effect, depends mainly on the particle characteristics: shape and size, degree of aggregation, surface characteristics, and also on the degree of dispersion of the particles in the PP matrix. Despite their strong tendency of agglomeration hindering the dispersion process, the formation of microparticle clusters was limited (Figure 8). Moreover, there were no visible cavities on the fiber surface. This denotes a relatively good adhesion between the polymeric matrix and the particles. Furthermore, the Young's modulus variations have been correlated to the work of adhesion, and thus, for a low value of work of adhesion, the Young's modulus is lower due to a loss in compatibility between the matrix and the microparticles. It has also been observed that the damage zone in the PP composite is composed of some crazes and matrix/filler debonding due to poor adhesion. The fracture mechanism is undoubtedly due the formation of voids around the microparticles and to the propagation by repeated debonding or cavitation (Figure 9). Cone calorimeter experiments have been conducted to characterize the degradation of the knitted structure. It was observed that the presence of microparticles containing DAHP in PP decreased the HRR values, by 10 to 14%, when compared to the neat PP. In all cases, this decrease was accompanied by a shorter ignition time and lower value of total heat evolved (THE). The decrease in ignition time may be due to the fact that the microcapsules decompose earlier than the neat polymer matrix. They concluded that the mechanical properties are improved upon addition of the filler particles, and were directly influenced by the quality of the matrix/particle interactions. One of the microparticles gave rise to an intrinsic intumescent flame retardant system.

Figure 8. SEM (longitudinal) and optical (cross-section) micrographs of PP composite fibers.

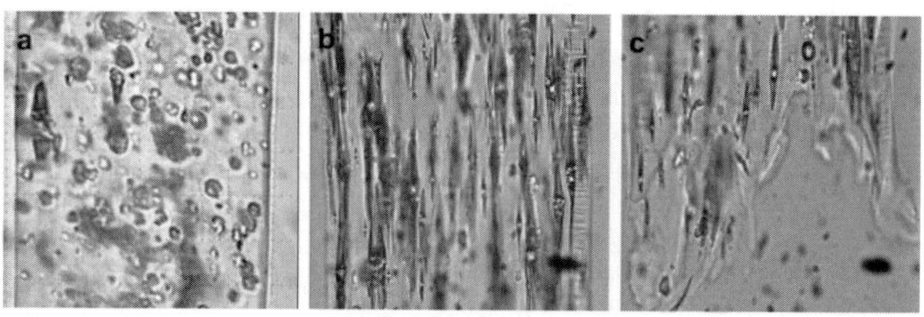

Figure 9. Optical micrograph of the deformed monofilament during the tensile test : (a) undamaged monofilament, (b) formation of voids in the matrix, (c) coalescence of voids and propagation of craze until the final fracture.

Coating

For a coating process, the wetted microencapsulated IFRs are dispersed throughout a polymeric binder, a surfactant, a dispersant, an antifoam agent and a thickener. All the common coating processes, such as knife over roll, screen printing, pad-dry-cure, knife over air, and gravure and dip coating should be used. Even if the method for manufacturing coating composition containing microcapsules was widely described in the patent literature, nevertheless few papers published in the literature give account to the formulation of coating, finishing of fabrics and therefore the evaluation of their characteristics more especially thermal and durability properties. Furthermore, the use of polymeric binder has some drawnbacks, since the incorporated amount should be enough to obtain permanent linkage which can alter the fabric properties such as drape, air permeability, breathability, thermal resistance, softness and tensile strength can be affected adversely as the percentage of binder add-on increases. Salaün et al. [122] proposed a thermodynamical method based on the comparison of the surface energy component to select the most suitable polymeric binder for melamine-formaldehyde microcapsules. From the 2000s, Giraud et al. demonstrated the FR effect of ammonium phosphate microcapsules on cotton fabric coated with polyurethane and polyurea binders [85-87]. From the cone calorimeter evaluation at 35 $kW.m^{-2}$, they found that the microcapsules containing DAHP showed better performance than raw DAHP compound. Vroman et al. evaluated the flame retardancy properties of PP fabrics with three types of microcapsules containing DAHP and having various polymeric shell, i.e. PVA-urethane-urea, melamine-formaldehyde and melamine-PHMA-formaldehyde [123]. Polypropylene fabric was treated by laboratory pad-dry process, in which the padding solution containing a commercial polyurethane and microcapsules in a ratio 1:2. The main degradation of the padded fabrics was very slightly affected by the presence of microcapsules, and all the samples have shown a residue at 800°C under nitrogen gas. From the cone calorimeter tests, it was expected that despite a decrease of ignition time, the DAHP-containing fabrics show decreased flame intensity with regard to based fabric. And, the best results were obtained for melamine-formaldehyde capsules even if they had the lowest loading content in DAHP.

Conclusion

The intumescent flame retardant system was usually composed of three active ingredients: an acid source, a carbon source, and a blowing agent. One of conventional intumescent flame retardant was composed from ammonium polyphosphate as acid source, pentaerythritol as carbon source and melamine as a blowing agent. Moreover, recent studies have shown that microencapsulation is an attractive approach for intumescent system. Among various shell materials presumably impermeable, amino resins, more specially melamine–formaldehyde resins play a main role in the patent literature. The polymerization *in situ* allows the formation of microcapsules containing water-immiscible dispersed phase, with improved mechanical properties and thermal stability. The properties of the membrane depend not only on its chemical structure but also on all the synthesis conditions. The polycondensation of the amino resin occurs in the continuous phase, and the phase separation is linked to the pH and the formaldehyde/melamine molar ratio. The shell polymer prevents the core from the attack of moisture since it decreases the polarity of the acid source, and thus enhances the water resistance and prevents the agglomeration of the particles.

Up to now, the manufacture of intumescent flame retardant textiles involves in the incorporation of microencapsulated intumescent formulation to fibers, fabrics and back-coating. To gain the desired intumescent effect during the thermal degradation, the researchers take care to select the appropriate compound formulation, and the most appropriated microencapsulation methods according to the shell choice to enhance PP protection.

When, microcapsules are incorporated in PP matrix, two polymorphic forms α and β occur in the composites when these particles have rough surface and contained DAHP, whereas only α-phase was detected in the other cases. The incorporation of rougher microparticles in the PP matrix led to a significant enhancement of the material's Young's modulus, as well as to a small improvement in the elongation at break. The thermal stability of PP matrix is substantially improved due to the presence of intumescent microparticles.

Future experimental studies on the testing of thermal and fire performances of intumescent flame retardant systems and their encapsulation, on the development and thermal properties of new intumescent compounds, on improvements in the manufacturing processes to apply them in or onto the surface of the textile substrate, on the development of new microencapsulation methods to enhance synergism with polymeric shell should expand the application range of microencapsulated intumescent systems in a wide range of applications fields such as automotive, household textiles, protective clothing and others.

REFERENCES

[1] E. M. Aizenshtein, *Fibre Chem.* 40, 399 (2008).

[2] R. G. Mansfield, *Success story : Polypropylene in the textile industry*. Wiley Hoboken, NJ, ETATS-UNIS (1999).

[3] C. Magniez, A. Dubois, M. Vouters, R. Delobel and F. Poutch, *J. Ind. Text.* 32, 255 (2003).

[4] V. Babrauskas, Heat of Combustion and Potential Heat. In S. J. G. V. Babrauskas (Ed.), *Heat Release in Fires* (pp. 207): Elsevier Applied Science, London (1992).

[5] T. J. Shields and J. Zhang, Fire hazard with polypropylene. In J. Karger-Kocsis (Ed.), *Polypropylene: An A-Z Reference* (pp. 248). Dordrecht: Kluwer Publishers (1999).

[6] Bourbigot S., Le Bras M. and Delobel R., Flame-retardant polypropylene compositions. In Karger-Kocsis J. (Ed.), *Polypropylene: An A-Z Reference* (pp. 254). Dordrecht: Kluwer Publishers (1999).

[7] S. Giraud, F. Salaün, G. Bedek, I. Vroman and S. Bourbigot, *Polym. Degrad. Stab.* 95, 315 (2010).

[8] S. Zhang and A. R. Horrocks, *Prog. Polym. Sci.* 28, 1517 (2003).

[9] E. D. Weil and S. V. Levchik, *J. Fire Sci.* 26, 5 (2008).

[10] A. R. Horrocks, Flame-retardant finishes and finishing. In Heywood DH (Ed.), *Textile finishing* (pp. 214). Bradford: Society of Dyers and Colourists (2003).

[11] J. Pitts, *Fire and Flammability* 3, 51 (1972).

[12] P. R. Hornsby and R. N. Rothon, Fire retardant fillers for polymers. In R. S. o. Chemistry (Ed.), *Fire retardancy of polymers: new applications of mineral fillers* (pp. 19) (2005).

[13] J. T. Lutz, *Thermoplastic polymer additives: theory and practice*. M. Dekker (1989).

[14] S. Bourbigot and M. Le Bras, Flame Retardants. In J. Troitzsch (Ed.), *Plastics flammability handbook: principles, regulations, testing, and approval*. Munich: Hanser Publishers (2004).

[15] F. Rault, E. Pleyber, C. Campagne, M. Rochery, S. Giraud, S. Bourbigot and E. Devaux, *Polym. Degrad. Stab.* 94, 955 (2009).

[16] S. Zhang, A. R. Horrocks, R. Hull and B. K. Kandola, *Polym. Degrad. Stab.* 91, 719 (2006).

[17] A. R. Horrocks, B. K. Kandola, G. Smart, S. Zhang and T. R. Hull, *J. Appl. Polym. Sci.* 106, 1707 (2007).

[18] F. Rault, C. Campagne, M. Rochery, S. Giraud and E. Devaux, *J. Polym. Sci., Part B: Polym. Phys.* 48, 1185 (2010).

[19] A. F. Vargas, V. H. Orozco, F. Rault, S. Giraud, E. Devaux and B. L. López, *Composites Part A: Applied Science and Manufacturing* 41, 1797 (2010).

[20] A. Fina, D. Tabuani and G. Camino, *Eur. Polym. J.* 46, 14 (2010).

[21] S. Boubigot, X. Flambard, M. Rochery, M. Le Bras, E. Devaux and J. D. Lichtenhan, Polyhedral oligomeric silsesquioxanes : application to flame retardant textile. In M. Le Bras, C. A. Wilkie & S. Bourbigot (Eds.), *Fire Retardancy of polymers* (pp. 189). London: Royal Society of Chemistry (2005).

[22] S. Bellayer, S. Bourbigot, X. Flambard, M. Rochery, J. W. Gilman and E. Devaux, "Polymer/MWNTs nanocomposite yarns and fabrics: processing, characterization, flammability and thermal properties". 4th AUTEX Conference, 22, Jun. 2004 (ENSAIT, Roubaix (France)).

[23] A. K. Sen, B. Mukherjee, A. S. Bhattacharya, L. K. Sanghi, P. P. De and A. K. Bhowmick, *J. Appl. Polym. Sci.* 43, 1673 (1991).

[24] M. Le Bras and S. Bourbigot, Intumescent fire retardant polypropylene formulations. In J. Karger-Kocsis (Ed.), *Polypropylene: an A-Z reference* (pp. 357). London: Kluwer Academic Publishers (1999).

[25] C. Cazé, E. Devaux, G. Testard and T. Reix, New intumescent systems: an answer to the flame retadant challenges in the textile industry. In M. Le Bras, G. Camino, S. Bourbigot & R. Delobel (Eds.), *Fire Retardancy of Polymers : The Use of Intumescence* (pp. 363). Cambridge: The Royal Society (1998).

[26] S. Bourbigot, M. Le Bras, S. Duquesne and M. Rochery, *Macromol. Mater. Eng.* 289, 499 (2004).

[27] L. Morice, S. Bourbigot and J.-M. Leroy, *J. Fire Sci.* 15, 358 (1997).

[28] R. Delobel, M. L. Bras, N. Ouassou and F. Alistiqsa, *J. Fire Sci.* 8, 85 (1990).

[29] S. Bourbigot, S. Duquesne and J.-M. Leroy, *J. Fire Sci.* 17, 42 (1999).

[30] G. Camino, L. Costa and L. Trossarelli, *Polym. Degrad. Stab.* 7, 25 (1984).

[31] H. L. Vandersall, *Journal of Fire and Flammability* 2, 97 (1971).

[32] S. Bourbigot and M. Le Bras, Flame Retardants *Plastic Flammability Handbook* (2004).

[33] S.-H. Chiu and W.-K. Wang, *Polymer* 39, 1951 (1998).

[34] P. Lv, Z. Wang, K. Hu and W. Fan, *Polym. Degrad. Stab.* 90, 523 (2005).

[35] G. Liu, J. Q. Zhao, Y. H. Zhang, S. M. Liu and H. Ye, *Polymers & Polymer Composites* 15, 191 (2007).

[36] J. D. Zuo, S. M. Liu and Q. Sheng, *Molecules* 15, 7593 (2010).

[37] X. Almeras, M. Le Bras, P. Hornsby, S. Bourbigot, G. Marosi, S. Keszei and F. Poutch, *Polym. Degrad. Stab.* 82, 325 (2003).

[38] X. Almeras, F. Dabrowski, M. Le Bras, F. Poutch, S. Bourbigot, G. Marosi and P. Anna, *Polym. Degrad. Stab.* 77, 305 (2002).

[39] H. Demir, E. Arkış, D. Balköse and S. Ülkü, *Polym. Degrad. Stab.* 89, 478 (2005).

[40] N. H. Huang, Z. J. Chen, J. Q. Wang and P. Wei, *eXPRESS Polymer Letters* 4, 743 (2010).

[41] S. Bourbigot, M. Le Bras, R. Delobel, P. Bréant and J.-M. Tremillon, *Polym. Degrad. Stab.* 54, 275 (1996).

[42] Y. Tang, Y. Hu, S. Wang, Z. Gui, Z. Chen and W. Fan, *Polym. Int.* 52, 1396 (2003).

[43] Y. Tang, Y. Hu, B. Li, L. Liu, Z. Wang, Z. Chen and W. Fan, *J. Polym. Sci., Part A: Polym. Chem.* 42, 6163 (2004).

[44] Q. Wu and B. Qu, *Polym. Degrad. Stab.* 74, 255 (2001).

[45] X. Wang, Y. Song and J. Bao, *J. Vinyl Add. Tech.* 14, 120 (2008).

[46] Jing Wu, Yuan Hu, Lei Song and Wenjie Kang, *J. Fire Sci.* 26, 399 (2008).

[47] Y. Li, B. Li, J. Dai, H. Jia and S. Gao, *Polym. Degrad. Stab.* 93, 9 (2008).

[48] X. Chen, J. Yu, M. He, S. Guo, Z. Luo and S. Lu, *J. Polym. Res.* 16, 357 (2009).

[49] P. Wei, J. Hao, J. Du, Z. Han and J. Wang, *J. Fire Sci.* 21, 17 (2003).

[50] Y. Chen, Y. Liu, Q. Wang, H. Yin, N. Aelmans and R. Kierkels, *Polym. Degrad. Stab.* 81, 215 (2003).

[51] Y. Liu and Q. Wang, *Polym. Degrad. Stab.* 91, 2513 (2006).

[52] Y. Chen and Q. Wang, *Polym. Degrad. Stab.* 92, 280 (2007).

[53] G. Nelson, *Rev. Prog. Color. Relat. Top.* 21, 72 (1991).

[54] G. Nelson, *Int. J. Pharm.* 242, 55 (2002).

[55] G. Erkan and M. Sariisik, *Colourage* 51, 61 (2004).

[56] S. Siler-Marinkovic, D. Bezbradica and P. Skundric, *Chemical Industry & Chemical Engineering Quaterly* 12, 58 (2006).

[57] R. F. Kover, B. Farrel, T. Tiano, R. Levi and J. W. H. Harpel, "Microencapsulated Fire Retardants for Flame Resistant Tentage Textiles". Eigth Annual BCC Conference on Flame Retardancy, 1997

[58] R. E. Aseeva and G. E. Zaikov, *Plastickeskie Massy* 6, 46 (1984).

[59] I. S. Reshetnikov, N. S. Zubkova, A. V. Antonov, E. V. Potapova, V. S. Svistunov, M. A. Tuganova and N. A. Khalturinskij, *Mater. Chem. Phys.* 52, 78 (1998).

[60] N. Zubkova, M. Tyuganova, I. Reshetnikov and N. Khalturinskii, *Fibre Chem.* 29, 166 (1997).

[61] N. Zubkova, N. Butylkina, S. Chekanova, M. Tyuganova, N. Khalturinskii, I. Reshetnikov and Y. Naganovskii, *Fibre Chem.* 30, 11 (1998).

[62] K. Wu, L. Song, Z. Wang and Y. Hu, *Polym. Adv. Technol.* 19, 1914 (2008).

[63] K. Wu, Z. Wang and Y. Hu, *Polym. Adv. Technol.* 19, 1118 (2008).

[64] K. Wu, L. Song, Z. Wang and Y. Hu, *J. Polym. Res.* 16, 283 (2009).

[65] K. Wu, Z. Wang and H. Liang, *Polym. Compos.* 29, 854 (2008).

[66] S. Benita, *Microencapsulation: methods and industrial applications*, Marcel Dekker Inc., New York (1996).

[67] R. Arshady, *Polymer Engineering & Science* 30, 905 (1990).

[68] G. Nelson, *Rev. Prog. Color. Relat. Top.* 31, 57 (2001).

[69] R. Cox, *Chem. Fibers Int.* 48, 475 (1998).

[70] Y. G. Bryant, *Advances in Heat and Mass Transfer in Biotechnology* HTD-Vol.363/BED-VOl.44, 225 (1999).
[71] D. P. Colvin. (2000). *Encapsulated phase change materials*. Paper presented at the The 2nd International Conference on Safety & Protective Fabrics.
[72] D. P. Colvin and Y. G. Bryant, US Patent 5, 637, 389, 1996
[73] R. J. Pushaw, *United States Patent* No. 5,677,048, Oct. 14, 1997
[74] H. E. Praetzel and H. Jenkner, *United States Patent* No. 3,660,321, May 2, 1972
[75] H. J. Eichel, *United States Patent* No. 3,317,433, Feb. 5, 1967
[76] D. Vincent and R. Golden, *United States Patent* No. 4,138,356, Feb. 6, 1979
[77] D. Vincent and R. Golden, *United States Patent* No. 3,968,060, Jul. 6, 1976
[78] H. Hahn, U. Guhr, H. Hintz, D. Wittenberg and P. Gleich, *United States Patent* No. 5,043,218, Aug. 27, 1991
[79] W. Pieper, H. Staendeke and G. Elsner, EP0180795, May 14, 1986
[80] R. Cipolli, R. Oriani, E. Masarati and G. Nucida, EP0542359, May 19, 1993
[81] R. Cipolli, R. Oriani, E. Masarati and G. Nucida, *United States Patent* No. 5,576,391, Nov. 19, 1996
[82] G. Elsner, H. Staendeke and G. Heymer, EP0131097, Aug. 29, 1990
[83] G. Elsner, H. Staendeke and G. Heymer, *United States Patent* No. 4,639,331, Jan. 27, 1987
[84] C. Fukumura, M. Iwata, N. Narita, K. Inoue and R. Takahashi, *United States Patent* No. 5,599,626, Feb. 4, 1997
[85] S. Giraud, S. Bourbigot, M. Rochery, I. Vroman, L. Tighzert and R. Delobel, *J. Ind. Text.* 31, 11 (2001).
[86] S. Giraud, S. Bourbigot, M. Rochery, I. Vroman, L. Tighzert and R. Delobel, *Polym. Degrad. Stab.* 77, 285 (2002).
[87] S. Giraud, S. Bourbigot, M. Rochery, I. Vroman, L. Tighzert, R. Delobel and F. Poutch, *Polym. Degrad. Stab.* 88, 106 (2005).
[88] T. Schwarzbach and J. Mcknight, *United States Patent* No. 7,485,370, Feb. 3, 2009
[89] D. Saihi, I. Vroman, S. Giraud and S. Bourbigot, *React. Funct. Polym.* 66, 1118 (2006).
[90] J. Ni, Q. Tai, H. Lu, Y. Hu and L. Song, *Polym. Adv. Technol.* 21, 392 (2010).
[91] S. Zhou, H. Lu, L. Song, Z. Wang, Y. Hu, J. Ni and W. Xing, *Journal of Macromolecular Science, Part A* 46, 136 (2008).
[92] B. Wang, Q. Tang, N. Hong, L. Song, L. Wang, Y. Shi and Y. Hu, *ACS Appl Mater Interfaces* 3, 3754 (2011).
[93] D. Saihi, I. Vroman, S. Giraud and S. Bourbigot, *React. Funct. Polym.* 64, 127 (2005).
[94] M. Lin, Y. Yang, P. Xi and S.-l. Chen, *J. Appl. Polym. Sci.* 102, 4915 (2006).
[95] F. Salaün and I. Vroman, *Eur. Polym. J.* 44, 849 (2008).
[96] Z. Wang, K. Wu and Y. Hu, *Polymer Engineering & Science* 48, 2426 (2008).
[97] K. Wu and Z. Wang, *Polymer-Plastics Technology and Engineering* 47, 247 (2008).
[98] W.-j. Luo, W. Yang, S. Jiang, J.-m. Feng and M.-b. Yang, *Polym. Degrad. Stab.* 92, 1359 (2007).
[99] H. Wang, X. Meng, B. Wen, X. Gao, S. Zhang and M. Yang, *Mater. Lett.* 62, 3745 (2008).
[100] Y. Liu and Q. Wang, *Polym. Degrad. Stab.* 91, 3103 (2006).
[101] M. Gao, Y. Wo and W. Wu, *J. Appl. Polym. Sci.* 119, 2025 (2011).
[102] K. Wu, M.-M. Shen and Y. Hu, *J. Polym. Res.* 18, 425 (2011).

[103] K. Wu, L. Song, Z. Wang, Y. Hu, E. Kandare and B. K. Kandola, *Journal of Macromolecular Science, Part A* 46, 837 (2009).

[104] S. Nie, M. Zhang, S. Yuan, G. Dai, N. Hong, L. Song, Y. Hu and X. Liu, *J. Therm. Anal. Calorim.*, 1 (2011).

[105] X. Chen, C. Jiao and J. Zhang, *J. Therm. Anal. Calorim.*, 1 (2011).

[106] J. Goldshtein and S. Margel, *Eur. Polym. J.* 45, 2987 (2009).

[107] Y.-q. Liu, F.-x. Zhang, Y.-y. Sun and G. Zhao, *J. Coat. Technol. Res.* 5, 465 (2008).

[108] R. Arshady, *J. Microencapsulation* 6, 13 (1989).

[109] J. Lukaszczyk and P. Urbas, *React. Funct. Polym.* 33, 233 (1997).

[110] K. Hong and S. Park, *React. Funct. Polym.* 42, 193 (1999).

[111] A. M. Pensé, C. Vauthier and J. P. Benoit, *Colloid & Polymer Science* 272, 211 (1994).

[112] Y. Liu and G. Zhao, *Chin. J. Chem. Eng.* 15, 429 (2007).

[113] B. Šumiga, E. Knez, M. Vrtačnik, V. F. Savec, M. Starešinič and B. Boh, *Acta Chimica Slovenica* 58, 14 (2011).

[114] F. Salaün, E. Devaux, S. Bourbigot and P. Rumeau, *Chem. Eng. J.* 155, 457 (2009).

[115] W. Li, X.-X. Zhang, X.-C. Wang and J.-J. Niu, *Mater. Chem. Phys.* 106, 437 (2007).

[116] M. Le Bras, M. Bugajny, J.-M. Lefebvre and S. Bourbigot, *Polym. Int.* 49, 1115 (2000).

[117] Q. Wu, J. Lü and B. Qu, *Polym. Int.* 52, 1326 (2003).

[118] F. Salaün, I. Vroman, G. Bedek and M. Lewandowski, *J. Polym. Sci., Part B: Polym. Phys.* 46, 2566 (2008).

[119] K. Wu, M.-M. Shen, Y. Hu and H.-G. Ji, *Journal of Macromolecular Science, Part B* 50, 1508 (2011).

[120] M. Leskovšek, G. Jedrinović and U. S. Elesini, *Acta Chimica Slovenia* 51, 699–715 (2004).

[121] F. Salaün, M. Lewandowski, I. Vroman, G. Bedek and S. Bourbigot, *Polym. Degrad. Stab.* 96, 131 (2011).

[122] F. Salaün, E. Devaux, S. Bourbigot and P. Rumeau, *Text. Res. J.* 79, 1202 (2009).

[123] I. Vroman, S. Giraud, F. Salaün and S. Bourbigot, *Polym. Degrad. Stab.* 95, 1716 (2010).

In: Polypropylene
Editors: L. P. Silva and E. F. Barbosa

ISBN: 978-1-62417-142-0

Chapter 11

NATURAL AND SYNTHETIC FIBER REINFORCED POLYPROPYLENE COMPOSITES

Quazi T. H. Shubhra*

Research Institute of Chemical and Process Engineering,
Faculty of Information Technology, University of Pannonia,
Veszprem, Hungary

ABSTRACT

Polypropylene (PP) is a thermoplastic material that is produced by polymerizing propylene molecules into long polymer molecules or chains. It has many properties which widen its applications. Polypropylene is characterized by high tensile strength, high stiffness and high heat deflection temperature under load, transparency, flame resistance and dimensional stability. It has low density, high melting point and relatively good resistance to impact. A typical composite material is a system of materials consisting of two or more materials on a macroscopic scale. The matrix holds the reinforcement to form the desired shape, while the reinforced material improves the overall properties of the matrix. There has been a continuous effort to develop polypropylene-based composites to replace many types of engineered thermoplastic materials in high-performance applications. Polypropylene composites have many advantages over metal such as corrosion resistance, low weight, easily molded complicated shapes, and low cost. As a matrix material, PP is widely used, because it has some excellent characters for composite fabrication. PP is also very suitable for filling, reinforcing and blending. Reinforcement of natural fibrous polymers in PP is one of the most promising routes to create natural-synthetic polymer composites. In this chapter, fabrication, mechanical properties and improvement of mechanical properties of fiber reinforced PP by various methods are discussed.

* E-mail: shubhro.du@gmail.com.

INTRODUCTION

A typical composite material is a system consisting of two or more materials (mixed and bonded) on a macroscopic scale. A composite material is generally composed of reinforcement (fibers, particles, fillers, etc) embedded in a matrix (polymers, metals, ceramics, etc). The matrix holds the reinforcement to form the desired shape while the reinforced material improves the overall mechanical properties of the matrix. Fibers reinforced polypropylene composites (FRPCs) are composed of fibers and polypropylene matrix with distinct interfaces (boundaries) between them. In such composites, fibers are the reinforcement and the main source of strength (principal load-carrying members) while the polypropylene matrix keeps them in the desired location and orientation and protect them from environmental damages.

Composites were first used near the time 1500 B.C. when early Egyptians and Mesopotamian settlers used a mixture of mud and straw to create strong and durable buildings. Straw continued to provide reinforcement to ancient composite products including pottery and boats. The Mongols invented the first composite bow in 1200 A.D. A combination of wood, bone, and "animal glue", were used in bows and were pressed and wrapped with birch bark. These bows were not only powerful, but also accurate. Composite Mongolian bows provided Genghis Khan with military dominance. Because of the composite technology, this weapon was the most powerful weapon on earth until the invention of gunpowder. The modern era of composites began after the development of plastics by scientists. Until then, the only source of glues and binders were natural resins derived from plants and animals. In the early 1900s, various plastics like polyester, polystyrene, vinyl, and phenolic were developed and became superior to natural resins. But plastics alone could not provide sufficient strength for structural applications. Reinforcement was needed to improve the strength, rigidity and other properties. In 1935, Owens Corning introduced the first glass fiber, fiberglass. Fiberglass was made almost by accident in 1930, when an engineer became intrigued by a fiber that was formed during the process of applying lettering to a glass milk bottle. Fiberglass, when combined with plastic polymers results incredibly strong, lightweight structure. This is the beginning of the fiber reinforced polymers (FRPs) industries, as we know it today.

Depending on the source, fibers are largely divided into two categories: natural and synthetic. A lot of work has been done by many researchers on the composites containing natural and synthetic fibers [1-3]. But both these fibers have advantages and drawbacks. Synthetic fiber reinforced thermoplastic composites give better mechanical properties than that made by natural fibers, but they are not environmentally friendly. The natural fibers such as cellulose fiber [4], wood fiber [5], flax [6], hemp, [7], silk [8], jute [9], sisal [10], kenaf [11], cotton [12], etc are being used to reinforce thermoplastics and thermosets by many researchers. Some advantages of natural fibers are low abrasion resistance, low density, high toughness, acceptable specific strength properties, good thermal properties, enhanced energy recovery, biodegradability, etc [13-14]. Natural fibers produce composites that offer advantages like environmental friendliness, renewability of the fibers, good sound abatement capability and improved fuel efficiency [15-16]. On the other hand, synthetic fiber reinforced PP composites attracted much attention due to their high strength, better durability and moisture resistance properties [17]. The most used synthetic fibers in composites are glass

[18], carbon [19] and aramid [20]. Among the synthetic fibers, glass fibers are widely used due to their low cost (compared to carbon and aramid) and better physical-mechanical properties.

PP is one type of thermoplastic engineering polymer. It possesses several useful properties like high heat distortion temperature, transparency, flame resistance, dimensional stability and high impact strength. As a matrix material, PP is widely used because it has some excellent characters for composite fabrication. PP is also very suitable for filling, reinforcing and blending. Reinforcement of fibrous polymers in PP is one of the most promising routes to create natural-synthetic polymer composites [21]. The performance of FRPCs depends on the constituent materials. The length of reinforcing fiber [22] and fiber content [23] are the two most important factors affecting the mechanical properties of a fiber reinforced composite.

FRCs find applications in construction industries [24], decking, window and door frames [25], sports equipment such as bicycle frames, baseball bats, exercise equipment, etc [26]. They are also suited for many automotive applications [21].

REINFORCING AGENT

Reinforcing agents are special classes of fillers that have better 'reinforcement' properties than the matrix and after incorporation into the matrix, influence the matrix properties. Usually composites consist of a reinforcing material embedded in a matrix. In the broadest sense, composites are the result of a combination of substantial volume fractions of high strength, high stiffness reinforcing components with a lower modulus matrix. Generally, the reinforcing component in a composite structure can be divided into two classes:

a) Discontinuous: It can be in the form of dispersions/particles, flakes, whiskers and discontinuous short fibers with different aspect ratios.
b) Continuous: It can be long fibers and sheets.

In practice, the most commonly employed reinforcing component is in particulate or in fibrous form. On the basis of these two types of reinforcing agents, sometimes composites are divided into two groups: particulate reinforced composites and fiber reinforced composites. In addition to these two types of composites, existence of another group of composite system should be noted which is called "laminar composites (or simply laminates)". In lamina composites, the reinforcing agents are in the form of sheets bonded together and are often impregnated with more than one continuous phase in the system [27].

Particulate (or particle) reinforcing agents have aspect ratios around '1' and possess similar length and breadths. They cover particulates of both regular shapes (such as spheres) and irregular shapes that may have extensive convolution and porosity. Particulate reinforcing agents are the most common and the cheapest reinforcing agent. They are divided into two groups: organic (e.g., powdered cellulose, powdered rubber, a wide variety of starches and particulate carbon [28] and inorganic (e.g., Silica and some simple metal oxides like alumina, glass etc.).

Fibers are stronger and stiffer materials than any other form of reinforcements which is the main reason for the overwhelming attraction for fibrous reinforcements. Fibers are an important class of reinforcements, as they satisfy the desired conditions with strengthening the matrix constituent influencing and enhancing their properties as desired. Generally fiber means a continuous single material whose length is at least 200 times of its width or diameter and filaments are endless or continuous fibers. Fiber reinforcing agents can be either natural or synthetic. Fibers can be used in as woven, non-woven fabrics, yarn or rovings.

FIBERS

Fiber is a class of material that is continuous filaments or is in discrete elongated pieces, similar to lengths of thread. They can be spun into filaments, rope or string. Fibers are of two types: natural and synthetic. Figure 1 shows general classification of fibers.

Natural Fibers

Natural fiber-reinforced polymers have created interest in the recent years in the automotive industry. Besides, other applications of natural fiber composites include door inner panel, seat back, roof inner panel, and so on. Natural fibers are being used in automotive industry due to the following reasons:

a) Natural fibers are comparatively of very low cost than glass and carbon fibers;
b) They are biodegradable and environment friendly. Unlike glass and carbon fibers, the energy consumption to produce them is very small;
c) The density of natural fibers is in the range of 1.25-1.5 g/cm^3 whereas that of E-glass fibers and carbon fibers is 2.54 g/cm^3 and 1.8–2.1 g/cm^3 respectively;
d) Natural fiber composites provide higher acoustic damping than glass or carbon fiber composites that makes natural fibers more suitable for noise attenuation which is an increasingly important requirement in automotive applications;
e) The modulus–weight ratio of some natural fibers is greater than that of E-glass fibers indicating that they can be very competitive with E-glass fibers in stiffness-critical designs.

However, there are several limitations of natural fibers such as:

a) The strength of natural fibers is relatively low;
b) They absorb moisture;
c) They start to degrade at a comparatively lower temperature than synthetic fibers like glass or carbon fibers.

The two main sources of natural fibers are plants and animals. The main component of animal based fibers is various proteins [29]: examples include mohair, wool, silk, alpaca, angora, etc.

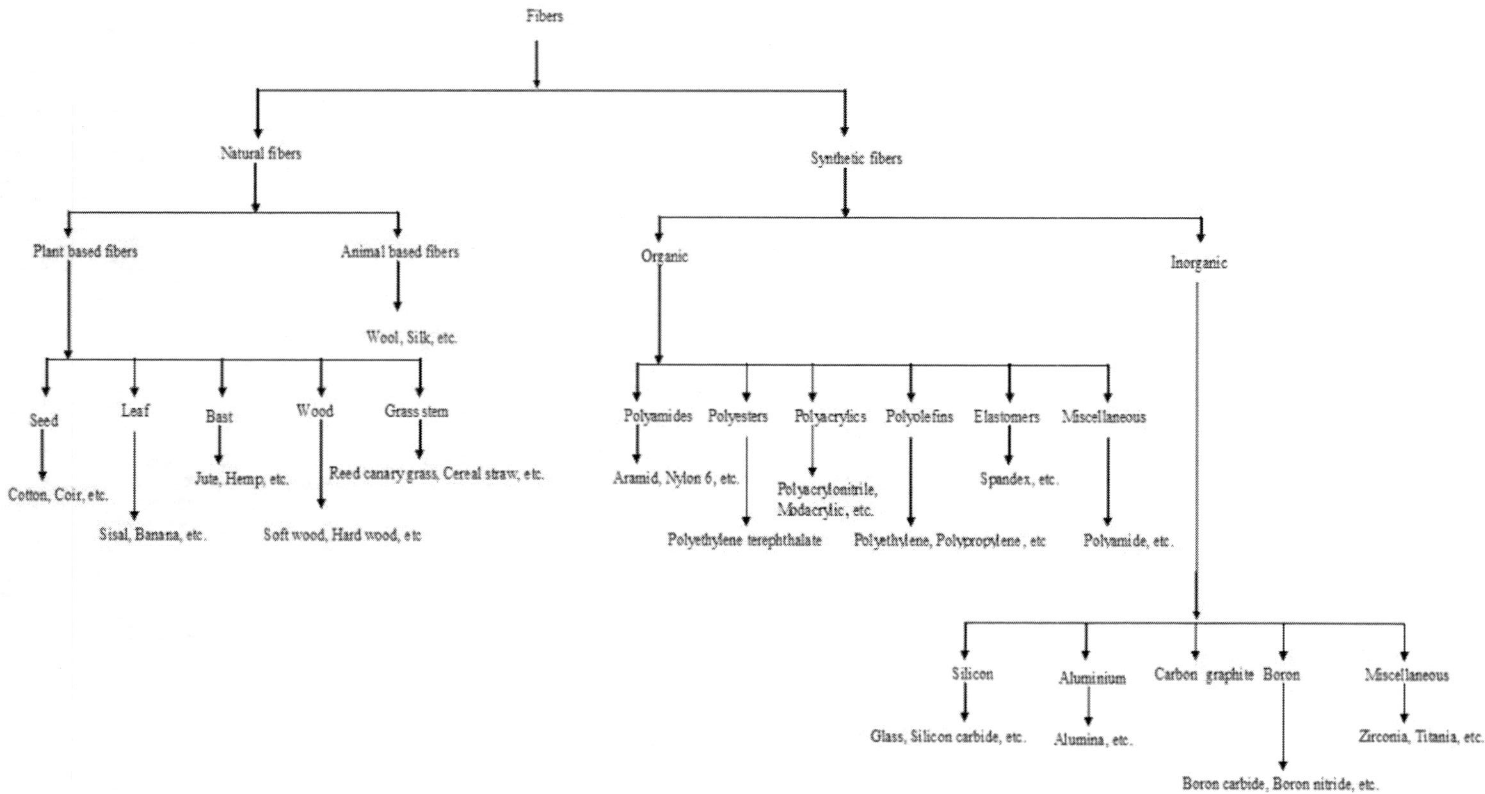

Figure 1. General classification of fibers.

The components of plant fibers are cellulose microfibrils dispersed in an amorphous matrix of lignin and hemicellulose [30]: examples include cotton, jute, flax, ramie, sisal, hemp, etc.

Plant Based Natural Fibers

Plant fibers are generally comprised mainly of cellulose: examples include cotton, jute, flax, ramie, sisal and hemp. These fibers are abundantly available throughout the world. Cellulose fibers, depending on the part of the plant from which they are taken, can be classified as [31]:

Grasses and reeds: These types of fibers collect from the stems of monocotyledonous plants such as bamboo and sugar cane.

Leaf fibers: Leaf fibers are fibers that run lengthwise through the leaves of most monocotyledonous plants such as abaca, sisal, henequen, esparto, etc. These fibers are also referred to as 'hard fibers'.

Bast fibers: Bast fibers (bundles) are those fibers that come from the inner bark (phloem or bast) of the stems of dicotyledonous plants. Common examples are hemp, jute, flax, kenaf, etc.

Seed and fruit hairs: These are fibers that come from seed-hairs and flosses. Common examples of these fibers are cotton and coconut.

Wood fibers: Wood fibers are those fibers that come from the xylem of angiosperm (hardwood) and gymnosperm (softwood) trees. Examples include yellow poplar, pine, maple, spruce, etc. Wood fiber is the most abundantly used cellulosic fiber due to their extensive use in pulp and paper industries.

Jute

Jute is the common name given to the fiber extracted from the stems of plants belonging to the genus *Corchorus* and family *Tiliaceae*. Jute fibers can be extracted from the bast of stems of plants. Favorable conditions for cultivation of jute are found in the deltas of the great rivers of the tropics and subtropics where alluvial soils and irrigation are combined with long day lengths to provide an opportunity for considerable vegetative growth before flowering. Optimum growing temperature of jute is 18 to 33°C with a minimum annual moisture requirement of 250 mm in a soil pH between 6.6 and 7.0. Jute is the complex lignocelluloses based polymeric fiber. It is not uniform in the chemical composition as found by the multi-cellular structure of jute fiber. Jute has three principle constituents: cellulose, hemicellulose and lignin. The approximate composition of raw jute materials is cellulose: 62-65%, hemicelluloses: 15-16% and lignin: 14-20%. The cellulose, hemicelluloses and lignin all exist in the form of long chain molecules. In addition to these major components, it contains minor constituents such as fats and waxes, inorganic (mineral) matters, nitrogenous matter, vitamins, pectinious materials and traces of pigments like–carotene and xanthophylls. The main structural element of jute fiber is cellulose which is held entirely within the cell units where the lignin is and hemicellulose are distributed throughout the entire body of the fiber serving as a cementing material. Cellulose is the key component of the jute fiber. Basically it is the major component of the vegetable kingdom and the cellulose content varies from plant to plant. The term cellulose applies only to the pure plant cell materials consisting of high polymer of the hydro glucose unit as monomer which is liner and the degree of polymerization usually varies from several hundred to several thousands. Cellulose is a 1→4-linked linear polymer of β-D-glucopyranose as shown in Figure 2.

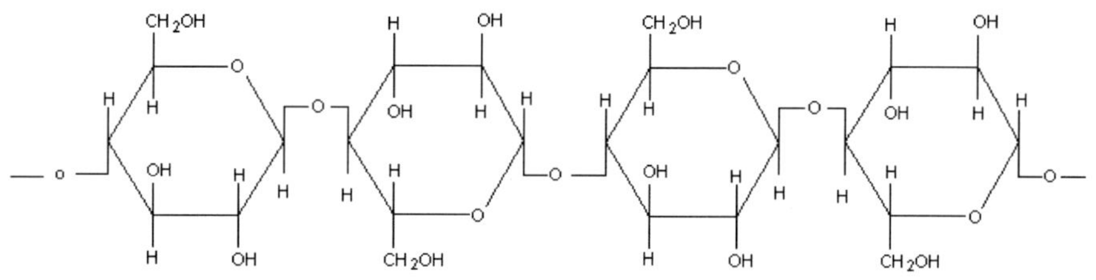

Figure 2. Segment of cellulose chain.

These monomers are linked together by elimination of one molecule of water between the hydroxyl groups attached to carbon number 1 of one glucose molecule and the number 4 carbon atom of another glucose molecule. Since the condensation reaction to make the cellulose polymer eliminates a molecule of water per glucose, the monomers are often referred to as anhydroglucose units or as glucose residues. Jute fibers present some disadvantages such as high moisture sorption, poor dimensional stability, intrinsic polarity, low thermal resistance, anisotropic fiber resistance and variability.

Cotton

Cotton fibers are seed hairs from plants of the order *Malvales*, family *Malvaceae*, tribe *Gossypieae*, and genus *Gossypium*. Each cotton fiber is a single, elongated, complete cell developing in the surface layer of cells of the cottonseed. The mature cotton fiber is actually a dead, hollow, dried cell wall. In the dried out fiber, the tubular structure is collapsed and twisted to give cotton fiber convolutions which differentiate cotton fibers from all other forms of seed hairs and are partially responsible for some of the unique characteristics of this fiber. Raw cotton fiber, after ginning and mechanical cleaning, is approximately 95% cellulose. The structure of cotton cellulose is nothing but a linear polymer of β-D-glucopyranose. Noncellulosic constituents of cotton fibers are proteins, amino acids, other nitrogen-containing compounds, wax, pectic substances, organic acids, sugars, inorganic salts, and a very small amount of pigments. These noncellulosic constituents are located principally in the cuticle, in the primary cell wall, and in the lumen. Cotton is the only significant textile fiber whose strength increases with humidity while most others are weakened by increased moisture. For most cotton, elongation at break is in the range 6–9%, initial Young's moduli for cotton range from about 80 g/den for Sea Island cotton down to 40 g/den for Asian cottons. Tenacity of the Upland cottons ranges from approximately 20 to 30 gf/tex. Cotton being a strong fiber is used in composite fabrication especially with polymeric matrices.

Flax

Flax is among the oldest crop plants cultivated for the purpose of oil and fiber. It belongs to the genus *Linum* and family Linaceae. It is an annual herbaceous plant with shallow root system. Flax has been cultivated for nearly 10,000 years and its fibers, i.e., linen, are reported to be the oldest textile known, dating back to early Mesopotamian times. The flax plant consists of the root, stalk, and branches carrying the seed capsules. Flax is dicotyledonous. Only the central portion of the plant (up to 75% of the plant height) can be used to produce fibers. Flax cell wall generally consists of ~70−75% cellulose, 15% hemicelluloses, 10−15% pectic material, about 2% waxes and 2% lignin. Color of the flax fibers varies from light blond to gray. Flax fibers are very strong and flexible. It is a particularly inextensible fiber. Within the small degree of extension, flax is an elastic fiber. Flax fibers lose strength gradually on exposure to sunlight. Flax has good resistance to alkaline solutions and can withstand dilute weak acids. Hot dilute acids or cold concentrated acids attack it. Due to many labor-intensive production steps, flax fibers are quite expensive. Flax is used as reinforcement for high value-added composite products especially for interior automotive components.

Hemp

Hemp (*Cannabis sativa*) is another famous bast fiber. Hemp plants have impressive growth rates so that they quickly cover the ground and, therefore, suppress weeds and some soil-borne pathogens and needs almost no or minimal herbicides, pesticides, fungicides, and fertilizers. Hemp can be whitish to yellow in color. They have good tensile strength and are water-resistant. Hemp fibers are coarser as compared to flax. The fibers have an excellent moisture resistance. These fibers have high tenacity but low elongation at break. Hemp fibers are used is specialty paper, textiles, construction materials, plastics and composites etc.

Kenaf

Kenaf is the name given to a fiber similar to jute and obtained from the stems of plants belonging to the genus *Hibiscus*, family *Malvaceae*, especially the species *H. cannabinus* L. Kenaf is a pale-colored fiber and this fiber contains less noncellulosic materials than jute. Kenaf fibers are coarse, brittle and they are difficult to process. They have a breaking strength similar to that of low-grade jute and are weakened only slightly when wet. Now, various new applications for kenaf products are emerging, including those for paper products, building materials, absorbents, etc. Kenaf fibers are used as a reinforcing fiber in thermoplastic composites because of its superior toughness and high aspect ratio in comparison with other fibers.

Animal Based Natural Fibers

Animal based natural fibers can also be used as alternatives for producing biodegradable and environment friendly composite materials for bio-engineering and bio-medical applications. Animal based fibers are mainly composed of various proteins, like wool, spider and silkworm silk.

Silk

Silk is a solidified viscous fluid excreted from a special gland by a number of insects and spiders. It is a protein fiber. It is composed of amino acid residues. Silk protein is a kind of protein like collagen, elastin, keratin, fibroin, sporgin etc. Insects mainly belong to two families, viz., Saturnidae and Bombycidae which spins silk fibers. There are two main types of silkworms: 'mulberry silks' (*Bombyx mori*) also called 'cultivated silk' and 'wild silk' of which Tussah silk is the most important representative. Silk proteins are natural polymers and are biodegradable with reactive functional groups that open up possibility to be cross-linked with many other polymers. The mechanical properties of silk fibers consist of a combination of high strength, extensibility, and compressibility [29]. The mature silkworm builds its cocoon by extruding a viscous fluid from two large glands in the body of the silkworm. This solution is extruded through two ducts in the head of the silkworm into a common spinneret. The viscous part (fibroin) is covered by another secretion (sericin) which flows from two other symmetrically placed glands. These two components are cemented together by emerging into the air, coagulating and producing a firm continuous filament. As a consequence of this spinning process, the fiber has two main parts called sericin and fibroin. Sericin, called silk gum is a minor component of the fiber and it also has some impurities such as waxes, fats and pigments. Sericin is a yellow, brittle and inelastic substance. It acts as an adhesive for the twin fibroin filaments and conceals the unique luster of fibroin. Sericin dissolves in a hot soap solution. The greatest sericin content is present in the outer layer of a cocoon whereas the least sericin proportion is present in the innermost layer of a cocoon. Fibroin is the principal water insoluble protein. Fibroin has a highly oriented and crystalline structure. Microfibrils are packed together to form the fibril bundle and several fibril bundles produce a single strand. Fibroin has high proportions of alanine, glycine and serine. A small amount of cysteine residue gives a very small amount of sulfur in the fiber. Fibroin contains only a very small amount of amino acid residues which have acid side chains. The amount of basic side groups in fibroin is relatively low. Because of this, the amounts of acids and alkali that can be absorbed by silk are relatively lower than those absorbed by wool.

Wool

Wool is a natural protein fiber that grows from the follicles of the sheep's skin. Like human hair, wool is also composed of keratin-type protein. Chemically these proteins contain five elements: carbon, hydrogen, oxygen, nitrogen and sulfur. These five elements are combined into 19 amino acids linked together in ladder-like polypeptide chains. Wool fibers consist of two types of cells: cuticle (the outer layer composed of scales) and cortex (the main part of the fiber composes about 90% of the fiber mass) cells. In coarse fibers, a central medulla also exists. Each cuticle cell consists of three layers with different cystine and isodipeptide contents: epicuticle, exocuticle and endocuticle. On the other hand, the cortex consists of spindle shaped cells. The medulla in coarser wool fibers consists of hollow cells with a skeleton of amorphous proteins and fine filaments. Wool fiber is a hygroscopic fiber which means that it takes up moisture readily. Wool can absorb moisture almost one-third of its own weight. Wool has moderate abrasion resistance, good resilience, medium density, lower electrical conductivity and poor dimensional stability.

Synthetic Fibers

Synthetic fibers are usually more uniform in size, more economical to use and behave in a more predictable manner. For engineering applications the most commonly employed significant fibers are synthetic fibers which includes glass fibers, metallic fibers, organically derived synthetic fibers, etc. Most strong and stiff fibers (e.g., ceramic, fibers of glass, graphite-carbon, boron carbide and silicone carbide) are usually difficult to use as structural materials in bulk. However, embedding such materials in a ductile matrix (such as a polymer or metal) enables them to behave as a stronger, stiffer, tougher material.

Carbon Fibers

Carbon fibers are generally manufactured from two types of precursors (starting materials) which are: textile precursors and pitch precursors. The most common textile precursor is polyacrylonitrile (PAN). The molecular structure of PAN contains highly polar CN groups that are randomly arranged on either side of the chain. Pitch is a by-product of petroleum refining or coal coking and comparatively cheaper than PAN. The carbon atoms in pitch are arranged in low molecular weight aromatic ring patterns. Structurally, carbon fibers contain a blend of amorphous carbon and graphitic carbon. Carbon fibers are typically carbonized between 1200–1400ºC and contain 92%–95% carbon. After carbonization, tensile strength values of 3000 MPa and moduli of 250 GPa or even higher are usually obtained. Tensile moduli can even be improved up to 350 GPa, at the expense of some drop in strength by the post treatment. Their high tensile modulus results due to graphitic form in which carbon atoms are arranged in a crystallographic structure of parallel planes. In each plane, carbon atoms are arranged at the corners of interconnecting regular hexagons. Although the bond between the planes is much weaker (van der Waals-type), strong covalent bonds exist between the carbon atoms in each plane. This results in highly anisotropic physical and mechanical properties for the carbon fiber. There are different types of carbon fibers depending on the origin of precursor, such as (a) PAN-based, (b) isotropic pitch based, (c) anisotropic pitch-based, (d) rayon-based, or (e) gas phase grown. Depending on mechanical properties, carbon fibers are classified into four categories: (a) PAN-based high modulus (HM)-low strain to failure- type (b) PAN-based high tensile (HT)-high strain to failure-type and (c) PAN-based intermediate modulus (IM) type and (d) mesophase (pitch)-based. Among the advantages of carbon fibers, their exceptionally high tensile strength-weight ratios as well as tensile modulus-weight ratios are of special interest. Carbon fibers also possess very low coefficient of linear thermal expansion, high fatigue strengths and high thermal conductivity. Main disadvantages of carbon fibers are their low strain-to-failure, low impact resistance and high electrical conductivity that may cause ''shorting'' in unprotected electrical machinery. Their high cost has so far excluded this fiber from widespread commercial applications. They are used mostly in the aerospace industry, where weight saving is considered more critical than cost.

Glass Fibers

Glass fibers are the most common of all synthetic fibers for fiber-reinforced composites (FRCs). Glass fibers are normally divided into three classes: E-glass, S-glass and C-glass. The E-glass is designated for electrical use, the S-glass for high strength and C-glass is for high corrosion resistance. Of the three fibers, the E-glass is the most common reinforcement material used in civil structures. It is produced from lime-alumina-borosilicate which can be easily obtained from abundance of raw materials like sand. Main chemical constituents of E-glass are SiO_2, Al_2O_3, CaO, MgO, B_2O_3 and Na_2O, etc. The fibers are drawn into very fine filaments with varying diameters. The glass fiber strength and modulus can degrade with increasing temperature. Although the glass material creeps under a sustained load, it can be designed to perform satisfactorily. The fiber itself is regarded as an isotropic material and has a lower thermal expansion coefficient than that of steel. The average tensile strength of freshly drawn glass fibers may exceed 3.45 Gpa. The principal advantages of glass fibers are low cost, high strength, high chemical resistance and excellent insulating properties. The disadvantages are relatively low tensile modulus and high density (among the commercial fibers), sensitivity to abrasion during handling which frequently decreases its tensile strength, relatively low fatigue resistance, and high hardness (which causes excessive wear on molding dies and cutting tools).

Aramid Fibers

These are synthetic organic fibers consisting of highly crystalline aromatic polyamides. The aramid fibers have excellent fatigue and creep resistance. Although there are several commercial grades of aramid fibers available, the most common one used in structural applications is Kevlar 49. When a solution of Kevlar is extruded into fiber form, a structure with an exceptional degree of alignment of straight polymer chains parallel to fiber axis develops. Weak hydrogen bonds between hydrogen and oxygen atoms in adjacent molecules hold them together in the transverse direction. The resulting filament is highly anisotropic with much better physical and mechanical properties including higher strength and modulus in the longitudinal direction than in the radial direction. Aramid fibers are used as reinforcement in many marine and aerospace applications where high tensile strength, lightweight, and resistance to impact damage are important. Aramid fibers have a negative coefficient of thermal expansion in the longitudinal direction which is used in designing low thermal expansion composite panels. The major disadvantages of aramid fiber-reinforced composites are their low compressive strengths and difficulty in cutting or machining.

Ceramic Fibers

Ceramic fibers are polycrystalline refractory materials. These fibers are composed of various metal oxides, metal nitrides, metal carbides and their mixtures. Ceramic fibers are available in different fiber lengths and as fabrics. Continuous alumina ceramic fibers offer good compressive strength rather than tensile strength. It's important property is it's high melting point of about 2000°C and the composite can be successfully used at temperature up to about 1000°C. Magnesium and aluminum matrices frequently use alumina fiber as they do not damage the fiber even in the liquid state. Alumina silicate ceramic fibers with significant quantities of boria as continuous filament yarns are produced in braided and woven forms. These fibers are the most advanced high performance fibers available commercially and are known as Sumitomo and Nextel ceramic fibers. A considerable amount of zirconium oxide exists in Nextel ceramic fiber and is called zirconiasilica ceramic fiber. Other ceramic fibers include silicone nitride, silicone carbidethoria, aluminium nitride, boron nitride, potassium titanate, high silica, quartz and a recently introduced polymeric material, polycarbosilane.

Silicone carbide fibers are low cost fibers and are used mostly for ceramic matrix composites and metal matrix composites.

Metallic Fibers

Metallic fibers can be in the form of whiskers, metal wool and filament length fibers. The term whisker applies to single crystals with fibrous or fibrillar characteristics having higher tensile strength and moduli as compared with polycrystalline continuous or discontinuous counterparts. As reinforcement, metal fibers have many advantages. They are easily produced using several fabrication processes and are more ductile, apart from being not too sensitive to surface damage. They also possess high strengths and temperature resistance. However, their weight and the tendency to react each other through alloying mechanisms are major disadvantages. Better flexural properties are observed in some metal fibers reinforced plastic composites that also offer improved strength and weight, than glass fibers. Metals in wire or filament form exhibit considerable high strength. Metal wires, of the continuous version, also reinforce plastics like polyethylene and epoxy. Such combinations ensure high strength, lightweight and good fatigue resistance. Besides, continuous metal fibers are easily handled, unlike glass fibers. The main disadvantage of wires is their high densities. Tungsten wire is used in some nickel and cobalt-based super alloys, while steel wire is commonly used for concrete and tires.

Boron Fibers

Boron fiber is a synthetic fiber with extremely high tensile modulus which is in the range of 379–414 GPa. Boron fibers are manufactured by chemical vapor deposition (CVD) of boron onto a heated substrate (either a tungsten wire or a carbon monofilament). By the reaction of boron chloride with hydrogen, vapor of boron is produced:

$$2BCl_3 + 3H_2 = 2B + 6HCl$$

The most common substrate used in the production of boron fibers is tungsten wire. It is continuously pulled through a reaction chamber in which boron is deposited on its surface at 1100°C–1300°C. During boron deposition, the tungsten substrate is converted into tungsten boride by diffusion and reaction of boron with tungsten. The major disadvantage of boron fiber is its high cost which is even higher than that of many forms of carbon fibers. For this reason, its use is at present restricted to a few aerospace applications. Boron fiber is primarily used in space and aerospace applications.

Polypropylene (PP)

As stated earlier, the matrix holds the reinforcement in a composite. Matrix supports the fibers and keeps them together in the composite material. If load is applied to composite, matrix transfers that to the fibers, thereby keeping the fibers in their position and chosen orientation. The matrix also gives the composite environmental resistance. Some important functions of the matrix in FRCs are: (1) to hold the fibers in place in the structure; (2) to protect the fibers from environment; (3) to help to distribute or transfer loads; (4) to enhance transverse properties of a laminate; (5) to improve impact and fracture resistance; (6) to avoid propagation of crack growth through the fibers by providing alternate failure path along the interface between the fibers and the matrix; (7) to carry interlaminar shear, etc. Matrix must have some desired properties which are important for a composite structure. Some of these properties are:

a) Matrix must wet the fibers with bonding;
b) Minimization of moisture absorption;
c) Easily processable into the desired composite shape;
d) Must flow to penetrate the fiber bundles completely and eliminate voids;
e) Must have reasonable strength, modulus and elongation (elongation should be greater than fiber);
f) Have low shrinkage and coefficient of thermal expansion;
g) Must be elastic to transfer load to fibers;
h) Have dimensional stability to maintain its shape.

There are many matrix choices available and different types have distinct impact on the processing techniques, physical and mechanical properties and environmental resistance of the finished composites. In selecting matrix material, some factors may be taken into consideration like: (1) the matrix must be easy to use in the selected fabrication process, (2) the resultant composite should be cost effective, (3) the matrix is be able to withstand service conditions, viz., temperature, humidity, exposure to ultra-violet environment, exposure to chemical atmosphere, abrasion by dust particles, etc. Matrix can be classified in different ways. Generally it can be divided into three classes: (1) ceramic, (2) metal and (3) organic.

Polypropylene is one kind of polymer matrix. Polypropylene is a thermoplastic polymer and can be made by polymerizing propylene molecules (Figure 3). The macromolecule of PP contains about 10,000 to 20,000 monomer units. The steric arrangement of the methyl groups attached to every second carbon atom in the chain may vary in different ways as shown in Figure 4. If all the methyl groups are on the same side of the winding spiral chain molecule, then the product is called isotactic PP. If pendant methylene groups are attached to the polymer backbone chain in an alternating manner, the PP is known as syndiotactic PP. The PP structure where pendant groups are arranged randomly on the polymer backbone is called atactic PP.

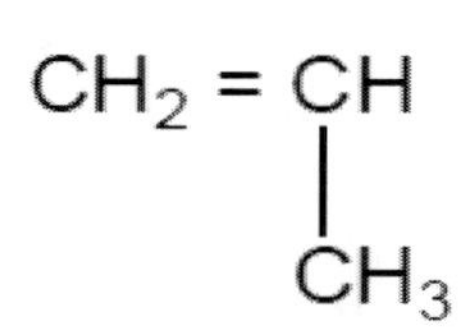

Figure 3. Propylene monomer.

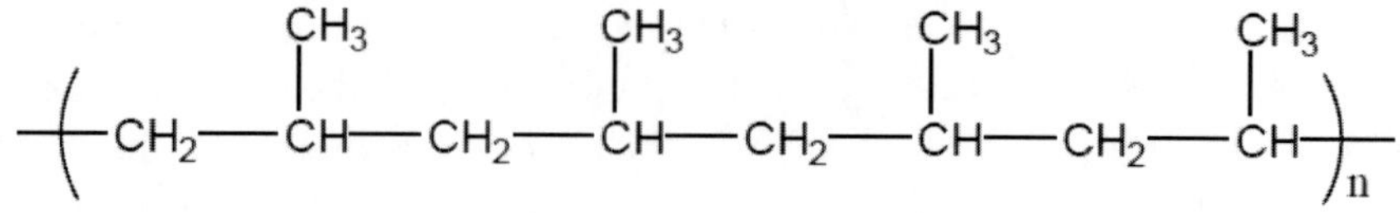

Isotactic polypropylene

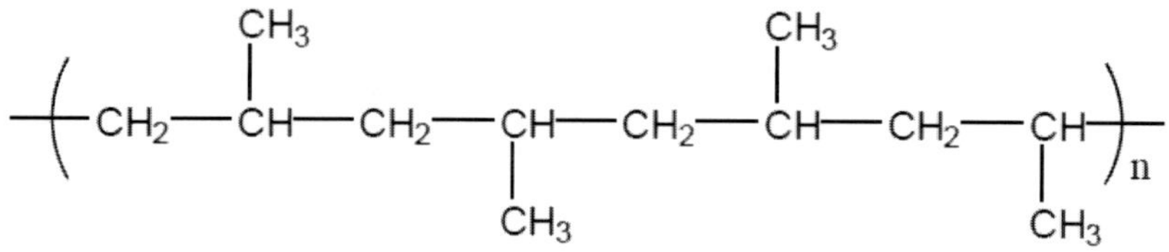

Syndiotactic polypropylene

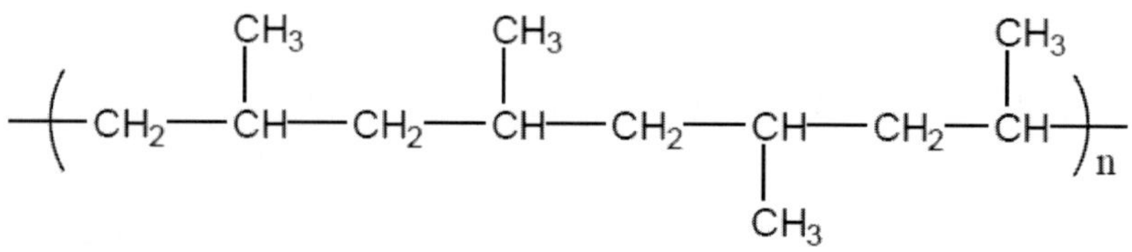

Atactic polypropylene

Figure 4. PP polymer molecule in isotactic, syndiotactic and atactic forms.

PP is derived from three major sources. Globally, most propylene monomer comes from the steam cracking process using naphtha which is a valuable fraction of crude oil. Usually, the target product of naphtha cracker's is ethylene monomer. Propylene is a byproduct of the cracking process and is produced at various ratios depending on the crude oil feedstock. Many cracking processes have a propylene plant intimately connected to effectively collect the propylene that comes from naphtha cracking. The second largest production of propylene comes from the gasoline refining process. Finally, and most recently, a new process by which propane is dehydrogenated to propylene monomer is being used to produce propylene.

The melting of polypropylene occurs as a range. Most commercial PP has an intermediate level of crystallinity which ranges between 40-60% [32]. Perfectly isotactic PP has a melting point of 171 °C (340 °F). Depending on atactic material and crystallinity, commercial isotactic PP has a melting point ranging from 160 to 166 °C (320 to 331 °F) [33] whereas, syndiotactic PP with a crystallinity of 30% has a melting point of 130 °C (266 °F). Some mechanical and thermal properties of commercial PP is shown in Table 1 [34].

Why PP?

PP possesses several useful properties like high heat distortion temperature, transparency, flame resistance, dimensional stability and high IS which widen its application. As a matrix material, PP is widely used because it has some excellent characters for composite fabrication. PP is also very suitable for filling, reinforcing and blending. PP with natural fibrous polymers is one of the most promising routes to create natural–synthetic polymer composites.

Table 1. Some mechanical and thermal properties of commercial PP [34]

Property	Test method	Homopolymer		
Melt flow index (MFI)	(a)	3.0	0.7	0.2
Tensile strength	(b)	500 (lb in^{-2})	4400 (lb in^{-2})	4200 (lb in^{-2})
		34 (MN/m^2)	30 (MN/m^2)	29 (MN/m^2)
Elongation at break (%)	(b)	350	115	175
Flexural modulus	-	190000 (lb in^{-2})	170000 (lb in^{-2})	160000 (lb in^{-2})
		1310 (MN/m^2)	1170 (MN/m^2)	1100 (MN/m^2)
Brittleness temperature (°C)	I.CI./ASTM D746	+15	0	0
Vicat softening oint (°C)	BS 2782	145-150	148	148
Rockwell hardness	-	95	90	90
Impact strength (ft lb)	-	10	25	34

(a) Standard polyethylene grader: load 2.16 kg at 230°C.
(b) Straining rate 18 in/min.

Competitive Materials

PP is most frequently compared with polyethylene (PE). Other competitive materials are polystyrene derivatives along with polystyrene, cellulose acetate butyrate (CAB), cellulose acetate (CA) and polyvinyl chloride (PVC). PP is used to replace engineering plastics like polycarbonate (PC), polyethylene terephthalate (PET), acrylonitrile butadiene styrene (ABS), polyamide (PA), etc., in kitchen appliances and domestic appliances. In non-plastics, PP faces competition from glass and metal.

Major competitive materials for PP and their advantages and disadvantages over PP are given in Table 2. This is a broad comparison only. In many cases, polymers are filled or modified to improve properties or to reduce cost which makes the distinction between the properties of two polymers. Choice of a particular material for a given application will require a careful study of the material properties, product requirements, commercial issues, environmental issues and legislative issues.

Table 2. Comparative advantages/disadvantages of other thermoplastics to PP

Polymer	Advantages	Disadvantages
Polypropylene	High impact resistance Transparency Flame resistance Low cost	Higher mould shrinkage Higher thermal expansion
High density polyethylene	Lower brittle temperature Low cost High impact resistance	Lower heat distortion temperature Lower strength
Low density polyethylene	Lower brittle temperature High impact resistance	Lower strength Lower heat distortion temperature
Acrylonitrile butadiene styrene	Better gloss Better stiffness	High cost High weight Lower solvent resistance
Polyethylene terephthalate	Higher clarity Better impact properties	High cost Unsuitable for hot fill
Polyvinyl chloride	Higher clarity Good weather resistance	Lower heat deflection temperature Worse environmental acceptance

PROPERTIES

The mechanical and thermal properties of PP are dependent on the molecular weight and its distribution, isotacticity, the crystallinity and the type and the amount of comonomer. Additionally, PP is thermoplastic and a viscoelastic material. Mechanical properties of PP are strongly dependent on time, temperature and stress.

The typical density of PP is 0.9 g/cm^3 and it is one of the lightest among the widely used thermoplastics. Polymethylpentene (PMP), a commercially available semi-crystalline transparent thermoplastic, has a lower density (0.84 g/cm^3) than PP. For PE, changes in the degree of crystallinity result in quite large variations in density whereas for PP, the density changes little over the whole range of homopolymer and copolymers. Elastomer-modified, filled or reinforced grades PP might have significantly higher density depending on their formulation. For example, a 40% talc-filled grade has a density of 1.2 g/cm^3.

Unlike metals, plastics are extremely sensitive to changes in temperature. The chemical, electrical or mechanical properties of plastics cannot be considered without knowing the temperature at which the values are derived. The thermal properties of a polymer typically determine its low- and high-temperature applications, impact properties and processing characteristics. The applications where high-temperature properties of PP are of particular interest include sterilization, particularly steam sterilization, parts for dishwashers and washing machine which are subjected to hot water in the presence of detergents, microwave oven proof containers, etc. Typical low-temperature applications for PP are in refrigerator parts and food packaging for refrigerated shelves. Shubhra et al. studied mechanical properties of PP and determined tensile strength (TS), Young's modulus (YM), elongation at break (Eb%), bending strength (BS), bending modulus (BM), impact strength (IS) and hardness of the PP which is shown in Table 3.

Table 3. Mechanical properties of polypropylene [29]

Tensile strength (MPa)	Young's modulus (MPa)	Elongation at break (%)	Bending strength (MPa)	Bending modulus (MPa)	Impact strength (kJ/m^2)	Hardness (Shore A)
22	545	322	27	2050	4.2	95

MANUFACTURING PROGRESS

FRPCs are mainly fabricated by different extrusion [35], injection [36] or by compression molding [37-38] methods. Figure 5 shows these three types of equipment.

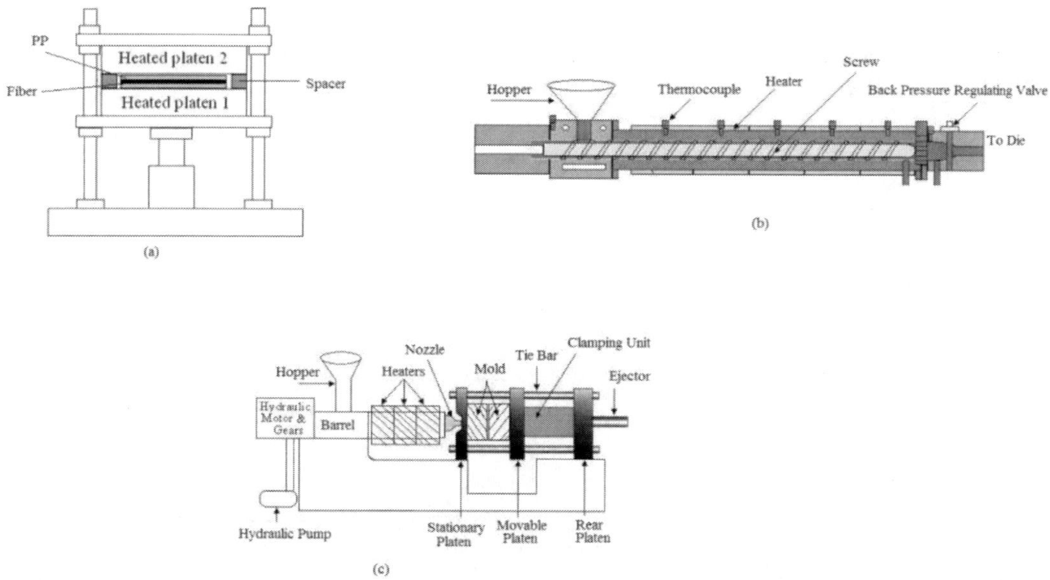

Figure 5. (a) compression, (b) extrusion and (c) injection molding machines.

Compression Molding

Compression molding is one of the oldest manufacturing techniques that use large presses to compress the prepreg material between two matched steel dies [39]. The recent development of high strength, fast cure, sheet molding compounds and advancement in press technology has made this process very popular for mass production of composite parts. In comparison with the injection molding process, better physical and mechanical properties can be obtained in compression molding. This process utilizes large tonnage presses wherein the curing occurs between two matched steel dies under pressure and high temperature. Compression molding basically involves the pressing of a deformable materials charged between the two heated mold, and its transformation into a solid product under the effect of the elevated mold temperature. After placing the laminate to be cured, the cavity is closed. The molds are heated to a high temperature which causes the charge viscosity to be reduced. With increasing mold pressure, the charge flows towards the cavity extremities, forcing air out of the cavity. High pressure helps to eliminate the problem of development of voids. The primary advantage of the compression molding is its ability of producing large number of parts with little dimensional variations. Various shapes, sizes and complexity can be achieved by compression molding. This process has high tooling cost and not practical for low volume production. It is very important to control the cure time otherwise cracking, blistering or warping may occur.

Compression molding processes [40] is the process used for the fabrication of unidirectional FRPCs. Shubhra et al., [39] fabricated unidirectional silk/PP composite using compression molding method. Similarly, Avik et al., fabricated Ca-alginate fiber/PP composite [41], Ruhul et al., phosphate glass/PP [42] composite, Mubarak et al. jute/PP [43] composites in the same way using compression process. Shubhra et al., fabricated PP sheets by using granules of PP (about 6 gm) by placing the granules into two steel plates [29]. The steel plates were then placed into the heat press at 185°C under 6 bar pressure for 1 minute. Then it was cooled for 1 minute in a separate press under 6 bar pressure at room temperature. The PP sheets were cut into desired size for composite fabrication. Composites were prepared by sandwiching three layers of silk fibers between four sheets of PP. Fibers were placed unidirectionally between PP sheets. This was placed against two steel plates and heated at 185°C for 1 minute and then cooled in another press to 6 bar pressure for 1 minute [29] to get unidirectional silk fiber reinforced polypropylene composite.

Injection Molding

Injection molding refers to a process that generally involves forcing or injecting a fluid plastic material into a closed mold of desired shape. The molding compound is fed into injection chamber through the feed hopper. In the injection chamber, the molding compound is heated wherein it changes into liquid form. It is forced into the injection mold by the plunger. This method is normally used for high-volume and low-cost component manufacturing.

Both thermoplastic and thermoset are subjected to injection molding [44]. A thermoplastic material is first melted and then forced through an orifice into the mold which is kept relatively cool. The material solidifies in the mold from which it can then be removed. But in thermoset injection molding, high temperature is required for solidification. Therefore, a reaction material is forced into a generally warm mold in which the material further polymerizes into a solid part. This method is suitable for high-volume and low-cost component manufacturing. But the method is limited to materials with short fibers.

Many researchers fabricated fibers reinforced polypropylene composite by injection molding method. For example, abaca/PP, jute/PP and flax/PP composites were fabricated by Bledzki et al., using injection molding method [45]. Similarly, Arzondo et al., fabricated sisal/PP [46], Thi-Thu-Loan et al., jute/PP [47], Bledzki et al., wood fiber/PP [48], Suhara et al., hemp-glass/PP [49] and Abraham et al., fabricated nylon/PP [50] composites by using injection molding method. For fabrication of abaca/PP composite by injection method, Bledzki et al., dried abaca fibers at 80°C in an oven for 24 hours and then mixed with PP by high speed cascade mixer. Then the fiber and PP at different proportions were placed into hot mixer and heated up to the melting temperature of polypropylene (173°C). Hot agglomerate granules were then transferred to the cool mixer where hot agglomerate granules cools down to room temperature due to application of cold water. Cold agglomerate granules were then dried again at 80°C (24 hours) before the sample preparation by injection molding process. Test samples were prepared from dried agglomerate by injection molding process (temperature zone 150–180°C) maintaining mold temperature of 80°C under an injection pressure of 20 kN/mm^2 [45].

Extrusion

To fabricate FRPCs with unidirectional fibers, extrusion method is not used as this process can only result short fibers reinforced polypropylene composites (few millimeters at most). In an extruder, a screw pushes mixture of natural fiber and PP through a heated cylinder, changing PP from solid state to liquid and mixing with the fibers as it moves through the barrel. This is followed by the die which will give the mixture a constant cross section area. Extrusion method is widely used to fabricate short fiber reinforced polypropylene composites. Many researchers fabricated FRPCs by extrusion method. For example, Mora´n et al., fabricated flax/PP composite [51], Masudul et al., Betel nut short fiber/PP composite [52], Van der over et al., jute/PP [35] and Fu et al., fabricated glass fiber/PP and carbon fiber/PP [53] using extrusion method. For fabrication of flax/PP composite, Mora´n et al., cut flax fiber to an approximate length of 10 mm to make the fiber's incorporation easier in the extruder. He dried the fibers (vacuum dry) in an oven at 70ºC for 24 hours. Fibers and PP were blended in a twin screw extruder. Fibers and polypropylene were extruded and pelletized five times. Materials extracted after each extrusion step were compression molded to get flax/PP composite [51].

MECHANICAL PROPERTIES OF FRPCS

For fiber reinforced composites, mechanical properties are of major concern as reinforcement affects the mechanical properties of FRPCs. Tensile, bending, impact strength and hardness are some mechanical properties that are considered significantly for a fiber reinforced PP composite. The mechanical properties of FRPCs depend on properties of fibers. Some properties of different natural and synthetic fibers are listed in Table 4.

For fiber reinforced composites, mechanical properties are of major concern as reinforcement affects the mechanical properties of FRPCs. Tensile, bending, impact strength and hardness are some mechanical properties significantly considered for a fiber reinforced PP composite material. The mechanical properties of FRPCs depend on properties of fibers.

The strength of a unidirectional fiber reinforced composite, in the direction of fibers is given by the rule of mixture:

$$\sigma_{cL} = \sigma_f V_f + \sigma_m^{'} (1-V_f) \quad (1)$$

where V_f is the volume fraction of fibers, σ_f is the tensile strength of fibers and $\sigma_m^{'}$ is the stress developed in the matrix. For composites with long parallel fibers, the Young's modulus (E_c) in the fiber direction is calculated using the following equation:

$$E_c = E_f V_f + E_m V_m \quad (2)$$

where E_c, E_f, E_m are the Young's moduli of the composite, fiber, and matrix respectively, and V_f and V_m are the volume fractions of the fiber and matrix, respectively.

Table 4. Properties of different fibers [34]

Fiber	**Tensile strength (MPa)**	**Young's modulus (GPa)**	**Density (gcm^{-3})**
Cotton	330-585	4.5–12.6	1.5–1.54
Flax	345-1035	27.6–45.0	1.43–1.52
Hemp	690-1000	50.0	1.47–1.50
Jute	393-800	13-26.5	1.3-1.45
Silk	650-750	16	1.3-1.38
Kenaf	930	53.0	1.5
Ramie	400-1000	61.5	1.5–1.6
Sisal	511-635	9.4–15.8	1.16–1.5
Banana	500-700	7-20	1.4
Softwood	100-170	10-50	1.4
Hardwood	90-180	10-70	1.4
E-glass	1800	69.0–73.0	2.5
HM carbon	2400	380	1.95
HS carbon	3400	230	1.75
Kevlar 49	3000	130	1.45

Tensile strength [38] is the maximum stress that a material can withstand without tearing apart. It is a measurement of the ability of a material to withstand forces that tend to pull it apart. Tensile strength is measured in unit of force per unit area. Tensile properties of FRPCs can be determined according to ASTM D638 [50, 54] or according to DIN 53455 method [55]. Tensile strength of FRPCs is determined by the following equation:

$$\sigma_{fu} = F_u/A_f \quad (3)$$

where, F_u is the force at failure, A_f is the average filament cross sectional area.

Bending strength [56], also known as flexural strength is defined as a material's ability to resist deformation under load. The bending strength represents the highest stress experienced within the material at its moment of rupture. There are two methods that cover the determination of bending properties of material: three-point loading system and four point loading system. Prior to 1997, one ASTM standard covered both loading modes which is ASTM D 790. However since 1997, only three-point loading is included in ASTM D 790 and a new ASTM standard has been created for four-point loading which is ASTM D 6272. Without ASTM standard, DIN 53452 method [55] can be used also to determine bending properties of FRPCs. For a rectangular sample under a load in a three-point bending setup, the bending strength is calculated by the following formula:

$$\sigma = 3FL/2bd^2 \quad (4)$$

where F is the load (force), L is the length of the support span, b is width and d is thickness. For a rectangular sample under a load in a four-point bending setup where the loading span is one-third of the support span, bending strength is calculated by the following formula:

$$\sigma = FL/bd^2 \quad (5)$$

where F, L, b, d have the same meaning as indicated above. But if the loading span is 1/2 of the support span, bending strength is calculated by the following formula:

$$\sigma = 3FL/4bd^2 \quad (6)$$

where F, L, b, d have the same meaning as indicated above.

Impact strength [57] is the ability of a material to resist suddenly applied loads. In most cases, impact test of FRPCs are carried out by two different types of testing: Charpy test and Izod test. Izod impact strength of FRPCs can be determined according to ASTM D256 [58] or DIN EN IS0 180 and Charpy impact strength can be determined according to ASTM D 6110 or DIN EN IS0 179 [55].

Mechanical properties of different fibers reinforced polypropylene composites are studied by many researchers. Shubhra et al., worked on silk fiber/PP composites [59], Avik et al., worked on calcium alginate fiber/PP [41], Barkoula et al., flax/PP [60], Sushanta et al., bamboo fiber/PP [61], Ajay et al., wood fiber/PP [62], Mominul et al., abaca fiber/PP [63], Haydaruzzaman et al., coir fiber/PP [64], Mominul et al., palm fiber/PP [65] and Shubhra et al., phosphate glass fiber/PP [66] composites and their studies are discussed in this section. They all used 10% fibers for composite fabrication. Besides, Shubhra et al., worked on 20% jute fiber reinforced PP [29] and Ruhul et al., worked on 50% E-glass fiber reinforced PP composites [67] and for convenience, these two studies are also included in the discussion of this section. Tensile strength (TS), tensile modulus (TM), bending strength (BS), bending modulus (BM) and impact strength of different FRPCs are shown in Figure 6-10, respectively.

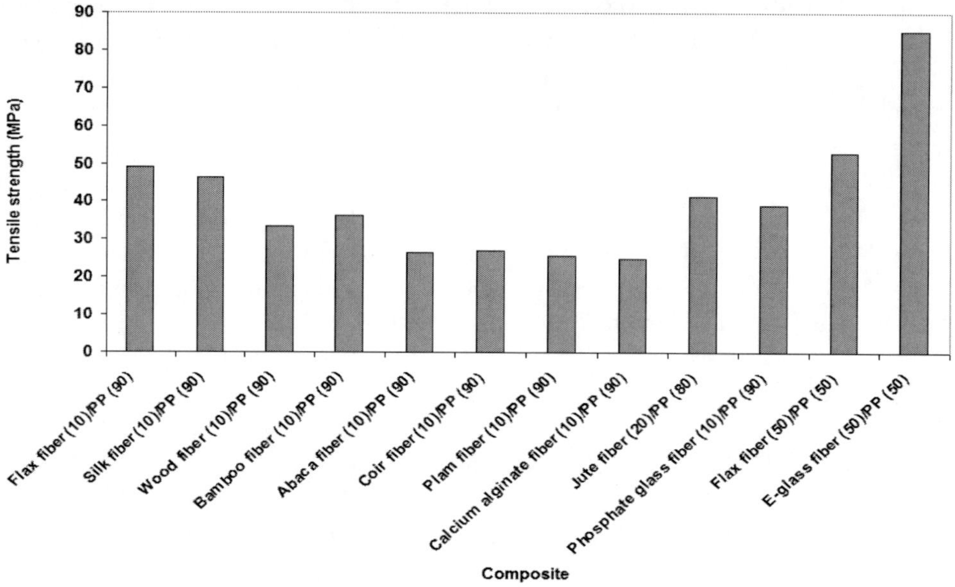

Figure 6. Tensile strength of different FRPCs [34]. [Reprinted from Journal of Thermoplastic Composite Materials, Shubhra et al., Mechanical properties of polypropylene composites: A review, with permission from SAGE Journals].

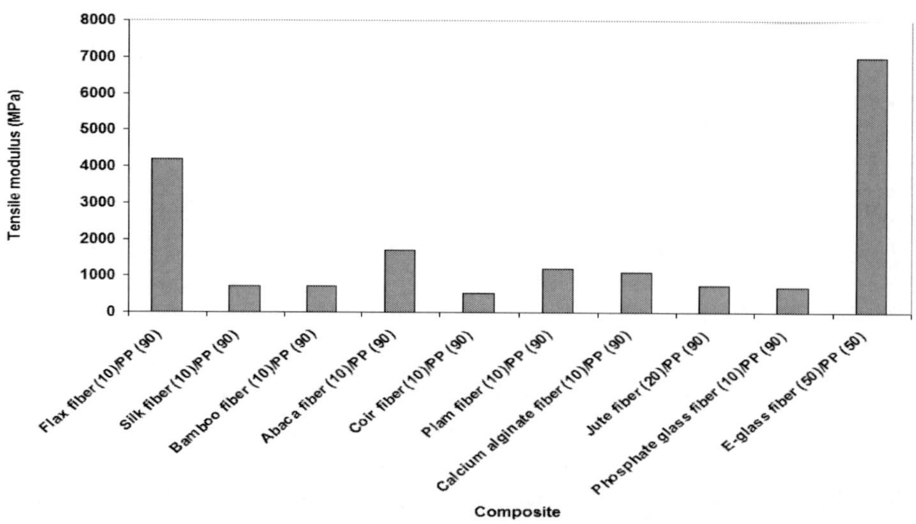

Figure 7. Tensile modulus of different FRPCs [34]. [Reprinted from Journal of Thermoplastic Composite Materials, Shubhra et al., Mechanical properties of polypropylene composites: A review, with permission from SAGE Journals].

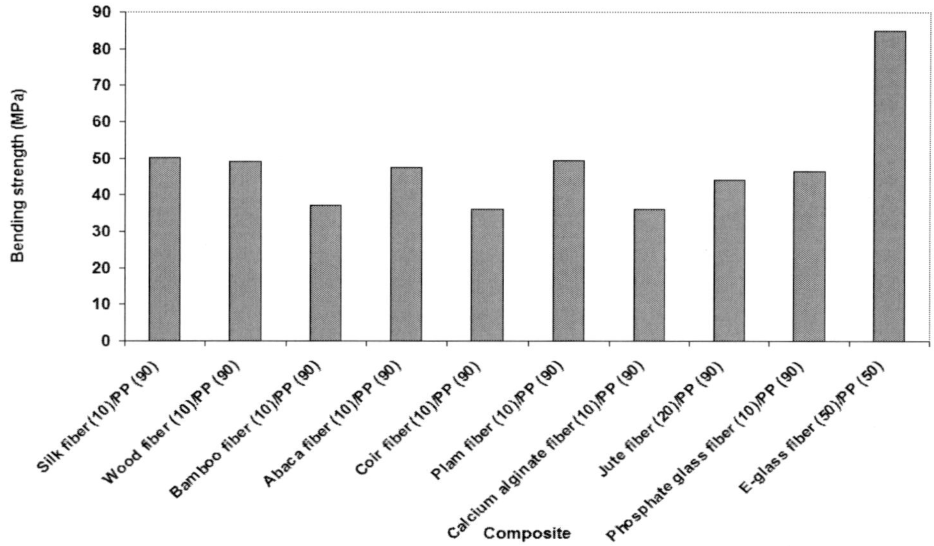

Figure 8. Bending strength of different FRPCs [34]. [Reprinted from Journal of Thermoplastic Composite Materials, Shubhra et al., Mechanical properties of polypropylene composites: A review, with permission from SAGE Journals].

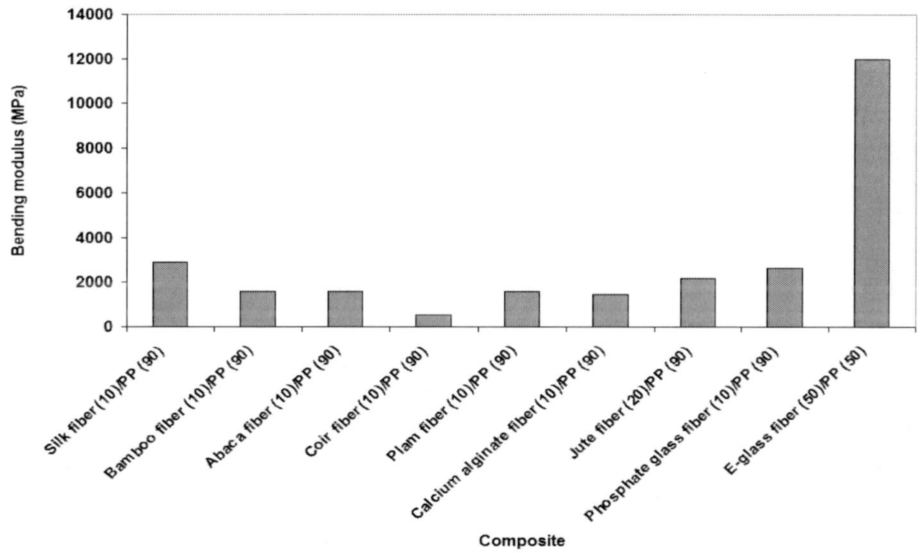

Figure 9. Bending modulus of different FRPCs [34]. [Reprinted from Journal of Thermoplastic Composite Materials, Shubhra et al., Mechanical properties of polypropylene composites: A review, with permission from SAGE Journals].

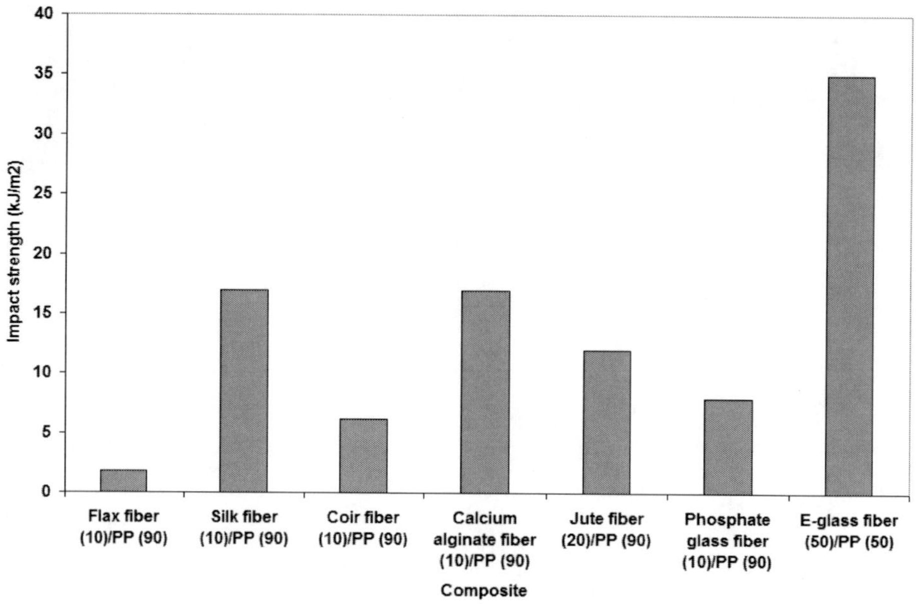

Figure 10. Impact strength of different FRPCs [34]. [Reprinted from Journal of Thermoplastic Composite Materials, Shubhra et al., Mechanical properties of polypropylene composites: A review, with permission from SAGE Journals].

From the figures it is clear that synthetic E-glass/PP composites have greater mechanical properties than that of natural fibers/PP composites. It is observed from the figures that, among natural fibers/PP composites, flax fiber/PP composites have highest tensile and bending properties. For convenience, tensile strength of 50% flax fiber reinforced PP composite is shown in Figure 6 to compare the strength with that of E-glass fiber/PP composite containing same percentage of fiber content. It is clear from Figure 6 that 50% flax fiber/PP composite have tensile strength below 60 MPa [34] whereas, 50% E-glass fiber reinforced PP composites have tensile strength more than 80 MPa. These values make it clear that E-glass fiber/PP composites have superior tensile strength than that of any other natural fiber/PP composites. Among natural fibers studied in this article, flax fiber offers highest reinforcement effect. A reason behind this may be the morphology of flax fiber.

It is a generally accepted fact that the strength of a fiber reinforced polymer matrix composite is mainly dependent on the fiber's strength. From Table 4, it is interesting to note that, among all natural fibers listed in that table, flax has highest tensile strength. Thus high strength is preferred for flax containing PP composite over other natural fiber's composites which is also found in practice and reflected in the results shown in Figure 6. Flax fiber/PP composites showed a better reinforcing effect due to the high matrix ductility and high strength/modulus ratio of flax fibers.

Utilization of fibers with a higher fineness generally leads to improved properties for composite materials [68]. Fiber diameter is also important since increase in fiber diameter after a certain value results decrease in strength of composites as found for many fibers like coir, banana, sisal, silk, jute, etc., [69-70]. Since, with the increase in fiber diameter, fiber strength decreases, fibers with more diameters when reinforced with PP for composite fabrication will result lower strength. Flax fibers show a higher fineness and possess less fiber diameter compared to hemp or kenaf. Strength of flax/PP is greater than hemp/PP or kenaf/PP [68]. High fiber fineness and small diameters generally result better fiber embedment and therefore composites with higher strength will be obtained from those fibers. Moreover, higher fiber fineness and diameter should lead to an improved ratio between surface and volume and therefore an increased contact surface between fiber and PP matrix will be obtained [68].

Impact strength of FRPCs is a measure of the ability of the composites to resist the fracture failure under stress applied at high speed and is directly related to the toughness of the composites. It is generally accepted that the toughness of a fiber composite is mainly dependent on the stress-strain behavior of fiber. Strong fibers with high failure strain impart high work of fracture on the composites. Fibers play an important role in the impact resistance of FRPCs as they interact with the crack formation and act as stress-transferring medium [71]. According to Ruhul et al., the virgin PP shows very low impact strength (4.47 kJ/m^2). He observed that, addition of 50 wt % of jute fiber increased the impact strength by 302% whereas addition of 50 wt % of E-glass fiber increased the impact strength by 683% [67]. Microfibrillar angle in plant fibers plays an important role in determining the impact behavior of FRPCs and this effect should be taken into account along with the other parameters.

Generally, strength of any fiber reinforced polymer matrix composites depends on: (1) strength of the fiber used as reinforcement and (2) the interfacial interaction between fiber and the matrix. The major drawbacks associated with the use of natural fibers as reinforcement in polypropylene matrix are the poor wettability and weak interfacial bonding with the PP due to the inherently poor compatibility as well as dispersability of the hydrophilic cellulose fibers with the hydrophobic PP [34]. When load is applied to a FRPC, PP transfers the stress to the reinforced fibers. Effective transfer of stress and load distribution throughout the interface is possible when strong adhesion exists at the interfaces.

FIBER MATRIX ADHESION

For FRPCs, the stress transfer at the interface between the matrix and fiber phases is determined by the degree of adhesion. Effective transfer of stress and load distribution throughout the interface is possible when strong adhesion exists at the interfaces. The adhesion phenomenon is described in several theories including mechanical interlocking, absorption, primary bonding, interdiffusion, electronic theories, etc.

Testing of Adhesion in FRPCs: Micromechanical Characterization

Interfacial shear strength (IFSS) is measured by the pullout test. In the first step of this test, one end of the fiber is embedded in a block of a polymer matrix and in the second step a force is applied to the free end to pull it out of the matrix. Applied force is continuously monitored and recorded and the force at which debonding occurs is obtained. The average IFSS can be calculated from using the following equation:

$$\tau = F / \pi \times d \times l \quad (7)$$

Here,

F= Maximum load measured prior to debonding of the fiber,
d = The diameter of used fiber, and
l = The length of embedded portion of fiber.

Single-fiber fragmentation test (SFFT) is also used to calculate the IFSS. A single fiber is embedded in the matrix which is then converted to a dogbone-shaped tensile coupon using appropriate method (e.g. cutting). The specimen is then subjected to a tensile load and tensile forces are transferred from the matrix to the fiber. Depending on the level of fiber-matrix adhesion, build up of tensile stresses occur within the fiber. After certain level of tensile strength, stress concentration reaches to a certain high level at which debonding occurs resulting fracture. This loading process continues until the fiber fragmentation process ceases. The final fiber fragment length is termed as the fiber critical length and denoted by lc. The critical length (l_c) can be measured using the following formula: $l_c = 4/3\ l_f$, where l_f indicates average fragment length [72]. Critical length is a good indicator of the ability of the interphase to transmit loads between the matrix and fiber. Kelly and Tyson [73] developed the equation by which IFSS is calculated using this length. The equation is as follows:

$$\tau_i = d \times \sigma_f / 2 \times l_c \quad (8)$$

In equation (8), d is the diameter of fiber used, σ_f is the tensile strength of single fiber at the critical length. The ratio (l_c/d) can be used as an indicator of the fiber-matrix bond strength.

Drzal et al [74] altered the above equation reflecting Weibull statistics and rearranged the Kelly and Tyson's equation, proposing the following modification to calculate the IFSS:

$$\tau_i = \sigma_f \times \Gamma \times [1\text{-}1/\alpha]/ 2\beta \quad (9)$$

In equation (9), β is the scale and α is the shape parameters in the Weibull distribution and Γ is the Gamma function. The strength of fiber can be calculated from the extrapolation gauge length using Weibull chart [75]. At the critical fragment length, the strength of fiber is found by the following equation [72]:

$$\sigma_f = \sigma_{f0}\ (l_c / l_0)\text{exp-}1/\rho \quad (10)$$

In equation (10), σ_{f0} is the fiber strength at gauge length l_0 and ρ is the shape parameter of the Weibull distribution for the tensile strength of fiber.

Karlsson et al., [76] used the Wilhelmy plate method to measure the apparent perimeter of regenerated cellulose. He used this parameter to calculate the IFSS. Due to inherent variability of the henequen fibers (*Agave fourcroydes*), equations (8) and (9) were rearranged to use the perimeter P_f instead of the diameter to calculate the IFSS. Following this approach the calculation of IFSS for the pullout test was done using the following equation:

$$\tau_i = F_f \ / P_f \times l_e \quad (11)$$

In equation (11), F_f is the debonding force, P_f is the measured perimeter of the henequen fiber and l_e is the fiber-embedded length. In the case of the single fiber fragmentation test the IFSS is calculated with the following equation:

$$\tau_i = 2F_f / \beta \times \Gamma \times P_f \times [1+1/\alpha] \quad (12)$$

In equation (12), F_f is the tensile load at break for the henequen fiber, P_f, α, β and Γ have the same meaning as indicated above.

Ruhul et al., [72] studied IFSS of phosphate glass/PP composites. He prepared single fiber composite samples using one single filament of fiber between two sheets of polypropylene. He performed fragmentation test using universal testing machine and he counted the number of fragments by microscope. The author found the critical length 3030 μm for the stated system. He measured tensile strength of the fiber at the critical length from the Weibull weakest chart rule. The author found the IFSS of phosphate glass fiber/PP system 5.93 MPa. He concluded that, his result suggests excellent fiber matrix adhesion between phosphate glass fiber and PP matrix.

IMPROVEMENT OF MECHANICAL PROPERTIES OF FRPCS

By Fiber Treatment

Generally fiber treatment is carried out for natural fibers so that their reinforcement increase properties of fiber reinforced composites. The major drawbacks associated with the use of natural fibers as reinforcements in polypropylene matrix are poor wettability and weak interfacial bonding with the polypropylene due to the inherently poor compatibility as well as dispersability of the hydrophilic cellulose fibers with the hydrophobic polypropylene. So in order to improve the fiber-matrix adhesion a pre-treatment of the fiber surface or the incorporation of surface modifier is required. Several studies have been reported based on the influence of various types of chemical modification on the physical and mechanical properties of natural fiber filled polypropylene composites.

Various fiber surface treatments such as alkali/mercerization [77-78], silane [79-80], combination of alkali and silane [81-82], monomer grafting under UV radiation [83-84], acetylation, benzyolation, etc. [85-86] have been reported by several authors.

A lot of work has been done by many researchers on fiber treatment to improve the mechanical properties of fiber/PP composites. In this section few of them are discussed. TS and BS of treated and untreated palm fiber/PP [65], abaca fiber/PP [63], coir fiber/PP [63] and calcium alginate fiber/PP [41] composites are shown in Figure 11 and 12, respectively. In those composites, fiber content was 10 wt%. From Figure 11 and 12, it is clear that fiber treatment sharply increases TS and BS of composites.

Haydar et al., has studied the effect of surface treatment of jute with $KMNO_4$ and found increased mechanical properties of jute/PP composites due to that treatment [87]. For 0.03 % $KMnO_4$ treated jute/PP composite at 2 minutes soaking time, the author found maximum 20% and 24% increase in TS and TM values, respectively than that of untreated one. In another study [55], the same author treated jute with starch solutions of different concentrations varying treating times. Maximum TS, BS and IS was found for 5% starch treated composites for 5 min soaking time and the values were 57.25 MPa, 65.34 MPa and 27.5 kJ/m^2, respectively which were 19, 27 and 60 % higher than the TS, BS and IS values of untreated composites [55]. Mubarak et al., studied surface modification of jute with acrylic monomers

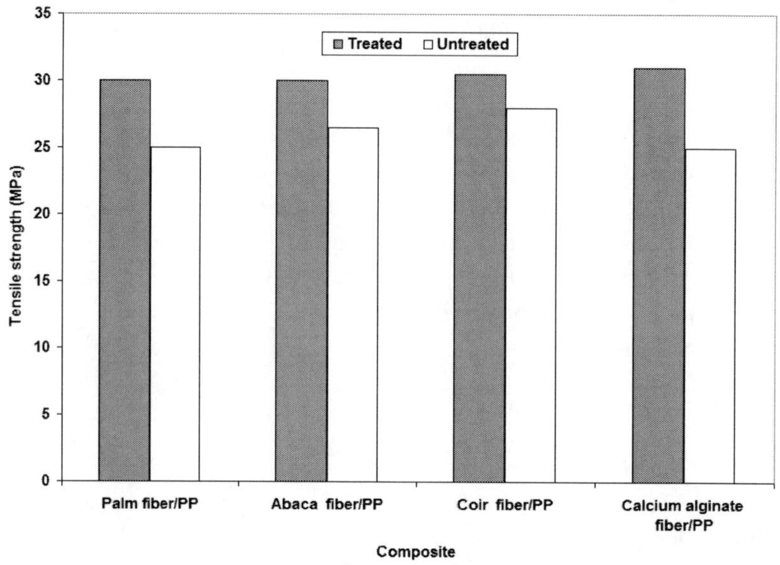

Figure 11. Tensile strength of treated and untreated FRPCs [34]. [Reprinted from Journal of Thermoplastic Composite Materials, Shubhra et al., Mechanical properties of polypropylene composites: A review, with permission from SAGE Journals].

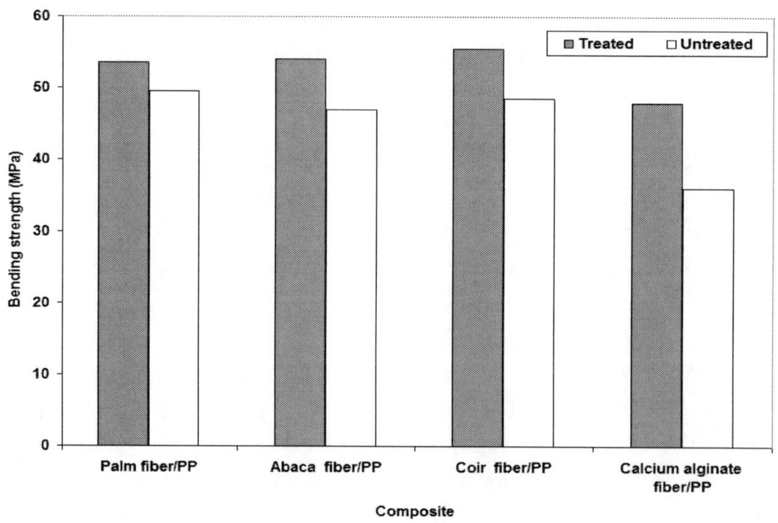

Figure 12. Bending strength of treated and untreated FRPCs [34]. [Reprinted from Journal of Thermoplastic Composite Materials, Shubhra et al., Mechanical properties of polypropylene composites: A review, with permission from SAGE Journals].

[88]. He prepared different formulations using urethane diacrylate oligomer, 2-ethyl hexyl acrylate (EHA) monomers, 1-vinyl 2-pyrrolidone (NVP) and photo-initiator in methanol to modify the surface of jute. After soaking in the prepared formulations, he cured jute by UV radiation source of varying intensities. The author found that, the formulation containing 5% NVP, 5% EHA, 2.5% oligomer and 2% photo-initiator and cured at 15 UV passes showed the highest mechanical properties. For the stated treatment of jute, the author found composites with 20 and 15% more TS and BS values respectively than that of untreated one [88].

Haque et al., chemically treated palm fiber using benzene diazonium salt. According to him, the hydroxyl groups in the raw palm is responsible for high water absorption and weak interfacial bonding between the palm fiber and PP matrix. Diazonium salt converts two hydroxyl groups into diazo group and results azo product [65]. Due to the elimination of most of the hydroxyl groups from the treated palm fiber, the interfacial bonding between the palm fiber and PP matrix increased resulting composites of high tensile, bending and impact strength [65]. Haque et al., also studied the chemical treatment of abaca and coir fiber with benzene diazonium salt. Due to the treatment, the interfacial bonding between the abaca and coir fiber with PP matrix increased resulting composites of high tensile, bending and impact strength [63].

Guillermo et al treated flax fibers by esterification. He also treated the fiber with silane. The author found that treated fibers result composites of higher mechanical properties than the untreated one. Flax fibers were esterified with maleic anhydride (MA) dissolved in boiling acetone (T=50±5±°C) with a fiber: solvent ratio 1:25 (w/v) [14]. For silanization, he dissolved vinyl trimethoxy silane (VTMO) in acidified water (pH=3.5) for 10 minutes to get a better functionalization. Then the fibers were added and kept 1 hour in the solution for obtaining a 2.5 wt% silane with respect to the weight of flax fiber [14]. Arbelaiz et al., treated flax fiber with alkali and found increased mechanical properties for treated composites. For alkalization, he soaked flax fibers in a 20wt% aqueous solution of sodium hydroxide for 1 hour at room temperature followed by washing. He washed fibers several times in distilled water followed by neutralization with a few drops of acetic acid. The fibers were then washed again and finally dried in an oven. The ratio of fiber weight to alkali solution volume was 1:20 [89].

Joseph et al., worked on sisal fiber/PP composites [90]. Treatments of sisal fibers with chemicals such as sodium hydroxide, maleic anhydride and potassium permanganate were carried out to improve the bonding at the fiber–polymer interface. It was observed that all the treatments enhanced the tensile properties of the composites. Untreated composite had TS of 36.5 MPa whereas alkali treated, $KMnO_4$ treated and maleic anhydride treated composites had TS of 44.35, 40.3 and 42 MPa, respectively [91].

Avik et al., treated calcium alginate fibers with vinyl triethoxy silane [H_2=CH-Si-$(OC_2H_5)_3$] (5%, v/v) to reduce strong hydrophilic nature of the fiber and to improve the mechanical and interfacial properties of calcium alginate fiber reinforced PP composites [41]. The author found that TS, BS and IS of untreated calcium alginate fiber/PP was 25 MPa, 36 MPa and 17 kJ/m^2 respectively whereas corresponding values for treated calcium alginate fiber/PP were 31 MPa, 48 MPa and 21 kJ/m^2 [41].

Anuar et al., employed sulfuric acid surface oxidative treatment on carbon fiber. Firstly he heated sulfuric acid of 1M concentration to 70°C and then soaked the fibers into the acid solution for 90 minutes. Later, he refluxed the carbon fiber using 1M sodium hydroxide for 60 minutes. Then, the carbon fibers were re-soaked in distilled water for 5 days. Finally, rinsing of carbon fibers were carried out using distill water. Then the fibers were dried for 3 hours at 110°C. Dried carbon fibers were kept in a desiccator to minimize the moisture content prior to use. Comparing with untreated carbon fiber composites, the author found 11% increase in tensile strength for treated carbon fiber composites [92].

In all these studies improved mechanical properties of FRPCs were found due to use of treated fiber as reinforcement. Thus it can be concluded that, to improve mechanical properties of FRPCs, proper fiber treatment is an effective way.

By Using Coupling Agent

Polypropylene is very versatile polymer. There are many ways through which the mechanical properties of polypropylene can be increased. Fillers, such as talc and calcium carbonate, are often used with PP to produce a low cost material. Most reinforcements used for PP are polar in nature. Polar groups affect the fiber–matrix adhesion to a large extent. Polypropylene on the other hand is nonpolar. In absence of polar groups on PP, only physical and/or mechanical forces like friction influences the interface properties.

To overcome these problems, modification of the chemistry of the PP is done by attaching polar groups onto the molecular backbone, such as acrylic acid (AA) or maleic anhydride (MA). Lots of coupling agents are used in the research of FRPCs. Among them, maleated polypropylene (MAPP) is the most popular as well as the most effective one [93]. Recent works suggest that the use of maleated polypropylene significantly improves the fiber-matrix adhesion [94]. This section is devoted to the study of the widely used polymeric coupling agent, maleated polypropylene and mechanical properties of fiber/PP composites containing the coupling agent MAPP.

Without MAPP, the tensile and bending strengths of FRPCs are lower than that containing MAPP which suggests that there is comparatively little stress transfer from the matrix to the fibers for FRCPs containing no MAPP than that containing MAPP [94]. The use of MAPP improves interaction and adhesion between the fibers and matrix leading to better stress transfer from matrix to fiber. Incorporation of polar maleic anhydride grafted polypropylene, improves the fiber dispersion and fiber/matrix interfacial interaction through hydrogen bonding between hydroxyl groups of natural fibers and carbonyl groups of the maleic anhydride segment of the MAPP [34]. Due to hydrogen bonding, the rate of moisture absorption reduces and the mechanical strength in the composites increases [34].

Xiang et al., studied glass fiber/PP composites. In his study, he used three different compatibilizers including octane–ethylene copolymer, maleic anhydride grafted octane–ethylene copolymer and maleic anhydride. The author found that these compatibilizers had different adhesion effects in the order: PP-g-MAH>POE-g-MAH>POE i.e., glass fiber/PP containing MAPP resulted highest mechanical properties as it showed most adhesion between fiber and matrix [95]. Barkoula et al., studied flax fiber/PP composites (PP with and without 3 wt.% MA-PP). The author found sharp increase in the strength of flax/PP composites when he added MAPP to the composites [60].

Sanjay et al., found that the tensile strength and modulus of hybrid glass and sisal (15% glass and 15% sisal) fiber reinforced PP composites containing 2wt% MAPP increases 17.74 and 14.52%, respectively than that containing no MAPP [96]. According to him, this is primarily due to the covalent bonding between the anhydride groups of MAPP and hydroxyl groups of the sisal fiber. Chain entanglement between MAPP and PP chains also facilitates the stress transfer at the interface. He also claimed that, incorporation of 2wt% MAPP to the hybrid composites prepared using 15: 15 wt% of sisal: glass fibers enhances the tensile strength and modulus due the formation of strong covalent linkage between the –OH groups of sisal fiber and SiO groups of glass fibers with MAPP which results an improved interfacial adhesion between the matrix and both type of fibers [96]. Mohanty et al., also studied flax/PP without and with MAPP and found that the later one gives greater mechanical properties [97]. Sushanta et al., worked on hybrid bamboo/glass fiber reinforced PP composites and he also found that MAPP content upto 2% increases the mechanical properties of composites. But 3% MAPP content in the composite results lower mechanical strength than that containing 2% MAPP [61].

Kazayawoko et al. [98] studied the effectiveness of MAPP in bleached kraft pulp (BKP) and unbleached thermo-mechanical pulp (TMP)-PP composites. The author found that surface treatment of BKP and TMP increased the tensile strength of the composites by 28 and 27%, respectively. Yuan et al. [99] used MAPP as a coupling agent for PP-waste news paper flour composites and reported improved compatibility and hence mechanical properties of his prepared composite. Sanadi et al. [100] reported that improvement in properties by using MAPP depends on the amount of MA in the graft copolymer and the molecular weight of the copolymer. Similarly Mariano et al., worked on hemp/PP composites [101], Arbelaiz et al., worked on flax/PP [89], Sam-Jung et al., worked on cotton/PP [102] composites and all of them found that MAPP results greater mechanical properties.

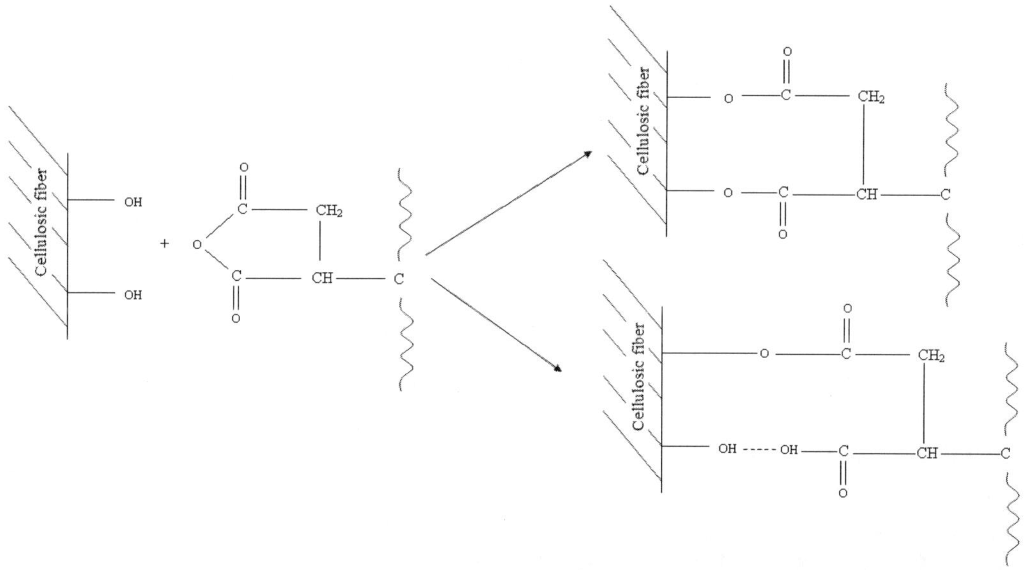

Figure 13. Bond formation between hydroxyl groups of natural fiber and anhydride ring of MAPP [34]. [Reprinted from Journal of Thermoplastic Composite Materials, Shubhra et al., Mechanical properties of polypropylene composites: A review, with permission from SAGE Journals].

When MAPP is used as coupling agent, bonding between the anhydride groups of MAPP and hydroxyl groups of the cellulosic fibers occurs as shown in Figure 13 [34]. This is the situation for plant based cellulosic fibers reinforced PP composites containing MAPP. For proteinous animal based fibers, bonding between amino group of proteinous fibers and MAPP can result increased mechanical properties of FRPCs. In case of synthetic fibers, bonding between MAPP and synthetic fibers occur. Reaction occur between MAPP and carbon fiber [92], the anhydride groups of MAPP and SiO groups of glass fibers [96] resulting improved mechanical properties when MAPP is used with synthetic fibers reinforced PP composites.

Thus MAPP increase the mechanical properties of FRPCs but the content of MAPP should not be high. Because at higher MAPP content, a marginal decrease in mechanical properties occur which may be explained due to self-entanglement of MAPP resulting from the migration of excess MAPP around the fiber surface rather than causing inter-chain entanglement and contributing to the mechanical continuity of the system [99]. Table 5 shows effect of concentration of MAPP on mechanical strength of 30% sisal fiber/PP composites [34].

Table 5. Effect of concentration of MAPP on mechanical strength of 30% sisal fiber/PP composites [34]

MAPP Grade	Concentration of MAPP (in %)	Tensile strength (MPa)	Bending strength (MPa)	Impact strength (J/m)
G-3015	0	29.25	48.96	51.79
	0.3	32.35	50.13	52.45
	0.5	35.44	52.44	57.16
	1	43.66	62.42	68.66
	2	34.55	51.16	55.30
HC_5	0	29.25	48.96	51.79
	0.3	32.82	51.10	57.43
	0.5	41.07	60.25	74.40
	1	43.84	63.66	81.57
	2	33.95	52.13	57.38

By Changing Fiber Loading

It is often observed that increase in fiber content leads to the increase in the strength and modulus of a fiber reinforced polymer matrix composite [34,103]. Joseph et al., [104] have studied the influence of interfacial adhesion on the mechanical and fracture behavior of short sisal fiber reinforced polymer composites of several thermoset resin matrices and a thermoplastic matrix with respect to fiber length and fiber loading. He observed that all the composites showed a general trend of increasing properties with fiber loading.

Generally with the increase in fiber content, TS and BS of FRPCs increases upto a certain limit and then decreases with further increase in fiber content. Sushanta et al., has worked on bamboo fiber reinforced PP composites [61]. He observed that the tensile strength increases with the increase in bamboo fiber content up to 30wt% bamboo fiber content in PP with a subsequent decrease in tensile strength at 40wt% of fiber loading. According to Sushanta, this increase in the tensile strength is due to the increased wt% of the fiber loading within the PP matrix leading to an efficient stress transfer from the PP to the fiber [61]. Masudul et al., worked on betel nut fiber/PP composite and found maximum TS and BS for 10% fiber content in PP. The author observed that increase in fiber content from 3 to 5, 5 to 10 also increase TS and BS of that composite but when the fiber content is 20 wt%, those strengths decrease [52]. Haydaruzzaman et al., studied the effect of coir fiber loading on the mechanical properties of coir/PP composite [64]. The author found that TS and BS increase with increasing the percentage of coir yarn up to 20wt% in the composite and then starts decreasing with the increase in fiber loading. Mubarak et al., observed that TS and BS of jute/PP composite increases with increasing the percentage of jute content up to 40wt% in the composites and after that increase in fiber content results decrease in TS and BS values [88]. Mominul et al., observed that TS and BS of palm fiber/PP composites increase with the increase in percentage of palm fiber content up to 10% in the composites and after that both TS and BS decreased [65]. Figure 14 and 15 show effect of fiber content on the TS and BS of different FRPCs. From these two figures, it is clear that tensile and bending strength of FRPCs increase with the increase in fiber content upto a certain value of fiber content and after that further increase in fiber content results decrease in those strength. This is due to fact that, after a certain range, agglomeration of fibers result and they cannot share the stress and hence reduce the strength of composites.

At lower levels of fiber content, the composites show poor mechanical properties due to poor fiber population and low load transfer capacity to one another. As a result, stress gets accumulated at certain points of the composites and highly localized strains occur in the matrix [64]. At intermediate levels of fiber loading, fibers actively participate in stress transfer [64]. Decrease in strength of fiber/PP composite at higher fiber content is a direct consequence of poor fiber/matrix adhesion which leads to micro-crack formation at the interface under loading and non-uniform stress transfer due to the fiber agglomeration in the matrix [61]. Higher wt% of fiber content also leads to an increase in fiber–fiber interaction which results difficulties in dispersion of the fibers within the PP matrix [61].

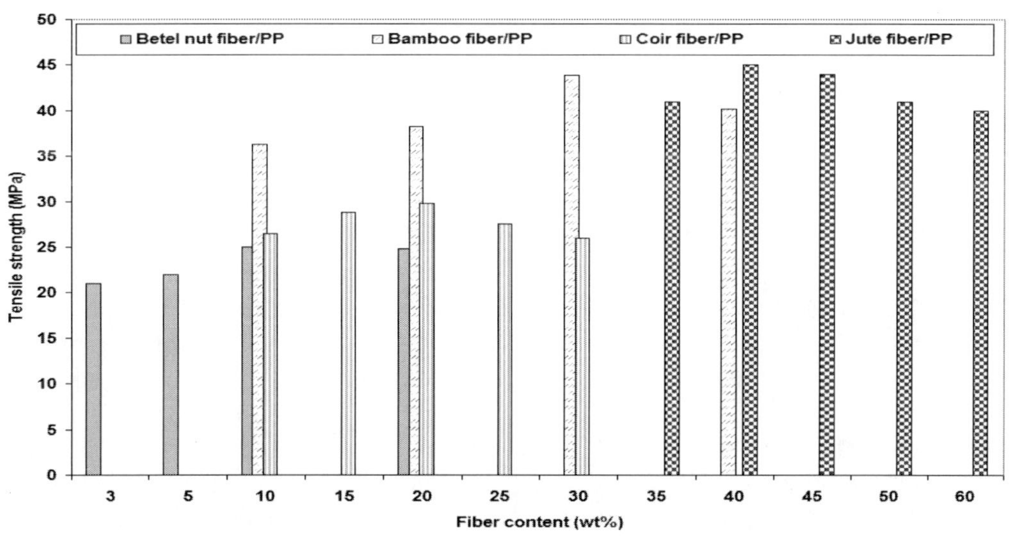

Figure 14. Effect of fiber content on the tensile strength of different FRPCs [34]. [Reprinted from Journal of Thermoplastic Composite Materials, Shubhra et al., Mechanical properties of polypropylene composites: A review, with permission from SAGE Journals].

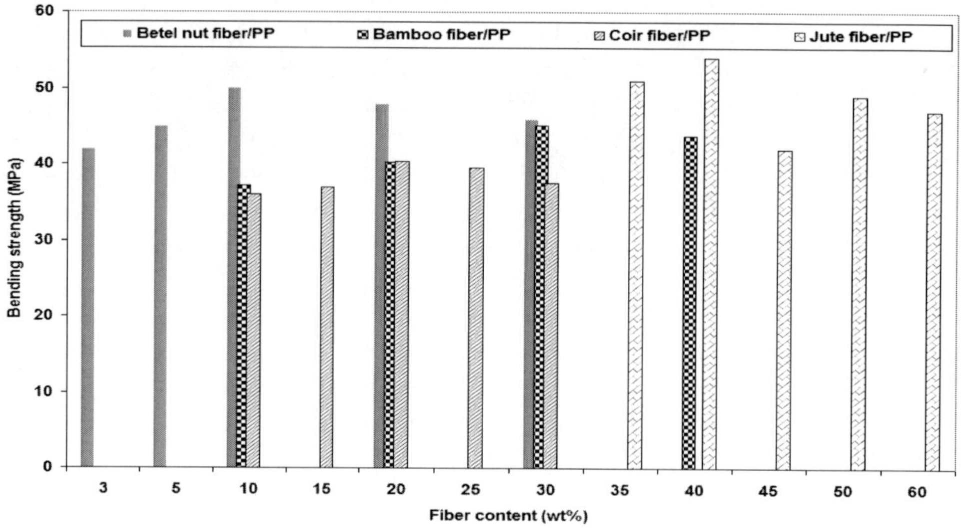

Figure 15. Effect of fiber content on the bending strength of different FRPCs [34]. [Reprinted from Journal of Thermoplastic Composite Materials, Shubhra et al., Mechanical properties of polypropylene composites: A review, with permission from SAGE Journals].

CONCLUSION

Among different types of reinforcements, fibers are stronger and stiffer materials than any other form which is the main reason for the overwhelming attraction for fibrous reinforcements. Different types of fibers used in composite have different properties and so affect the properties of the composite in different ways. The mechanical properties of most reinforcing fibers are considerably higher than those of un-reinforced resin systems. The mechanical properties of the fiber/resin composite are therefore dominated by the contribution of the fiber to the composite. Between synthetic and natural fibers, synthetic fibers are usually more uniform in size, more economical to use and behave in a more predictable manner. On the other hand, natural fiber-reinforced polymer composites have created interest in the automotive and other industries in recent years due to their lower cost than glass and carbon fibers, biodegradability and environment-friendliness. Unlike glass and carbon fibers, natural fibers are CO_2 neutral. The modulus–weight ratio of some natural fibers is greater than that of E-glass fibers indicating that they can be very competitive with E-glass fibers in stiffness-critical designs. Polypropylene matrix composites are very popular due to their low cost, simple fabrication methods, lightweight and desirable mechanical properties. Composite materials are now well established all over the world and used in almost every type of advanced engineering structure, with their usage ranging from aircraft, helicopters and spacecraft through to ships, boats and offshore platforms and to automobiles, sports goods, chemical processing equipment and civil infrastructure such as bridges and buildings. The development of fiber reinforced polypropylene composite materials has not only been a great motivating factor for scientists, but also an important provider of opportunities to improve the standard of living of people around the world. Along with researchers and entrepreneurs, we also expecting the expansion of research for new fiber reinforced polypropylene composite materials with high performance at affordable costs.

REFERENCES

[1] E. Rodriguz, R. Petrucci, D. Puglia,M. K. Jose, and A. Vazauez, *J. Comp. Mat.* **39,** 265 (2005).

[2] S. O. Han, S. M. Lee, W. H. Park and D. Cho, *J. App. Polym. Sci.* 100, 4972 (2006).

[3] S. Mohanty, V. K. Sushil, N. K. Sanjay and T. S. Sudhansu, J. App. Polym. Sci. 94, 1336 (2004).

[4] J. Ganster, H. P. Fink and M. Pinnow, *Comp.: Part A* 37, 1796 (2006).

[5] P. Srikanth, G. Shaoqin, E. O'Neill, Y. Liqiang and M. R. Roger, *J. App. Polym. Sci.* 111, 37 (2009).

[6] C. Baley, *Comp. Part A* 33, 939 (2002).

[7] K. L. Pickering, G. W. Beckermann, S. N. Alam and N. J. Foreman, *Comp. Part A* 38, 461 (2007).

[8] R. Somashekar, R. U. Gopalkrishne, *Polym.* 34, 2711 (1993).

[9] O. A. Khondker, U. S. Ishiaku, A. Nakai and H. Hamada, *Comp. Part A* 37, 2274 (2006).

[10] K. L. Fung, X. S. Xing, R. K. Y. Li, S. C. Tjonga and Y. W. Mai, Comp. Sci. & Tech. 63, 1255 (2003).

[11] M. Zampaloni, F. Pourboghrat, S. A. Yankovich, B. N. Rodgers, J. Moore, L. T. Drzal, A. K. Mohanty and M. Misra, *Comp. Part A* 38, 1569 (2007).

[12] L. Y. Mwaikambo, E. Martuscelli and M. Avella, *Polym. Test.* 19, 905 (2000).

[13] T. A. Bullions, R. A. Gillespie, J. Price-O'Brien and A. C. Loos, J. App. Polym. Sci. 92, 3771 (2004).

[14] G. Cantero, A. Arbelaiz, R. Llano-Ponte and I. Mondragon, Comp. Sci. & Tech. 63, 1247 (2003).

[15] M. A. Khan, M. M. Hasan and L. T. Drazal, *Comp. Pat A* 36, 71 (2005).

[16] A. Nick, U. Becker and W. Thoma, *J. Polym. & Environ.* 10, 115 (2002).

[17] Haydaruzzaman, A. K. Ruhul, A. K. Mubarak, A. H. Khan and M. A. Hossain, *Rad. Phy. & Chem.* 78, 986 (2009).

[18] J. L. Thomason and M. A. Vlug, *Comp. Part A* 27, 477 (1996).

[19] Y. Tomohiro, O. Toshio and I. Takashi, *Comp. Sci. & Tech.* 65, 2140 (2005).

[20] J. Bijwe, S. Awtade and A. Ghosh, *Wear* 260, 401 (2006).

[21] A. K. Bledzki and J. Gassan, *Prog. Polym. Sci.* 24, 221 (1999).

[22] B. Madsen and H. Lilholt, *Comp. Sci. & Tech.* 63, 1265 (2003).

[23] W. de Van, I. Ivens, J. De Coster, A. B. Kino, E. Baetens and I. Verpoest. *Comp. Sci. & Tech.* 63, 1241 (2003).

[24] J. A. G. Thomas, *Comp.* 3, 62 (1972).

[25] S. K. Rai and P. S. Padma, *J. Reinf. Plast. & Comp.* 25, 565 (2006).

[26] H. James and H. Dan, *J. J. Min. Met. & Mat. Soc.* 58, 80 (2006).

[27] L. Hollaway, Polymer Composites for Civil and Structural Engineering, Chapman and Hall Inc., Glasgow, UK (1993).

[28] R. P. Sheldon, Composite Polymeric Materials. Applied Science Publishers, London, UK (1982).
[29] Q. T. H. Shubhra, A. K. M. M. Alam, M. A. Gafur, S. M. Shamsuddin, M. A. Khan, M. Saha, D. Saha, M. A. Quaiyyum, J. A. Khan, and M. Ashaduzzaman, *Fib. & Polym.* 11, 725 (2010).
[30] D. N. Saheb and J. P. Jog, *Adv. in Polym Tech.* 18, 351 (1999).
[31] P. S. Mukherjee and K. G. Satyanarayana, *J. Mat. Sci.* 21, 51 (1986).
[32] P. V. Joseph, R. S. Marcelo, L. H. C. Mattoso, K. Joseph and S. Thomas, *Comp. Sci. & Tech.* 62, 1357 (2002).
[33] R. H. Somani, L. S. Yang, I. Hsiao, B. S. Pogodina, N. V. Winter, H. H. Agarwal, P. H. Fruitwala and A. Tsou, *Macromol. Symp.* 185, 105 (2002).
[34] Q. T. H. Shubhra and A. K. M. M. Alam, *J. Therm. Comp. Mat.* doi: 10.1177/0892705711428659.
[35] M. J. A. van den Oever and M. H. B. Snijder, *J. App. Polym. Sci.* 110, 1009 (2008).
[36] G. Morales, M. I. Barrena, J. M. S. de Gómez, C. Merino and D. Rodríguez, *Comp. Struc.* 92, 1416 (2010).
[37] A. K. Mohanty, A. Wibowo, M. Misra and L. T. Drzal, *Comp. Part A* 35, 363 (2004).
[38] K. Oksman, App. Comp. Mat. 7, 403 (2000).
[39] Q. T. H. Shubhra, M. Saha, A. K. M. M. Alam, M. D. H. Beg, and M. A. Khan, *J. Reinf. Plast. & Comp.* 29, 3338 (2010).
[40] A. Magurno, Die Ang. Makrom. Che. 272, 99 (1999).
[41] K. Avik, H. Tanzina, M. Saha, R. A. Khan, M. A. Khan and M. A. Gafur, *J. Comp. Mat.* 44, 2875 (2010).
[42] R. A. Khan, A. J. Parsons, I. A. Jones, G. S. Walker and C. D. Rudd, *J. Reinf. Plast. & Comp.* 29, 1838 (2010).
[43] M. A. Khan, R. A. Khan, Haydaruzzaman, A. Hossain and A. H. Khan, J. Reinf. Plast. & Comp. 28, 1651 (2009).
[44] M. A. Khan, G. Johannes and F. Hans-Peter, *Comp. Part A,* 40, 846 (2009).
[45] A. K. Bledzki, A. A. Mamun and O. Faruk, *eXPRESS Polym. Let.*, 1, 755 (2007).
[46] L. M. Arzondo, C. J. Pérez and J. M. Carella, *Polym. En.g & Sci.* 45, 613 (2005).
[47] A. C. Karmaker and J. A. Youngquist, *J. App. Polym. Sci.* 62, 1147 (1996).
[48] A. K. Bledzki and F. Omar, *J. Cell. Plast.* 42, 63 (2006).
[49] R. G. Venkata, N. S. Venkata and R. T. Shobha, *J. Reinf. Plast. & Comp.* 27, 1789 (2008).
[50] N. A. Thomas and K. E. George, *J. Therm. Comp. Mat.* 22, 5 (2009).
[51] J. Mora´n, V. Alvarez, R. Petrucci, J. Kenny and A. Vazquez, *J. App. Polym. Sci.* 103, 228 (2007).
[52] M. M. Hassan, M. H. Wagner, H. U. Zaman and M. A. Khan, *J. Nat. Fib.* 7, 165 (2010).
[53] S. Y. Fu, B. Lauke, E. Ma¨der, C. Y. Yue and X. Hu, *Comp. Part A* 31, 1117 (2000).
[54] R. Rahman, M. M. Hasan, M. Huque and M. N. Islam, *J. Reinf. Plast. Comp.* 29, 445 (2010).
[55] Haydaruzzaman, A. H. Khan, M. A. Hossain, M. A. Khan, R. A. Khan and M. A. Hakim, *J. Comp. Mat.* 44, 1945 (2010).
[56] M. J. A. van den Oever, H. L. Bos and M. J. J. M. van Kemenade, *App. Comp. Mat.* 7, 387 (2000).

[57] S. K. Garkhail, R. W. H. Heijenrath, and T. Peijs, *App. Comp. Mat. 7,* 351 (2000).
[58] Q. Yuan, D. Wu, J. Gotama and S. Bateman, *J. Therm. Comp. Mat. 21,* 195 (2008).
[59] Q. T. H. Shubhra, A. K. M. M. Alam, M. A. Khan, M. Saha, D. Saha and M. A. Gafur, *Comp. Part A* 41,1587 (2010).
[60] N. M. Barkoula, S. K. Garkhail and T. Peijs, *J. Reinf. Plast. & Comp.* 29, 1366 (2010).
[61] K. S. Sushanta, M. Smita and K. N. Sanjay, *J. Reinf. Plast. & Comp.* 28, 2729 (2009).
[62] K. Ajay, S. S. Chauhan, M. M. Jayant, and M. Chanda, *Comp. Part A* 38, 227 (2007).
[63] M. Haque, R. Rahman, N. Islam, M. Huque and M. Hasan, *J. Reinf. Plast. & Comp.* 29, 2253 (2010).
[64] Haydaruzzaman, A. H. Khan, M. A. Hossain, R. A. Khan and M. A. Khan, *J. Comp. Mat.* 44, 401 (2010).
[65] M. Haque, S. Islam, S. Islam, N. Islam, M. Huque and M. Hasan, *J. Reinf. Plast. & Comp.* 29, 1734 (2010).
[66] Q. T. H. Shubhra, A. K. M. M. Alam, M. D. H. Beg, M. A. Khan, M. A. Gafur, *J. Comp. Mat.* 45, 1305 (2011).
[67] R. A. Khan, M. A. Khan, U. Z. Haydar, P. Shamim, K. Nuruzzaman, S. Sabrina, M. Saha, A. I. Mustafa, *J. Reinf. Plast. & Comp.* 29, 1078 (2010).
[68] D. H. Mueller, *Int. Nonwoven J.* 13, 31 (2005).
[69] C. Pavithran, P. S. Mukherjee, M. Brahmakumar and A. D. Damodaran, *J. Mat. Sci. Let.* 6, 882 (1987).
[70] N. M. Sergio, G. S. Kestur and P. D. L. Felipe, *Mat. Sci. Forum* 638-642, 961 (2010).
[71] P. Suhara and S. Mohini, *J. App. Polym. Sci.* 103, 2432 (2007).
[72] R. A. Khan, M. A. Khan, S. Sultana, K. Nuruzzaman, Q. T. H. Shubhra and G. N. Farhana, *J. Reinf. Plast. & Comp.* 29, 466 (2010).
[73] A. Kelly and W. R. Tyson, *J. Mech. & Phy. Sol.* 13, 329 (1965)..
[74] L. T. Drzal, M. J. Rich, M. F. Koeng and P. F. Lloyd, *J. Adh.* 16, 133 (1983).
[75] H. F. Wu, and A. N. Netravali, *J. Mat. Sci.* 27, 3318 (1992)..
[76] J. O. Karlsson, P. Gatenholm, J.-F. Blachot and A. Peguy, *Polym. Comp.* 17, 300 (1996).
[77] D. Ray, B. K. Sarkar, S. Das, A. K. Rana, *Comp. Sci. & Tech.* 62, 911 (2002).
[78] D. Ray, B. K. Sarkar, R. K. Basak and A. K. Rana, *J. App. Polym. Sci.* 85, 2594 (2002).
[79] S. Díez-Gutiérrez, M. A. Rodríguez-Pérez, J. A. De Saja, J. I. Velasco, *Polym.* 40, 5345 (1999).
[80] M. Z. Rong, M. Q. Zhang, Y. Liu, Z. W. Zhang, G. C. Yang and H. M. Zeng, *Polym. & Polym. Comp.* 10, 407 (2002).
[81] L. A. Pothan and S. Thomas, *Comp. Sci. & Tech.* 63, 1231 (2003).
[82] P. J. Herrera-Franco and A. Valadev-Gonzales, *Comp. Part A* 35, 339 (2004).
[83] M. A. Khan, M. M. Rahman and K. S. Akhunzada, Polym. Plast. Tech. & Eng. 41, 677 (2002).
[84] J. Gassan and V. S. Gutowski, *Comp. Sci. & Tech.* 60, 2857 (2000).
[85] J. George, S. S. Bhagawan and S. Thomas, *Comp. Interf.* 5, 201 (1998).
[86] L. A. Pothan, J. George and S. Thomas, *Comp. Interf.* 9, 335 (2002).
[87] U. Z. Haydar, M. A. Khan, R. A. Khan, M. A. Rahman, L. R. Das and M. Al-Mamun, *Fib. & Polym.* 11, 455 (2010).
[88] M. A. Khan, R. A. Khan, U. Z. Haydar, M. Noor-A-Alam and M. A. Hoque, *J. Reinf. Plast. & Comp.* 29, 1195 (2010).

[89] A. Arbelaiz, B. Ferna´ndez, G. Cantero, R. Llano-Ponte, A. Valea and I. Mondragon, *Comp. Part A* 36, 1637 (2005).

[90] P. V. Joseph, J. Kuruvilla and T. Sabu, *Comp. Sci. & Tech.* 59, 1625 (1999).

[91] S. Mukhopadhyay, B. L. Deopura and R. Alagiruswamy, *J. Therm. Comp. Mat.* 16, 479 (2003).

[92] H. Anuar, S. H. Ahmad, R. Rasid, S. Zakaria, *American J. App. Sci.* (Special Issue), 34 (2005).

[93] J. Z. Lu, Q. Wu and S. M. Harold, *Wood & Fib. Sci.* 32, 88 (2000).

[94] A. Alireza and N. Amir, *J. App. Polym. Sci.* 111, 1684 (2009).

[95] F. Xiang, H. Bobing and C. Xian, *J. Reinf. Plast. & Comp.* 29, 936 (2010).

[96] K. N. Sanjay and M. Smita, *J. Reinf. Plast. & Comp.* 29, 1551 (2010).

[97] S. Mohanty, S. K. Nayak, S. K. Verma and S. S. Tripathy, *J. Reinf. Plast. & Comp.* 23, 2047 (2004).

[98] M. Kazayawoko, J. J. Balatinecz and L. M. Matuana, *J. Mat. Sci.* 34, 6189 (1999).

[99] X. Yuan, Y. Zhang and X. Zhang, *J. App. Polym. Sci.* 71, 333 (1999).

[100] A. R. Sanadi, R. A. Young, C. Clemons and R. M. Rowell, *J. Reinf. Plast. & Comp.* 13, 54 (1994).

[101] M. Pracella, D. Chionna, I. Anguillesi and Z. Kulinski, *Comp. Sci. Tech.* 66, 2218 (2006).

[102] K. Sam-Jung, M. Jin-Bok, K. Gue-Hyun and H. Chang-Sik, *Polym. Test.* 27, 801 (2008).

[103] H. Bijsterbosch and R. J. Gaymans, *Polym. Comp.* 16, 363 (1995).

[104] J. Kuruvilla, V. Siby, G. Kalaprasad, S. Thomas, L. Prasannakumari, P. Koshy and C. Pavithran, *European Polym. J.* 32, 1243 (1996).

In: Polypropylene
Editors: L. P. Silva and E. F. Barbosa

ISBN: 978-1-62417-142-0

Chapter 12

HIGH LOAD POLYPROPYLENE COMPOSITES

***L. Fambri**[*1], **D. Lorenzi**[1], **E. Masarati**[2] **and E. Costantini**[2]*
[1]Department of Industrial Engineering, University of Trento,
via Mesiano, Trento, Italy
[2]Basell Poliolefine Italia srl a company of Lyondellbasell,
"G. Natta" R&D, P.le L. Donegani, Ferrara, Italy

ABSTRACT

Polypropylene composites with high content of inorganic filler or short glass fiber could be produced after compounding in a single screw Buss technology or in a twin-screw extruder. Specific formulations were properly prepared to guarantee an adequate processability for injection molding or extrusion. Polypropylene matrix was a proper combination of low and high melt flow grades in the range of about 3-12 g/10min and 500-2300 g/10min respectively. High content polypropylene composite containing up to 60% by wt. of talc, 80% of calcium carbonate or 65% of short glass were produced using a relatively low viscosity matrix.

Good flow-ability and processability for all compositions were evidenced. ISO specimens and large plates obtained after injection molding evidenced the increase of stiffness, strength, toughness at low temperature, and reduction of shrinkage. Deep investigation was conducted on glass fiber composites at 40-50-60%. (DSC, TGA, SEM, DMA, fiber distribution, flammability evaluation, and comparison). Results have been discussed and correlated with the compositions.

INTRODUCTION

Reinforced polypropylene (PP) materials represent one of the high volume production compounds for a large variety of applications where improvement of stiffness, strength and increase of heat deflection temperature are required. Moreover, the significant decrease of the relative price becomes a strategic factor. Three fillers are the most considered due to

[*] E-mail address: Luca.Fambri@ing.unitn.it

favorable combination of properties, processing [1] and costs: talc, calcium carbonate, and glass fiber.

However, some limitations must be considered, especially in polypropylene composite with a high percentage of filler, due to the reduction of other properties such as the impact strength, deformation at break, and critical problems in processing conditions like injection molding. Hence, because of the increase in melt viscosity for general applications, typical percentages reach about 20-40% by wt. for particulate filler and about 30% by wt. of glass fiber [2-4].

Concerning production and applications, the market of glass fiber polypropylene composites represents a significant and technical sector with respect to polypropylene resins. According to recent data, in 2010 the European polypropylene resins sales reached 9.9Mtons, as result of 7.5 Mtons in Western Europe and 2.4 Mtons in Central/Eastern Europe [5]. And concerning glass fiber reinforced polypropylene, European sales in 2009 required 160 ktons as reported in Table 1 and are expected to largely increase in the coming years. The main application sector is automotive, followed by electrical and electronics, furniture and other segments (Non automotive).

It is interesting to underline that the production of polypropylene composite granule is distinguished in short glass fiber up to about 2-4 mm (SGF) and long glass fibers up to 8-15 mm (LGF), with 2009 consumptions in the proportions of about 67% and 33%, respectively.

Literature reveals papers and publications of researchers dealing with the properties of short glass fibers polypropylene composite [6-20] in dependence on the characteristics of the matrix, the type and content of fibers, their average length and distribution, the presence of any adhesion promoter, the processing conditions, and the final shape of the required/desired item.

Table 1. Glass reinforced PP compound consumption in Europe by ktons [5]

Application sector	Year	2006	2009	2014
Automotive		110	100	160
Non Automotive		80	60	75
	Total	190	160	235

On the other hand, recent developments suggested the use of long glass fiber in reinforced polypropylene to improve mechanical properties, such as impact resistance and creep behavior [21-28]. It should be taken into account that several factors affect the final properties, as result of a combined effect of the composition and processing conditions (injection molding and mold design).

Moreover, attention is specifically dedicated to the fiber orientation in dependence on the processing conditions in extrusion and injection, and its direct effects on the composite properties [29-50]. In all cases, a critical problem, especially for long glass fibers, is the severe fiber breakage occurring in injection molding, determining processed composites with shorter fiber length distribution [51-56].

Following these general considerations, the study of preparation, processing, and the characterization of high load polypropylene composites will be described/treated in this chapter with particular relevance on the short glass fibers polypropylene composites up to 40-60% by wt. It should be underlined for this purpose that a specific selection of polypropylene matrix was necessary to overcome the effects of high viscosity after filler addition in compounding and specimens processing.

MATERIALS AND SAMPLE PREPARATION

Isotactic polypropylene (iPP) is a product of LyondellBasell. Four polymers with different melt flow rate (MFR) at high and low grade were selected and coded, as in the following. PP2300 is a homopolymer with a MFR of 2300 g/10 min; PP550 is a homopolymer with a MFR of 550 g/10 min; PP12 is a homopolymer with a MFR of 12 g/10 min; PP3 is a random copolymer with a MFR of 3 g/10 min and containing 1.6 % of ethylene.

Polybond 3200, a polypropylene homopolymer, grafted with about 1% of maleic anhydride (PP-g-MA) and MFR 115 g/10 min, and commercialized by Chemtura, was used as coupling agent (CA) in glass fiber composites with a GF/CA ratio of 20, in line with what also described elsewhere [57-58].

Two type of talc at medium and fine size particles fillers were: Luzenac HAR T84 from Imerys-France (T1), a High Aspect Ratio Talc with aspect ratio of about 10-20 vs. 3-5 of a standard talc and HM05 from IMI Fabi SpA-Italy (T2) with more than 95% of particles lower than 5 μm.

Calcium carbonate (CC) Omyacarb 2T-UM was supplied by Omya SpA-Italy (more than 32% have particles lower than 2 μm, and more than 99% lower than 45 μm).

Short glass fiber (SGF) with density 2.65 g/cm^3, diameter of 13 μm and average length of 3 mm was selected (Fibre Glass ECS O3T 480, from NEG–Japan).

Composites with particulate fillers (Table 2) were prepared using a single screw extruder, Buss kneader technology model MDK70 (oscillating screw shaft), with screw diam. 70 mm and L/D = 17. Inorganic fillers were fed into the melt polypropylene through second vertical feeding ports. Running conditions were screw speed = 270 rpm, capacity = 50-60 kg/h and barrel temperature = 200-230 °C.

Table 2. Composition of particulate polypropylene composites

Matrix and Filler (%)	T40	T50	T60	CC80
PP 2300 Homopolymer	-	25	27	13
PP 12 Homopolymer	40	-	13	-
PP 3 Random Copolymer	-	25	-	7
PP 550 Homopolymer	20	-	-	-
Talc (T1)	-	50	60	-
Talc (T2)	40	-	-	-
Calcium Carbonate (CC)	-	-	-	80

Glass fiber composites (Table 3) were prepared using a twin screw extruder model Werner and Pfeiderer ZSK40SC with a screw diam. 40 mm and L/D = 43 provided with gravimetric feeders; polypropylene and a coupling agent were fed into the first barrel, and the short glass fiber was fed into the fifth barrel via forced side feeding. Running conditions were screw speed = 200 rpm, capacity = 50-60 kg/h and barrel temperature = 200-220 °C.

Table 3. Composition of fibrous polypropylene composites

Matrix and Filler (%)	GF40	GF45	GF50	GF55	GF60	GF65
PP 2300 Homopolymer	-	17.6	23.75	14.1	18.5	15.85
PP 12 Homopolymer	58	17.6	-	14.1	-	-
PP 3 Random Copolymer	-	17.6	23.75	14.1	18.5	15.85
Polybond 3200	2	2.2	2.5	2.7	3	3.3
Glass Fiber (GF)	40	45	50	55	60	65

For determination of materials properties, T-bar specimens ISO 527-1 type 1A (length: 170 mm; thickness: 4.0 mm; gauge length: 80 mm; gauge width: 10 mm) were prepared by injection molding following ISO 1873-2 (2007) by using a Negri-Bossi injection molding machine (melt temperature: 230-270°C; injection pressure: 30-60 MPa; mold temperature: 50°C).

Selected composites were also injection molded to produce large plates (length: 250 mm; width: 150 mm; thickness: 3.2 mm) in a specific mold provided by a film gate 1.5 mm thick and 150 mm wide.

CHARACTERIZATION METHODS

Density was measured according to ISO 1183-2:2004 (Part 2: Density gradient column method).

Melt flow rate (MFR) was performed on the produced composites by using the instrument Melt Flow Index LMI 4000 (Dynisco) according to ISO 1133:2005 with a load of 2.16 kg at 230°C.

Spiral flow test were performed using an injection molding machine by Sandretto Model 190 with a clamping force of 1900 kN. Melt temperature of 230°C, injection pressure between 2 and 10 MPa, a single cavity endless spiral flow mold with 2.5 mm depth, and mold temperature of 40°C were selected as experimental conditions.

Heat deflection temperature (HDT) and Vicat softening temperature (VST) were determined by using a HDT-Vicat apparatus model MP3 (product of ATS-FAAR, Italy). HDT test was performed on T-bar specimens at 1.8 MPa following ISO 75A. VST was measured according to ASTM D1525-00 with a load of 10 N and a heating rate of 120°C/h. The dimensions of test specimens prepared from the plates were 12 mm x 12.8 mm x 3.2 mm.

Mechanical properties were determined at room temperature (23°C) by using a dynamometer Instron mod.4502 in bending and tensile mode at 50 mm/min, according to ISO178 and ISO527, respectively. Tensile modulus (secant) was determined with gauge length of 1 mm/min and span of 50 mm.

Charpy tests were performed on unnotched and notched (Notch A) rectangular specimens 80x10x4 mm from T-bars following ISO 179 (type 1; edgewise).

Thermal degradation and composition of pure and GF reinforced polypropylene were studied in thermogravimetric analysis (TGA) using thermobalance Mettler TG 50 at a heating rate of 10°C/min in air flow of 100 ml/min in the range 50-700°C.

Differential scanning calorimetry (DSC) analysis was performed using a Mettler DSC30 calorimeter by using a cycle of heating-cooling-heating in the range 0-230°C at ±10°C/min flushing nitrogen at 100 ml/min. Melting temperature and crystallization temperature were reported. The crystallinity content was evaluated from the melting peak and referred to the value of 209 J/g for 100% crystalline polypropylene reported in literature [59].

Scanning electron microscopy (SEM) pictures were obtained by using a Philips® XL30 Environmental Scanning Electron Microscope at an acceleration voltage between 15 and 30 kV. Samples were fractured in liquid nitrogen and then fracture surface was metalized.

Measurements were performed using a Dynamic mechanical thermal analysis DMTA Mk II (Polymer Laboratories Ltd, Loughborough, UK). The storage E' and loss E" moduli were measured in the single cantilever bending mode at frequency 1 Hz in the range –50 to 150°C (test pieces 19 mm x 5 mm x 3.2 mm were cut of the large plates; heating rate 3°C/min; dynamic displacement 0.064 mm). The glass transition temperature of the PP component was identified with the position of the maximum respective loss modulus peak on temperature scale. Effect of orientation was evaluated after tensile DMTA performed on rectangular specimens (19 mm x 2 mm x 3.2 mm) longitudinally and transversally cut of the large plate (static stress of 1 MPa). Longitudinal and transversal deformation (D) were calculated as $D = 100x\ \Delta L/L_0$, in dependence on the initial specimen length L_0 and the length variation ΔL. The coefficient of (linear) thermal expansion $CTE = (\Delta L/L_0)/\ \Delta T$ was calculated for three temperature intervals ΔT, by using the initial specimen length L_0 and the length variation ΔL over the selected interval ΔT.

RESULTS AND DISCUSSION

Preliminary analysis of compounded composites evidenced a suitable flowability for injection molding. Melt flow rate, as reported in Table 4 and 5, was found in the range between 1-82 g/10 min for particulate composites and 3-7 g/10 min for fibrous composites in dependence on the matrix and the filler content.

The density of composite increased linearly with the percentage of filler, reaching 1.51 g/cm^3 for 60% of talc, 1.82 g/cm3 for 80% of calcium carbonate, and 1.56 g/cm^3 for 65% of SGF.

See Tables 2 and 3 for the composition, and Tables 4 and 5 for the properties of particulate and fibrous composites, respectively.

Table 4. Properties of particulate polypropylene composites

Properties	T40	T50	T60	CC80
Density (g/cm^3)	1.244	1.360	1.514	1.820
Melt Flow Rate (g/10min)	22.8	8.6	1.8	82
Tensile modulus (GPa)	5.17	7.50	9.60	6.65
Tensile strength (MPa)	37	39	31	20
Deformation at break (%)	2.2	1.5	1.3	1.0
Charpy unnotched (kJ/m^2)	13	10	3.4	3.4

Table 5. Properties of fibrous polypropylene composites

Properties	GF40	GF45	GF50	GF55	GF60	GF65
Density (kg/dm^3)	1.216	1.275	1.320	1.404	1.475	1.557
MFR (dg/min)	3.0	5.3	6.8	4.7	6.8	3.8
Tensile Modulus (GPa)	8.90	10.04	12.20	12.96	14.75	16.06
Strength at Break (MPa)	116	121	131	126	130	121
Elongation at Break (%)	2.4	2.5	2.4	1.7	2.2	1.6
Charpy unnotched (kJ/m^2)	59.5	61.2	63.4	49.4	44	40
Charpy unnot.-30°C (kJ/m^2)	63	67.5	67.6	57	57	41
HDT @1.8MPa (°C)	151	150	148	149	148	150

Glass fiber polypropylene composite revealed higher mechanical properties with respect to the correspondent filled with talc and calcium carbonate as shown in Table 5.

Moreover spiral flow test was performed on selected PP composites and evidenced a good processability for T40, and GF40 GF50, as depicted in Figure 1.

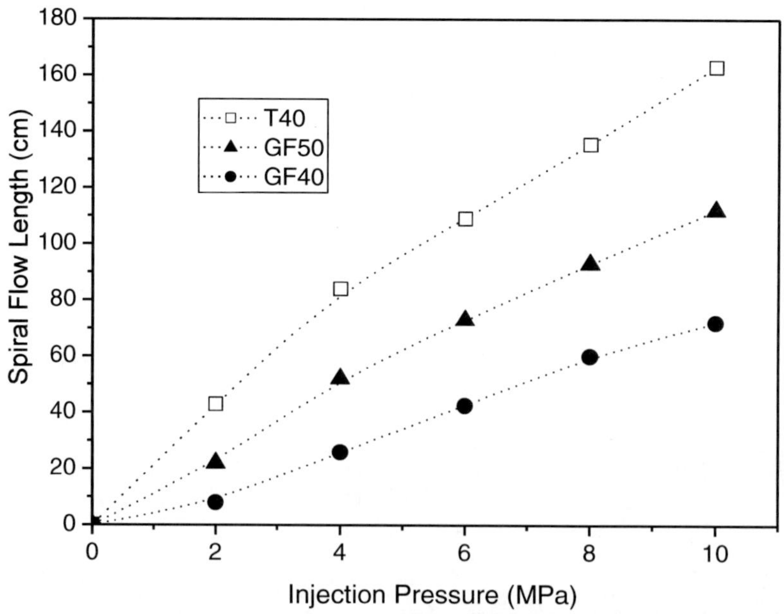

Figure 1. Effect of injection pressure at 230°C on spiral flow length of PP composite containing talc at 40% by wt. (□); short glass fiber at 50% (▲) and 40% by wt. (●).

It is well evident the effect of polypropylene matrix and the correspondent MFR.

As expected, higher values of flexural modulus and Notched Charpy test at 23°C were evidenced for short glass fiber composites with respect to particulate composites, as shown in Figure 2 and 3.

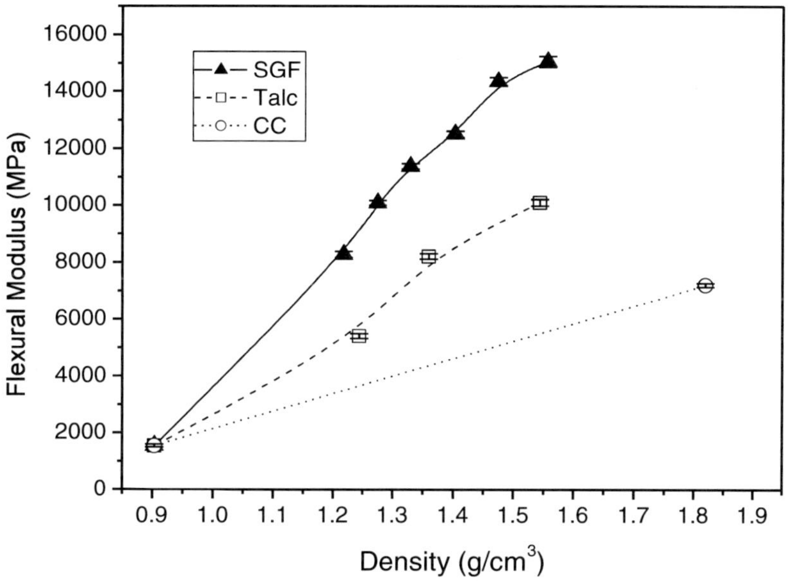

Figure 2. Effects of various filler content and density of PP composite on flexural modulus. Short glass fiber 40-65% by wt. (▲); talc 40-60% by wt. (□); calcium carbonate 80% by wt. (O).

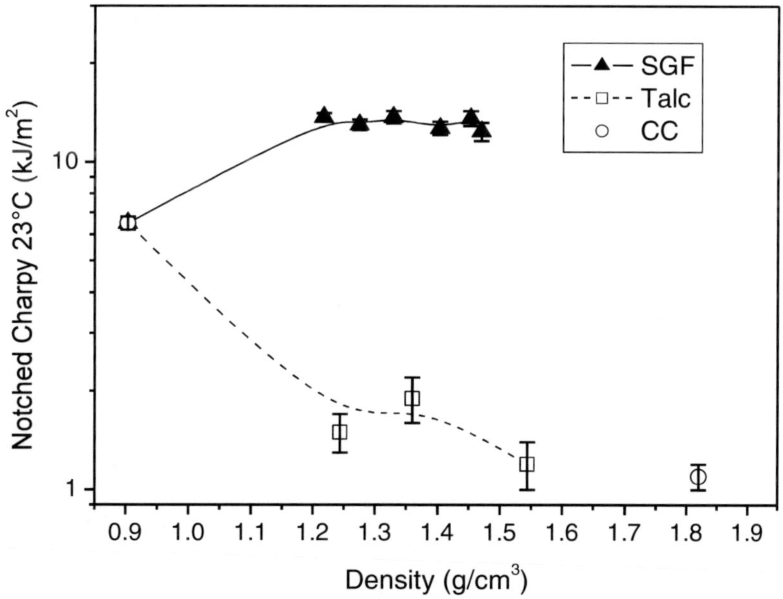

Figure 3. Effects of various filler content and density of PP composite on notched Charpy. Short glass fiber 40-65% by wt. (▲); talc 40-60% by wt. (□); calcium carbonate 80% by wt. (O).

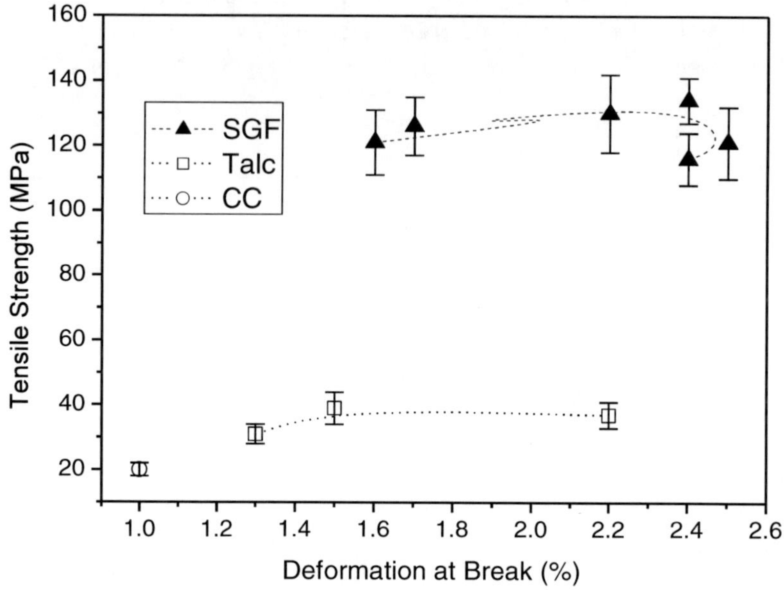

Figure 4. Effects of various filler content on tensile strength and deformation at break of PP composite. Short glass fiber 40-65% by wt. (▲); talc 40-60% by wt. (□); calcium carbonate 80% by wt. (O).

To better understand elastic properties, the tensile modulus (E_C) was evaluated using the traditional model for random oriented short fiber composites [60-61] according to equation I:

$$E_C = \frac{3}{8}E_L + \frac{5}{8}E_T \tag{I}$$

where E_L and E_T are elastic moduli of the composite with longitudinal and transversal aligned fibers, respectively. These moduli can be calculated following the semi-empiric model Halpin-Tsai with the equations II-IV:

$$E_L = \frac{1+2(l/d)(\eta_L\phi)}{1-\eta_L\phi} E_M \tag{II}$$

$$E_T = \frac{1+2\eta_T\phi}{1-\eta_T\phi} E_M \tag{III}$$

and

$$\eta_L = \frac{(E_F/E_M)-1}{(E_F/E_M)+2(l/d)} \tag{IVa}$$

$$\eta_T = \frac{(E_F / E_M) - 1}{(E_F / E_M) + 2} \tag{IVb}$$

where l/d is the aspect ratio of the fibers, Φ is the volume fraction of the fibers, and E_F and E_M are the fiber and matrix moduli (70 GPa and 1.68 GPa, respectively).

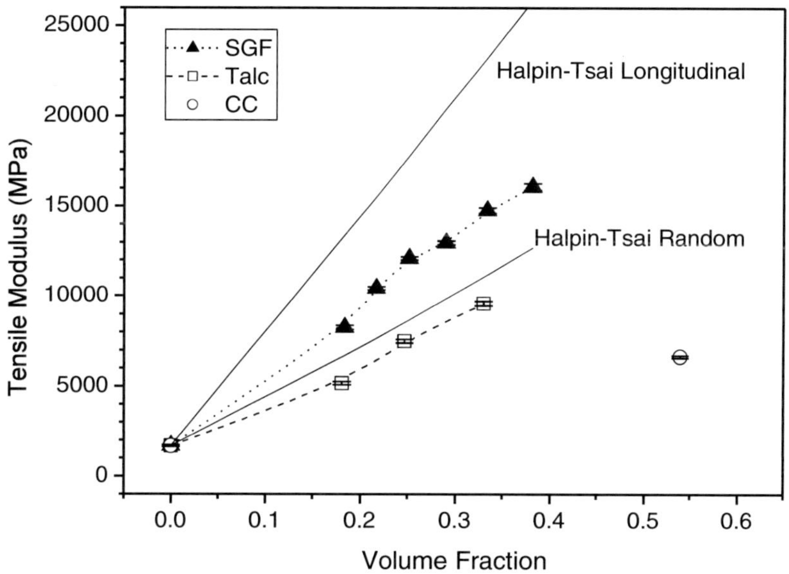

Figure 5. Tensile modulus of PP composites containing short glass fiber 40-65% by wt. (▲); talc 40-60% by wt. (□); calcium carbonate 80% by wt. (O), as function of volume fraction. Continuous lines represent the prediction according to Halpin-Tsai model for longitudinal and random oriented fiber composites.

Note that the Charpy test on notched specimens revealed at room temperature an almost constant energy to break of about 12-14 kJ/m^2 for all glass fiber polypropylene composites. On the other hand, at -30°C the energy to break is much higher and is directly proportional to the percentage of fiber content up to 65% by wt. (Figure 6).

Due to the higher properties, PP composites based on 40, 50 and 60% by wt. of short glass fiber were more deeply characterized. To avoid asymmetrical flow in injection molding, as shown by Karian and Stoops [38] in the case of single gate mold, the large rectangular plates were produced filling the mold cavity through a film gate 1.5 mm thick and 150 mm wide. These plates were used for TGA, DSC, DMTA analysis, and some tests to compare the material flammability.

DSC analysis evidenced in Figure 7 similar behavior of PP matrix and GF50 and GF60, for both crystallinities (44-45%), melting temperature (157-158°C) and crystallization temperature (112-115°C). Higher values were found for GF40, in dependence on the polypropylene composition.

TGA analysis confirmed the nominal composition, as shown in Figure 8, revealing a residual mass of about 37, 49 and 58% of glass fibers respectively. Two tests after longitudinal and transversal break were reported for each composition.

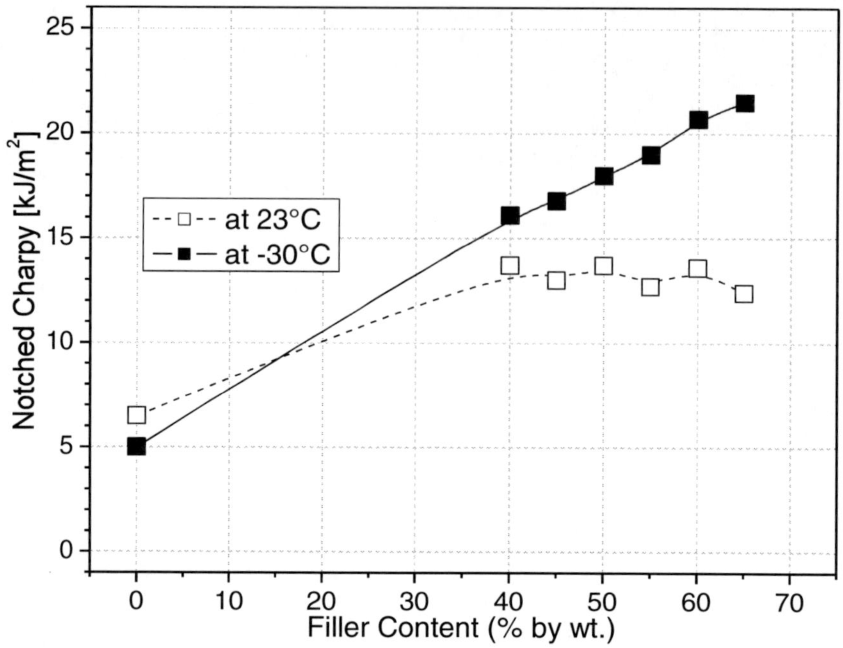

Figure 6. Notched Charpy test of SGF-PP composites at 23°C and -30°C.

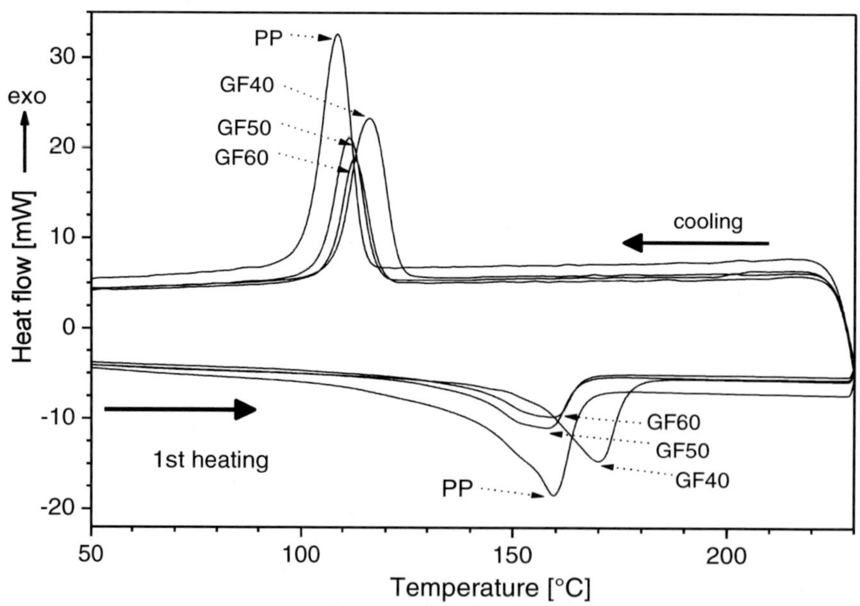

Figure 7. DSC thermograms of SGF-PP composites.

Table 6 reports some results of mechanical characterization at high temperature (flexural test at 80°C, creep experiments at 105°C), shrinkage evaluation according to standard or internal procedures, and thermal data.

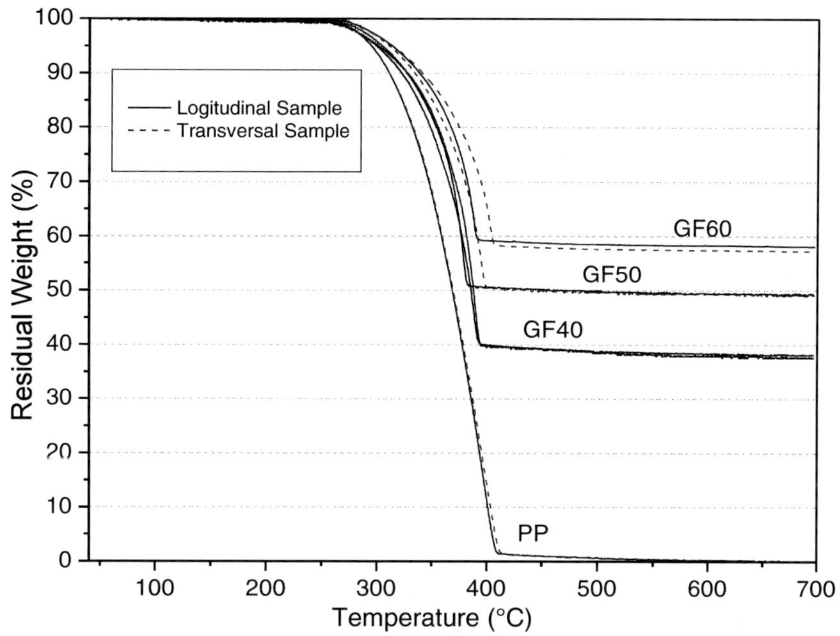

Figure 8. TGA thermograms of SGF-PP composites in air flow.

Table 6. Selected properties of SGF polypropylene composites

Properties	Method	GF40	GF50	GF60
Flexural Modulus @80°C (GPa)	ISO 178	5.05	7.20	8.80
Flexural Strength @80°C (MPa)	ISO 178	76	95	98.5
VST (°C)	ASTM D1525	158	146	152
Longitudinal Shrinkage (%)	Internal	0.21	0.09	-
Transversal Shrinkage (%)	Internal	0.87	0.55	-
Tensile Creep 105°C, 17MPa Time to Fracture (h)	ISO 899	41.5	58.5	-
Elongation to Fracture (%)	ISO 899	2.13	1.45	-
Polypropylene crystallinity (%)	DSC	48	44	45
Melting Temperature 1 scan (°C)	DSC	168	157	158
Crystallization Temperature (°C)	DSC	119	114	115
Melting Temperature 2 scan (°C)	DSC	166	154	155
Glass fiber content (%)	TGA	37	49	58
CLTE @ -40/-20°C (10^{-6}/°C)	DMTA	9-24	13-25	13-18
CLTE @ 5/35°C (10^{-6}/°C)	DMTA	15-52	34-54	16-33
CLTE @ 60/90°C (10^{-6}/°C)	DMTA	11-61	46-84	20-69
Glass transition temperature (°C)	DMTA	7-8	5-6	7-8

Fiber Length Analysis and Distribution

About 30 mg of each GF/PP composition were thermally degraded in TGA crucible after heating in air flow up to 700°C. The residue was collected and undergone to optical image analysis. The fiber length measurements were made with the help of polarized light microscopy (Leitz Ortholux II POL-BK) and computer aided image analysis (ImageJ), as shown in Figure 9. The fiber length distribution (FLD) was determined by the average fiber length, which calculated from a minimum of 500 length measurements on fibers recovered from the incineration (Figure 10). As expected the curve distribution evidenced a much higher length reduction for the 50-60% GF PP composite with respect to the 40% GF content. Values of D50 and D90 were found in accordance to literature [56].

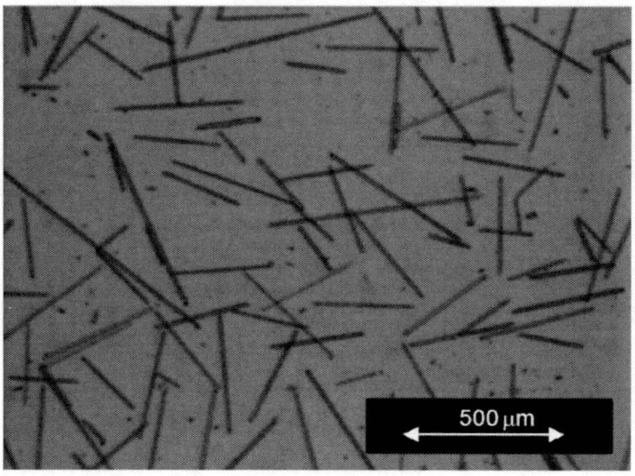

Figure 9. Fiber length analysis (Optical microscope).

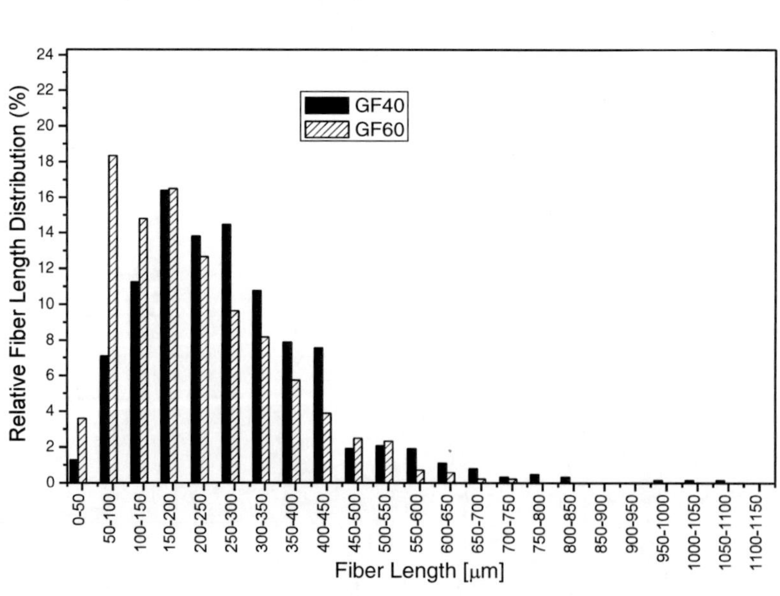

Figure 10a. Relative fiber length distribution in SGF PP composites.

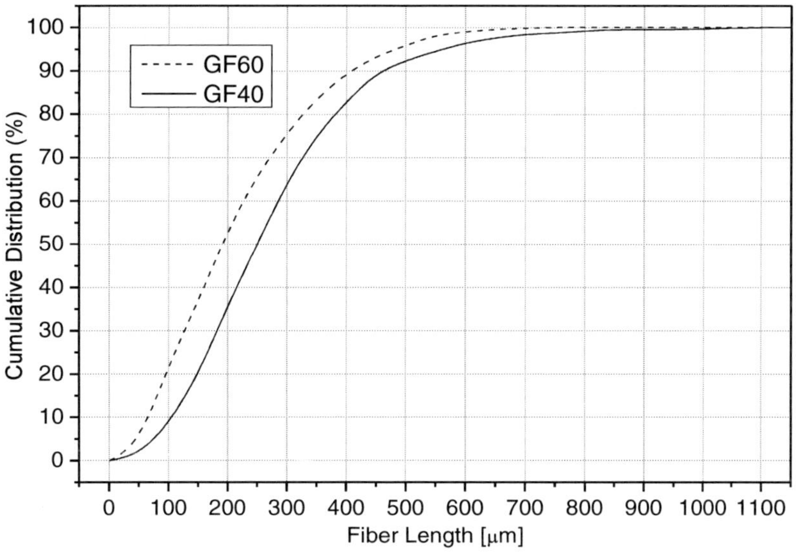

Figure 10b. Cumulative fiber length distribution in SGF PP composites.

Vicat softening temperature (50 N, 2°C/min) was increased from 106±5°C of the i-PP matrix to 146±4°C and 152±3 °C of iPP composites, at 50 and 60% SGF respectively, whereas higher VST 158±2°C was found for 40%, directly dependent on matrix composition, as shown in Figure 11.

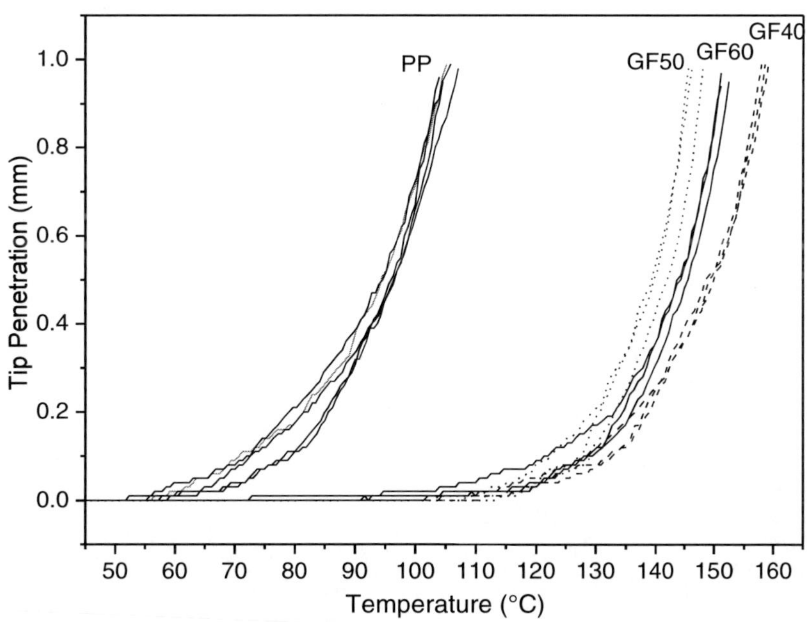

Figure 11. Vicat Test on PP and SGF-PP samples.

Higher stiffness and strength, and reduced shrinkage with better creep resistance were found as the most interesting improvements of the high-content SGF-iPP composites. On the other hand, deformation at break, HDT and Charpy tests remained almost unchanged versus 40 % GF PP. For example, flexural modulus increased from about 9 GPa up to 12 and 14 GPa, for i-PP composites containing 40, 50 and 60% by wt of SGF respectively.

Bending DMTA analysis showed significant differences in storage modulus (Figure 12), depending on both SGF content and injection molding direction. Loss modulus data (Figure 13) reported a glass transition peak at about 5°C independently with a shoulder peak at around 60°C.

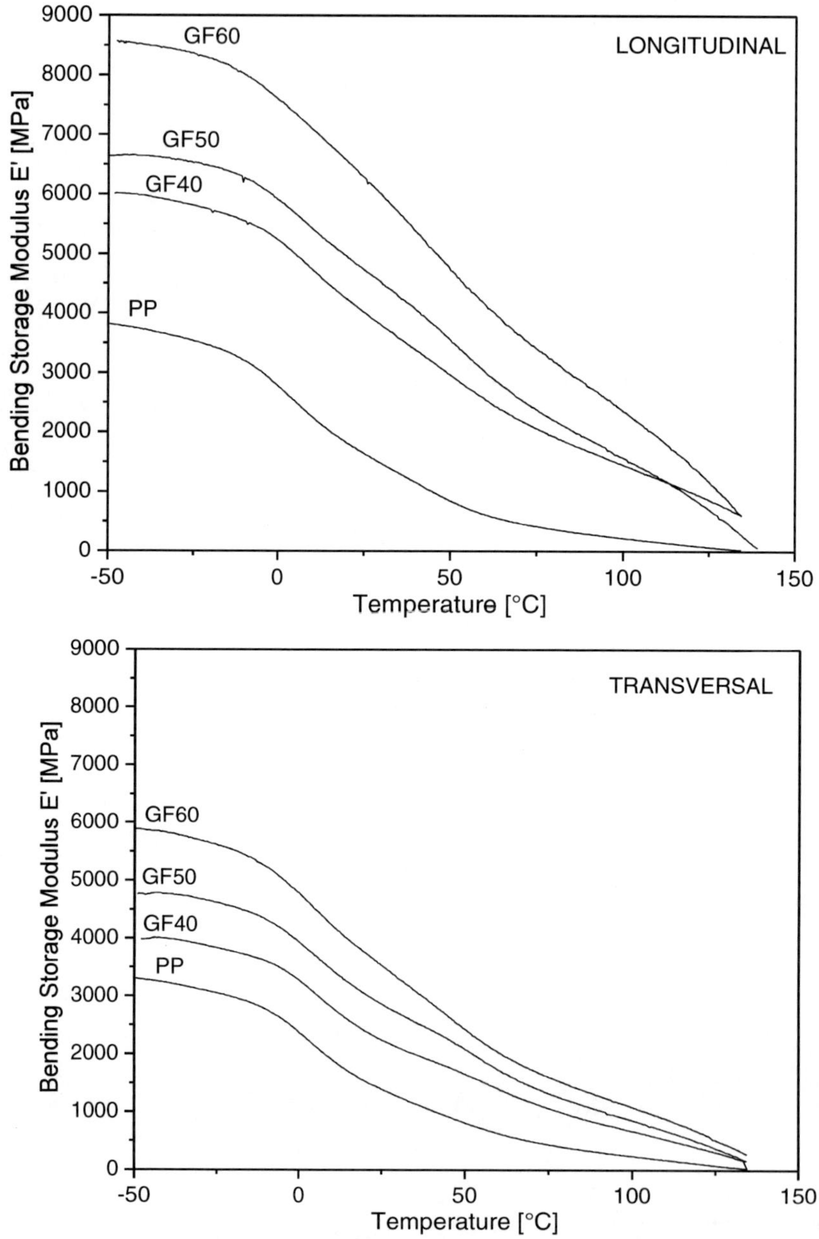

Figure 12. Bending storage modulus of PP composites as function of composition and longitudinal/transversal direction.

Specific analysis of the anisotropic properties of injection molded plates was obtained by comparing characterization of specimens machined in longitudinal and transverse directions, evidencing higher storage and loss modulus and lower tensile thermal deformation (Figure 14).

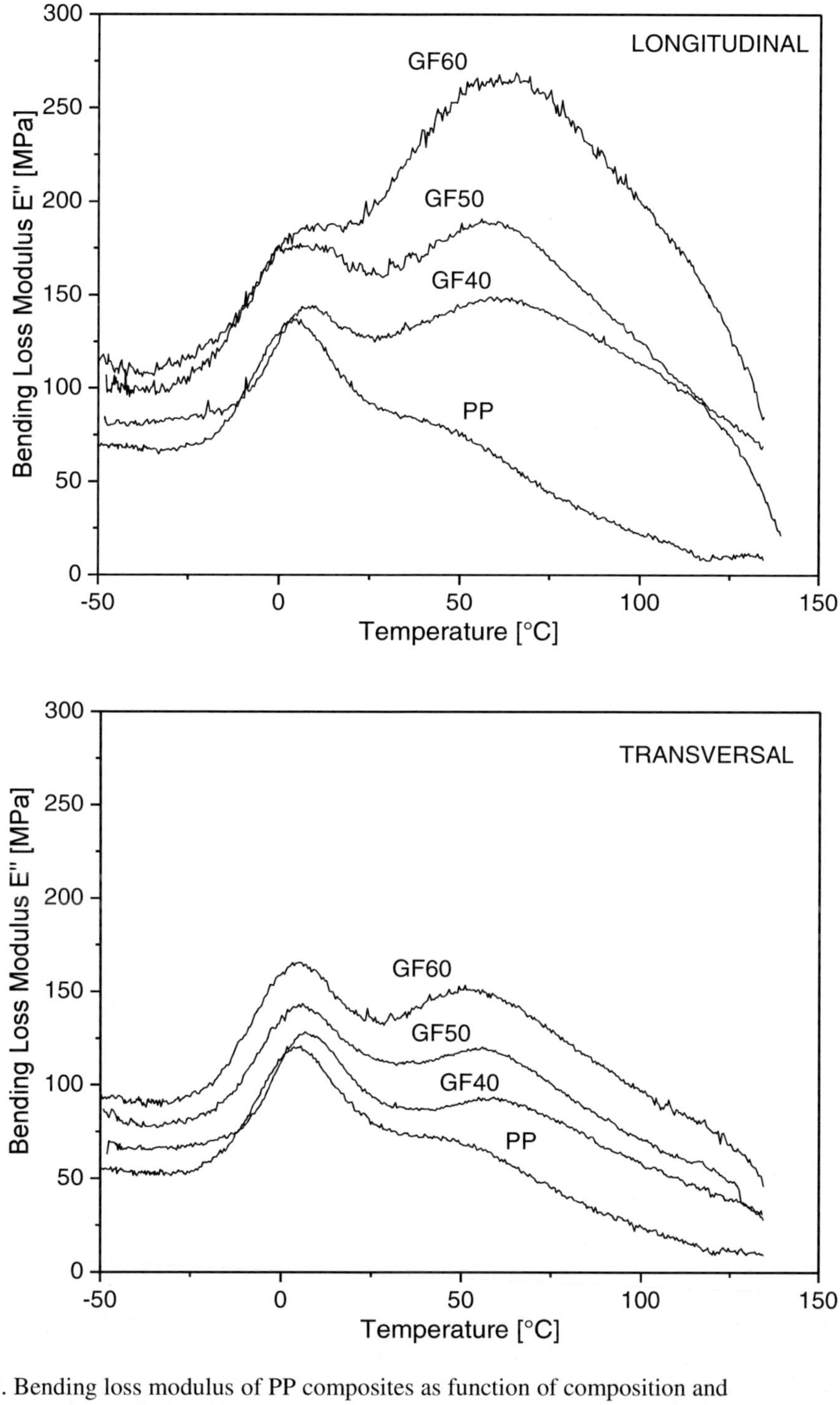

Figure 13. Bending loss modulus of PP composites as function of composition and longitudinal/transversal direction.

The anisotropy factor evaluated as the ratio between longitudinal storage modulus and transversal storage modulus is reported in Figure 15 as function of temperature.

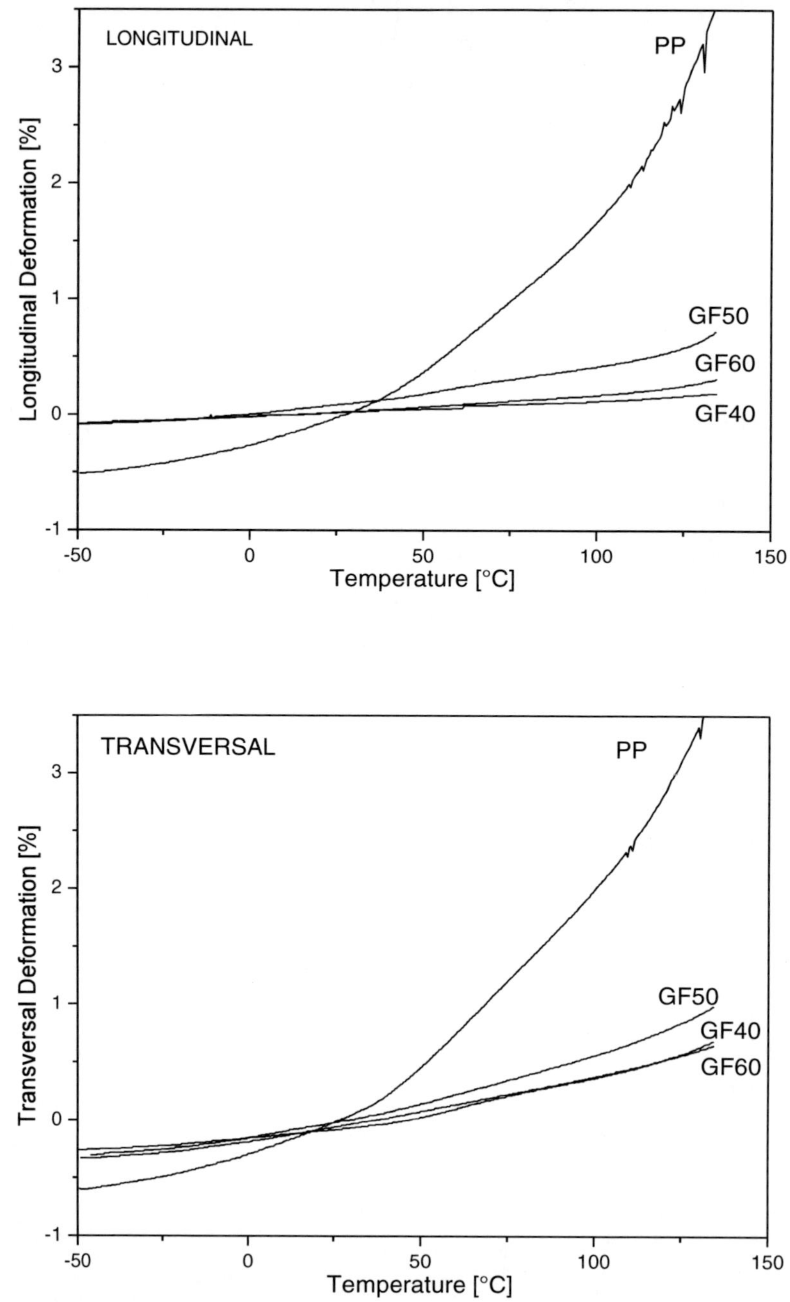

Figure 14. Thermal deformation during tensile DMTA test in longitudinal/transversal direction.

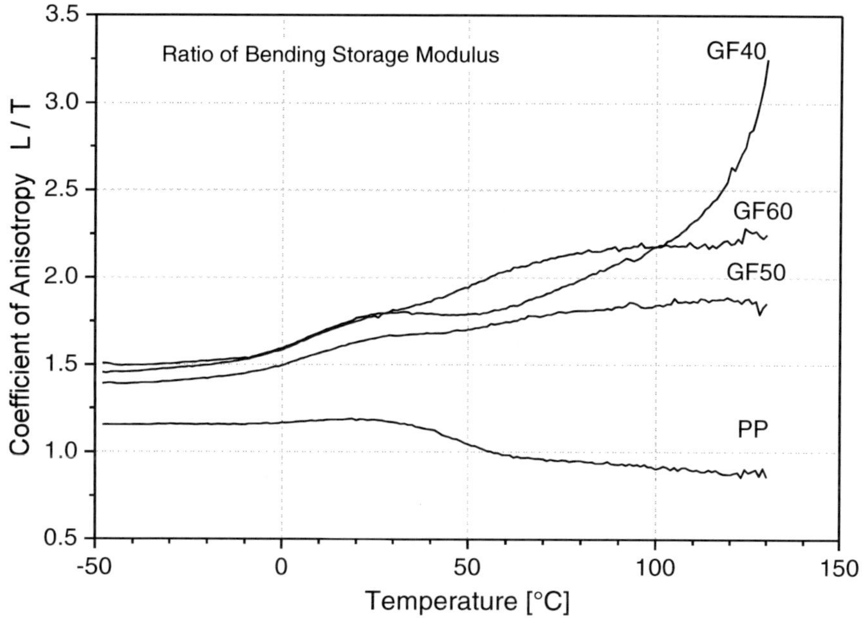

Figure 15. Anisotropy factor (E'_L/E'_T) of PP composites from DMTA.

Morphology and Orientation

SEM images show that fibers are oriented in the direction of flow in the skin zones and are more randomly oriented in the core zone (skin/core effect in the fountain flow of the injection molding process).

Figure 16a evidences the fiber orientation in the skin and core of GF50 after a transversal fracture in the direction of cross flow. In the Figure 16b, the fracture surface of GF50 after longitudinal break in the direction of the flow injection is depicted.

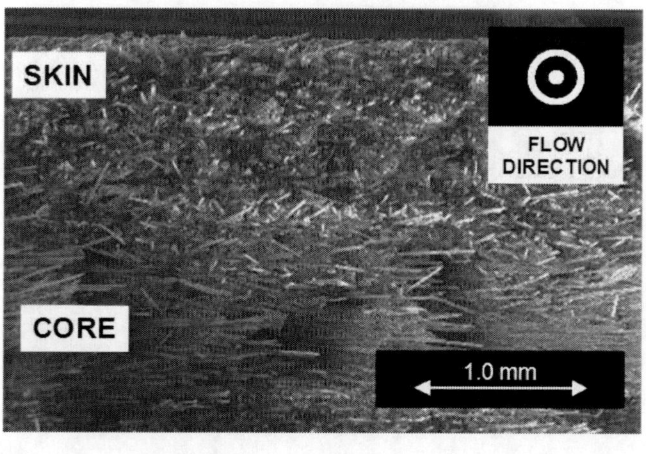

Figure 16a. SEM micrograph of GF50 (transversal fracture - cross flow direction).

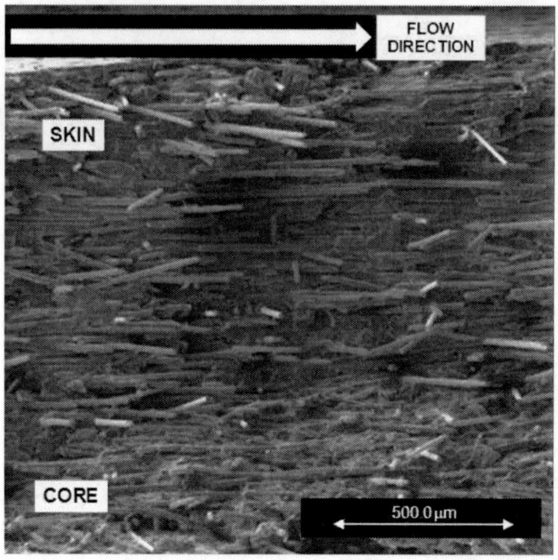

Figure 16b SEM micrographs of GF50 (longitudinal fracture-along flow direction).

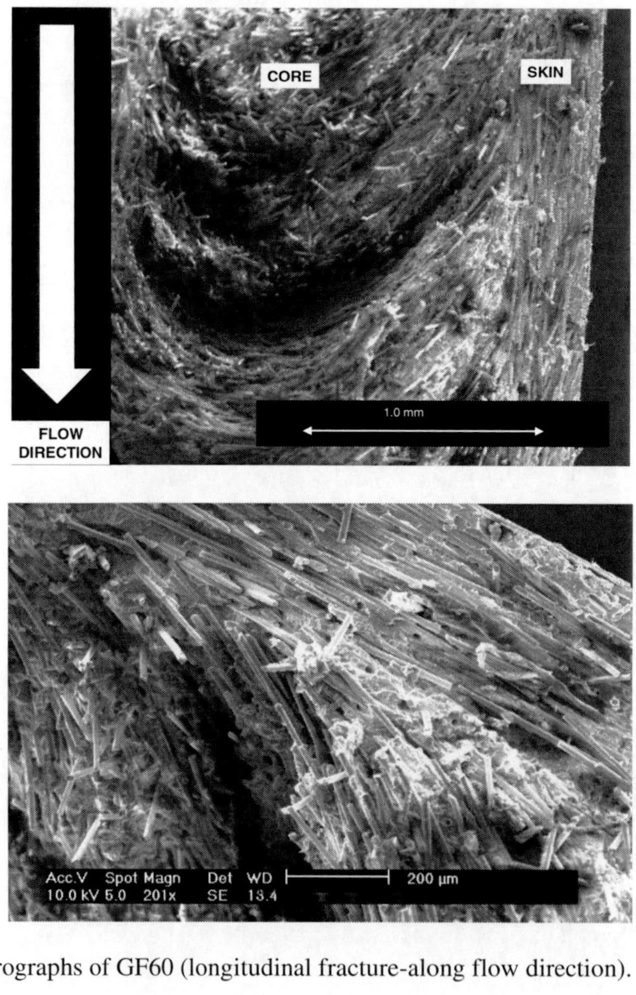

Figure 17. SEM micrographs of GF60 (longitudinal fracture-along flow direction).

Figures 17 show the fracture surface of GF60 after a longitudinal fracture in the direction of flow injection.

Fiber orientations in the core and in the skin layer are evidenced at different magnifications.

Flammability Test

To compare the effect of short glass fiber on the combustion of PP composite, ISO specimens were properly tested using an apparatus for LOI Analysis according to ASTM D-2863-06. Both polypropylene and its composites evidenced a LOI value lower than 0.19, as reported in literature [62-63].

Three different atmospheres with oxygen content of 19-21-23% were selected. Figure 18 shows the specimens of PP matrix and glass fiber composites after 180 seconds of the combustion test at 19% of oxygen.

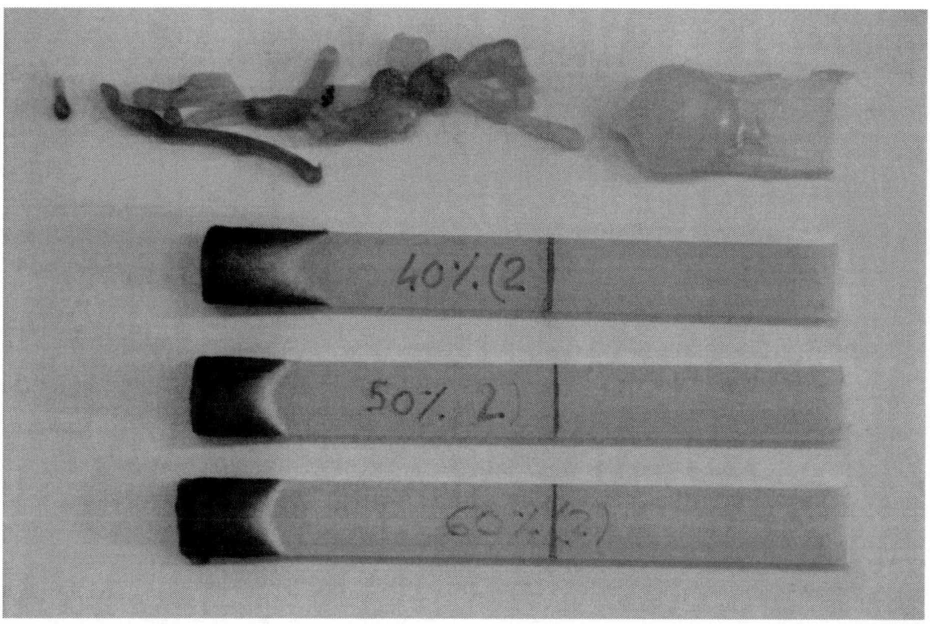

Figure 18. PP and GF-PP (40-50-60 % by wt.) specimens after 180 seconds of burning test in atmosphere with 19% of oxygen.

The rate of propagation of the advancing flame was evaluated for all materials, as shown in Figure 19 in the case of PP and GF40.

It is well evident that the propagation rate increases with the oxygen content and decreases with the presence of glass fibers. In particular, the flame propagation rate in short glass fiber composite decreases at about one-third of the polypropylene in various conditions of oxygen content, almost independently on the glass content (Table 7).

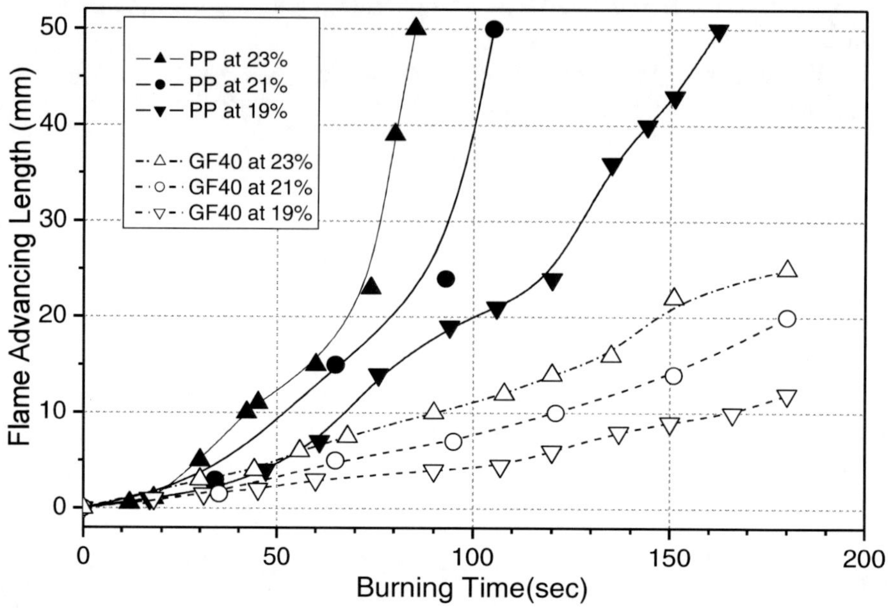

Figure 19. Length of the burned specimen vs. time of burning in flammability test for PP and PP-GF40 at various oxygen content.

Table 7. Propagation rate of PP matrix and GF composites as function of the percentage of oxygen in candle-like combustion (ASTM D2863-06)

Propagation rate (mm/sec) Oxygen Content (%)	PP	GF40	GF50	GF60
19	0.28	0.06	0.07	0.07
21	0.36	0.10	0.11	0.11
23	0.39	0.13	0.13	0.15

CONCLUSION

High load polypropylene composite containing talc, calcium carbonate and short glass fibers up to 60%, 80% and 65% by wt., respectively, were produced by compounding with a relatively low viscosity polymer matrix. All materials exhibited adequate flowability for injection molding processes. Significant increase of stiffness and strength, with a reduction of deformation at break was found for all the composite composition. According to literature data, short glass fibers were progressively chopped during processing in both compounding and injection, reaching average length in between 0.2-0.5 mm, nonetheless keeping outstanding composite properties in particular stiffness and strength. In the same time, short glass fiber composite exhibited a higher toughness (Charpy impact values) at low temperature and heat deflection temperature in the range 148-152°C. Finally, anisotropy factor was evaluated for injection molded large plate.

Following these results, high load polypropylene composite recipes described here are produced in almost standard compounding machines (that is not true for LGF) and appear suitable for the production of articles via traditional injection molding equipment without any special tooling or gating (e.g. hot runner) that are required in the case of long fiber polypropylene composites.

REFERENCES

[1] B. Pukansky, *Fillers for polypropylene, Particulate filled polypropylene, in "Polypropylene. An A-Z Reference"*, J. Karger-Kocsic ed., Kluwer Acad. Publish, Dordrecht – The Netherlands (1999).

[2] P.F. Chu, Glass fiber-reinforced polypropylene, in "*Handbook of polypropylene and polypropylene composite*", H.G. Karian ed., Marcel Dekker, New York, (1999).

[3] A.G. Gibson, Processing and properties of reinforced polypropylene, in "Polypropylene. Structure, blends and composites. Composites" Vol 3, J. Karger-Kocsic ed., Chapman Hall, London (1995).

[4] J.I. Kroschwitz, A. Seidel, *Kirk-Othmer, Encyclopedia of Chemical Technology*, V ed., Wiley-Interscience, Hoboken, NJ 2004-2007.

[5] Lyondellbasell data (2011).

[6] M. Akay, and D. Barkley, *Compos. Struct.* 3, 269 (1985).

[7] V.B. Gupta, R.K. Mittal, P.K. Sharma, G. Mennig, and J. Wolters, *Polym. Comp.* 10,16 (1989).

[8] M. Akay, and D. Barkley, *J. Mater. Sci.* 26, 2731 (1991).

[9] S. Hashemi, and M. Koohgilani, *Polym. Eng. Sci.* 35, 1124 (1995).

[10] J.L. Thomason, and M.A. Vlug, *Composites Part A* 27, 477 (1996).

[11] J.L. Thomason, and M.A. Vlug, *Composites Part A* 28, 277 (1997).

[12] S.Y. Fu, B. Lauke, E. Mader, X.Hu, and C.Y. Yue, *J. Mater. Proc. Techn.* 89, 501 (1999).

[13] A. Pegoretti, and T. Riccò, *J. Comp. Mater.* 34,1009 (2000).

[14] S.Y. Fu, B. Lauke, E. Mader, X.Hu, and C.Y. Yue, *Composites Part A* 31,1117 (2000).

[15] J.L. Thomason, *Compos. Sci. Technol.* 62, 1455 (2002).

[16] J.L. Thomason, *Composites Part A* 33, 1283 (2002).

[17] A.P. Gupta, U.K. Saroop, G.S. Jha, and Minakshi Verma, *Polym.-Plast. Technol. Eng.* 42, 297 (2003).

[18] C. Mobuchon, P.J. Carreau, M.C. Heuzey, M. Sepehr, and G. Ausias, *Polym. Comp.* 26, 247 (2005).

[19] J.L. Thomason, *Composites Part A* 38, 210 (2007).

[20] I. Taha, and Y.F. Abdin, *J. Composite Material* 45, 1805 (2011).

[21] J. Denault, T. Vu-Khanh, and B. Foster, *Polym. Compos.* 10, 313 (1989)

[22] P. Peltonen, E.J. Paakkonen, P.K. Jarvela, and P. Tormala, *Plast. Rubber Compos. Proc. Appl.* 23, 111 (1995).

[23] J.L. Thomason, *Composites Part A* 33, 1641 (2002).

[24] J. Hartikainen, M. Lindnerm, T. Harmia, and K. Friedrich, *Plast. Rubber Compos.* 33, 77 (2004).

[25] J.L. Thomason, *Composites Part A* 36, 995 (2005).
[26] K. Senthil Kumar, A.K. Ghosh, and N. Bhatnagar, *Polym. Compos.* 28, 259 (2007).
[27] M. Rohde, A. Ebel, F. Wolff-Fabris, and V. Altstadt, *Int. Polym. Proc.* 26, 292 (2011).
[28] J. Wang, C. Geng, F. Luo, Y. Liu, K. Wang, Q. Fu, and B. He, *Mater. Sci. Eng. A* 528, 3169 (2011).
[29] B. He, H. Liu, J. Leng, B. Yang, X. Chen, J. Fu, and Q. Fu, *J. Reinf. Plast. Compos.* 30, 222 (2011).
[30] D.V. Rosato, and D.V. Rosato, Injection Molding Handbook: The Complete Molding Operation Technology, Performance, Economics, 2nd Edition, Chapman and Hall, New York, (1995).
[31] P.F. Bright, R.J. Crowson, and M.J. Folkes, *J. Mater. Sci.* 13, 2497 (1978).
[32] M. Sanou, B. Chung, and C. Cohen *Polym. Eng. Sci.* 25, 1008 (1985).
[33] P. Singh, and M.R. Kamal, *Polym. Compos.* 10, 344 (1989).
[34] M. Gupta, and K.K. Wang, *Polym. Compos.* 14, 367 (1993).
[35] S.Y. Fu, and B. Lauke, *Compos. Sci. Technol.* 56, 1179 (1996).
[36] L. Averous, J.C. Quantin, A. Crespy, and D. Lafon, *Polym. Eng. Sci.* 37, 329 (1997).
[37] J.P. Green, and J.O. Wilkes, *Polym. Eng.Sci.* 37, 1019 (1997).
[38] H.G. Karian, and K. Stoops, *Polym. Compos.* 19, 31 (1998).
[39] S.E. Barbosa, and J.M. Kenny, *J. Reinf. Plast. Compos.* 18, 413 (1999).
[40] A. Larsen, *Polym. Compos.* 21, 51 (2000).
[41] N. Sombatsompop, and W. Chaiwattanpipat, *Polym. Test.* 19, 713 (2000).
[42] E.G. Kim, J.K. Park, and S.H. Jo, *J. Mater. Proc. Technol.* 111, 225 (2001).
[43] P.J. Hine, and R.A. Duckett, *Polym. Compos.* 25, 237 (2004).
[44] S. Patcharaphun, and G. Mennig, *Polym. Compos.* 26, 823 (2005)
[45] W.N. Ota, S.C. Amico, and K.G. Satyanarayana, *Compos. Sci. Technol.* 65, 873 (2005).
[46] C.A. Silva, J.C. Viana, F.W.J. van Hattum, and A.M. Cuhna, *J. Mater. Sci.* 42, 5203 (2007).
[47] M. Benhadou, and A. Haddout, *J. Reinf. Plast. Compos.* 26, 1357 (2007).
[48] J.P. Qu, X. Zhang, G. Jin, *Polym.-Plast. Technol. Eng.* 47, 186 (2008)
[49] P.Onishi, and S. Hashemi, *J. Mater. Sci.* 44, 3445 (2009).
[50] C. Yang, H.-X. Huang, and K. Li, *Polym. Compos.* 31, 1899 (2010).
[51] R. Turkovic, and L. Erwin, *Polym. Eng. Sci.* 23, 743 (1983).
[52] B. Franzen, C. Klason, J. Kubat and T. Kitano, *Composites* 20, 65 (1989)
[53] V.B. Gupta, R.K. Mittal, P.K. Sharma, G. Mennig, and J. Wolters, *Polym. Compos.* 10, 8 (1989).
[54] J.P. Hernandez, T. Raush, A. Rios, S. Strauss, and T.A. Osswald, *Eng. Anal. Bound. Elem.,* 26, 621 (2002).
[55] G. Zhang, and M.R. Thompson, *Compos. Sci. Technol.* 65, 2240 (2005).
[56] S. Patcharaphun, and G. Opaskornkul, *Kasetsart J. (Nat. Sci.)* 42, 392 (2008).
[57] K.A. Geddes, R.A. Guyer, and R.C. Miller, High melt flow fiber reinforced propylene polymer compositions, *United Stated Patent* No. 4,997,875, Mar. 5, 1991.
[58] Basell Poliolefine Italia srl, E. Masarati, E. Costantini, and M. Consalvi, Filled polyolefin compositions, *Patent WO 2008/074713,* June 26, 2008.
[59] D.W. Van Krevelen, "Properties of Polymers", 3rd Ed., Elsevier, Amsterdam (1990).
[60] J.C. Halpin, and J.L. Kardos, *Polym. Eng. Sci.* 18, 496 (1978).
[61] P.K. Mallick, "*Fiber Reinforced Composites*", Marcel Dekker Inc., New York (1988).

[62] H. Dvir, M. Gottlieb, S. Daren, and E. Tartakovsky, *Compos. Sci. Technol.* 63, 1865 (2003).

[63] B. K. Kandola, and R. Toqueer-Ul-Haq, *Fire Mater.* (2011) - doi/10.1002/fam.1120/full

In: Polypropylene
Editors: L. P. Silva and E. F. Barbosa

ISBN: 978-1-62417-142-0

Chapter 13

FOAMED POLYPROPYLENE FOR INDUSTRIAL APPLICATIONS

Marcelo Antunes and José Ignacio Velasco*
Centre Català del Plàstic,
Departament de Ciència dels Materials i Enginyeria Metal·lúrgica,
Universitat Politècnica de Catalunya, BarcelonaTech (UPC),
Barcelona, Spain

ABSTRACT

There is a great deal of industrial interest in the development of increasingly lighter materials based on polymer foams with improved specific properties for the most varied purposes, from flexible foams for packaging to rigid ones for structural applications. Due to its good combination of properties and reduced cost, foamed polypropylene (PP) could come as a good option. Depending on the base material, foam density and developed cellular structure and microstructure, PP foams may cover a wide range of characteristics and properties which, combined with the incorporation of functional fillers, could ultimately result in the development of multifunctional lightweight materials. This chapter presents a review of the most recent developments in polypropylene foams, starting out by the commercially available PP grades thought out for foaming applications, the different types of PP-based foams depending on their characteristics and final uses, from low-density flexible foams for cushioning and packaging to medium-high density foams for structural purposes, and the main industrial foaming processes, going from melt-like foaming to solid-state chemical and physical foaming. Recent developments in the field of PP foams are also considered, focusing on the combination of cellular structure control via foaming and the incorporation of micro and nanosized functional fillers, with the ultimate objective of developing PP-based foams with a wide range of properties and unique characteristics, from electrically conductive foams through the incorporation of conductive carbon-based nanofillers to foams with direction-dependent transport properties.

---

* E-mail address: jose.ignacio.velasco@upc.edu

INTRODUCTION

Owing to its good set of properties and reduced cost, polypropylene (PP) has been considered in the development of foams to be used in the most varied applications, from soft and flexible foams to more rigid and tougher ones, depending on the initial polymer molecular characteristics, foaming process, amount of blowing agent, foam density, cellular structure and microstructure, among others [1].

Together with these, there is a recent interest in further extending the properties and thus applications of PP foams by adding functional fillers, from more conventional ones such as magnesium hydroxide to nanometric-sized such as montmorillonite or carbon-based nanofillers (carbon nanotubes and nanofibers, graphene, etc). In the case of nanometric-sized fillers, the particular characteristics of the filler, with at least one of its dimensions in the nanometric range and hence high possible specific surface interaction with the polymer molecules, enable the addition of low amounts (< 5 wt.%) and even very low amounts (< 0.5 wt.%) to result in important improvements in the final properties of the polymer foam [2]. Also, it has been proven that there is a synergistic effect between these nanofillers and foaming, the first for instance promoting the formation of smaller cells (i.e. higher cell densities) by a heterogeneous cell nucleation effect, and foaming favouring the dispersion of the filler particles throughout the polymer matrix, thus promoting particle-polymer interaction [2-3]. Also, not only these nanofillers are characterized by their high specific surface area, but are also added taking into mind some of their intrinsic characteristics, which in the case of PP foams could lead to the development of multifunctional lightweight materials. For instance, carbon-based nanofillers such as carbon nanofibers or nanotubes, or the recently considered graphene, are known to have intrinsically high thermal and electrical transport properties [4], which could, in combination with cellular structure control or other foaming characteristics, result in PP foams with tailor-made properties.

Polymer foams commonly display non-uniform geometrical cell orientations, direct result of the high stretching forces inherent to gas expansion and foaming parameters [5]. These cellular anisotropies have been shown to directly affect the characteristics of PP foams, from mechanical to thermal transport properties [6-8]. In the particular case of thermal conductivity, cellular anisotropy, and more specifically preferential cell orientation, has been shown to affect the thermal conductivity of low density foams, recently proven by the analysis of the thermal conductivity anisotropy in PP foams prepared by supercritical CO_2 dissolution [9], demonstrating that there is a direct influence of cell orientation on the thermal conduction behaviour of these foams.

This chapter not only presents a review of the most recent developments in PP foaming regarding its industrial use and applicability, focusing on the different available PP-based grades specifically developed for preparing from flexible to rigid foams for the most varied applications and new foaming processes, but also covers the new developments in the field of PP (nano)composite foams and particularly the idea of developing multifunctional PP-based lightweight materials by the incorporation of different types of functional (nano)fillers.

POLYPROPYLENE GRADES FOR FOAMING

Although being considered a "commodity plastic", i.e. one of the high consumption plastics, and thus being well-known and vastly characterized, due to its inherent poor melt strength and extensibility derived from its linear molecular structure [10], the preparation of PP foams presents some peculiarities. In order to overcome some of its processing limitations, several strategies have been considered in terms of modifying the rheology of the polymer to the specific conditions required when using conventional foaming technology. As previously said, this is the direct consequence of the linear molecular structure of conventional PP, which ultimately results in melts with melt strengths below the required values for foaming, particularly when using conventional melt-like foaming methods, such as extrusion or injection moulding [11-12].

Nowadays, the use of PP in the preparation of low density foams is rather limited when compared to polyethylene (PE). Polyethylene is cheaper and easier to foam at typical foaming conditions, i.e. above polymer's melting point, due to its inherently higher melt strength and extensibility and the higher number of available melt strengthening strategies, which include branching and crosslinking, ultimately broadening the optimum foaming temperature window [13].

Different strategies have been considered depending on the final foam characteristics, and have led to the development of specific PP grades thought for foaming applications:

For the preparation of flexible and soft PP-based foams for sectors such as the automotive and packaging, where a combination of high foam expansion and material flexibility is required, PP homopolymer is blended with low melting point ethylene copolymers such as ethyl vinyl acetate copolymer (EVA) or ethylene-octene copolymers [14], or alternatively low ethylene content random or block copolymers [15] are used. As will be shown later on, there is a new grade of PP homopolymer specifically developed by Borealis that combines the good thermal stability and strength characteristics of PP homopolymer with the flexibility and softness of polyethylene, enabling its use in the production of "soft-like" 100% PP-based low density foams.

There are several polyolefin manufacturers that have PP-based grades specifically thought for the production of low density flexible foams for the packaging sector. Under the generic name of *Isplen*, Repsol offers two specific PP-based grades for the preparation of foamed sheets by means of a physical foaming extrusion process: *Isplen PB-110 H2E*, a very low MFI block copolymer, which can also be used in blow-moulding applications, and *Isplen PR-240 G1E*, a random copolymer. While the first tends to be used when higher impact strengths are required [16], the random copolymer is commonly used in combination with other PP homopolymers, contributing to add flexibility to the foam (low density flexible packaging sector) [17].

In this way, LyondellBasell, under the generic trade name *Moplen*, offers a wide range of PP-based random copolymers which can be used alone or in combination with conventional PP homopolymers for the preparation of low density flexible foams.

Similarly to Repsol and LyondellBasell, other polyolefin manufacturers such as Sabic offer a wide range of PP-based random copolymers which can be used in extrusion foaming. In the case of medium-high density rigid PP foams, due to the particular linear-like molecular structure of PP, specific PP homopolymers with improved melt strengths, usually obtained by post-polymerization or during polymerization chemical modification [18-21], are used. These are commonly known as high melt strength polypropylenes.

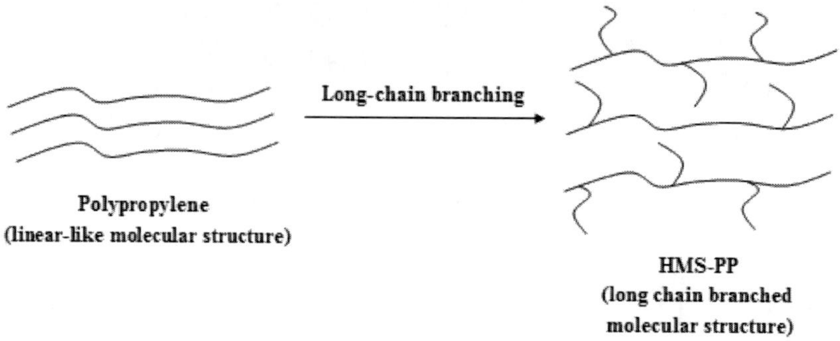

Figure 1. Schematic view of HMS-PP long chain-branched molecular structure.

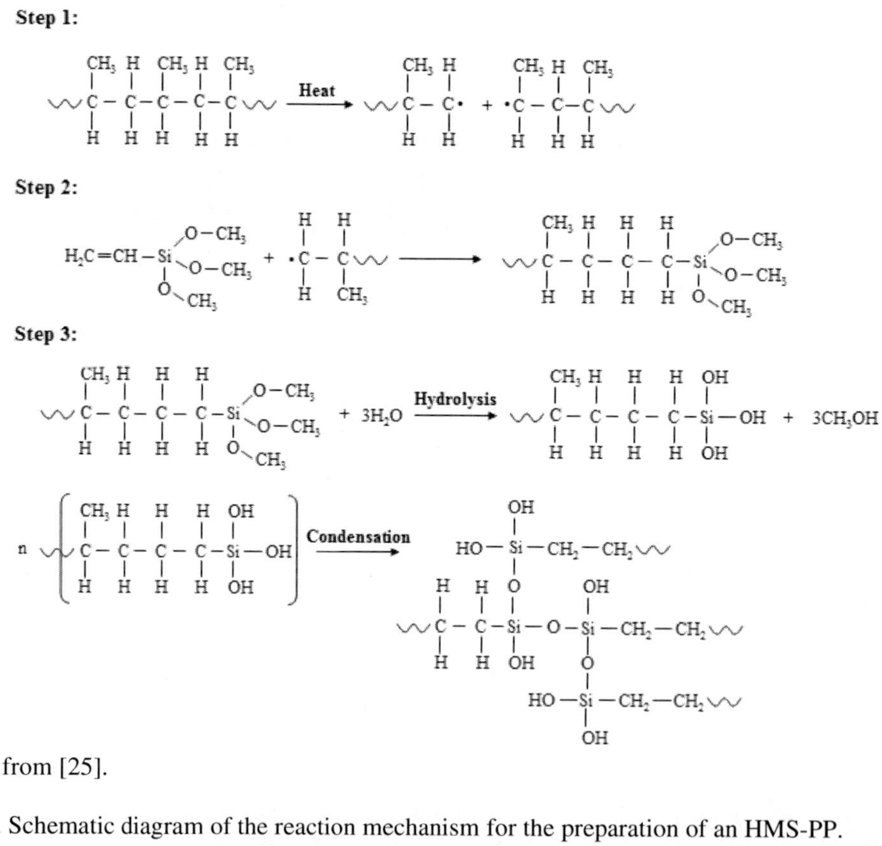

Adapted from [25].

Figure 2. Schematic diagram of the reaction mechanism for the preparation of an HMS-PP.

High melt strength polypropylene (HMS-PP) is a propylene-based polymer with improved melt strength and extensibility, thus characterized by its improved foaming processability, where considerable polymer stretching occurs during cell growth. As previously stated, standard polypropylene is a semi-crystalline polymer with a linear molecular structure and inherently poor melt resistance, limiting its melt strength and drawability, thus resulting in a material that cannot withstand the extensional forces generated during foaming, leading to foams with considerably high densities and poor cellular structures [22-23]. This limitation can be solved by chemically introducing long chain branches into the linear PP backbone, resulting in the development of the so-called high melt strength polypropylene grades (HMS-PP's), their name deriving from the improved melt strength resulting from branching (see the schematic of long-chain branched HMS-PP displayed in Figure 1 and an example of the reaction mechanism for the preparation of an HMS-PP in Figure 2). These specially modified polypropylenes have been shown not only to increase the expansion ratio of PP foams, but also to improve cell uniformity and retard cell coalescence, ultimately broadening the optimum foaming processing window [3, 24]. Particularly, HMS-PP grades are specially thought for physical foam extrusion, where, depending on the specific foam technology, foams can reach density values as low as 30 kg/m^3.

Among the few manufacturers that offer high melt strength polypropylene homopolymers, Borealis and LyondellBasell are the two most important ones.

Borealis is one of the pioneers in the industrial implementation of specially-modified high melt strength polypropylenes for foaming applications. Among the *Daploy HMS* PP grades offered by Borealis, some types are specifically thought for foaming, being divided into homopolymers, *Daploy WB130HMS*, *Daploy WB135HMS* and *Daploy WB140HMS*, and copolymers, *Daploy WB260HMS*. The first two are especially thought for the production of low density PP foams by means of physical foaming by direct extrusion. *Daploy WB140HMS* is a recent development in long chain-branched polypropylenes with improved melt resistance, enabling to decrease even more the density of the produced PP foams to the very low density range (see section "Recent Developments in Materials" for further details). So, Borealis *Daploy HMS* grades are long chain-branched PP-based materials, which combine two of the most important characteristics required for stable low density closed-cell foam production: high strength and high extensibility/drawability in melt-like conditions, i.e. at high temperatures, which help control cell growth and limit cell coalescence through a strain hardening mechanism [26-27]. As previously mentioned, they have opened up the use of PP in the production of low density foams, with all the benefits associated to PP, such as wide mechanical property range and high thermal stability. Some of the main characteristics of *Daploy HMS* are, besides their high melt strength and high melt extensibility, making it suitable for physical foam extrusion, high stiffness, high service temperature and 100% recyclability (non-crosslinked).

Deriving from the above-mentioned characteristics, *Daploy HMS* PP is used in the production of thermoformable foamed sheets and films, lightweight packaging trays and containers, microwaveable lightweight food packages, foamed components for the automotive sector, cushioning and protective packages and thermal and acoustic insulation elements, among others.

LyondellBasell offers high melt strength polypropylenes through the commercial name of *Pro-fax*. Particularly, *Pro-fax PF814* PP homopolymer is specially designed for the production of low density foamed products using conventional physical foaming extrusion methods, since its particular rheological behaviour (improved melt strength and resistance properties) supports stable cell growth using common physical blowing agents.

HMC Polymers, an associate of LyondellBasell, offers several PP grades prepared by a new polymerization patented technology called Spherizone using the Multizone Circulating Reactor (MZCR) [28] (see Figure 3). Since this dual reactor technology allows to carefully control the polymerization reaction of PP, it widens the range of PP uses, allowing even to prepare high melt strength PP for foaming and thermoforming applications. One of the new grades, called *Higran RS1684*, displays very high melt strength, enabling its use in the production of low density PP foams mainly for the packaging sector by means of both physical and/or chemical foaming sheet extrusion, or even through the expansion of PP beads (EPP).

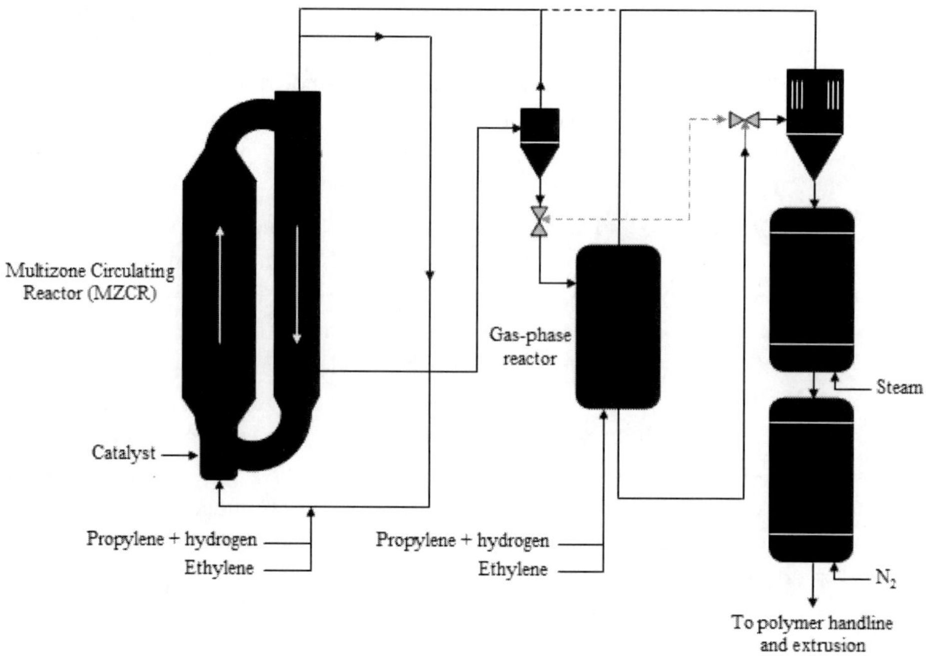

Adapted from [28].

Figure 3. Simplified Spherizone-Multizone Circulating Reactor (MZCR) flow diagram for PP polymerization.

Depending on the application, HMS-PP can be used on its own or blended to some extent with more conventional lower melt strength polypropylenes, as can be seen in Figure 4. As increasingly higher proportions of linear PP are added to the long chain-branched PP, the rheological properties of the blend start to drop. Nevertheless, a high amount of linear PP (maximum around 60 wt%) can still be added to HMS-PP before the characteristics of the blend fall outside the optimum foaming window.

An adequate control of the foam properties is a crucial necessity in order to produce PP foams with properties that meet the performance demands of particular end-use applications. The scheme shown in Figure 5 displays how an appropriate choice of blend polymer to the HMS-PP may regulate the impact strength vs. stiffness characteristics of the foam, enabling to obtain from high impact strength foams (using for instance random or block copolymers) to foams with improved stiffness (as for instance when using 100% HMS-PP or blending it with linear PP homopolymers). *Borsoft*, a Borealis-patented low tensile modulus random heterophasic copolymer, can also be used to produce soft PP foams.

As previously stated, in order to overcome some of the limitations related to the high stiffness of conventional HMS-PP, excluding its use in the production of soft PP foams for sectors such as the automotive or packaging, Borealis has recently developed a new HMS-PP Daploy grade, *Daploy WB260HMS*, which enables PP to be competitive in applications usually dominated by polyurethane, polyethylene and PVC foams. This enables the use of an all-PP product (100% *Daploy WB260HMS* or blended with other PP grades) for applications where low density PP-based foams (with densities as low as 30 kg/m^3) were usually based on blends of standard HMS-PP with polyethylenes, TPOs or low melting point copolymers such as EVA.

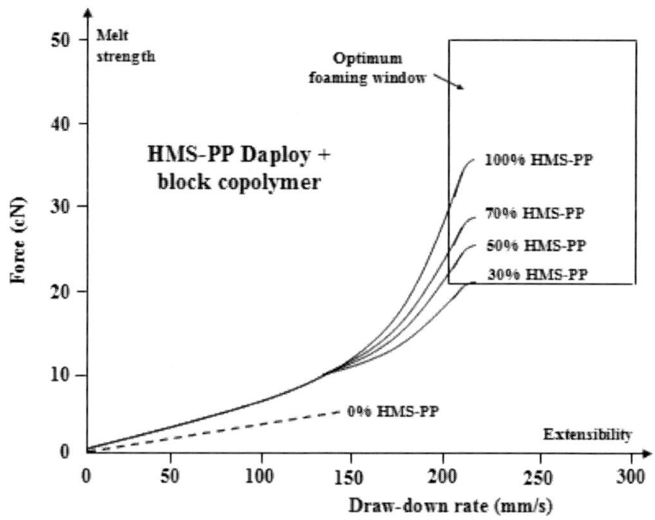

Adapted from [29].

Figure 4. Extensional rheology curves for HMS-PP (*Daploy WB130HMS* + block copolymer blends).

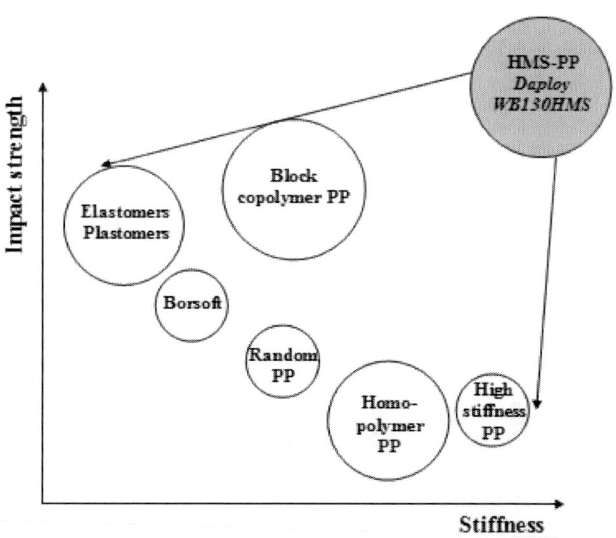

Adapted from [29].

Figure 5. Influence of different PP-based blend partners on the impact strength vs. stiffness relation of PP foams.

Due to the particular characteristics of these low density PP foams, formed by a very low volume percentage (< 5%) of a 100% recyclable non-crosslinked polymer, they make a significant contribution not only towards material savings (weight reduction) but also in terms of final product recyclability, mainly compared to other non-foamed/high density foamed products or crosslinked systems. This has obvious environmental advantages and brings clear cost-efficiency advantages to the manufacturer.

TYPES OF POLYPROPYLENE FOAMS AND APPLICATIONS

Although polymer foams and thus polyolefin foams are commonly classified according to their density and cellular structure in closed-cell, open-cell, partially open-cell and interconnected, and/or according to the base polymer in elastomeric, thermoplastic and thermoset [30-31], a quite used classification considers the mechanical characteristics of the foam and is determined by a combination of the nature of the base polymer and foam density: flexible and rigid foams.

Due to their combination of good energy absorption and low thermal conductivity, low density flexible PP-based foams are vastly used in the packaging and transport sectors, as well as thermal insulators in sectors such as construction. More specifically, there is an increasing interest in developing soft flexible PP-based foams by controlling the polymer formulation, as well as open-cell PP foams, commonly prepared by post-foaming open-cell operations. These would allow to extend the range of possible applications of low density flexible foams for instance to 'soft-touch' applications (automotive sector) or, in the case of open-cell foams, to acoustic insulation. Rigid PP foams stand out by their considerably high density, typically above 180 kg/m^3, and high specific mechanical properties, hence being mainly used for structural purposes in sectors such as construction or automotive.

Flexible PP Foams: Packaging and Thermal Insulation

It is well known the interest behind the study of heat transfer in polymer foams, as one of their main applications is as thermal insulators, especially in the case of low density foams, where the volume percentage of gas, a very good thermal insulator due to its low thermal conductivity, surpasses 90% [30]. Particularly, the use of PP in the low density foam range is somewhat limited compared to PE, mainly due to difficulties related to the already mentioned poor 'foamability' of PP (inherently weak melt resistance and extensibility). This is often overcome by blending PP with low melting point ethylene copolymers, or alternatively using low ethylene content random copolymers.

Limited work has been dedicated to the thermal conductivity analysis of PP-based flexible foams [13]. For instance, Rodríguez-Pérez et al. [32] have investigated the thermal conductivity of physically crosslinked closed cell polyolefin foams, with the objective of optimizing the thermal insulating properties of the foams. One of the considered foams, formed by a blend of PP and LDPE, displayed better insulating properties than the rest, related to the intrinsically lower thermal conductivity of PP when compared to LDPE [33].

Polypropylene-based formulations specially developed for the production of soft-like closed-cell PP foams with excellent flexibility and heat resistance have also been considered [34-35]. These low density foams, usually obtained using PP random or block copolymers or blends of linear or high melt strength polypropylenes with polyethylene or other low melting point ethylene copolymers such as EVA, can be used in shock absorbing applications, heat and/or sound insulation, electric insulation, among others. PP-based soft foams have also been considered as possible replacers of soft thermoset foams, with obvious advantages in terms of recyclability. Maximum foam expansion has been achieved through a proper control of the content of the low melting point soft copolymer, in this case poly(ethylene co-octene), added to the polymer matrix [36].

Recently, soft PP foams have awakened a great deal of interest in the field of piezoelectric sensors and actuators, where polymer foams are also known as ferroelectrets. The concept of ferroelectret as piezoelectric active material has been investigated in foamed PP films with different cellular structures and elastic properties, and their elastic behaviour has been correlated with the piezoelectric properties of the developed foams [37-38].

Among the flexible foams, open-cell PP foams, obtained by post-foaming operations or by inducing inhomogeneous 'softer' domains in the polymer matrix and carefully controlling the foaming conditions in order to promote cell rupture during expansion, are gaining a great deal of interest as they are capable of absorbing acoustic energy and at the same time support good mechanical loads, which is critical for packaging and protection.

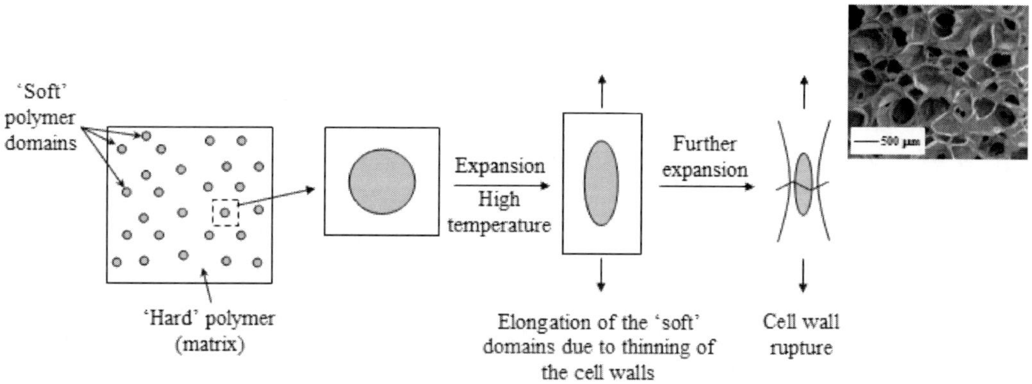

Adapted from [40] and [42].

Figure 6. Schematic view of possible cell opening mechanism during foaming by using a 'hard/soft' inhomogeneous PP/PE blend and typical micrograph of an open-cell polyolefin foam.

Open-cell PP foams have been prepared using a combined compression-moulding chemical foaming and particulate leaching processes [39]. It was demonstrated that foamed samples with smaller cell sizes displayed better sound absorption properties. Nevertheless, as foam thickness increased, foams with medium cell sizes offered a better acoustic insulation. As expected, the compressive strength of the PP foams increased with increasing relative density. PP/PE blends have also been used in order to prepare flexible open-cell PP-based foams [40-41]. The basic strategy behind the use of these blends consisted in inducing a 'hard/soft' inhomogeneous melt structure in which the 'softer polymer' could be opened at the high temperature foaming conditions, resulting in open-cell foams (see schematic view displayed in Figure 6). Both HMS-PPs as well as more conventional linear PPs have been used with metallocene and low-density polyethylenes, and it has been demonstrated that an almost 100% open-cell content may be obtained by carefully optimizing the polymer blending ratios, foaming temperature and pressure drop rate using a CO_2 high-pressure dissolution physical foaming process [40].

Rigid PP Foams: Structural Applications

Some of the advantages of PP when compared to PE, such as higher stiffness or higher impact strength, start to be relevant for foam densities above 100 kg/m^3. For that reason, a PP homopolymer-based material, in its linear form blended with other polyolefinic block copolymers, or more conventionally with a long-chain branched molecular architecture (see HMS-PP explained in section "Polypropylene Grades for Foaming"), or used in blends of linear and long-chain branched homopolymers, is used in order to produce rigid PP-based foams with a closed-cell structure, mainly thought for structural purposes (in sectors such as construction), where the component must fulfil certain specific mechanical requirements. As explained, density is important as it is well known that foaming globally reduces the mechanical performance of the material (stiffness, strength, etc). Generally speaking, rigid PP foams are prepared with densities above 100 kg/m^3.

So, alongside the always important requirement of low cost, a polymer foam thought for structural applications must present a good balance between mechanical properties and lightness, i.e., high specific mechanical properties, if possible comparable to the specific properties of the respective solid material or even higher (please see the concept of 'microcellular foam' presented in sections "Continuous/Semi-continuous Foaming Processes" and "New Foaming Processes and Innovations"). Alongside a careful control of the developed cellular structure, where for instance microcellular foams have been reported to have improved mechanical properties compared to foams with a more conventional cellular structure [43], one strategy that could be used to improve even further the mechanical performance of polymer foams consists in mechanical reinforcing the base material formulation by adding rigid functional reinforcements, leading to foamed materials that are also known as cellular polymer composites or polymer composite foams [44] (see schematic displayed in Figure 7).

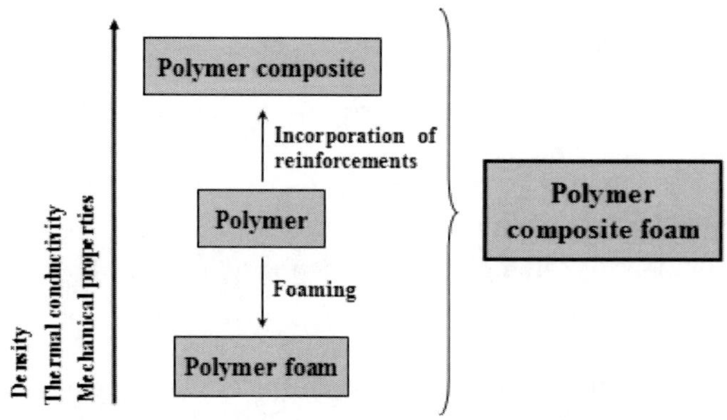

Figure 7. Schematic view displaying the concept of polymer composite foam by combining foaming with the incorporation of reinforcements.

For all the previously mentioned reasons (good balance of mechanical properties/low cost), together with the development of the already mentioned high melt strength PP grades, vastly extending its use in the foam industry, PP could come as a very good option for the development of rigid lightweight components for the most varied applications, from the construction sector (thermal insulation of floors, walls, ceilings, sub-layer of wood floors, etc, or used as cellular cores of sandwich-like structures) to the automotive one (bumpers, lateral panels, dashboards, etc).

As mentioned, there is a great interest in extending the properties and thus possible applications of rigid PP foams, mainly considering that all sectors where weight reduction is critical set increasingly higher mechanical and functional requirements. Regulation of the foam properties, mainly mechanical, could be possible by a combination of cellular structure control (foaming process and foaming parameters) and/or incorporation of functional reinforcements, leading to what have been named as 'multifunctional foams' [45].

Microcellular foams, characterized by a micrometric-sized isotropic-like cellular structure, have been initially thought in the industry for the weight reduction of different moulded parts keeping the mechanical performance of the solid base material [43]. Microcellular PP injection-moulding, already industrially used (please consult section "Foaming Processes" for further details), only results profitable in the case of big production series with significant material and process cost savings and cycle time reductions (for instance in the automotive sector). In the case of small part series, batch foaming processes, such as the high-pressure gas dissolution in specially designed autoclaves, could come as a good alternative. Nevertheless, these batch foaming processes are time consuming, and should only be industrially considered for the production of very specific components. From a scientific point of view, it constitutes a more versatile process than injection-moulding, since it enables to study the possible synergistic effects between foaming and the incorporation of functional reinforcements, with the final objective of developing new multifunctional foams [46].

Though many advantages could come from the preparation and use of microcellular rigid PP-based foams, some problems normally arise from the use of PP, limiting the formation of a microcellular structure using conventional foaming technologies. The main one is related to the fact that PP is a semi-crystalline polymer. Crystallization of PP while cooling in order to stabilize the cellular structure may result in cellular structure inhomogeneities, as the blowing agent does not dissolve well in the crystal domains [47]. This can be overcome by carefully controlling the foaming conditions in a pressure-quench CO_2 dissolution foaming process [48] or blending PP with other polyolefins such as HDPE [49-52]. Other works have considered the preparation of PP microcellular foams by incorporating heterogeneous nucleating agents such as talc or more recently clay nanoparticles [53-55] (see section "Polypropylene Nanocomposite Foams" for further details).

The loss in mechanical properties inherent to foaming may also be counteracted by incorporating mechanical reinforcements to the polymer. Though there are many possibilities for mechanical reinforcing PP foams, from conventional macroscopic or micrometric to nanometric-sized reinforcements, these last ones have aroused a great deal of interest in the scientific community, mainly due to the possible improvements achieved at low reinforcement concentrations and synergies with foaming [45]. Particularly, it has been shown that these nanoreinforcements promote finer and more isotropic cellular structures, due to a heterogeneous cell nucleation effect, which, combined with their inherently higher mechanical properties, could contribute to mechanically reinforce the foam. Also, foaming has been shown to add to a better dispersion of the nanoreinforcements which, due to their extremely small sizes, could result in cell wall reinforcement during foaming, improving the foaming behaviour through local enhancement of the melt strength.

FOAMING PROCESSES

Though open-cell polyolefin foams may be obtained, commonly using post-manufacturing operations, the industrially available polyolefin foaming processes give way to predominantly closed-cell cellular structures [13]. Also, although PE-based foams are normally crosslinked prior or during expansion, PP-based foams do not require crosslinking. Instead, especially melt strengthen PP grades (HMS-PP, PP-based copolymers or blends of PP with other olefinic copolymers) are used (please consult section "Polypropylene Grades for Foaming").

There are two basic foaming processes used to prepare closed-cell PP-based foams: continuous and semi-continuous processes for producing foamed sheets and batch-foaming for preparing foamed products with higher thicknesses. Figures 8 and 9 respectively present the main industrial continuous/semi-continuous and batch foaming processes, as well as their basic stages, used to produce closed-cell PP foams.

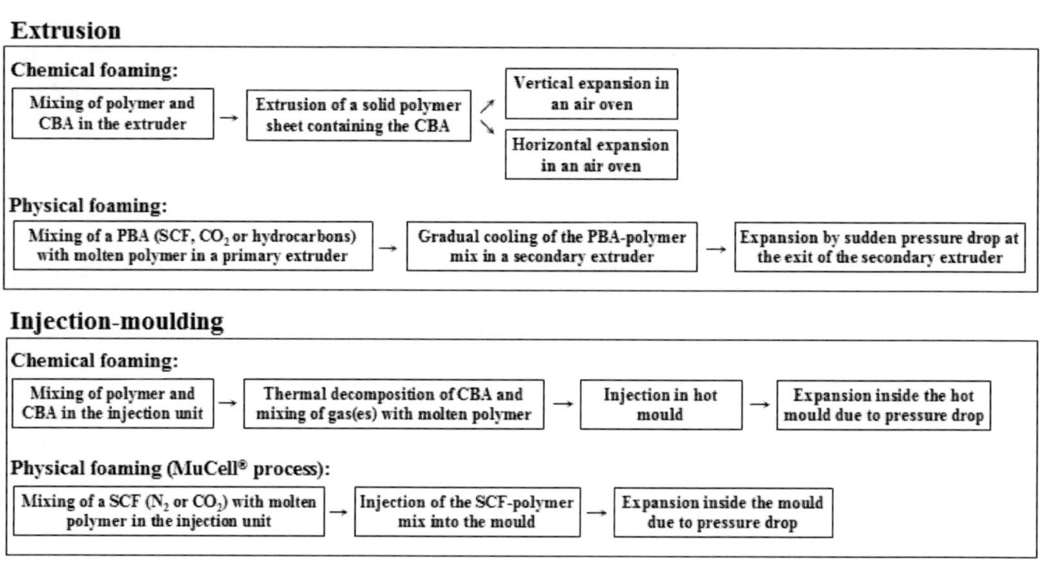

Figure 8. Scheme displaying the main industrial continuous/semi-continuous foaming technologies used to produce closed-cell PP-based foams.

Continuous/Semi-Continuous Foaming Processes

As can be seen in Figure 8 for the continuous/semi-continuous foaming processes, each process (extrusion and injection-moulding) has been divided in terms of the type of blowing agent used into chemical and physical foaming.

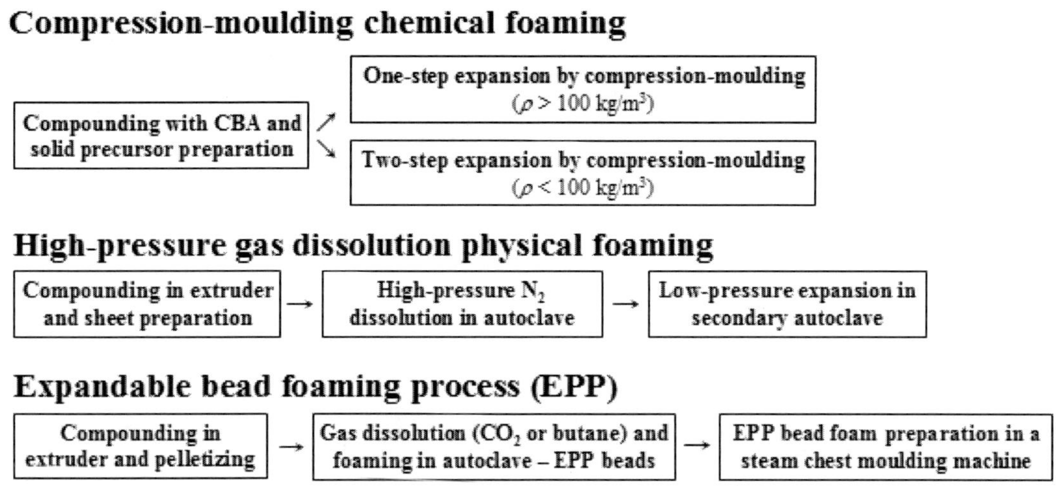

Figure 9. Scheme displaying the main industrial batch-foaming technologies used to produce closed-cell PP-based foams.

Extrusion chemical foaming was first introduced in 1958 by Dow Chemical and developed in Japan by companies such as Sekisui, currently Sekisui Alveo, halfway through the 1960s for producing crosslinked PE foams. This process is limited to the preparation of foams with relatively reduced thicknesses and is commonly used for the production of flexible foams in the form of sheet rolls for packaging and transportation.

As this foaming process has been mainly used in the production of crosslinked PE-based polyolefin foams, it has normally been classified according to the type of crosslinking process used: physical by means of high energy irradiation or chemical by the addition to the extruder of a crosslinking agent [13]. The usual crosslinking agent for PE-based foams is dicumyl peroxide (DCP) [1]. In the case of polyolefins, the chemical blowing agent (CBA) is normally azodicarbonamide (ADC) or other azo-based compounds, well in its non-activated form (PP-based formulations) or thermally activated, for instance using oxide metals such as zinc oxide (PE-based formulations) [1].

The foaming process basically consists in compounding the polymer with the different additives, mainly the CBA and, in the case of chemical crosslinking, the crosslinking agent, as well as other processing additives, using an extruder. After exiting the extruder in the form of a solid sheet, the material is expanded in specially designed air ovens, normally placed in-line with the extruder. Special care must be taken to avoid any thermal decomposition of the CBA inside the extruder during the compounding stage.

Another classification considers the way in which the material is foamed: vertical expansion, where the solid compounded sheet is foamed by passing it vertically through a hot air oven, and horizontal expansion, in which foaming occurs horizontally.

In the case of vertical expansion (see Figure 10), the sheet is stretched in the foaming direction, hence resulting in foamed products with elongated cells in said direction [32, 56]. After exiting the oven, the foamed sheet is cooled between a series of chill rolls. The main advantage of this chemical foaming process lies in the fact that very thin foamed products may be obtained, with thicknesses as low as 0.5 mm till up to 8 mm and variable densities (25-250 kg/m^3).

In horizontal expansion, the sheet is moved horizontally using a conveyor belt. Although stretching of the sheet in the transport direction is not as significant as in vertical expansion, a certain degree of cell elongation in the foaming direction still occurs. Once again, the foamed sheet is cooled using chill rolls. The main advantage of the horizontal expansion is that it enables the production of thicker foamed sheets with thicknesses as high as 13 mm with similar densities as the ones obtained using vertical expansion.

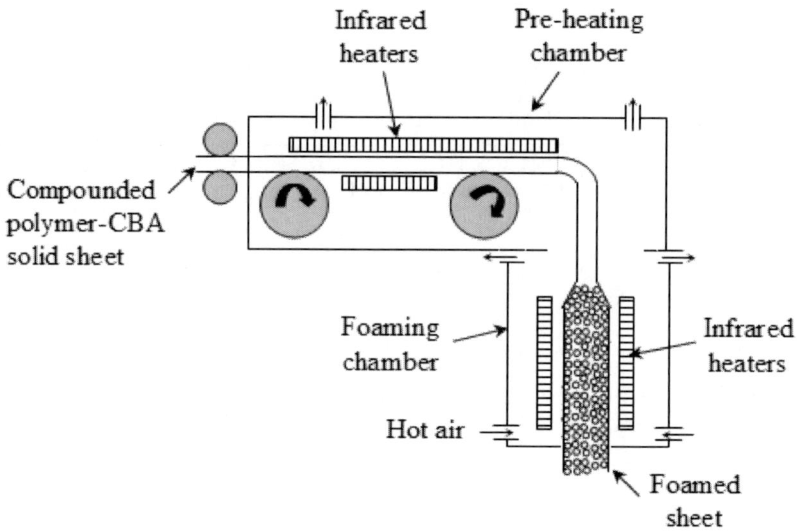

Adapted from [1].

Figure 10. Scheme of a vertical air oven used to prepare PP-based foams by means of chemical extrusion.

In the case of physical foaming by direct extrusion, the foam is directly obtained by the sudden pressure drop generated at the exit of an extruder die, normally using a physical blowing agent (PBA) such as CO_2 or hydrocarbons as iso-butane [1, 57], although gases (N_2 or CO_2) in supercritical conditions may also be used. Although there are industrially available physical foaming extrusion processes that only use one extruder to mix the polymer with the blowing agent(s) and later on promote the formation of the foam at the exit of the die (please see *MicroZote* PP foams produced by Zotefoams plc), normally the mix of the polymer with the PBA is done in a single or twin-screw primary extruder, with the polymer-gas mix being progressively cooled in a secondary extruder. This enables to increase the melt strengthening of the polymer and promote cell nucleation by sudden depressurization at the exit of the die, as well as facilitate foam stabilization by cooling [1]. The system formed by two in-line extruders is known as tandem-extrusion system. Physical foaming by tandem-extrusion using CO_2, and mainly hydrocarbons such as iso-butane or n-pentane (see Figure 11), is widely used in the production of low density PP foams in the form of tube, rod, sheet or film, depending on the geometry of the die, reaching productions as high as 450 kg/h, though typical productions are usually between 120 and 150 kg/h [58-59].

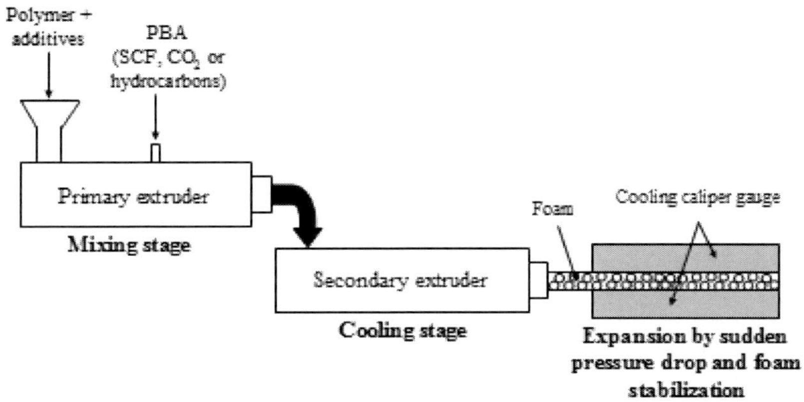

Figure 11. Basic scheme of a tandem-extrusion physical foaming process.

In terms of their final cellular structure, physically foamed extruded products tend to show a rather anisotropic cellular structure, with cells elongated in the extrusion flow direction due to the high temperatures and stresses applied during extrusion.

As a wide range of PP foams may be prepared in terms of density and final form, from densities as low as 20 kg/m^3 to medium-high densities of 500 kg/m^3 and forms that can go from tubes to sheets, physically foamed extruded PP foams are used in the most varied applications: automotive sector (door seals, trunk and sealing insulation components, among others), packaging (protection during transport, thermoformed foamed trays, etc), construction (acoustic insulation of laminated floors, thermal insulation elements, …).

Some of the companies that use this foaming process are OK Company S.A. from Spain, which sells, under the trade name *Propilen*, foamed PP rolls with a density of 45 kg/m^3 and thicknesses from 1.5 to 3 mm; or Sekisui Alveo, which, under the trade name *Alveocel*, sells low density PP-based foams with densities going from 20 to 500 kg/m^3 and a wide interval of thicknesses that can go from as low as 0.6 mm to as high as 13 mm.

Physically foamed PP films and sheets made by extrusion meet growing market demand for various components which require food contact regulatory clearance, such as food trays, cups and goblets, food packaging, among other applications [60].

Under the name *MicroZote*, Zotefoams plc has recently brought out to the market a new range of closed-cell, non-crosslinked, microcellular PP foams applying the patented *MuCell*® technology to extrusion (please consult this section on *MuCell*® foaming technology). The process uses supercritical N_2 or CO_2 to foam PP in a single extrusion stage. Foamed sheets display variable thicknesses with densities that go from 55 to 250 kg/m^3. As PP is physically foamed, no solid residues coming from CBA decomposition are present, making *MicroZote* suitable for medical and direct food contact applications. Other potential markets include seals and gaskets, automotive, construction, etc.

Another semi-continuous foaming process used to prepare PP-based foams, mainly medium to high density rigid foams for the automotive and construction sectors, is injection-moulding. As shown in Figure 8, both chemical as well as physical foaming are industrially used.

Chemical foaming typically occurs by the thermal decomposition of a blowing agent, commonly known as chemical blowing agent or CBA, which at a given temperature thermally decomposes releasing one or more gases that act as foaming agent (see Table 1 for the most commonly used chemical blowing agents for PP foaming). In the case of injection-moulding, the CBA, chosen according to the characteristics of the polymer and foaming parameters, is typically endothermic, although exothermic or endothermic/exothermic blends are also used. In the case of PP, a mix of citric acid and sodium bicarbonate is often used as endothermic blowing agent (see in Figure 12 the decomposition reaction of this mix to give CO_2 and H_2O), as it has been shown to facilitate cellular control during foaming at the high injection-moulding temperatures [61]. Nonetheless, exothermic blowing agents based on ADC or other azo compounds are also commonly used, alone or in combination with endothermic systems [1, 60].

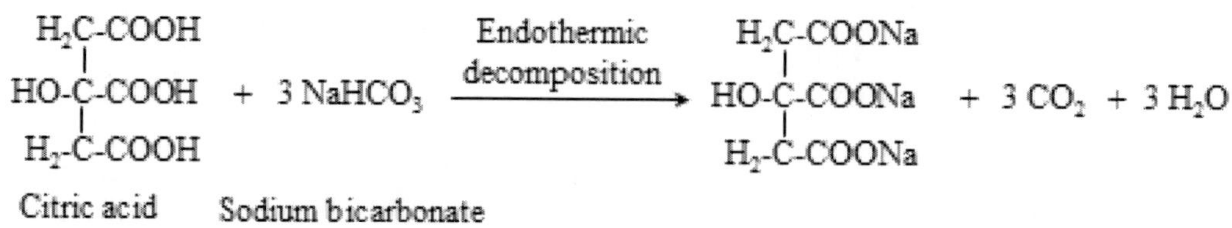

Figure 12. Decomposition reaction of citric acid + sodium bicarbonate chemical blowing agent to give carbon dioxide and water.

Table 1. Commonly used chemical blowing agents for PP foaming

Blowing agent	Exothermic/ endothermic	Decomposition temperature in air (°C)	Gas yield (cm^3/g)	Blowing gas
ADC	Exothermic	205-215	220	N_2, CO (NH_3, CO_2)
ADC, activated*	Exothermic	155-215	150-220	N_2, CO (NH_3, CO_2)
Citric acid/sodium bicarbonate	Endothermic	150-230	185-240	CO_2, H_2O

* Normally activated with metal oxides, such as ZnO.
Adapted from [60].

Some of the companies that provide these sort of chemical blowing agents for the preparation of PP foams using extrusion, injection or other chemical foaming processes, include Clariant Masterbatches, which sells, under the generic trade name *Hydrocerol*, sodium bicarbonate formulations, alone or in combination with other exothermic azo-based blowing agents; Bergen International, provider of a mix of citric acid and sodium bicarbonate (*Foamazole*); Lanxess Energizing Chemistry, which sells ADC or other formulations containing ADC or other azo compounds under the trade names *Porofor* and *Genitron*; Hughes Polymer Additives Corp., which in 2009 bought the *Ficel* chemical blowing agents branch of Lanxess; Arkema and its *Azobul* ADC-based formulations, etc. Other Asia-based companies that provide exothermic blowing agents for PP foaming include Chemtura (*Celogen* grades), Otsuka Chemical (*Unifoam*) and High Polymer Labs (*Microfine*).

The foams obtained using this chemical foaming process are known as integral foams, as they are formed by a cellular core, resulting from the thermal decomposition of the CBA, and solid surface skins. They present a cellular structure which is characterized by a gradient of cell sizes along the thickness, with smaller cell sizes closer to the surface skins due to cooling by contact with the inner walls of the injection mould (see Figure 13). Panels produced using this foaming method are designed to replace plywood and other lightweight polymer composites in a variety of applications, from the transportation market to outdoor upholstery and whiteboards, among others.

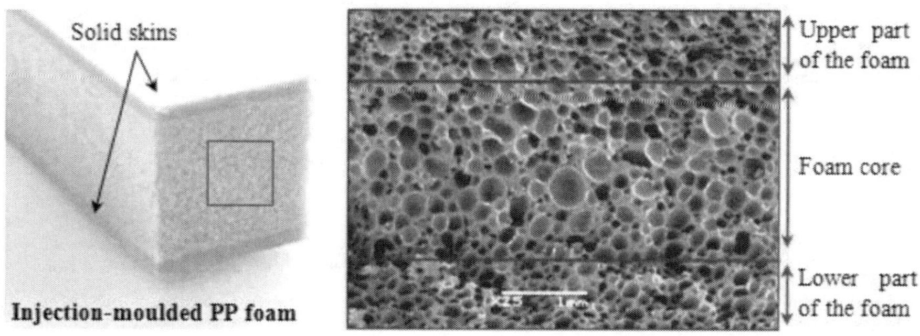

Figure 13. View of a chemically foamed injection-moulded PP foam (VarioLine Systems) and SEM micrograph showing typical cell size gradient.

German-based company VarioLine Systems provides chemically foamed injection moulded PP panels with high densities ranging from around 400 to up to 600 kg/m^3 and thicknesses up to 20 mm (trade names *PPH Standard*, *PPH Ultralight* and *PPH Ultrastiff* – see Figure 13), which offer a unique balance of high specific stiffness and strength, enabling their use as replacers of plywood or other wood and plastic lightweight structures in high mechanically demanding sectors, such as building or construction.

Alongside injection-moulding chemical foaming, there is also an industrially available method to prepare structural PP foams by means of injection-moulding: the *MuCell*® process. This process has the advantage that it does not require the addition of a chemical additive to the polymer, since it uses a physical blowing agent to foam the material. Its name derives from the particular microcellular structure of the produced foams (*Mu* – micrometer and *Cell* – cellular), with typical cell sizes between 5 and 50 μm.

In this process, a PBA, typically N_2, although CO_2 may also be used, is introduced under supercritical conditions (-147 ºC and 34 bar for N_2, and around 31 ºC and 74 bar in the case of CO_2 [62]) using a special supercritical fluid (SCF) metering system patented by Trexel [63] into the injection unit and dissolved with the melted polymer. A specially designed screw must be used in order to guarantee a sufficient back pressure at all times, preventing premature foaming by loss of pressure. The mix is then injected into the mould and the material is allowed to expand filling the mould cavity (see Figure 14).

Unlike injection-moulding chemical foaming, this physical foaming process does not have significant temperature limitations and does not leave residues resulting from the thermal decomposition of CBAs, enabling the foams to be used in consumer good applications in direct contact with food. Also, as SCFs act as plasticizers of the melted polymer, lower temperatures (reduced energy consumption) and injection pressures are required to fill the injection mould cavity, allowing for faster production of lightweight parts using smaller injection machines. Due to the high control of process variables, enhanced dimensional stability and reduced warpage of the produced parts, it is a process used in the production of thin wall moulded parts (thickness < 3 mm) with weight reductions higher than 20%.

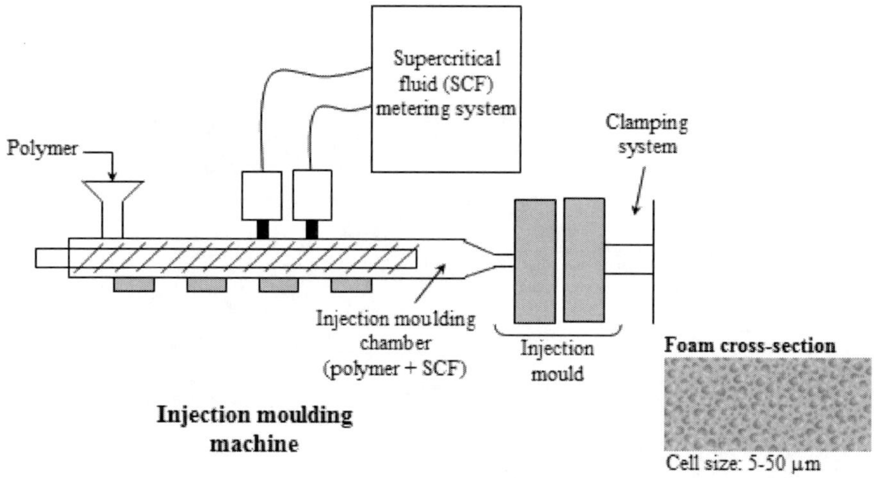

Figure 14. Schematic view of the *MuCell*® injection-moulding physical foaming process.

Some of the applications of PP foams prepared using this foaming technology include business equipment such as printers, and important sectors such as the automotive (rear door carriers, trunk liners, etc), industrial and electrical, medical (clamps, syringe plungers, etc) and packaging (rigid storage bins).

As a patented technology, Trexel has sold along the years to several injection-moulding manufacturers the rights to use the SCF metering system and injection-unit modifications. As a widely extended injection-moulding technology, the *MuCell*® process is used by companies such as Hewlett Packard, Canon, Audi, BMW, Porsche, Ford, Nissan, and Samsung, just to mention some.

Batch Foaming Processes

Chemical foaming by means of press moulded foaming is a type of batch-foaming process, first introduced at the beginning of 1970s by BASF in Germany, in which the polymer is foamed by the thermal decomposition of a CBA by applying heat and pressure on a solid precursor previously compounded with the CBA and pre-moulded to a desired shape, typically by means of extrusion. The solid compounded foaming precursor is foamed by the thermal decomposition of the CBA and sudden release of pressure. This leads to a quick cell nucleation and growth, with fast cooling of the surface being required in order to avoid possible gas loss and foam collapse.

Compression-moulding chemical foaming of polyolefins may be classified in one-step and two-step expansion processes (see Figure 15). In one-step expansion, the compounded polymer is expanded by applying heat and pressure for a certain time to guarantee the decomposition of the CBA, and afterwards applying a sudden release of pressure. This process is used to prepare medium-high density foams with densities higher than 100 kg/m^3, i.e., maximum expansion ratios of 9-10. In the two-step expansion process, the material is initially pre-expanded either by carefully controlling the degree of blowing agent decomposition or by quenching the pre-foam. Foam expansion is completed by heating the pre-foam in a mould at atmospheric pressure, resulting in foams with densities below 100 kg/m^3 and thicknesses as high as 120 mm.

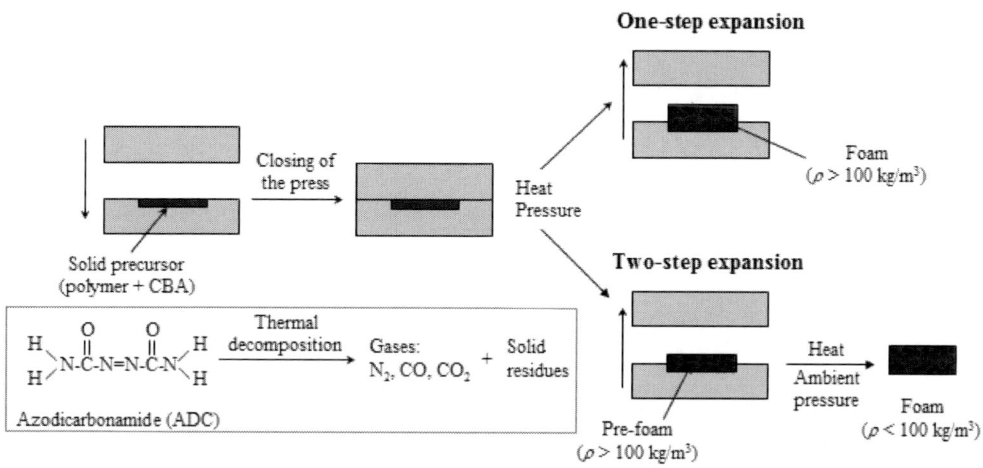

Figure 15. Schematic view of one-step and two-step compression-moulding chemical foaming processes. Also shown is azodicarbonamide's thermal decomposition.

Although dicumyl peroxide (DCP) is often added to low density polyethylene (LDPE) as crosslinking agent in order to increase the melt strength of the polymer during foaming, PP foams produced by this method do not require any crosslinking agent. Instead, HMS-PP or melt strengthen PP grades (PP-based copolymers or blends of PP with other copolymers) are used (please consult section "Polypropylene Grades for Foaming"). Azodicarbonamide or azo-based non-activated or activated formulations are commonly used as chemical blowing agents in polyolefins when using compression-moulding (consult Table 1). As previously stated, no activation is required in the case of PP foams.

Polyolefin foams prepared by compression-moulding chemical foaming display closed-cell structures with typical cell sizes around 300-600 μm, though almost microcellular foams may also be obtained under specific conditions [44]. Depending on the amount of blowing agent, polymer matrix characteristics and foaming parameters, this process can be used to prepare foams with a very broad range of densities, from low density foams (20-40 kg/m^3) to medium-high density foams (300-700 kg/m^3). Due to its versatility, PP foam blocks produced by this method may be used in applications going from protection and packaging (low density foams) to construction and automotive (medium-high density foams).

Some of the companies that presently use this process in order to prepare polyolefin-based block foams include OK Company S.A. from Spain, Sanwa Kako from Japan and

Switzerland-based Sekisui Alveo. In the case of the later, the leading European manufacturer of crosslinked polyolefin foams, compression-moulded PP foam blocks, under the trade name *Alveolux*, are provided with densities ranging from 110 to 330 kg/m^3.

The high-pressure gas dissolution physical foaming process, developed and industrially implemented by Zotefoams plc, initially under the name BXL Plastics [64], a company seeded in the UK, is a type of batch-physical foaming process where a gas is added and dissolved into a polymer, promoting the later formation of the polymer foam by generating a thermodynamic instability in the polymer-gas mixture through heating [1]. The main difference between this process and other physical foaming methods lies on the fact that gas dissolution is done below the melting temperature of the polymer, i.e., in a semisolid-like state [48]. A schematic view of the three basic stages involved in the gas dissolution physical foaming process used by Zotefoams is shown in Figure 16.

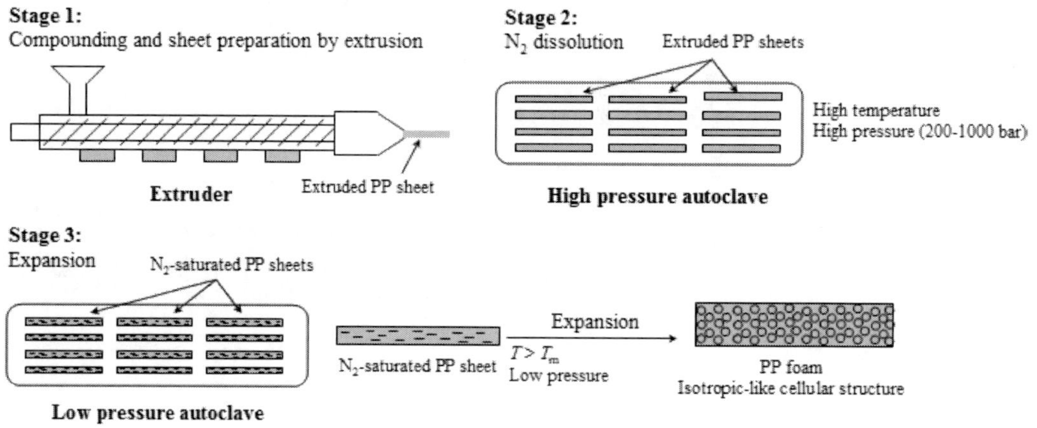

Figure 16. Schematic view of the three stages involved in the gas dissolution physical foaming process used by *Zotefoams*. Note: T_m – Melting temperature.

Stage 1 considers mixing of the different formulation components and preparation of a solid sheet foaming precursor by means of extrusion. The produced polymer sheet is then foamed in two stages: in Stage 2, N_2, acting as physical blowing agent, is dissolved into the extruded sheets at high temperatures and pressures during long times inside specially designed high pressure autoclaves; after removal of the N_2-saturated sheets from the high pressure autoclave, these are placed in a second autoclave and expanded by heating at a temperature above the glass transition temperature of the gas-saturated polymer maintaining a low pressure (Stage 3). In the case of semi-crystalline polymers such as PP, this final stage is done above the melting temperature of the material (T_m).

As can be seen in Figure 16, one of the main advantages of this foaming process lies in the very accurate control of the individual parameters governing foam production due to the separation of the different stages involved in foaming (cell nucleation, foam growth and foam stabilization [44]). As foam expansion occurs after a prior cell nucleation stage and expansion is done under well controlled temperature at isostatic pressure conditions, foams are obtained with a characteristic isotropic-like cellular structure. It even offers the possibility of preparing foams with a uniform microcellular structure, i.e., foams with micrometric-sized cells [44,

46]. However, it is a time consuming and thus expensive process, as it is limited by the long times required to dissolve N_2 in the PP sheets.

Under the generic trade name *Azote*, Zotefoams offers a wide variety of polyolefin foams prepared using this physical foaming method, from crosslinked LDPE, HDPE and blends, EVA and EMA copolymers, to non-crosslinked PP-based foams. In the specific case of PP, the name *Propozote* is used, being a closed-cell 100% PP foam with applications in the automotive and returnable packaging industries.

Besides the already mentioned chemical and physical foaming processes, there is one that can be used to easily obtain PP foams with densities ranging from as low as 18 kg/m^3 to foams with medium-high densities around 250-300 kg/m^3: the expandable bead foaming process. This process allows obtaining a type of PP foam known as Expanded Polypropylene (EPP). EPP is a closed-cell bead foam in which individual expanded PP beads initially impregnated with a blowing agent inside autoclaves and later expanded (typical final diameters of 4 to 6 mm), are fused together to give the final product by applying hot steam and pressure inside especially designed moulds, known as "steam chest moulds".

Unlike Expanded Polystyrene (EPS), where EPS beads still contain a certain amount of blowing agent, EPP beads are already fully expanded when used to prepare the final product. A conventional manufacturing process of EPP bead foams includes three basic stages (see Figure 17): a first one for compounding PP with the required additives (nucleating agent(s), compatibilizer(s), etc) and later pelletizing, commonly by means of extrusion (see Stage 1); the compounded PP pellets are then placed in an autoclave reactor with a dispersion medium, usually water, and mixed with a PBA (commonly CO_2 or butane) at an elevated pressure and at a temperature above the melting point of PP for approximately 1 hour to guarantee full saturation of the pellets [65-66]. At the end of saturation, the pressure is released, allowing the material to expand (Stage 2). Finally, these expanded beads are cooled and washed in order to be used in the last stage of the process, where EPP parts are produced by fusing pneumatically compressed EPP beads with the help of superheated steam in a steam chest moulding machine (Stage 3).

Due to its excellent energy absorption, thermal insulation and high specific strength properties, EPP is used for a range of applications in the automotive industry for components such as energy absorbers in bumpers, side impact protection, seating structures, consoles, among many other elements; packaging (containers, boxes, etc); construction materials (as insulation material and noise absorber, for drainage, etc); and consumer goods (sports protective gear, outer consumer goods casting, etc).

Among the main EPP foam manufacturers are Japanese-based JSP, global leader and responsible for the initial development of this product in the 1980s [67], and Kaneka Corporation, with its European branch Kaneka Belgium N.V., as well as BASF in a worldwide basis and Porexpan in Spain.

JSP provides EPP beads as well as already moulded EPP bead foams in the form of sheets to be used in multiple industries for the protection of products during storage and transportation. In terms of EPP beads, JSP products are named *Miradrain*, which due to their good noise absorbing performance, drainage and thermal insulation are used for surface and subsurface drainage and insulating applications in the automotive sector; *P-Block* is a specially developed EPP bead grade with improved impact properties, enabling its use in the packaging sector. Last but not least, JSP also provides EPP bead foams in the form of panels under the trade name *P-Mat*.

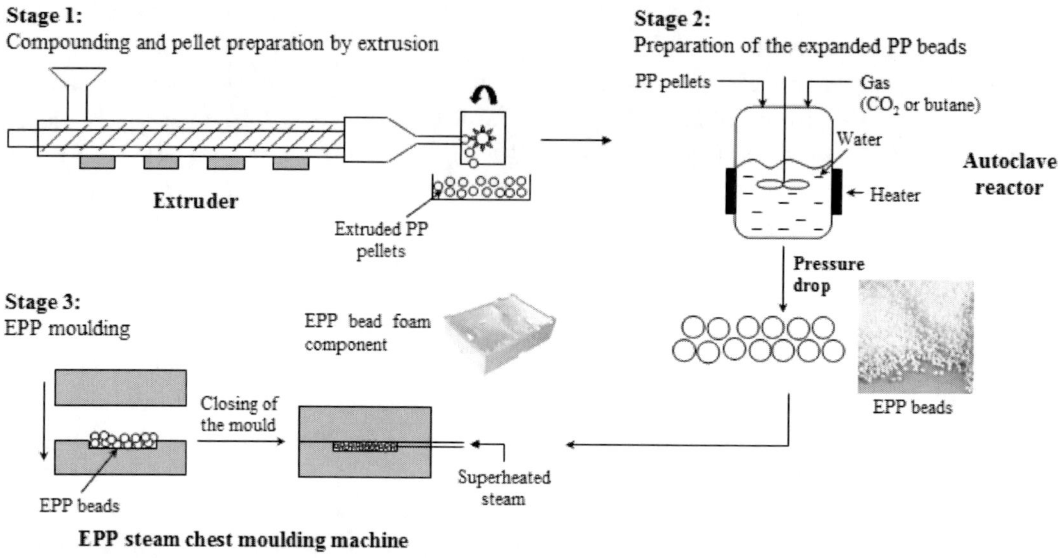

Figure 17. Conventional manufacturing process of EPP beads and EPP bead foams.

Kaneka Corporation, through its European branch Kaneka Belgium N.V., provides EPP beads or already prepared EPP bead foams (under the trade name *Eperan-PP*), normally used as shock absorbers for components in the automotive industry and packaging sector. Other special grades are also available, such as coloured, dissipative and very small bead grades, this last one enabling the production of extremely thin-walled products. Kaneka has also developed *Eperan-PP "FGX"*, especially thought for the production of high-density EPP foams to meet with the increasing demands of the automotive sector.

BASF offers EPP beads under the generic trade name *Neopolen-P*. The different available grades (white or coloured) of these EPP beads allow to prepare from very low density foams (18 kg/m^3) to what are considered as medium-high density EPP foams ($\rho > 100$ kg/m^3) for the automotive industry.

In Spain, Porexpan fabricates and sells already moulded EPP foams with densities up to 150 kg/m^3 in the form of boxes or containers to be mainly used in the packaging and automotive sectors.

Recent Developments

Recent developments in the field of PP foams, mainly related to the arrival of the new long chain-branched PP grades with improved melt strengths and hence improved foaming behaviours (consult section 2 Polypropylene grades for foaming), may be divided according to base material developments and modifications; new foaming processes and control of the developed cellular structure; and the incorporation of (nano)reinforcements/(nano)fillers (polymer nanocomposite foams).

Recent Developments in Materials

The extended use of PP-based and even 100% PP homopolymer materials in foaming has arisen from the emergence of the long-chain branched high melt strength polypropylenes (see section "Polypropylene Grades for Foaming" for further details). Nevertheless, as this is a somewhat new material, improvements are still being done regarding the base material in the particular field of foaming, from the development of new PP-based polymer blends or copolymers, to new polymerization processes, always with the objective of covering the widest possible range of applications. As these are mainly related to foam density (from low density, $\rho < 50$ kg/m^3, mainly for packaging and insulation purposes, to high density, $\rho > 100$ kg/m^3, for structural applications), and balance of mechanical properties (from flexible and soft to rigid foams), innovations in the base material have been done with these two goals into mind.

Generally speaking, low density PP-based foams, mainly thought for the packaging, automotive and construction sectors, where they are vastly used as insulators, are obtained by using specially designed long-chain branched grades with improved melt strengths, i.e., through controlling the molecular architecture of PP, or blending HMS-PP with other polymers or copolymers. High-density PP foams ($\rho > 100$ kg/m^3) for structural applications (automotive and construction sectors) are usually obtained by using 100% HMS-PP homopolymers or blends of these stiffer grades with more conventional linear PP homopolymers. In the case of high-density PP foams, high specific mechanical properties are usually achieved through a careful control of the developed cellular structure. Until now, soft and flexible PP-based foams, where PP has mainly been used due to its higher thermal stability when compared to PE, have been obtained by blending PP with more flexible polymers and copolymers. Nevertheless, as will be shown later on, Borealis has introduced new HMS-PP grades that allow preparing all-PP 'soft-like' low density foams.

The Spherizone PP polymerization technology patented by LyondellBasell (see Figure 3), which allows to carefully control the molecular architecture of PP during its polymerization, has opened up a whole lot of new possibilities for PP in foaming applications. One of the first commercially available examples is the already mentioned *Higran RS1684*, a 100% PP material that displays extremely high melt resistance and extensibility, making it useful for the production of very low density PP foams by means of conventional melt-like foaming processes such as direct extrusion. An example of a thermoformed foamed tray produced using *Higran RS1684* can be seen in Figure 18(a).

Borealis, one of the pioneers in the industrialization of high-melt strength polypropylenes for the extruded PP foam market with the *Daploy HMS* PP grades, has recently extended their product offer to the production of very low density PP foams and even to 'soft'-foam applications. Besides *Daploy WB130HMS* and *Daploy WB135HMS* high-melt strength PP homopolymers, especially designed for the production of PP foams with densities as low as 45 kg/m^3 by means of direct extrusion, Borealis has recently introduced in the market *Daploy WB140HMS*, a next-generation long-chain branched PP homopolymer with improved melt strength and extensibility for the production of very low density extruded PP foams ($\rho < 45$ kg/m^3). Also, and to meet the growing demand for 'soft-like' foams, Borealis has introduced *Daploy WB260HMS*, a high-melt strength PP copolymer especially thought for the production of 'soft' PP foams by extrusion (see Figure 18(b) for example of an extruded foamed roll made of *Daploy WB260HMS*). As mentioned previously, this enables the use of a 100% PP-based product for applications where low density soft PP foams were usually obtained by blending standard high-melt strength polypropylenes with softer polymers such as metallocene polyethylenes, thermoplastic olefin elastomers and ethyl vinyl acetate copolymers [36].

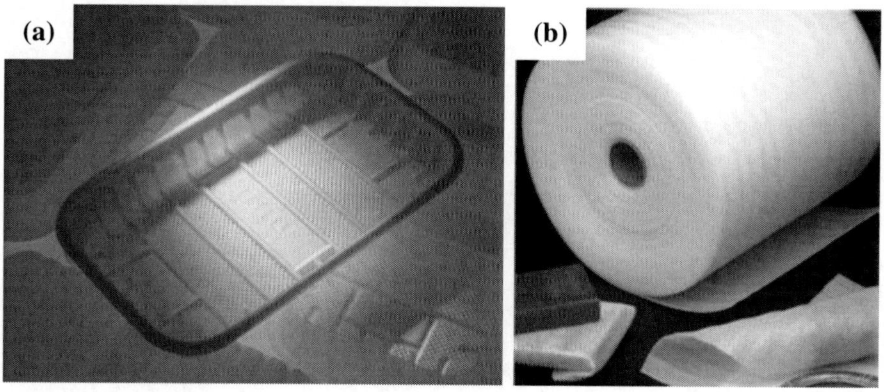

Figure 18. (a) *Higran RS1684* thermoformed foamed tray and (b) extruded roll of low density 'soft-like' PP foam.

Concerning the production of low density EPP foams, BASF has developed new types of expanded foams with reduced brittleness and improved crack-resistance, flexibility and thermal insulation by blending different polymers with low amounts of blowing agents, improving fusion between the polymer beads.

New Foaming Processes and Innovations

In terms of new foaming processes and innovations in PP foaming, these have been focused in obtaining and commercializing new microcellular PP-based foams, mainly for structural applications, trying to take advantage of the extremely small cell sizes and thus high cell densities of these foams (cell sizes < 10 μm, i.e., cell densities > 10^9 cells/cm^3), which have been shown under specific conditions to result in foamed products with specific mechanical properties, i.e., related to density, even higher than that of the respective solid base materials [43]. A micrograph showing the typical structure of a microcellular PP foam is displayed in Figure 19(a).

As mentioned previously, there are some companies that over the last years have developed and commercialized microcellular PP foams, such as Zotefoams from the UK, with its *MicroZote* PP foam obtained by a physical foaming extrusion process based on *MuCell*® microcellular foaming technology (see Figure 19(b)). Due to their extremely fine and uniform cellular structure, these foams possess a unique balance of mechanical properties, which even removes the need for lamination or layer co-extrusion in some sealing applications. Also, several injection-moulded unfilled and filled PP structural foam manufacturers sell microcellular PP foams obtained using the injection-moulding *MuCell*® process, mainly for the automotive sector. The *MuCell*® microcellular foaming process not only is commonly used in the automotive sector for weight reductions up to 30%, but also allows for up to 60%

reduction in material viscosity due to mixing with the supercritical fluid, significantly reducing the processing temperatures and injection pressures, and hence time cycles [68].

Japanese Sumitomo Chemical Co., through its brand Sumika Plastech Co., which commercializes PP extrusion foam sheets under the trade name *Sumiceller*, with thicknesses ranging from 2 to 5 mm and expansion ratios of 1.3 (for protective materials used in floor and wall surfaces) and 3.0 (boxes, cushioning materials, etc), as well as in combination with other plastic films and nonwoven cloth, have developed new processes of lamination and integration of *Sumiceller* with different types of materials [69]. The particular small size of *Sumiceller* PP foam cells when compared to similar commercial products (10 to 100 times higher cell densities), have enabled the development of new foaming/moulding processes that extend the target area to high expansion ratios and thicknesses for automotive and construction purposes. Particularly, the so-called in-line foam expansive moulding technology has allowed overcoming the drop in surface quality intrinsic to high expansion ratio *Sumiceller* foams. Foam expansive moulding (FEM), a moulding technology developed by Sumitomo [70], has been incorporated in-line to the later stages of the extrusion foaming process, minimizing surface defects in highly expanded thick PP foams due to a higher control of their cellular morphology.

In terms of EPP bead foams, the worldwide leading company JSP has proposed new EPP foaming processes. Under the generic names of *Foamcore* and *Superfoam*, JSP combines simultaneous blow moulding and foaming in a single mould to manufacture sandwich-like products with solid outer layers and foamed cores that combine good specific mechanical strength and thermal insulation, mainly to be used in automotive parts [71]. EPP bead foam post-moulding operations are also being applied by BASF to improve the mechanical properties by means of improving fusion between EPP beads.

Figure 19. (a) Typical structure of a microcellular PP foam [48] and (b) *MicroZote* microcellular PP foam sheet commercialized by Zotefoams.

There are still some foam manufacturers trying to target the so-called 'tailor-made foam' market, where through a careful control of the foaming parameters and conditions, foams with specific densities and cellular structures, mainly cell size and distribution, as well as induced cell orientation, could be prepared. So, keeping the base material unaltered, it would be possible to obtain foams with properties that could go from almost isotropic (related to an isotropic-like cellular structure) and mainly density-dependent in terms of properties, to foams with specifically induced cell orientations and very different properties according to the foam's direction.

After showing that the cellular structure induced by a specific foaming process and foaming conditions may have an important effect in the mechanical properties and thermal conductivity of low density PP-based foams [6-8], with foams with similar relative densities showing lower thermal conductivities with reducing cell size and improving cell's isotropy, critical to regulate the characteristics of these foams for thermal insulation applications, we have studied the influence of induced cell orientation on the thermal transport properties of these foams by separating the axial and radial thermal conductivity contributions [9]. Results showed that though the global thermal conductivity of foams was mainly density-dependent, with only slight differences due to cell size, the highly oriented honeycomb-like cellular structure of the CO_2 dissolution foams resulted in thermal conductivity values that were considerably higher in the axial direction when compared to the radial one. These foams presented a thermal conductivity in the radial direction typical of foams with this density, while the thermal conductivity in the axial direction was similar as the one for solid PP, showing the possibility for the development of new PP foams with direction dependent thermal properties. This was taken a step further by studying the electrical conductivity of PP foams reinforced with carbon nanofibers and with similarly induced highly oriented cellular structures [72]. As with thermal conductivity but magnified by the presence of the inherently electrically conductive carbon nanofibers, these cell oriented foams displayed a higher through-plane electrical conductivity when compared to foams with a more isotropic and regular cellular structure, which was related to a partial reorientation of the nanofibers along the through-plane thickness direction of the foams promoted by the oriented cellular structure induced during foaming.

These results demonstrate that foams with a preferentially oriented cellular structure generated by carefully controlling the parameters during foaming may display direction-dependent transport properties, which could open up a whole new field of lightweight PP-based materials with tailor-made properties.

Polypropylene Nanocomposite Foams

Particularly relevant and interesting is the research field in polypropylene nanocomposite foams, and more specifically the idea of developing multifunctional PP-based lightweight materials by incorporating different types of functional nanoreinforcements/nanofillers.

The most used nanoscale reinforcements in polymer nanocomposites are layered silicate nanoclays such as montmorillonite [73] and carbon-based nanofillers (carbon nanotubes and nanofibers, graphene nanoplatelets, etc) [4, 74-76], though others have also been considered, such as layered double hydroxides (LDH) [77-78], nanosilica [79], polyhedral oligomeric silsesquioxanes (POSS) [80] or even nanosized reinforcements obtained from cellulosic fibres [81]. Layered silicate nanoclays have clearly been the most used and the ones that have shown the broadest commercial viability, mainly due to their considerably lower cost, especially when compared to carbon-based nanofillers. Comparatively, carbon-based nanoreinforcements tend to be used when specific transport properties such as thermal or electrical conductivity are required [4].

Although foaming is a useful tool for weight reduction and can be used to improve some of the material's properties (for instance, to reduce the thermal conductivity for thermal insulation or increase some specific properties in the case of microcellular foams), due to the nature of the base material, the use of polymer foams is somewhat limited. Apart from the regulation of properties through controlling the expansion ratio and developed cellular structure, it could be possible to further extend the range of properties of the material by incorporating inorganic functional phases, i.e., inorganic fillers/reinforcements that may add a specific functional characteristic to the material. Besides, if these phases were nanometric-sized, i.e., with at least one of their dimensions in the nanometer range (1 nm = 10^{-9} m), it could be possible to get multifunctionality results in the material even at low filler amounts due to a high surface interaction with the polymer matrix. The reduction in mechanical properties inherent to foaming could hence be counteracted by a local reinforcement effect of the cell walls by the nanoparticles, which could even add up to a local improvement of the melt resistance of the polymer cell walls during foaming, promoting higher expansions [54]. In the case of nanoparticles with a plate-like geometry such as layered silicates, these could even improve the 'foaming efficiency' by means of slowing down gas diffusion, as it has been shown that platelet-like particles have a gas barrier effect [2].

Nanoparticles have been shown to improve the cell nucleation rate during the earlier stages of foaming, promoting the formation of foams with finer and more homogeneous cellular structures by means of heterogeneous cell nucleation [23]. This could result in the formation of foams with a microcellular structure (cell sizes < 10 μm, i.e., cell densities > 10^9 cells/cm^3), which have been shown to have promising properties when compared to conventional foams (cell sizes > 100 μm and cell densities < 10^6 cells/cm^3). However, microcellular foams are very difficult to obtain as they require very stringent foaming conditions (high pressures and high pressure drop rates [43]). The addition of nanoparticles could therefore overcome this limitation.

Also, and as can be seen in the schematic view shown in Figure 20, as cell nucleation and subsequent expansion can happen between the platelets or fibres due to the reduced size of the dispersed gas molecules, foaming could come as a possible strategy to increase the exfoliation/dispersion of secondary inorganic phases in a polymer matrix. It has been shown that supercritical CO_2 (sc-CO_2) may help exfoliate and disperse layered nanoclays in PP [82-85], resulting in nanocomposites with improved mechanical properties. In the case of solid nanocomposites, this can be done by dissolving sc-CO_2 with the clay, rapid expansion and incorporation of the mix into an extruder to be dispersed in the PP [85]. The enhancement in mechanical properties related to improved clay exfoliation has been explained based on the fact that CO_2 quickly entered the interlayer gallery of the organically-modified clay and swelled the organic modifier, usually quaternary alkyl ammonium salts. The later release of pressure resulted in the expansion of the interlayer clay gallery, further exfoliating the particles.

So, there is a great synergistic possibility between foaming and the incorporation of nanosized reinforcements, as the first could improve the exfoliation/dispersion of the reinforcements, and the reinforcements could promote cell nucleation, and thus the formation of foams with finer and more homogeneous cellular structures and hence higher specific mechanical properties [2, 86].

The industrialization of polymer nanocomposite foams depends on reliable and affordable compounding and foaming processing methods. Though considerable advances have been made regarding the development of polymer nanocomposite foams, several industrial challenges have still to be overcome [2]. As mentioned, the most important one is related to the dispersion and especially exfoliation of the nanofillers in the polymer matrix in large quantities and at acceptable costs, i.e., an effective preparation of the nanocomposites prior to foaming. In the case of non-foamed polymer nanocomposites, General Motors has been the pioneer in the use of these materials in automotive parts. Particularly, a PP-based/clay nanocomposite appeared on the body side-moulding of the 2004 Chevrolet Impala, as well as in the centre bridge, sail panel and box-rail protector of the 2005 Hummer H2 SUT [87].

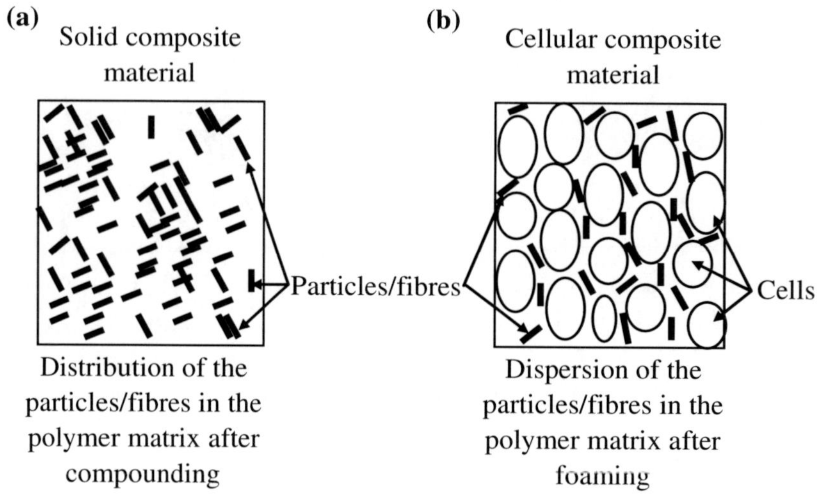

Adapted from [45].

Figure 20. Schematic view showing particle distribution/dispersion in a polymer (a) after compounding (solid composite material) and (b) after foaming (cellular composite material).

Due to difficulties related to nanofiller exfoliation in thermoplastic matrices, especially using conventional polymer melt-mixing techniques such as extrusion, related to the high melt viscosity of thermoplastics and, in the case of polyolefins, their hydrophobic nature, the majority of research works on PP nanocomposite foams have been done using the high-pressure gas dissolution physical foaming process with sc-CO_2 [44, 48, 53, 55].

The use of small amounts of platelet-like organically-modified montmorillonite clays (0.2-5 wt.%) has been show to result in foams with smaller cell sizes and higher cell densities, related to an effective heterogeneous cell nucleation effect, especially in the case where an adequate dispersion and partial or full exfoliation of the clay platelets was achieved [2, 86]. Likewise, these clay platelets stabilized the material during the expansion stage, limiting cell coalescence and collapse, as well as possible gas escape due to an improved gas barrier effect promoted by the nanoplatelets. This is of special importance in the case of PP, as it has been shown previously that conventional PP displays weak melt strength due to its linear molecular structure. The incorporation of variable concentrations of nanoclay has been shown to improve the foaming behaviour of linear PP-based materials during extrusion foaming (chemical foaming, for instance using ADC, as well as physical foaming using CO_2), promoting cell nucleation and suppressing cell coalescence at low temperatures [23, 36, 88]. Better results were achieved when compatibilizing PP with the clay using grafted copolymers, commonly with maleic anhydride (PP-g-MA) [54, 89], with an optimum value of clay concentration (as low as 3 wt%) being observed above which no significant changes were observed regarding the cell density of the resulting foams.

Alongside extrusion foaming, PP-clay nanocomposites have also been foamed using injection-moulding, both by physical means (*MuCell®* process), as well as by the thermal decomposition of a CBA [90-91]. These foams presented a characteristic microcellular structure with smaller cells and a more uniform cell distribution when compared to the respective unfilled PP foams, which resulted in globally higher tensile strengths, combination of the reinforcement effect promoted by the stiffer particles and the more uniform and finer cellular structure of the foams.

We have worked on the preparation of rigid PP-MMT nanocomposite foams by means of compression-moulding chemical foaming using ADC as chemical blowing agent [3, 92]. PP-MMT foams presented a wider foaming processing window and a more isotropic-like cellular structure when compared to the unfilled PP foams, which was related to an effective heterogeneous cell nucleation effect and cell wall stabilization induced by the partially exfoliated MMT platelets. The nanocomposite foams also presented higher specific storage moduli, direct result of their higher cell densities and the effective mechanical reinforcement of the MMT platelets.

Limited work has been done regarding the preparation and characterization of polymer foams reinforced with carbon-based nanofillers. Focusing on the analysis of the electrical conduction behaviour, a couple of works have considered the incorporation of carbon nanotubes and carbon nanofibers into amorphous polymer matrices and later foaming for electromagnetic interference (EMI) shielding applications [93-94]. Results show that there is a percolation threshold nanofiller concentration for electrical conduction in the foams similar to that of the respective solids. Likewise, it has been shown that the addition of these nanofillers to PS and PMMA and subsequent foaming resulted in foams with finer and more isotropic cellular structures [95-97], in some cases with cell size reductions in the order of 80% for carbon nanotube concentrations as low as 1 wt% [96], and enhancement of the tensile strength with increasing nanofiller concentration [98]. Nanomaterial Innovation Ltd., a company based in Ohio, US, have recently applied for a patent a process for the preparation of low density carbon-filled polymer nanocomposite foams with improved thermal insulation properties [99]. This process could boost the development and industrialization of novel nanocomposite foams based on carbon-based nanofillers such as carbon nanotubes.

We have recently published a series of articles that consider the preparation, by means of compression-moulding chemical foaming and high-pressure CO_2 dissolution, and characterization of PP-based foams with variable concentrations of carbon nanofibers. Regarding the foaming behaviour and cellular structure of the prepared foams, the incorporation of the carbon nanofibers resulted in foams with increasingly finer and more isotropic cellular structures with increasing the amount of nanofibers, related to a heterogeneous cell nucleation effect [92]. The combination of this finer cellular structure and inherently higher stiffness of the carbon nanofibers resulted in foams with specific storage moduli, i.e., relative to the foam's density, that increased with increasing the amount of added nanofibers, hence demonstrating their mechanical reinforcement effect [92, 100]. An increasingly higher electrical conductivity was assessed for both the solid nanocomposites and foams with increasing the concentration of carbon nanofibers, with a typical electrical conduction behaviour being observed for CNF concentrations above 10 wt%, the foamed nanocomposites showing promising electrical conductivities for electrostatic painting applications [72, 100]. Interestingly, we have shown that the foamed nanocomposites reached a high conductivity at lower CNF concentrations than the solids, demonstrating the synergistic effect between foaming and the addition of these conductive nanofillers. Also, we have proven that preferential cell orientation induced by foaming resulted in a partial reorientation of the nanofibers, increasing the through-plane electrical conductivity of the foams with regard to the in-plane one, hence demonstrating that control of the developed cellular structure may help produce foams with tailor-made transport properties [72].

Even more scarce are the results published in polymer foams reinforced with graphene, with the first successful preparation of graphene-reinforced polymer foams dating from 2008 [101-102]. In these works, variable concentrations of functionalized graphene sheets (0.10, 0.20 and 0.25 wt%) were mixed with an initially liquid silicone matrix and later foamed. Although cellular structure characterization revealed a certain cell coalescence and globally open-cell structure in the polymer nanocomposite foams due to the particular reactive foaming process of silicone, these foams showed great thermal stability improvement as well as remarkable increase in the normalized compressive modulus.

Recently, the incorporation of graphene to different amorphous polymers has been considered, mainly with the objective of developing new lightweight components for EMI shielding. It has been shown that like other nanofillers, the addition of small amounts of graphene to PS and subsequent foaming resulted in foams with higher cell densities and smaller cell sizes when compared to the unfilled counterparts [103]. Eswaraiah et al. [104] have developed new foamed PVDF-graphene composites and demonstrated that the incorporation of graphene concentrations as low as 0.5 wt% resulted in materials with a typical electrically conductive behaviour, while a 5 wt% of graphene resulted in foams with an EMI shielding effectiveness of around 20 dB and 18 dB, respectively in X-band (8-12 GHz) and broadband, demonstrating their possible use as lightweight elements for EMI shielding. Zhang et al. [105] have prepared microcellular PMMA-graphene nanocomposite foams, and shown that the minimum graphene concentration for electrical conduction was lower in the foams when compared to the solid nanocomposites. Adding to their high conductivity, foamed nanocomposites exhibited good EMI shielding efficiencies between 13 and 19 db in X-band, once again showing promising properties for EMI shielding applications.

We have recently prepared by means of melt-compounding and foaming by supercritical CO_2 dissolution polycarbonate foams reinforced with 0.5 wt% of graphene nanoplatelets and characterized them in terms of the CO_2 desorption behaviour and developed cellular structure [106]. The cellular structure was found to be highly dependent on the CO_2 saturation temperature and residual pressure, with the foams displaying higher cell densities with decreasing this temperature and uniform and regular-shaped cells with increasing the residual pressure.

CONCLUSION

This chapter presents the most recent developments in polypropylene foams both from an industrial point of view as well as from the main investigations carried out regarding these lightweight materials, with the main objective of extending their properties and hence possible applications.

In order to fully analyze the possible variables that may affect PP foaming and thus enable to improve its final characteristics, an extended review is presented regarding the polypropylene grades used for foaming (mainly focused on the new long chain-branched high melt strength polypropylenes), the industrially available foaming processes, from continuous/semi-continuous extrusion and injection-moulding to produce foamed sheets and rolls to batch-foaming processes used to prepare foamed blocks, and the main types of available PP foams, from low to high density and flexible and soft-like to rigid foams.

In terms of recent developments in the field of PP foams, it has been focused considering first of all the innovations in material formulation, secondly the new developed foaming processes and control of the foam's cellular structure, and last but not least the incorporation of nanosized reinforcements, all with the final goal of developing new PP foams that can cover the widest possible range of properties and functionalities, and thus applications.

In terms of material selection, the main innovations have been focused on increasing the inherently poor melt strength of PP in order to create 100% PP homopolymer-based foams for the most varied applications. This has been done by developing new polymerization processes, such as the Spherizone process, which enables to polymerize PP homopolymers with long chain-branching tailor-made characteristics, as well as blending different types of homopolymers and copolymers, even making possible the production of soft PP foams.

Also important are the new foaming processes, mainly focused on obtaining microcellular PP foams or combining PP foaming with other processes or materials to produce in-line complex lightweight components. A careful control of the cellular structure developed during foaming has also been shown to directly affect the foam properties. Particularly, some advances have been made in this field towards the fabrication of PP foams with tailor-made mechanical and transport properties by inducing preferential cell orientations.

Last but not least, there is an increasing interest in the scientific community in the concept of multifunctional foam by combining foaming with the incorporation of functional fillers, and especially nanosized reinforcements (polymer nanocomposite foams). It has been proven that nanoclays and carbon-based nanofillers have an important effect in foaming, promoting not only the formation of finer and more uniform cellular structures in PP foams, but also adding specific characteristics such as improved mechanical or transport properties, hence extending the field of applications of PP foams. Promising results have already been gathered and could enable the use of these materials for industrial applications.

ACKNOWLEDGMENTS

Financial assistance from the Spanish Ministry of Economy and Competitiveness for project MAT2011-26410 is gratefully acknowledged.

REFERENCES

[1] C. P. Park, *Polyolefin Foam.* In: Polymeric Foams and Foam Technology (eds. D. Klempner and V. Sendijarevic), 2nd edn., Hanser, Munich (2004).

[2] L. J. Lee, C. Zeng, X. Cao, X. Han, J. Shen and G. Xu, *Comp. Sci. Tech.* 65, 2344 (2005).

[3] M. Antunes, J. I. Velasco, V. Realinho and E. Solórzano, *Polym. Eng. Sci.* 49, 2400 (2009).

[4] M. Shaffer and J. Sandler, *Carbon nanotube/nanofibre polymer composites.* In: S. Advani, editor. Processing and properties of nanocomposites. World Scientific, p. 1-59 (2006).

[5] L. J. Gibson, M. F. Ashby, *Cellular Solids: Structure and Properties*, 2nd edn., Pergamon Press, Oxford (1999).

[6] M. Antunes, J. I. Velasco, V. Realinho, A. B. Martínez, M. A. Rodríguez-Pérez and J. A. de Saja, *Adv. Eng. Mater.* 11, 811 (2009).

[7] M. Antunes, V. Realinho, A. B. Martínez, E. Solórzano, M. A. Rodríguez-Pérez and J. I. Velasco, *Def. Diff. Forum* 297-301, 990 (2010).

[8] M. Antunes, V. Realinho, E. Solórzano, M. A. Rodríguez-Pérez, J. A. de Saja and J. I. Velasco, *Def. Diff. Forum* 297-301, 996 (2010).

[9] M. Antunes, V. Realinho, J. I. Velasco, E. Solórzano, M. A. Rodríguez-Pérez and J. A. de Saja. Thermal conductivity anisotropy in polypropylene foams prepared by supercritical CO_2 dissolution, *Mater. Chem. Phys.* (2012), doi: http://dx.doi.org/10.1016/j.matchemphys.2012.07.001.

[10] C. B. Park and L. K. Cheung, *Polym. Eng. Sci.* 37, 1 (1997).

[11] C. Y. Yu, Y. Wang, B. T. Wu, Y. H. Xie, C. Yu, S. H. Chen and W. G. Li, *Polym. Test.* 30, 887 (2011).

[12] T. Ishikawa and M. Ohshima, *Polym. Eng. Sci.* 51, 1617 (2011).

[13] M. A. Rodríguez-Pérez, *Adv. Polym. Sci.* 184, 97 (2004).

[14] Sekisui Alveo, *Product Guide*, Sekisui Alveo AG, Luzern, Switzerland (2007).

[15] S. Tokuda and T. Kemmotsu, *Rad. Phys. Chem.* 46, 905 (1995).

[16] S. Fukunaga, K. Ishikawa and K. Kondo, Patent JP2010150509-A (2009).

[17] K. Senda, Patent WO2010113471-A1 (2010).

[18] J. J. Park, L. Katz and N. G. Gaylord, U.S. Patent 5,149,579 (1992).

[19] E. M. Philips, K. E. McHugh and M. B. Bradley, *Kunststoffe* 82, 671 (1992).

[20] C. Liu, D. Wei, A. Zheng, Y. Li and H. Xiao, *J. Appl. Polym. Sci.* 101, 4114 (2006).

[21] D.-H. Han, J.-H. Jang, H.-Y. Kim, B.-N. Kim and B.-Y. Shin, *Polym. Eng. Sci.* 46, 431 (2006).

[22] J. Stange and H. Münstedt, *J. Cell. Plast.* 42, 445 (2006).

[23] W. T. Zhai, T. Kuboki, L. L. Wang, C. B. Park, E. K. Lee and H. E. Naguib, *Ind. Eng. Chem. Res.* 49, 9834 (2010).

[24] P. Spitael, C. W. Macosko and A. Sahnoune, in *Proceedings of the Annual Technical Conference 2002* 1792, Society of Plastics Engineers, San Francisco, USA (2002).

[25] Y. An, Z. Zhang, W. Bi, Y. Wang and T. Tang, *J. Appl. Polym. Sci.* 110, 3727 (2008).

[26] P. Spitael and C. W. Macosko, *Polym. Eng. Sci.* 44, 2090 (2004).

[27] C. Gabriel and H. Münstedt, *J. Rheol.* 47, 619 (2003).

[28] Licensed Polyolefin Technologies and Services – Spherizone, LyondellBasell. Available in www.lyondellbasell.com/technology.
[29] Borealis, Daploy HMS Polypropylene for Foam Extrusion (2004).
[30] N. C. Hilyard and A. Cunningham, *Low Density Cellular Plastics: Physical Basis of Behaviour*, 1st edn. Chapman and Hall, London (1994).
[31] D. Klempner and V. Sendijarevic, *Polymeric Foams and Foam Technology*, 2nd edn. Hanser Publishers, Munich (2004).
[32] M. A. Rodríguez-Pérez, O. Alonso, J. Souto and J. A. de Saja, *Polym. Test.* 16, 287 (1997).
[33] V. Shah, *Handbook of Plastics Testing Technology*, 2nd edn. John Wiley and Sons Inc (1998).
[34] S. Hashizume, K. Itakura and S. Yamamoto, Patent JP2009084304 (2009).
[35] D. Purcell, R. Willemse, T. Ross, M. R. Hennessy, F. T. Wolf, W. F. Thomas and T. Roth, Patent WO2007017250 (2008).
[36] W. T. Zhai and C. B. Park, *Polym. Eng. Sci.* 51, 2387 (2011).
[37] M. Wegener, Piezoelectric polymer foams: Transducer mechanism and preparation as well as touch-sensor and ultrasonic-transducer properties. In: Behavior and mechanics of multifunctional materials and composites 2010 (eds. Z. Ounaies and J. Li), Book Series: Proceedings of SPIE-The International Society for Optical Engineering, 7644, Article number 76441 (2010).
[38] M. Dansachmuller, R. Schwodiauer, S. Bauer-Gogonea, S. Bauer, M. Paajanen and J. Raukola, *Appl. Phys. Lett.* 86, Article number 031910 (2005).
[39] J. D. Mcrae, H. E. Naguib and N. Atalla, *J. Appl. Polym. Sci.* 116, 1106 (2010).
[40] P. C. Lee, J. Wang and C. B. Park, *Ind. Eng. Chem. Res.* 45, 175 (2006).
[41] C. Xin, Y. He, Z. Fu, X. Li, Q. Li, B. Yan, Z. Yang and Q. Ling, Patent CN 200810111671 (2008).
[42] M. A. Rodríguez-Pérez, M. Álvarez-Láinez and J. A. de Saja, *J. Appl. Polym. Sci.* 114, 1176 (2009).
[43] V. Kumar and K. V. Nadella, *Microcellular Foams*. In: *Handbook of Polymer Foams,* D. Eaves, ed., Rapra Technology 243 (2004).
[44] M. Antunes, J. I. Velasco and V. Realinho, *Polypropylene Foams. Production, Structure and Properties*. In: Advances in Materials Science Research. Volume 10 (ed. M. C. Wythers), NovaScience Publishers (2011).
[45] M. Antunes, *Preparación y caracterización de espumas multifuncionales a base de nanocompuestos de poliolefinas*, PhD Thesis (2010).
[46] J. I. Velasco, M. Antunes, V. Realinho and M. Ardanuy, *Polym. Eng. Sci.* 51, 2120 (2011).
[47] S. Doroudiani, C. B. Park and M. T. Kortschot, *Polym. Eng. Sci.* 36, 2645 (1996).
[48] M. Antunes, V. Realinho and J. I. Velasco, *J. Cell. Plast.* 46, 551 (2010).
[49] H.-X. Huang and J.-K. Wang, *J. Appl. Polym. Sci.* 106, 505 (2007).
[50] S. Doroudiani, C. B. Park and M. T. Kortschot, *Polym. Eng. Sci.* 38, 1205 (1998).
[51] P. Rachtanapun, S. E. M. Selke and L. M. Matuana, *J. Appl. Polym. Sci.* 88, 2842 (2003).
[52] P. Rachtanapun, S. E. M. Selke and L. M. Matuana, *J. Appl. Polym. Sci.* 93, 364 (2004).
[53] P. H. Nam, P. Maiti, M. Okamoto, T. Kotaka, T. Nakayama, M. Takada, M. Ohshima, A. Usuki, N. Hasegawa and H. Okamoto, *Polym. Eng. Sci.* 42, 1907 (2002).

[54] M. Okamoto, P. H. Nam, P. Maiti, T. Kptaka, T. Nakayama, M. Takada, M. Ohsima, A. Usuki, N. Hasegawa and H. Okamoto, *Nano Lett.* 1, 503 (2001).
[55] K. Taki, T. Yanagimoto, E. Funami, M. Okamoto and M. Ohshima, *Polym. Eng.Sci.* 44, 1004 (2004).
[56] M. A. Rodríguez-Pérez, S. Rodríguez-Llorente and J. A. de Saja, *Polym. Eng. Sci.* 37, 959 (1997).
[57] N. Mills, *Polymer Foams Handbook, Engineering and Biomechanics Applications and Design Guide*, 1st edn,, Elsevier, Oxford (2007).
[58] R. K. Senn and D. G. Shenefiel, *Mod. Plast.* 48, 66 (1971).
[59] F. Martelli, *Mod. Plast.* 46, 146 (1969).
[60] H. Hurnik, *Chemical Blowing Agents.* In: Plastics Additives Handbook (eds. H. Zweifel, R. D. Maier and M. Schiller), 6th edn., Hanser, Munich (2009).
[61] E. Bociaga and P. Palutkiewicz, *Polimery* 56, 749 (2011).
[62] R. C. Reid, J. M. Prausnitz and B. E. Poling, *Properties of gases and liquids*, 4th edn., McGraw Hill, New York (1987).
[63] J. E. Martine-Vvedensky, N. P. Suh and F. A. Waldman, U.S. Patent 4,473,665 (1984).
[64] D. E. Eaves, *Preprint of Cellular Polymers, 2nd International Conference* (March 23-25, 1993), Edinburgh; RAPRA Technology, Ltd.
[65] F. Braun, G. Gluck, K. Hahn, I. de Grave and H. Tatzel, U.S. Patent 6,677,040 (2004).
[66] H. Sasaki, M. Sakaguchi, M. Akiyama and H. Tokoro, U.S. Patent 6,077,875 (2000).
[67] K. Hirosawa and S. Shimada, U.S. Patent 4,379,859 (1983).
[68] *A guide to the MuCell® microcellular foam injection molding process – Fundamentals, Set-up, Optimization and Troubleshooting*, Trexel Inc.
[69] N. Usui, U.S. Patent 2003/0165672 (2003).
[70] J. Nambu, T. Hirota, A. Uotani, Y. Oomura and A. Sakamoto, *Development of Novel Polypropylene Foaming/Molding Technologies and Products for Sumiceller®*. Paper translated from R&D Report "Sumitomo Kagaku", vol. 2008-I, 1 (2008).
[71] Foamcore and Superfoam, JSP. Available in http://www.jsp.com/products/other.
[72] M. Antunes, M. Mudarra and J. I. Velasco, *Carbon* 49, 708 (2011).
[73] S. S. Ray and M. Okamoto, *Prog. Polym. Sci.* 28, 1539 (2003).
[74] J. R. Potts, D. R. Dreyer, C. W. Bielawski and R. S. Ruoff, *Polymer* 52, 5 (2011).
[75] E. T. Thostenson, Z. Ren andT.-W. Chou, *Comp. Sci. Tech.* 61, 1899 (2001).
[76] E. Hammel, X. Tang, M. Trampert, T. Schmitt, K. Mauthner, A. Eder and P. Pötschke, Carbon 42, 1153 (2004).
[77] T. Kuila, S. K. Srivastava, A. K. Bhowmick and A. K. Saxena, *Compos. Sci. Tech*, 68, 3234 (2008).
[78] V. Rives, *Layered Double Hydroxides: Present and Future.* Nova Science Publishers, New York (2001).
[79] H. D. Shu, X. H. Li and Z. J. Zhang, *Prog. Chem.* 20, 1509 (2008).
[80] S. W. Kuo and F. C. Chang, *Prog. Polym. Sci.* 36, 1649 (2011).
[81] G. Siqueira, J. Bras and A. Dufresne, *Polymers* 2, 728 (2010).
[82] M. Sauceau, J. Fages, A. Common, C. Nikitine and E. Rodier, *Prog. Polym. Sci.* 36, 749 (2011).
[83] S. M. Lee, D. C. Shim and J. W. Lee, *Macromol. Res.* 16, 6 (2008).
[84] M. A. Treece and J. P. Oberhauser, *J. Appl. Polym. Sci.* 103, 884 (2007).
[85] Q. T. Nguyen and D. G. Baird, *Polymer* 48, 6923 (2007).

[86] C. C. Ibeh and M. Bubacz, *J. Cell. Plast.* 44, 493 (2008).
[87] L. M. Sherman, *Plast. Tech.* 50, 56 (2004).
[88] W. G. Zheng, Y. H. Lee and C. B. Park, *J. Appl. Polym. Sci.* 117, 2972 (2010).
[89] A. K. Chaudhary and K. Jayaraman, *Polym. Eng. Sci.* 51, 1749 (2011).
[90] T. Abiko, *Idemitsu Giho* 46, 281 (2003).
[91] M. Guo, M. Heuzey and P. J. Carreau, *Polym. Eng. Sci.* 47, 1070 (2007).
[92] M. Antunes, V. Realinho and J. I. Velasco, *J. Nanomat.* Vol. 2010, Article ID 306384 (2010).
[93] Y. Yang, M. C. Gupta, K. L. Dudley and R. W. Lawrence, *Nano Lett.* 5, 2131 (2005).
[94] Y. Yang, M. C. Gupta, K. L. Dudley and R. W. Lawrence, *Adv. Mater.* 17, 1999 (2005).
[95] J. Shen, C. Zeng, and L. J. Lee, *Polymer* 46, 5218 (2005).
[96] C. Zeng, N. Hossieny, C. Zhang and B. Wang, *Polymer* 51, 655 (2010).
[97] L. Chen, R. Ozisik and L. S. Schadler, *Polymer* 51, 2368 (2010).
[98] J. Shen, X. Han and L. J. Lee, *J. Cell. Plast.* 42, 105 (2006).
[99] L. J. Lee, J. Yang, N.-R. Chiou and S.-K. Yeh, US Patent Application 2010/0099782 (2010).
[100] M. Antunes, J. I. Velasco, V. Realinho and D. Arencón, *J. Nanosci. Nanotech.* 10, 1241 (2010).
[101] R. Verdejo, F. Barroso-Bujans, M. A. Rodríguez-Pérez, J. A. de Saja and M. A. López-Manchado, *J. Mater. Chem.* 18, 2221 (2008).
[102] R. Verdejo, C. Saiz-Arroyo, J. Carretero-González, F. Barroso-Bujans, M. A. Rodríguez-Pérez and M. A. López-Manchado, *Eur. Polym. J.* 44, 2790 (2008).
[103] J. Yang, M. Wu, F. Chen, Z. Fei and M. Zhong, *J. Super. Fluids* 56, 201 (2011).
[104] V. Eswaraiah, V. Sankaranarayanan and S. Ramaprabhu, *Macromol. Mater. Eng.* 296, 894 (2011).
[105] H.-B. Zhang, Q. Yan, W.-G. Zheng, Z. He and Z.-Z. Yu, *Appl. Mater. Interfaces* 3, 918 (2011).
[106] G. Gedler, M. Antunes, V. Realinho and J. I. Velasco, *IOP Conf. Series: Mater. Sci. Eng.* 31, Article ID 012008 (2012).

In: Polypropylene
Editors: L. P. Silva and E. F. Barbosa

ISBN: 978-1-62417-142-0

Chapter 14

PHOTO-OXIDATIVE BEHAVIOR OF ISOTACTIC-POLYPROPYLENE/CLAY NANOCOMPOSITES PRODUCED VIA SINGLE-STEP EXTRUSION METHOD AT DIFFERENT COOLING CONDITIONS

Gordana Bogoeva-Gaceva[1*], Luljeta Raka[2] and Gjorgji Petruševski[3]

[1]Faculty of Technology and Metallurgy, Ss Cyril and Methodius University, Skopje, Macedonia

[2]Faculty of Natural Sciences and Mathematics, State University of Tetovo, Tetovo, Macedonia

[3]Research & Development Institute, ALKALOID AD, Skopje, Macedonia

ABSTRACT

The environmental durability of polypropylene/clay nanocomposites, as an important factor for their industrial application, is widely reviewed in the literature, however there are still controversial data concerning the role of the clay and matrix morphology on their behavior upon long-term photo-aging. In this chapter, we report the results of photo-oxidation on morphology, crystallization behavior and thermal stability of polypropylene/organo-modified clay nanocomposites prepared by single-step extrusion process, using maleic anhydride grafted polypropylene (PP-g-MA) as a compatibilizer. After extrusion, the nanocomposite samples for the analyses were produced by pressing, applying cooling rates of 1 and 15 ^{o}C min^{-1}, and as a result of the cooling regime, materials with different morphology were obtained, as revealed by wide-angle X-ray diffraction (WAXD). Predominant exfoliation was achieved for nanocomposites produced at lower cooling rate, and this effect was observed even in the absence of compatibilizer. Nanocomposites with exfoliated and hybrid exfoliated/intercalated structure (as confirmed by WAXD and scanning/transmission electron microscopy SEM/TEM investigation) were exposed to long wavelength radiation ($\lambda > 300$ nm) for prolonged time (50, 100 and 150 h) to investigate their photo-oxidative behavior. Fourier

* Corresponding author: gordana@tmf.ukim.edu.mk

transform infrared spectroscopy (FTIR) has been used to probe the photo-oxidation stability/degradation of PP and PP/clay nanocomposites. The results have shown that the nanocomposites, as well as the starting polymer matrix, produced by slow cooling, are less susceptible to UV-irradiation. Although the main portion of α-crystalline phase was not much affected by the presence of clay, certain amount of β-isotactic polypropylene (β-iPP) has been detected in the samples produced at higher cooling rate, slightly increasing with the time of irradiation. The overall crystallinity of all investigated nanocomposites was higher after exposure to irradiation. Drastically different behavior upon photo-oxidation was observed in DSC experiments: crystallization peak temperature and initial rate of crystallization were both increased for all nanocomposites, and this effect has been assigned to nucleation ability of clay platelets, especially in the sample with exfoliated structure, registered by TEM and WAXD. Multiphase crystal structure originate in irradiated materials upon nonisothermal crystallization (following the first melting run), with broader crystal size distribution seen for nanocomposites produced by fast cooling. The influence of crystal morphology on thermal stability of nanocomposites was analyzed by comparing oxidation induction time for samples containing 1% clay after 150 h photo-aging. The results revealed better thermal stability of nanocomposite, obtained at slow cooling, represented by lower mass loss after isothermal treatment at 200°C. However, as expected, due to the changes in chemical structure, the overall rate of degradation of photo-aged materials was decreased after irradiation.

INTRODUCTION

Polymer layered silicate hybrids have great potential as a new generation lightweight high performance nanocomposite materials, due to their improved mechanical, electrical, thermal, and optical properties, reported in the literature [1-6], in addition to their price-performance ratio, and as compared to their macro- and micro-counterparts [7, 8]. Moreover, polymer nanocomposites are known for excellent barrier properties against gases (e.g., oxygen, nitrogen and carbon dioxide), water and hydrocarbons, strongly dependant on the aspect ratio of clay platelets, but not on the chemical interactions and the type of gas or liquid molecules [9]. These improvements of the properties are mainly reported for addition of several percents of clay in polymer matrix, dispersed into single silicate layers with a thickness of about 1 nm and typically lateral dimensions of several hundred nanometers up to micrometers, which is in strong contrast to conventional polymer fillers or reinforcements that require much higher concentrations (>30 wt %) and usually providing only a fraction of these enhancements [1, 4].

Depending on a nanoscale dispersion of the layered silicate, the whole spectrum of structures ranging from intercalated to exfoliated nanocomposites can be observed [4, 10-13]. The resulting properties of nanocomposites are related to the modification of the structure and dynamics of the polymer at and near the particle surface. Because of the large surface area this fraction of the polymer contributes significantly to the properties of the whole nanocomposite, even at extremely low filler content.

Over the years, isotactic polypropylene (iPP) has been reinforced with various inorganic or organic fibers (e.g., carbon, glass, Kevlar, PET, PTFE, natural fibers), as well as other fillers (talc, mica, calcium carbonate, etc.) [see, for instance ref. 14] essentially aiming at improving its mechanical and functional performance. However, during the last decade much attention has been paid on methods for incorporation different nanoparticles (such as silicates, calcium carbonate and carbon nanotubes), in iPP, among which [1, 15-17] the most promising seems to be the melt intercalation and recently developed latex technology [16, 17]. The most important changes induced by the presence of nano-clay in PP are probably those reported recently [2] regarding drastically improved tensile modulus and elongation at break detected for PP/clay nanocomposites containing only 5% clay and specially designed acrylate type gradient copolymers as compatibilizers.

However, in spite of the promising properties, PP nanocomposites are still struggling to break into mainstream volume markets [18, 19], partly due to environmental durability concerns. Exposure to direct sunlight and environmental dew or rain can compromise the main properties of the generic polymer, due to the structural and morphological modifications [20-22]. Moreover, there is a clear agreement in the literature that the degradability of nanocomposites with different polymer matrices under UV light represent a serious problem, when their long-term outdoor application is concerned [9].

Although the environmental durability of PP/clay nanocomposites is reviewed in the literature, there are still controversial data concerning the role of the clay with respect to their photo-oxidative behavior. Generally, the rate of photo-oxidative degradation of polymer/clay nanocomposites seems to be higher than that of pure polymers [20], which in turn results in an unexpected decrease in the lifetime of the material. It has been mentioned that the presence of clay is the key parameter for the acceleration of the degradation, provided that the degree of the dispersion of the clay (exfoliation) has small influence and that conventional microcomposites with clay demonstrated almost similar oxidation rates to the rates found in nanocomposites materials [23]. The main reasons for the increased photo-oxidation rates when using organo-modified nanoclays include the presence of iron ions in silicate fillers, the degradation of the organo-modifier and the formation of acidic sites on the nanoclay layers [24, 25].

In addition to our previous study of the structure and basic properties of PP/clay nanocomposites, produced via melt extrusion and latex technology [26, 27], herein we report the study of photo-induced ageing on morphology, thermal stability and crystallization behavior of polypropylene/clay nanocomposites, obtained at different cooling rates applied during the pressing cycle performed after melt extrusion. Melting and crystallization behavior of UV-irradiated PP/clay nanocomposites might be of interest in terms of their recycling ability and further re-use.

MATERIAL AND METHODS

Materials

The isotactic polypropylene (iPP) matrix used was a homopolymer (Borclean HB300BF, Borealis) with $M_w = 608$ kg mol^{-1}, MFI=2.5 g/10min (230 °C, 2.16 kg) and a density of 0.91 g cm^{-3} supplied from COST Action P12: WG3-Polymer Bank [www.unirostock.de/fakult/manafak/physik/poly/COST_P12/Final Report 2007]. Maleic anhydride grafted PP (PP-*g*-MA, with MA content=0.6 wt%, MFI=115 at 190 °C/2.16 kg and a density of 0.95 g cm^{-3}), supplied by Aldrich was used as compatibilizer for promoting exfoliation of the organoclay,

Cloisite C93A (Southern Clay Products, Inc), with a cation-exchange capacity (CEC) of 90 meq/100 g clay, organically modified with ternary ammonium salt.

Preparation of iPP/Clay Nanocomposites

iPP nanocomposites were obtained by direct melt mixing the components in recirculating co-rotating twin-screw micro-extruder with a volume of 5 cm^3, made by DSM (The Netherlands), at 190 ^oC and 300 rpm for 30 min. (single step extrusion process). The clay concentration of 1 and 3 % wt was used for C93A based nanocomposites, and fixed concentration of 10 wt% PP-*g*-MA. PP/PP-g-MA and nanocomposite samples for further analyses were produced by pressing in a Collin Press 300G at 180^oC for 2 min without any applied pressure, and after this period, a pressure of 100 bars was applied for 3 min. The cooling rates of 1 ^oC min^{-1}(slow cooled samples, assign. "S") and 15 ^oC min^{-1} (plate cooled samples, assign. "P") were applied in order to produce polymer samples with different morphology. For further analyses by WAXD, SEM/TEM, DSC and TG, films of 100-300 μm average thickness were used.

Table 1. Assignation of the samples

Sample	Irradiation time/ h		
Fast (plate) cooled	0	50	150
PP/PP-*g*-MA (90/10 wt %)	P0-0	P0-50	P0-150
PP/PP-*g*-MA/1 wt % clay	P1-0	P1-50	P1-150
PP/PP-*g*-MA/3 wt % clay	P3-0	P3-50	P3-150
Slow cooled			
PP/PP-*g*-MA (90/10 wt %)	S0-0	S0-50	S0-150
PP/PP-*g*-MA/1 wt % clay	S1-0	S1-50	S1-150
PP/PP-*g*-MA/3 wt % clay	S3-0	S3-50	S3-150

Accelerated Weathering

The photo-induced degradation experiments were conducted using SUNTEST CPS+ photo-chamber. Small pieces of the tested materials were exposed directly to the light flux of 765 W m^{-2}, in the near ultraviolet and visible (300–800 nm) part of the electromagnetic spectrum, for a total time of 50-150 hours. The integrated cooling device was set to ambient temperature, preventing intense heating and thermal degradation of the samples. Before exposure to UV-radiation in photo-chamber, all samples underwent weathering in natural ambient (at temperature below 35 ^oC, during 40 days).

Characterization Methods

Fourier Transform Infrared (FT-IR) Spectroscopy

The Attenuated total reflectance (ATR) infrared spectra were obtained on Varian 660 FT-IR spectrometer (resolution 4 cm^{-1}, 16 scans per spectrum, 550–4000 cm^{-1}), using MIRAcle ZnSe ATR module (PIKE technologies) with low pressure micrometer clamp.

X-ray Analysis

Wide-angle X-ray diffraction (WAXD) analysis was performed on a Rigaku D/Max-B diffractometer (Japan), equipped with nickel filtered $CuK\alpha_1$ (λ =1.54056 Å) radiation source operated at 40 kV and 25 mA. Corresponding data were collected at step scan rates of 0.02 ^{o} min^{-1} and 1° min^{-1} in the range 2θ=1.1-27° for clay particles and nanocomposite samples. The measurements were made on thin samples prepared by compression molding. For each sample two measurements were performed and the results represent mean value (Table 2). Analysis of the WAXD patterns was done using basic peak fitting with Gaussian-Lorentzian function after background subtraction.

Morphological Characterization by TEM and SEM

Transmission electron microscopy (TEM) observation was performed on Tecnai 20 operated in bright-field mode at 200 kV. Ultrathin sections of the PP/clay nanocomposites with a thickness of approximately 90 nm stained with RuO_4 solution were prepared at room temperature using an ultramicrotome Reichert-Jung Ultracut E equipped with a diamond knife. The sections were transferred dry to carbon-coated Cu grids of 200 meshes.

Scanning electron microscopy (SEM) was performed using a Quanta 3D FEG (Fei Co.) equipped with a field-emission gun. High vacuum conditions were applied and a secondary electron detector was used for image acquisition. No additional sample treatment, such as surface etching or coating with a conductive layer, has been applied before surface scanning. Standard acquisition conditions for charge contrast imaging were used.

Differential Scanning Calorimetry

Crystallization and melting behavior before and after exposure of the samples to oxidative environment and UV-radiation were determined by differential scanning calorimetry (DSC). DSC analysis was performed under nitrogen using DSC-7 analyzer (Perkin Elmer) at scanning rate (heating/cooling) of 10 ^{o}C min^{-1} in the temperature range from 25 to 200 ^{o}C. From the obtained DSC crystallization curves (II scan: cooling performed after the first heating/melting scan, I) the initial slope of the exotherm at inflection on the high-temperature side, S_i, and the full width at half height of the exothermic peak, ΔW, are obtained to describe the nonisothermal crystallization kinetics [28]. Third heating/melting scan (III) was performed after nonisothermal crystallization (II scan).

Thermo-Gravimetric Characterization

Oxidation Induction Time (OIT) test [29] was performed using PerkinElmer TG/DTA Pyris Diamond system, by heating the composite samples to 200 ^{o}C at rate of 100 ^{o}C min^{-1} and afterwards treating them isothermally in oxygen (air) atmosphere for 180 min at 200 ^{o}C.

The intention of the present study was to investigate the role of the matrix morphology, created during production of nanocomposites by changing the cooling rate in the pressing cycle, on their behavior after exposure to long wavelength radiation for prolonged time. According to our previous experience [30], the melt extrusion method enables production of iPP nanocomposites (using the matrix modification as reported herein) with both exfoliated and hybrid exfoliated/intercalated structure, which is an additional subject of interest when environmental stability of PP nanocomposites is concerned.

RESULTS AND DISCUSSION

FTIR

ATR-FTIR spectra of nanocomposites and pristine polymer matrices, produced by fast (P) and slow (S) cooling process, before and after exposure to irradiation (50, 100 and 150 h) are shown in Figure 1(a and b) and Figure 2.

Nearly identical spectra were observed for all the investigated materials. Due to low relative intensity of the bands of interest (especially in the carbonyl region, where most significant changes were expected according to the literature data for photodegradation of PP-nanocomposites [20, 23, 31, 32]), further deconvolution of the spectra for estimation of the peak areas was not performed. An exception was the sample P0, produced by fast cooling procedure, for which deconvolution delivered predominant presence of C=O groups (band at 1730 cm^{-1}) (over the other bands in this region assigned to the formation of carboxylic acids and peracids [23]) after irradiation time of 50 and 150 h.

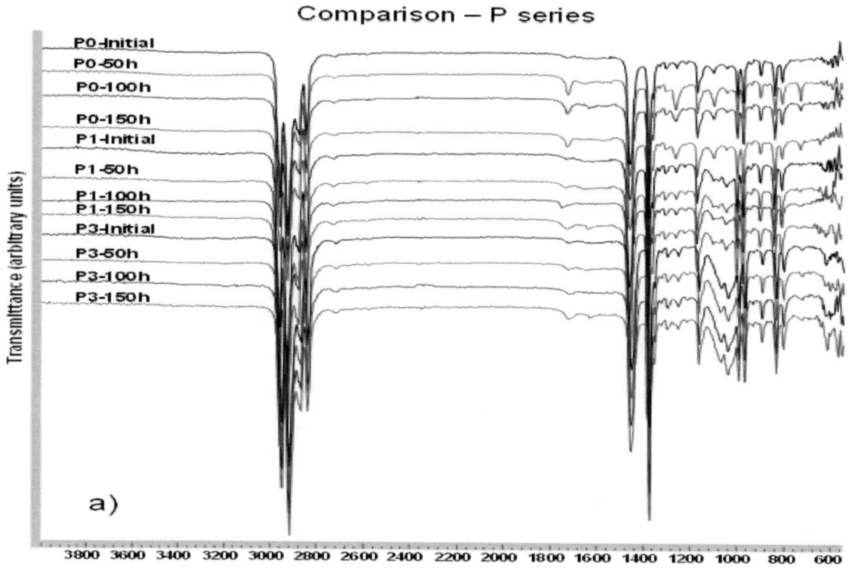

Figure 1. a) ATR-FTIR spectra of: matrix PP (P0-initial, P0-50h, P0-100h and P0-150 h), and nanocomposites P1 (P1-initial, P1-50h, P1-100h, P1-150h) and P3 (P3-initial, P3-50h, P3-100h, P3-150h), before and after exposure to UV-radiation, respectively.

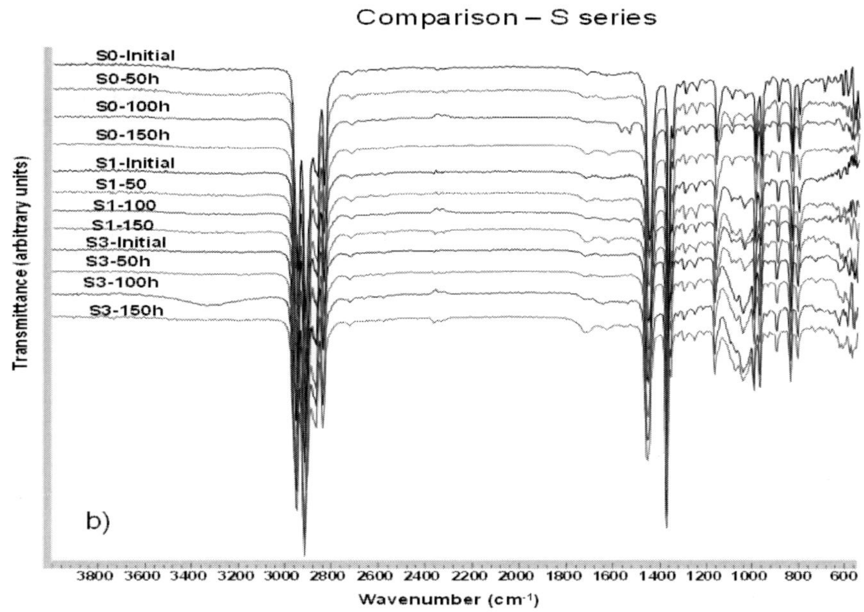

Figure 1. b) ATR-FTIR spectra of: matrix PP (S0-initial, S0-50h, S0-100h, S0-150h) and nanocomposites S1 (S1-initial, S1-50h, S1-100h, S1-150h) and S3 (S3-initial, S3-50h, S3-100h, S3-150h) before and after exposure to UV-radiation, respectively.

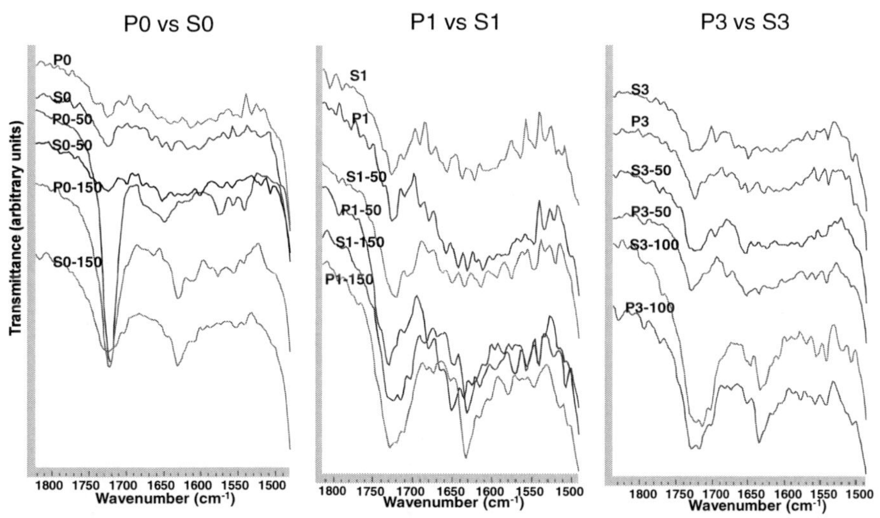

Figure 2. ATR-FTIR spectra of matrix PP and PP nanocomposites (slow, S and fast, P cooled) in the region 1500-1800 cm^{-1}, before and after irradiation.

The relative surface of the band in the region of 1770-1670 cm^{-1} is slightly increasing with the exposure time, and this trend is seen for both matrix polymers P0/S0 and for the nanocomposites. Figure 2 is focused in the frequency region of 1500-1800 cm^{-1}: the broad band with a peak maximum at approximately 1720-1735 cm^{-1}, attributed to the formation of carbonyl groups due to the oxidation of the matrix [23], is observed for all the irradiated samples, although with different intensity.

The appearance of a sharp carbonyl band in the spectra of pristine polymer P0 could be prescribed to a higher degree of photo-oxidative degradation as compared to S0; for nanocomposite samples, especially those with 1% of clay, similar tendency is seen: it appears that the nanocomposites produced by slow cooling are less susceptible to UV-radiation.

Morphological analysis by WAXD and SEM/TEM

Figure 3 shows WAXD patterns (2θ =1.3-10°) of clay (curve a) and P and S nanocomposites, containing 1 and 3 wt % clay (curve b and c, correspondingly), as well as of iPP/C93A 1 wt % nanocomposite prepared without compatibilzer, PP-*g*-MA (curve d). The peak of organically modified clay at 2θ~3.52° corresponds to basal spacing of silicate galleries of 25.1 Å. As can be seen from curve d, the characteristic peak of clay is intensive, although shifted to a lower value (2θ~2.84° for fast, and 2θ~2.40° for slow cooled sample), indicating intercalation of polymer chains into silicate galleries.

According to the peak position, better intercalation has been achieved for nanocomposites produced by slow cooling procedure, even in the absence of compatibilizer.

The results also clearly indicate that the extent of intercalation/exfoliation has been improved by addition of compatibilizer, in agreement with other reports [1, 4, 6, 33].

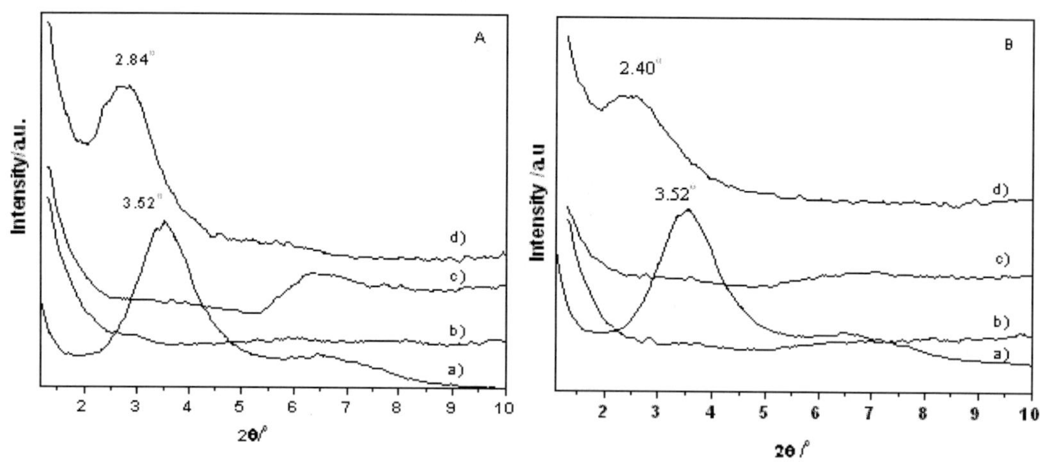

Figure 3. WAXD patterns of samples in the range 2θ = 1-10° for: A) clay (a), P1-0 (b), P3-0 (c), and iPP/clay (clay loading of 1 wt%; iPP matrix without compatibilizer) (d) (With kind permission from Springer Science+Business Media: Journal of Thermal Analysis and Calorimetry, *Characterization of polypropylene/layered silicate nanocomposites prepared by single-step method*, Vol. 100, Issue 2, 2010, pp. 629-639, Figure 1a), and B) clay (a), S1-0 (b), S3-0 (c) and iPP/clay (clay loading of 1 wt%; iPP matrix without compatibilizer) (d).

The absence of clay diffraction peak in the samples with iPP/PP-*g*-MA matrix (P and S) and 1 wt% clay shows that clay exfoliation occurred (curve b, Figure 3A, B). As confirmed by many authors, this is result of interactions between polar maleic anhydride groups and the silicate layer [34, 35]. In nanocomposite with 3% clay (curve c, Figure 3A), a broader diffraction peak appeared at higher 2θ-value, although the main clay peak disappeared. Considering the position of this peak, a possible coalescence of clay platelets can be supposed [34]. It should be mention that this peak is not seen in the pattern of slow cooled nanocomposite samples, indicating differences in the morphology of the resulting materials due to the different cooling/crystallization conditions.

The obtained results indicate the formation of predominant exfoliated structure of nanocomposites containing 1 wt % clay, and mixed intercalated/exfoliated structure at higher concentration of nanoclay (3 wt% clay), especially for fast cooled samples.

Figure 4 shows the effect of clay on the morphology of iPP nanocomposites (P and S) produced *via* two cooling regimes. As can be seen the main peak position of the crystal planes did not shift with addition of clay, although differences in relative intensities of the peaks α (110) and α (040) for the iPP matrix and the nanocomposites containing 1 and 3 wt% clay are observed. This indicates that the typical α-crystalline morphology of iPP was not much affected by the presence of nanoclay. However, in the patterns of P samples, the reflection peak at 2θ~15.77°, which corresponds to the (300) reflection of β-iPP [36], is clearly seen, and its intensity is changed by the addition of clay: namely, it appears as a shoulder peak in nanocomposites with higher amount of clay.

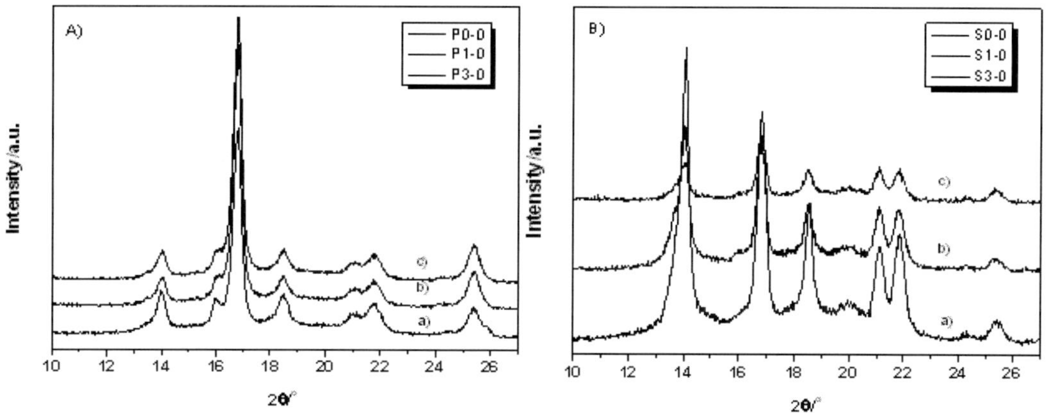

Figure 4. WAXD patterns of samples in the range 2θ = 10-27° for: A) P0-0 (a), P1-0 (b), and P3-0 (c), B) S0-0(a), S1-0 (b), and S3-0 (c).

WAXD patterns of the nanocomposites after exposure to UV-radiation (150h) are presented in Figure 5. For comparison, the patterns of samples before exposure are also given. The results of WAXD analysis are shown in Table 2.

As revealed by WAXD results, the overall crystallinity of all samples exposed to UV-radiation was increased (Figure 6), mostly certain as a result of the phenomenon of chemical crystallization [23]: upon exposure to UV radiation the polymeric matrix suffered chain scissions mostly of the amorphous fraction, resulting in small free polymeric moieties of high mobility, which then rearrange in the helical conformation. An overall crystallinity increase upon photo-oxidation of PP nanocomposites was also reported by Kandilioti and La Mantia [20, 23] The portion of β-phase was higher in starting nanocomposites produced by fast cooling, with tendency of increasing with prolongation of irradiation time (see Table 2). This finding is in agreement with the results obtained by FTIR, pointing out the role of crystalline morphology and different achieved degree of clay delamination in starting materials, S0 and P0.

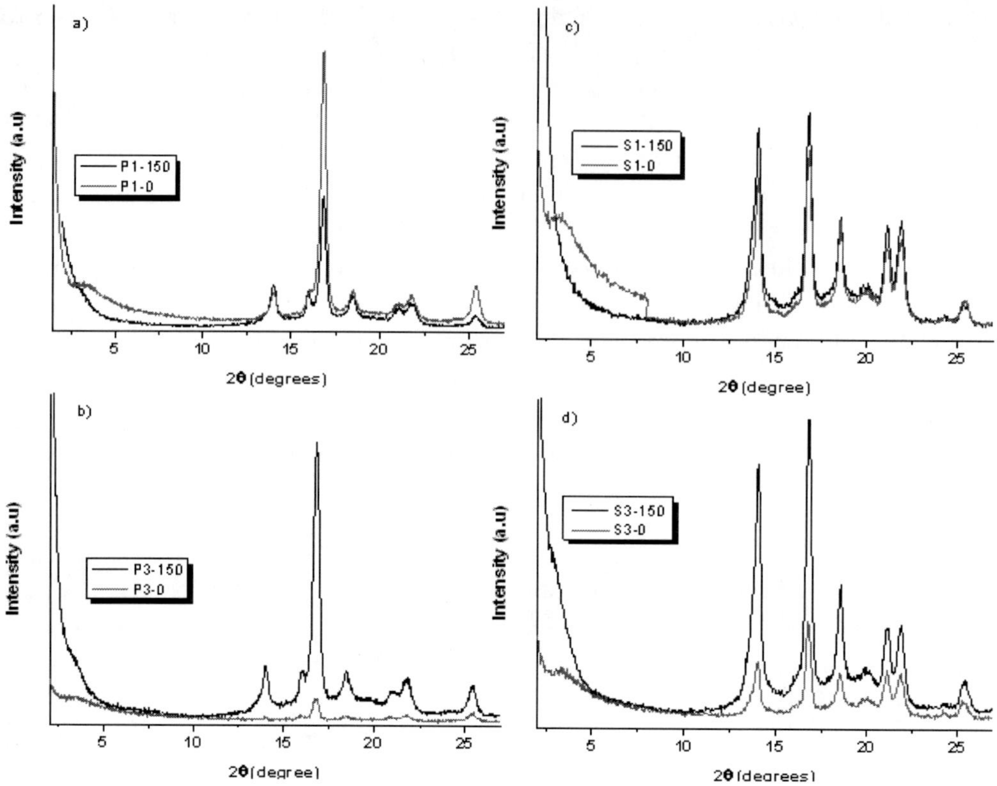

Figure 5. Comparison of WAXD patterns of nanocomposites P (a, b) and S (c, d) before and after irradiation of 150 h.

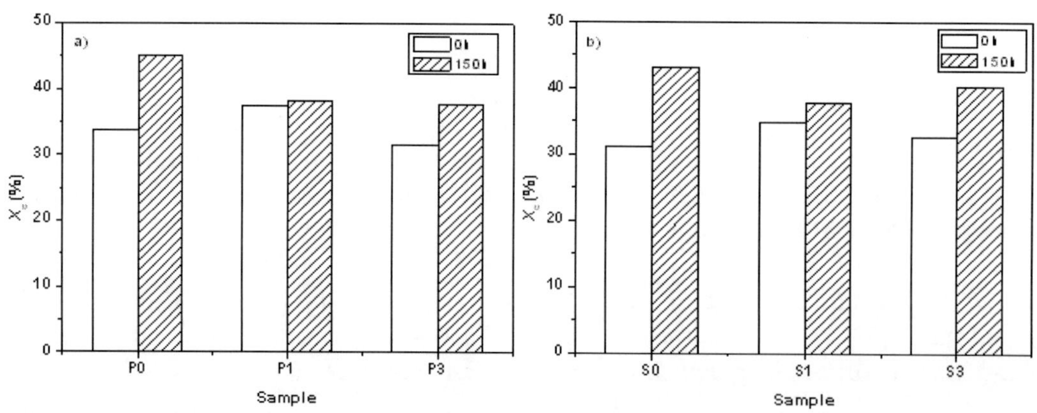

Figure 6. Overall degree of crystallinity for samples produced by different cooling regimes, before and after irradiation: a) fast (P) and b) slow (S) cooled.

Table 2. WAXD data obtained for P and S nanocomposites after irradiation time of 150 h

Sample	X_α /%	X_β /%	X_m /%	X_{am}/%	X_γ /%
P0-0	31.93	1.51	0.12	66.37	0.07
P0-150	43.44	1.35	0.24	54.83	0.14
P1-0	36.03	1.24	0.13	62.53	0.08
P1-150	35.80	2.11	0.20	61.77	0.12
P3-0	28.53	1.25	1.11	68.46	0.66
P3-150	35.77	1.54	0.22	62.35	0.13
S0-0	28.11	0.04	0.14	68.77	2.76
S0-150	41.64	0.05	0.18	56.80	1.34
S1-0	32.51	0.33	0.29	65.11	1.76
S1-150	37.28	0.08	0.28	62.19	0.17
S3-0	28.99	0.13	0.45	67.25	3.17
S3-150	37.84	0.06	0.22	59.66	2.22

Our previous investigation [30] has shown that good dispersion of clay is attained in samples with lower clay loading (see Figure 7a), without evidence of aggregation. Contrary, at higher clay content, existence of mixed intercalated/exfoliated structure was detected by SEM analysis (Figure 7b). The pores, seen on SEM micrographs of both samples, might be an indication of good adhesion, achieved between the matrix polymer and clay particles.

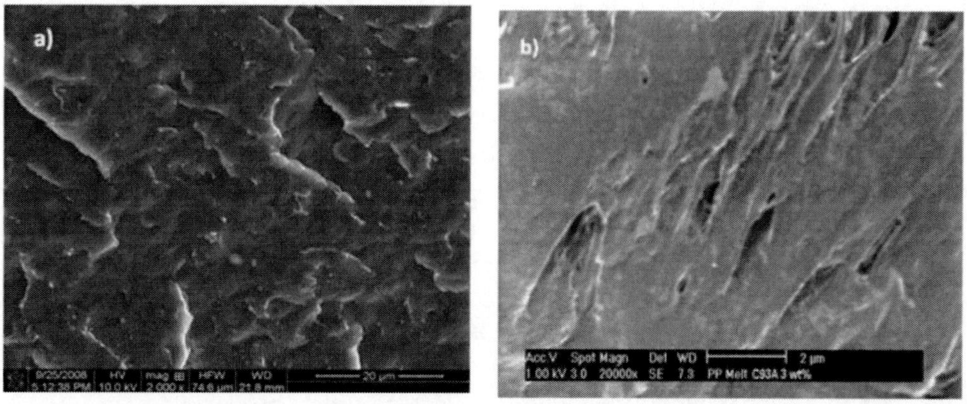

Figure 7. SEM micrographs of P1-0 (a) and P3-0 (b) (magnification 2000x and 20000x, respectively). (With kind permission from Springer Science+Business Media: Journal of Thermal Analysis and Calorimetry, *Characterization of polypropylene/layered silicate nanocomposites prepared by single-step method*, Vol. 100, Issue 2, 2010, pp. 629-639, Figure 3).

Transcrystalline organization of iPP lamellae around the clay layers, observable in TEM micrographs in the sample with 1% clay (Figure 8), shed light on the phenomenon of nucleation, induced by clay platelets [37].

Obviously, transcrystallization took place in the production (at cooling step) of nanocomposites, and it might be expected that during crystallization process performed after melting the clay could induce the same effect, depending on the conditions and tendency of clay platelets toward agglomeration. It is mostly certain that the formation of an ordered crystalline interface may be of prime important technological relevance, as it was for instance recognized that the nature of the interface plays a fundamental role in the reinforcement mechanism [38, 39].

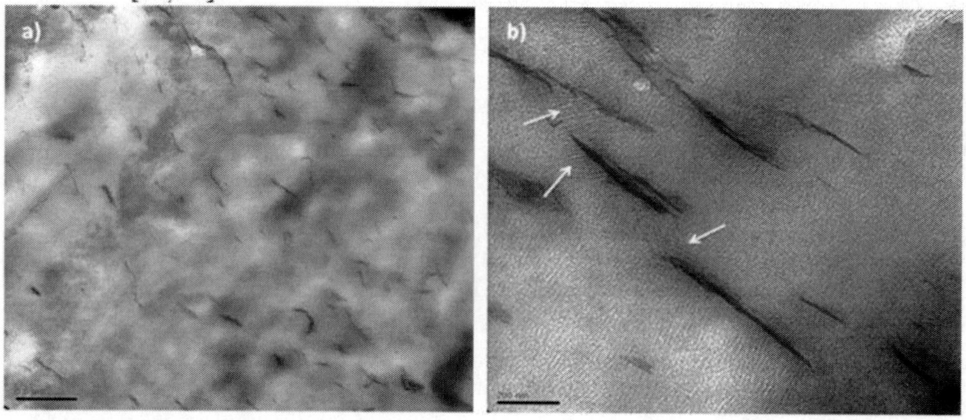

Figure 8. TEM bright-field micrograph of P1 nanocomposite at different magnification (a, b). White arrows indicate the presence of transcrystalline layer of iPP with different orientation around the clay platelets.
(With kind permission from Springer Science+Business Media: Journal of Thermal Analysis and Calorimetry, *Characterization of polypropylene/layered silicate nanocomposites prepared by single-step method*, Vol. 100, Issue 2, 2010, pp. 629-639, Figure 4).

Thermal Characterization by DSC and TGA

From the first melting scans of P0 and S0 samples (Figure 9) it is obvious that the presence of β-phase in P sample is indisputable, resulting in a very broad endotherms for starting and irradiated polymers. Additionally, more significant changes in the morphology of the samples exposed to UV-radiation are seen for fast cooled samples as compared to slow cooled: the melting endotherms of slow cooled polymers are almost identical before and after irradiation. Melting temperature (high melting peak) is slightly decreasing for P-samples with increasing time of irradiation (from 162.6 to 159.6 ^{o}C), while the value for S sample remains unchanged (164.4 ^{o}C) [see Appendix].

The behavior of the matrix polymer upon crystallization (II scan), regardless of the way of preparation (P and S), is similar, resulted in slightly decreased crystallization peak temperature with increased irradiation time (T_c is lowered for 2-3 ^{o}C); melting endotherms after nonisothermal crystallization are almost identical for P and S samples (T_m=161.4-163°C).

Similar conclusion concerning melting behavior of nanocomposites could be derived for samples P1 and S1 (containing 1 wt % clay) (Figure 10). Crystallization peak temperature is increased for both nanocomposites due to irradiation, and this effect is more pronounced for S sample (T_c=119.03 °C for starting nanocomposite, compared to T_c =121.6 °C for S1-50).

Nearly the same results upon DSC measurements (in I, II and III run) with very close values for T_m and T_c, as those obtained for P1/S1 nanocomposites, exposed to irradiation, were recorded for nanocomposites P3/S3 with 3 wt% clay (see Figure 11).

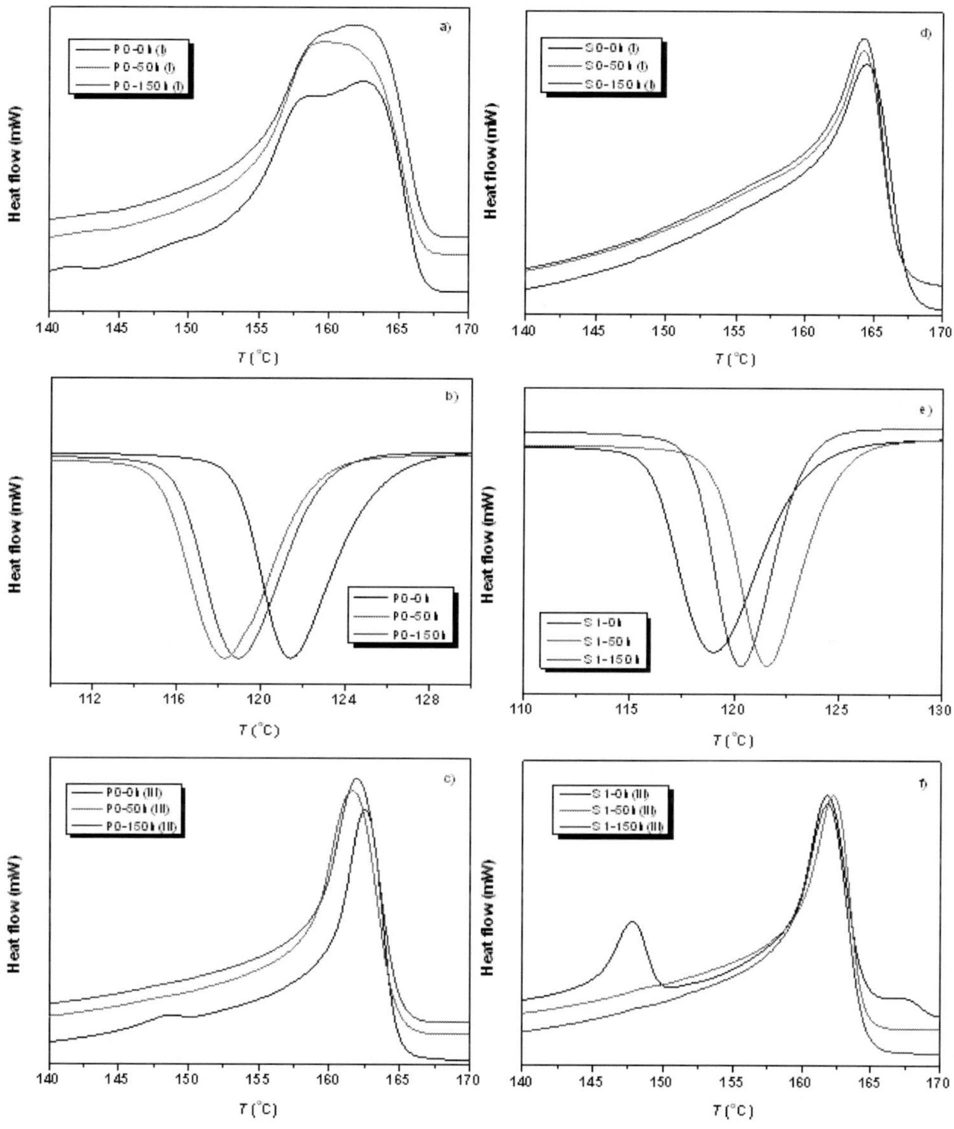

Figure 9. DSC curves of crystallization and melting of P0 (a, b, c) and S0 (d, e, f) samples.

The changes in crystallization behavior, detected for nanocomposite S1, is probably due to nucleation ability of clay platelets in this sample, for which exfoliated structure was predominant in the starting material, as well to the tendency of transcrystallization, which was in general observed at low clay content (see Figure 8b). Additional evidences for this consumption are the values for S_i, determined from nonisothermal crystallization curve, for nanocomposite S1 before and after irradiation: S_i = 2.1 for S1-0, S_i = 2.7 for S1-50 and S_i =

3.1 for S1-150 (Figure 12). The results obtained by two measurements were highly reproducible. The above mentioned results suggest that the interfacial phenomena [27] at polymer/clay interface are involved not only in the production step of PP nanocomposites, but also upon recrystallization performed after photo-aging of the material.

Multiphase crystal structure (detected previously by WAXD experiments), containing α- and β-iPP, has been created after crystallization from the melt of P1 and S1 samples before UV-radiation (the position of α- and β-melting peaks are at 161-162 °C and 148-149°C, respectively). The shoulder peak at high temperature side of the endotherms of pristine samples P1 and S1 could be assigned to melting of recrystallized portion of α-iPP (Figure 10).

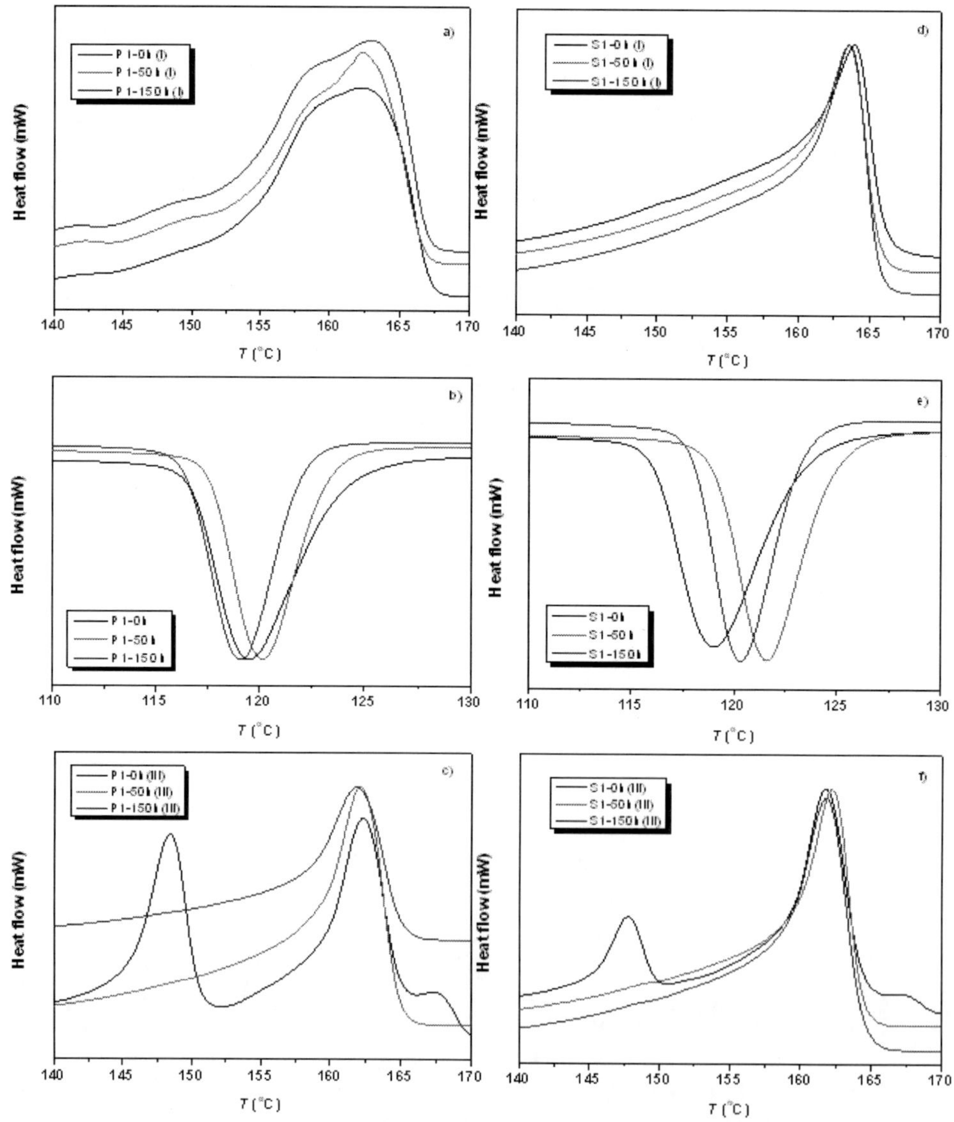

Figure 10. DSC curves of crystallization and melting of P1 (a, b, c) and S1 (d, e, f) samples.

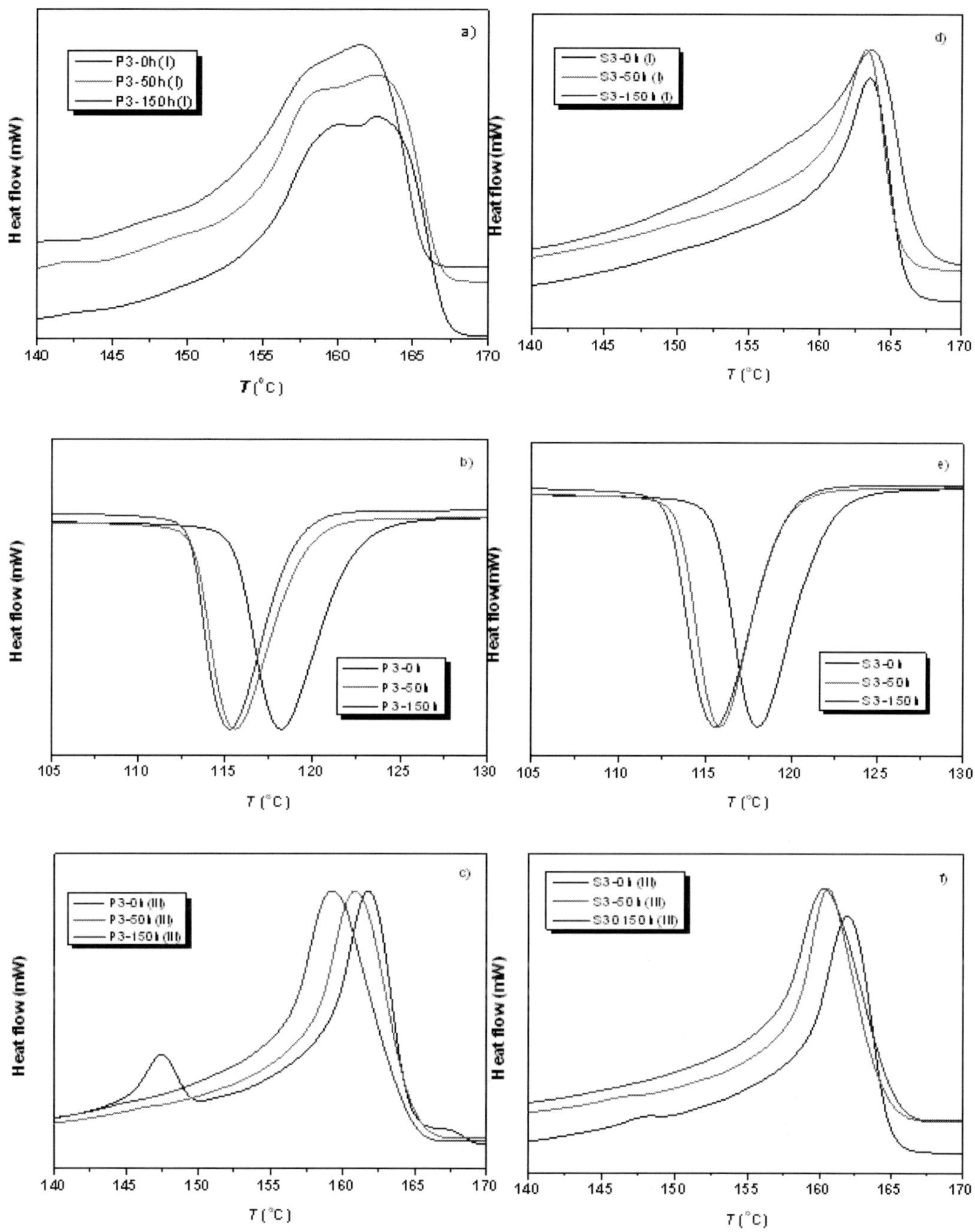

Figure 11. DSC curves of crystallization and melting of P3 (a, b, c) and S3 (d, e, f) samples.

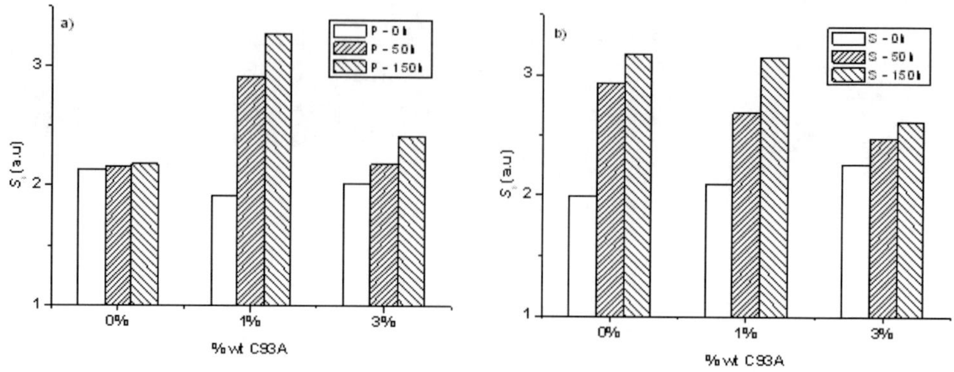

Figure 12. Dependence of S_i on clay loading for P (a) and S (b) samples before (0h) and after irradiation (50 and 150h).

It should be mentioned that the distribution of crystal dimensions, as determined from ΔW (Table 3) is quiet different for P and S series, and in general much broader distribution is seen for P nanocomposites; with prolongation of irradiation time ΔW is slightly increasing, whereas notable decreasing of ΔW was determined for S1-nanocomposites upon exposure to irradiation (for about 25%, after exposure time of 50h).

Different morphological changes induced by photo-oxidation of PP nanocomposites, such as increase in overall crystallinity, reduction in crystal size, multiphase morphology, etc., are reported in literature [20, 24, 40].

As revealed earlier [26], in air gas flow the nanocomposites with 1-3 wt% clay exhibited delay in weight loss, and this was assigned to barrier effect caused by diffusion of volatile decomposition products to the gas, and of oxygen to the polymer. At the same time, significant prolongation of oxidation induction time has been found for nanocomposites as compared to pristine polymer matrix.

Table 3. Full width half-height (ΔW) of endothermic peak determined from first and second DSC melting scan

$\Delta W/^o$					
Sample	**Melting scan**		**Sample**	**Melting scan**	
	I	**II**		**I**	**II**
P0-0	10.46	3.57	S0-0	8.07	3.98
P1-0	10.42	3.88	S1-0	4.51	3.94
P3-0	10.46	4.68	S3-0	5.71	4.53
P0-50	9.58	4.58	S0-50	7.26	3.73
P1-50	9.19	4.81	S1-50	5.30	4.05
P3-50	10.36	5.22	S3-50	5.48	4.53
P0-150	9.89	4.35	S0-150	6.95	3.83
P1-150	10.78	5.21	S1-150	5.08	4.39
P3-150	10.22	6.13	S3-150	5.96	6.00

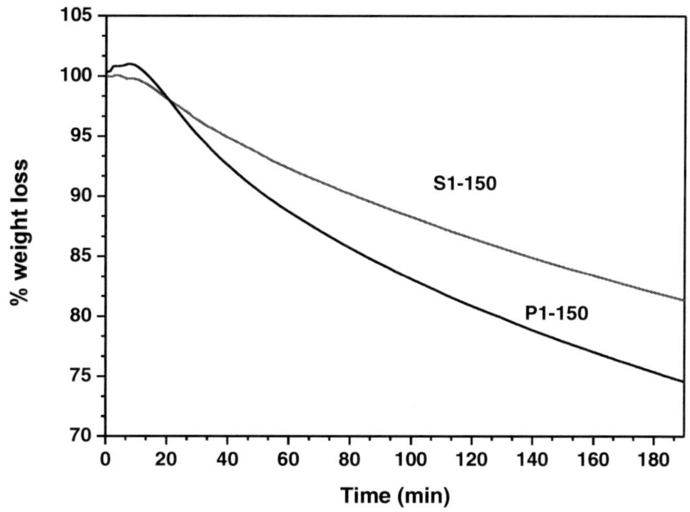

Figure. 13. OIT curves of nanocomposites P-1 and S-1 recorded after photo-aging (150 h).

Table 4. Time at weight loss during isothermal treatment at 200 °C

Sample	Time/min			
	5%	10%	15%	20%
P1	30.4	52.6	85.2	129
S1	38.9	81.9	138.8	-

After exposure to UV-radiation all nanocomposites, regardless of the morphology, degrade much faster under oxidative condition. Nevertheless, the outcome of OIT confirms the role of the material's morphology in oxidative degradation: delayed degradation and lower overall mass loss percent has been determined for the samples produced at slow cooling (see Figure 13 and Table 4). The remaining mass in OIT experiments after 180 min was 74.5% and 81.2% for P-1 and S-1, respectively. As confirmed by other authors, the dispersion state of nanosize particles (clay, calcium carbonate) in the polymer matrices has a little influence on the photo-oxidative degradation of the matrix [23, 41]. The dependence of photo-oxidation rate of PP/MMT composites on melt processing conditions and the resulting structure of the polymer has also been analyzed by other authors [42, 43].

CONCLUSION

Photo-oxidative behavior of two types iPP/clay nannocomposites, produced at different cooling rates applied in the phase of compression molding after extrusion, has been studied. As revealed by WAXD, predominant exfoliated clay structure was achieved for nanocomposites produced at lower cooling rate (1 °C min^{-1}), and this effect was observed even in the absence of compatibilizer, PP-g-MA. Nanocomposites with exfoliated and hybrid exfoliated/intercalated structure were exposed to long wavelength radiation for prolonged time to investigate their melting/crystallization behavior, morphology and thermal stability. The results of FTIR analysis have shown that the nanocomposites, as well as the starting polymer matrix, produced by slow cooling, are less susceptible to UV-irradiation. Although the main portion of α-crystalline phase has not much been affected by the presence of clay, certain amount of β-iPP has been detected in the samples produced at cooling rate of 15 °C min^{-1}, slightly increasing with the time of irradiation. The overall crystallinity of all investigated nanocomposites was higher after exposure to irradiation. Crystallization peak temperature and initial rate of crystallization (performed after first DSC melting run) were both increased for all nanocomposites, and this effect has been assigned to nucleation ability of clay platelets, especially in the sample with exfoliated structure, recorded by TEM and WAXD. Multiphase crystal structures originate in irradiated materials upon nonisothermal crystallization, with broader crystal size distribution seen for nanocomposites produced by fast cooling. According to oxidation induction time, determined by TGA, better thermal stability could be assigned to nanocomposites, obtained at slow cooling, pointing out the role of material's morphology originated in processing cycle. For long-term outdoor applications, it is very important that clay nanocomposite materials possess both thermal and photo-oxidative stability. By solving the insufficient photo-oxidative stability, extended application fields are opened which will help to turn nanocomposites into a real industrial opportunity beyond niches.

Acknowledgments

The part of the work presented in this chapter was supported by DAAD grant (under the programme: "International Masters and Postgraduate Programme in Materials Science and Catalysis" (MatCatNet). The authors are grateful to Professor Gert Klöß from University of Leipzig, Faculty of Chemistry and Mineralogy, for WAXD measurements. Thanks are due to Sonja Ugarkovic, pharm. spec. for enabling ageing experiments in "Alkaloid" – Research and Development.

APPENDIX

Table 1. DSC crystallization and melting data of nanocomposites prepared by fast cooling

Sample	I run melting			II run cooling			III run melting		
	T_m /°C	$T_{m,on}$ /°C	ΔH_m /J/g	T_c /°C	$T_{c,on}$ /°C	$-\Delta H_c$ /J/g	T_m /°C	$T_{m,on}$ /°C	ΔH_m /J/g
P0-0	162.49 (141.47; 159.51)	152.29	98.82	121.45	125.50	102.97	162.48	159.20	107.13
P0-50	159.60	153.11	96.85	118.34	122.42	101.62	161.58	157.58	104.93
P0-150	161.66	152.76	98.36	118.97	123.02	104.29	161.90	157.97	103.72
P1-0	162.30	152.42	94.42	119.53	123.88	98.42	148.40 (162.33)*	145.33 (144.91)*	106.92
P1-50	162.30	152.27	92.67	120.19	123.23	98.99	162.13	158.02	102.07
P1-150	162.98 (142.00)	151.70	94.21	119.11	122.04	100.69	161.88	157.04	100.52
P3-0	161.67	152.15	98.53	118.20	122.37	96.91	148.05(162.15)	145.32 (148.40)	101.07
P3-50	162.49	152.45	92.89	115.60	119.51	97.45	160.83	156.60	96.65
P3-150	161.44	150.57	97.02	115.28	118.96	98.29	159.24	154.72	101.35

Table 2. DSC crystallization and melting data of nanocomposites prepared by slow cooling

Sample	I run melting			II run cooling			III run melting		
	T_m /°C	$T_{m,on}$ /°C	ΔH_m /J/g	T_c /°C	$T_{c,on}$ /°C	$-\Delta H_c$ /J/g	T_m /°C	$T_{m,on}$ /°C	ΔH_m /J/g
S0-0	164.44	157.95	112.14	121.52	125.66	106.77	162.83	159.21	106.50
S0-50	164.23	158.82	113.45	119.27	122.34	99.32	161.42	158.39	102.23
S0-150	164.22	158.47	110.53	119.14	122.04	102.13	161.66	158.47	102.00
S1-0	163.86	158.02	109.39	119.03	123.37	108.11	148.07 (161.74)	158.36	105.11
S1-50	163.47	158.57	106.52	121.59	124.86	104.59	162.58	158.42	102.47
S1-150	163.53	158.84	107.69	120.33	123.33	101.04	160.27	155.86	101.38
S3-0	163.48	158.75	106.57	118.04	121.89	98.71	161.98	157.70	96.69
S3-50	163.25	158.53	104.39	115.83	119.40	102.11	160.57	157.18	100.04
S3-150	163.62	154.89	110.97	115.57	119.60	98.68	160.27	155.87	97.95

REFERENCES

[1] M. Alexandre and P. Dubois, *Mater. Sci. Eng.* 28, 1 (2000).

[2] R Pfaender, *Polym. Degrad. Stab.* 95, 363 (2010).

[3] J. Jordan, K. I. Jacob, R. Tannenbaum, M. A. Sharaf and I. Jasiuk, *Mat. Eng. Sci.* A393, 1 (2005).

[4] S. S. Ray and M. Okamoto, *Prog. Polym. Sci.* 28, 1539 (2003).

[5] J. A. Tarapow, C. R. Bernal and V. A. Alvarez, *J. Appl. Polym. Sci.* 111, 768 (2009).

[6] E. P. Giannelis, *Adv. Mater.* 8, 29 (1996).

[7] Y. Komori and K. Kurada. Layered silicate-Polymer Intercalation Compounds, In: T. J. Pinnavaia and G. W. Beal (Eds.), *Polymer Layered Silicate Nanocomposites*, Wiley, New York (2000).

[8] N. M. Hasegawa, M. Kawasumi, M. Kato, A. Usuki and A. Okada, *J. Appl. Polym. Sci.* 67, 87 (1998).

[9] R. Pfaendner, Stabilization of Nanocomposites. In: *Industry Guide to Nanocomposites*, G. Beyer (editor): Applied Market Information Ltd; 117 (2009).

[10] M. Zanneti, S. Lomakin and G. Camino, *Macromol. Mater. Eng.* 279, 1 (2000).

[11] X. Y. Ma, G. Z. Liang, H. L. Liu, J. Y. Fei and Y. Huang, *J. Appl. Polym. Sci.* 97, 1915 (2005).

[12] F. Perrin-Sarazin, M.T. Ton-That, M. N. Bureau and J. Denault, *Polymer*, 46, 11624 (2005).

[13] B. Lepoittevin, N. Pantoustier, M. Devalckenaere, M. Alexandre, C. Calberg, R. Jerome et al., *Polymer*, 44, 2033 (2003).

[14] J. Karger-Kocsis (Ed.), *Polypropylene: Structure, Blends and Composites*, Springer, London (1995).

[15] N. Grossiord, J. Loos and C. E. Koning, *J. Mater. Chem.* 15, 2349 (2005).

[16] O. Regev, P. N. B. El Kati, J. Loos and C. E. Koning, *Adv. Mater.* 16, 248 (2004).

[17] S. Pavlidou and C. D. Papaspyrides, *Prog. Polym. Sci.* 33, 1119 (2008).

[18] J. R. Salvetat-Delmotea and A. Rubio, *Carbon*, 40, 1729 (2002).

[19] D. Vink, *Eur. Plast. News*, May: 36-37 (2008).

[20] M. Morreale, N. T. Dintcheva and F. P. La Mantia, *Polym. Int.* 60, 1107 (2011).

[21] F. P. La Mantia and N. T. Dintcheva, *Plast. Rubber Compos.* 33, 184 (2004).

[22] R. Benavides, R. Gonzales-Hernandez, M. C. Gonzales-Cantu, B. Reyes-Vielma and N. C. Billingham, *J. Vinyl Addit. Technol.* 9, 41 (2003).

[23] G. Kandilioti, A. Elenis, K. A. Macchiarola and V. Gregoriou, *Appl. Spectroscopy*, 60, 1285 (2006).

[24] S. Morlat, B. Mailhot, D. Gonzales, J.-L. Gardette, *Chem. Mater.* 16, 377 (2004).

[25] H. Qin, Z. Zhang, M. Feng, S. Zhang and M. Yang, *J. Polym. Sci. Part B: Polym. Phys.* 42, 3006 (2004).

[26] G. Bogoeva-Gaceva, Lj. Raka and B. Dimzoski, *Adv. Compos. Lett.* 17, 161 (2008).

[27] L. Raka, G. Bogoeva-Gaceva, K. Lu and J. Loos, *Polymer*, 50, 3739 (2009).

[28] J. L. Wang, C. M. Dong, *Polymer*, 47, 3218 (2006).

[29] D. M. Price, J. Hourston and F. Dumont. Theromogravimetry of Polymers. In: *Encyclopedia of Analytical Chemistry*, R. A. Meyers (Ed.), John Wiley&Sons Ltd, Chichester, 2000

[30] L. Raka, G. Bogoeva-Gaceva and J. Loos, *J. Thermal Anal. Calorim.* 100, 629 (2010).

[31] S. Chmela, A. Kleinova, A. Fiedlerova, E. Borsig, D. Kaempfer, R. Thomann et al., J. *Macromol. Sci. A. Pure Appl. Chem.* 42, 821 (2005).

[32] H. Qin, C. Zhao, C. Zhang, G. Chen and M. Yang, *Polym. Degrad. Stab.* 81, 497 (2003).

[33] M. T. Ton-That, F. Perrin-Sarazin, K. C. Cole, M.N. Bureau and J. Denault, *Polym. Eng. Sci.* 44, 1212 (2004).

[34] L. Százdi, A. Abrányi, B. Pukánszky and G. J. Vancso, *Macromo. Mater. Eng.* 291, 858 (2006).

[35] M. Modesti, A. Lorenzetti, D. Bon and S. Besco, *Polymer*, 46, 10237 (2005).

[36] J. Varga, *J. Macromol. Sci.* B41, 1121 (2002).

[37] J. Zheng, X. Lu, C. L. Toh, T. H. Zheng and C. He, *J. Polym. Sci.* B42, 1810 (2003).

[38] H.E. Miltner, N. Grossiord, K. Lu, J. Loos, C.E. Koning and B. Van Mele, *Macromolecules*, 41, 5753 (2008).

[39] P. Maiti, P.H. Nam, M. Okamoto, N. Hasegawa and A. Usuki, *Macromolecules*, 35, 2042 (2002).

[40] J. Li, R. Yang, J. Yu and Y. Liu, *Polym. Degrad. Stab.* 93, 84 (2008).

[41] H. Qin, S. Zhang, H. Liu, S. Xie, M. Yang, D. Shen, *Polymer*, 46, 3149 (2005)

[42] M. S. Rabello and R. L. White, *Polym. Degrad. Stab.* 53, 55 (1997).

[43] S. Aslanzadeh, M. Haghighat Kish and A. A. Katbab *Polym. Degrad. Stab.* 95, 1800 (2010).

In: Polypropylene
Editors: L. P. Silva and E. F. Barbosa

ISBN: 978-1-62417-142-0

Chapter 15

NANOSCALE APPROACHES OVER POLYPROPYLENE PLASTIC SURFACES: AN ALTERNATIVE TOOL FOR QUALITY CONTROL CHECKING

Eduardo F. Barbosa and Luciano P. Silva*
University of Brasília (UnB),
Embrapa Genetic Resources and Biotechnology (EMBRAPA), Brasilia, Brazil

ABSTRACT

Materials and composites containing polypropylene (PP) can be considered as the most widely applied in industrial laboratories and also in scientific research labware. This wide range of uses is sometimes hindered by the drawbacks of plastic contamination or sample adhesion to the plastic surfaces. In this context, investigative approaches of this type of polymeric surface can reveal important information concerning the drawbacks. In this chapter, some surface properties of PP materials using atomic force microscopy (AFM) and Fourier transform infrared spectroscopy (FTIR) were tested. These tools allowed the acquisition of information about morphology, nanoroughness, interaction forces and chemical environment of the tested materials (plastic tips and microtubes internal surface). PP tips with different sizes were tested revealing considerably low nanoroughness values (Ra and Rq). However, statistically significant differences were found among the sample values for these parameters that can be related to manufacturing steps. Force spectroscopy (FS) results demonstrated an increased homogeneity between the two smaller samples, while the biggest plastic tip presented different patterns of interaction with the AFM probe, mainly related to adhesion forces. FTIR results showed a typical PP crystalline form in all evaluated samples. PP microtubes with different volume capacities were also analyzed, revealing high homogeneity (Ra and Rq). The largest microtube also presented discrepancies compared to the smaller ones (Rz, Rzjis, and Rv). FS results suggested a different scenario, highlighting a heterogeneous behavior of the plastic microtubes considering their patterns of interaction with the AFM probe. These experimental approaches can be seen as a potential route for quality control checking, applicable to materials derived from PP and this information can be largely

* E-mail address: luciano.paulino@embrapa.br, lucianopaulinosilva@gmail.com.

useful for further studies concerning molecules immobilization (e.g. enzymes) onto these surfaces.

INTRODUCTION

Chronologically, polypropylene (PP) materials emerged lately considering the synthetic polymer industry scenario. It appeared after celluloid, bakelite, polyacrylic acid, polystyrene, nylon, polyurethanes, polyethylene, and polyethylene terephthalate polymers [1]. Despite this, PP is known as the first synthetic stereo-regular polymer to achieve industrial importance [2], basically due to its technical end-uses combining high tensile strength with low cost [3]. PP based materials have been largely applied in several areas, namely the automotive industry as well as medical and food packaging films, and other diverse miscellaneous applications [4].

In addition to being applied to several useful plastic devices in research and industrial laboratories, contemporaneously to polymerase chain reaction (PCR) discovery (1980's), scientists developed a plastic microtube aiming to optimize heat transference during temperature gradients required for PCR protocols. This device should also minimize reagent volumes and allow automated handling for operations, such as solvent extraction and centrifugation [5]. The invention was described as a thin walled sample microtube, with a thin walled conical section with 17° angles relative to the longitudinal tube axis [5].

This kind of disposable plasticware became routinely used in most biological research laboratories [6]. The widespread use of these microtubes and their design variations prompted a hypothesis that this device type could act as a major source of contaminant, since its manufacturing steps include addition of lubricating agents, which could leach from the tubes while performing some assays [6]. Then, it had emerged out studies trying to elucidate some possible influence of plastic labware on experimental procedures.

Mc Donald et al reported the identification of two additives leaching from plastic microtubes that had the potential to effect enzymes and receptor proteins [6]. Watson et al reported a study containing important information about contaminants released from plastic labware upon DMSO exposure and their effects on specific biological assays [7]. However, Barbosa and Silva investigated possible molecular leaching from plastic microtubes by exposure to specific solvents (water, acetonitrile, trifluoroacetic acid, and ammonium bicarbonate) using mass spectrometry and they did not observe any polymer leaching [8]. They also characterized the nanometric surface features of internal plastic microtube wall, along with quantitative nanoroughness determination, establishing an alternative way for plastic tubes to be assayed for quality control [8]. Other types of plasticware commonly used in laboratories are disposable macro- and micropipette tips. The first tip design, that was attachable to adjustable pipets, had only a simple conical form [9]. Since then, the design was constantly improved to offer an easy removal and minimal interference on the experiments [10].

A low retention property is always desirable for micropipette tips in order to assure the accuracy. This feature is closely related to the nanoscale aspects of the internal and external tip surface in contact with solvents.

With this in mind the use of atomic force microscopy represents a very useful tool to scrutinize both nanoroughness and force spectroscopy (FS) parameters concerning surface properties. Nanoroughness can be seen as an important evaluation parameter concerning the interactions between the surface and the surrounding environment, relating directly to surface area [11-14]. Force spectroscopy parameters offer complementary information about how a surface interacts with other external molecular systems. It can point out some interaction forces, such as, snap in force, maximum load force, and some mechanical properties as Young's modulus. These parameters are closely related to attractive forces, stiffness, and elastic properties, respectively. Barbosa and Silva 2012 applied FS technique to investigate the pattern of interactions between some polymeric materials and an atomic force microscopy probe to evaluate the influence of each material over their attraction, adhesion, and elasticity covering a nanoscale approach [15].

In this chapter, an alternative way to investigate some nanoscale features of tips and microtube internal surfaces is proposed as a possible form of quality control check, relating the chemical composition of tips and tubes with surface morphology and physical properties. Beyond these technical applications, this approach can result in important and useful information in several other areas, such as, in enzyme immobilization and enzyme-linked immunosorbent assays (ELISA) approaches.

MATERIAL AND METHODS

Sampling

Two types of PP material surfaces were analyzed: microtubes (0.2, 0.6, and 1.5 ml of capacity) and micropipette tips (used for 0.5-10, 20-200, and 100-1000 μl capacity). Each tube (n=2) was segmented at three different areas along of the microtube, as illustrated in figure 1:

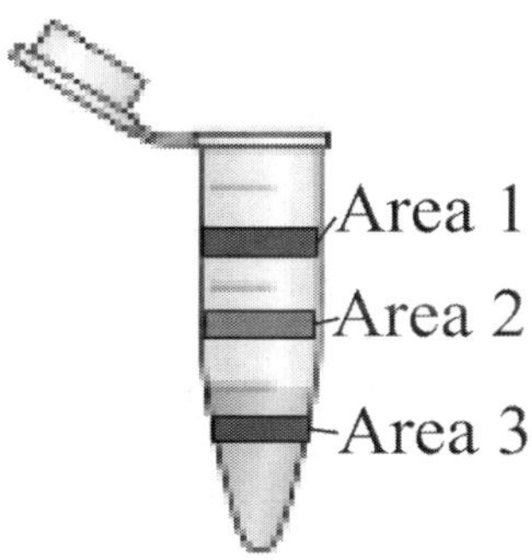

Figure 1. Scheme representing segmented areas along each PP microtube.

Small rectangular fragments of approximately 6 mm^2 were removed from each sectioned area to perform the experimental methodology. The analyzed surfaces were always performed on the internal side. Tip segments were removed from the middle portion of each tip, approximately the same length of the microtube segments (6 mm^2), however width was limited by the tip dimensions.

Atomic Force Microscopy (AFM) Analyses

Force Spectroscopy (FS)

To analyze the interaction forces between an AFM tip and PP surfaces, a commercial AFM instrument (SPM-9600, Shimadzu, Japan) was used operating in contact mode. A small segment of each sample was fixed on a metallic support using a double-sided adhesive tape. A pyramidal silicon nitride tip (Olympus, curvature radius < 20 nm) was used in this technique as well as a cantilever with a length of 200 μm (spring constant of 0.15 N/m and resonance frequency of approximately 24 kHz). The acquisition rate was set at 1 Hz and the method was loaded at 20 V of amplitude and 3 V of operating point. The software used to analyze acquired force curves was SPIP v. 5.1.5, set at 23°C, 0.15 N/m spring constant, cone indentation (Sneddon), 5 fitting points, and approach curve as baseline correction curve. At least 20 force curves analyses were performed to ensure the reproducibility of the assays. All force spectroscopy data were analyzed by ANOVA using Tükey test ($p < 0.05$).

Nanoroughness Characterization

Samples were analyzed using the same commercial AFM equipment previously described operating in dynamic phase mode. The scanner used has a travel of 125 μm in XY-directions and 7 μm in the Z-direction. Images were obtained in ambient air, at room temperature (23°C) and approximately 20-30% relative humidity. Scanned areas were squares of 50 μm × 50 μm with 1 Hz scan rate using an aluminum rectangular cantilever with 125 μm length, resonance frequency of 320 kHz, 42 N/m spring constant, and Si tip with curvature radius < 10 nm (Nanosensors PPP-NCHR-20). At least ten areas were scanned for each microtube surface to ensure the reproducibility of the assays. Trace and retrace procedures were performed in order to prove that the samples were not distorted during the scanning steps. All AFM images contained 512 × 512 data points and were first processed by SPM-9600 off-line software. The processing consisted of an automatic X-line, Y-line, and plane fit leveling of the surface, and two erase noisy lines steps. After processing, the images surface analysis were acquired and some nanoroughness values were obtained referring to arithmetic mean roughness (Ra), maximum height (Rz), ten point height of roughness profile (Rzjis), root-mean-square roughness (Rq), maximum profile peak height (Rp), and maximum profile valley depth (Rv). Finally, AFM 3D images were obtained using SPM-9600 off-line software. AFM nanoroughness data were analyzed by ANOVA and Tukey test ($p < 0.05$).

Fourier Transform Infrared Spectroscopy (FTIR)

Infrared absorption spectra were obtained using a Fourier transform infrared spectrometer - 4100 (Jasco, United Kingdom), with wavenumbers ranging from 500 to 4000 cm^{-1}. Reported spectra were averages of 32 scans. Samples were analyzed in the solid state, with small fragments of each sample deposited in the attenuated total reflectance system for subsequent acquisition.

RESULTS AND DISCUSSION

PP Micropipette Tips Nanoscale Characterization

Internal PP tip surface morphologies obtained by AFM analyses are showed in Figure 2. The presence of ordinate and regular parallel lines along each area is notable in all PP tips. This can be related to the production steps for the PP tips, including particular injection molding systems.

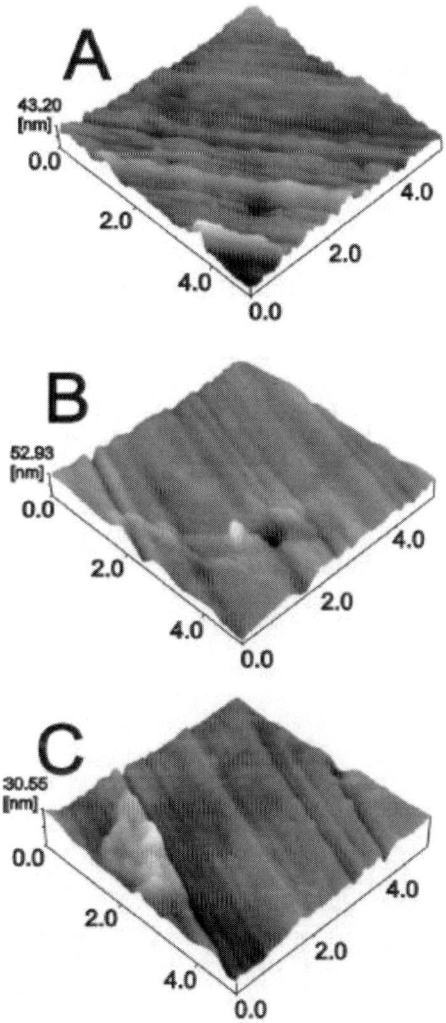

Figure 2. Representative three-dimensional images of PP tip samples. Images were obtained by atomic force microscopy analyses operating in dynamic mode. "A" represents the tip internal surface used for 0.5-10 μl micropipette, "B" represents the tip internal surface used for 20-200 μl micropipette, and "C" represents the tip internal surface used for 100-1000 μl micropipette.

In addition to the qualitative results, some quantitative outcomes for Ra, Rz, Rzjis, Rq, Rp, and Rv nanoroughness parameters are presented in Figure 3.

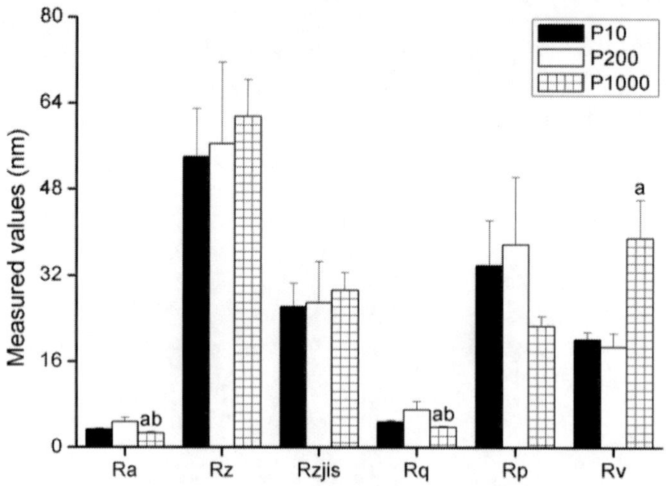

Presence of the symbol "a" means statistically significant difference from P10, and "b" from P200 ($p < 0.05$).

Figure 3. Nanoroughness results obtained by atomic force microscopy operating in dynamic mode. P10 represents the internal tip surface roughness (0.5-10 µl micropipette), P200 represents the internal tip surface roughness (20-200 µl micropipette), and P1000 represents the internal tip surface roughness (100-1000 µl micropipette).

In order to clarify the meaning of these results, we provide some theory concerning roughness parameters. Ra is the arithmetic average of absolute values of deviations, relative to the midline, in the measurement route. Ra may be calculated by the equation:

$$Ra = \frac{1}{n}\sum_{i=1}^{n}|y_i| \tag{1}$$

R_z is the arithmetic mean of five partial roughness values (z_i). Z_i is defined as the sum of absolute values of the 5 largest variations (both above and below the midline) existing within a sampling length. Graphically, this value represents the height between the maximum and minimum points of the profile within the sampling length. This is represented in Figure 4.

R_{zjis} is the sum of the fifth largest peak and mean value of largest valley to the fifth largest valley within a sampling length. R_q is a parameter related to R_a. It is the root mean square deviation. It is also called RMS (Root Mean Square). This parameter increases the effect of irregularities that deviate from the average. R_q may be represented by the equation:

$$Rq = \frac{1}{n}\sum_{i=1}^{n}y_i^2 \tag{2}$$

$$R_Z = \frac{Z_1 + Z_2 + Z_3 + Z_4 + Z_5}{5}$$

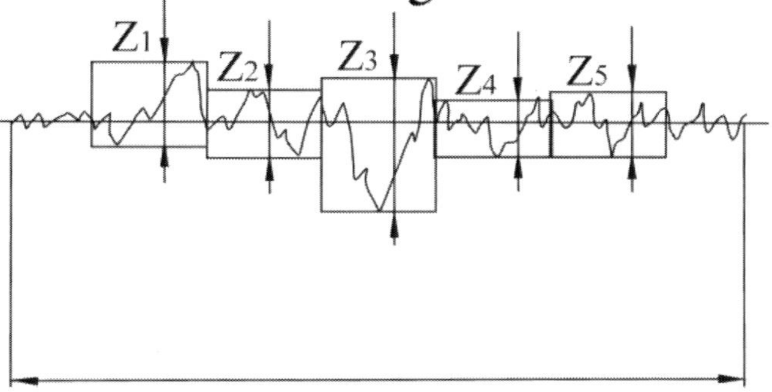

Figure 4. Equation and scheme representing R_z roughness parameter.

Finally, R_p and R_v correspond to the maximum profile peak height and maximum profile valley (depth) respectively.

Figure 3 shows the low nanoroughness values (for R_a and R_q) of each PP tips surface that suggests a high homogeneity and regularity of the morphology on nanoscale. R_q is often described as the best parameter to define surface irregularities [8]. Despite their low heterogeneity, it was possible to detect significant statistical differences among P1000, P200, and P10. This fact can be associated to the manufacturing steps associated with each tip size, involving injection molding systems. Another influencing factor on tips roughness can be the dyes applied to tip manufacturing.

Figure 3 shows that Rz values are notably higher than Ra and Rq, indicating a considerable influence of maximum vertical displacements on the PP tips topographical pattern organization. Rp and Rv exhibited approximately intermediate values that strongly suggest a similar effect of each one to the global surface irregularities, considering the vertical or z axis.

Moreover, it is important to highlight that some researchers have been investigating the direct effects of surface roughness on the contact mechanisms describing the relationships among deformations and loading/unloading adhesion forces of various polymeric surfaces [11, 15-18]. Then, this chapter assigns the use of AFM analysis as a promising tool to check the quality degree of common PP surfaces.

PP tips chemical environment was investigated, to verify some possible differences among the analyzed samples. In Figure 5, no difference among the samples was detected, but it showed characteristic functional groups, mainly referring to PP crystalline assignments. The frequencies: 805, 840, 898, 940, 972, 995, 1100, and 1166 cm^{-1} are related to CH_2 rocking; C-C stretching; and CH_3 wagging and rocking modes. The 1255 cm^{-1} refers to CH_2 twisting and CH bending; 1455 cm^{-1} refers to CH_2 bending and antisymmetric CH_3 bending; 2836 cm^{-1} refers to symmetric CH_2 stretching; 2867 and 2915 cm^{-1} refers to symmetric CH_3 stretching, antisymmetric CH_2 stretching and CH stretching; and 2960 cm^{-1} refers to antisymmetric CH_3 stretching [19]. All this information is compiled and summarized in table 1, shown together with each observed peak intensity.

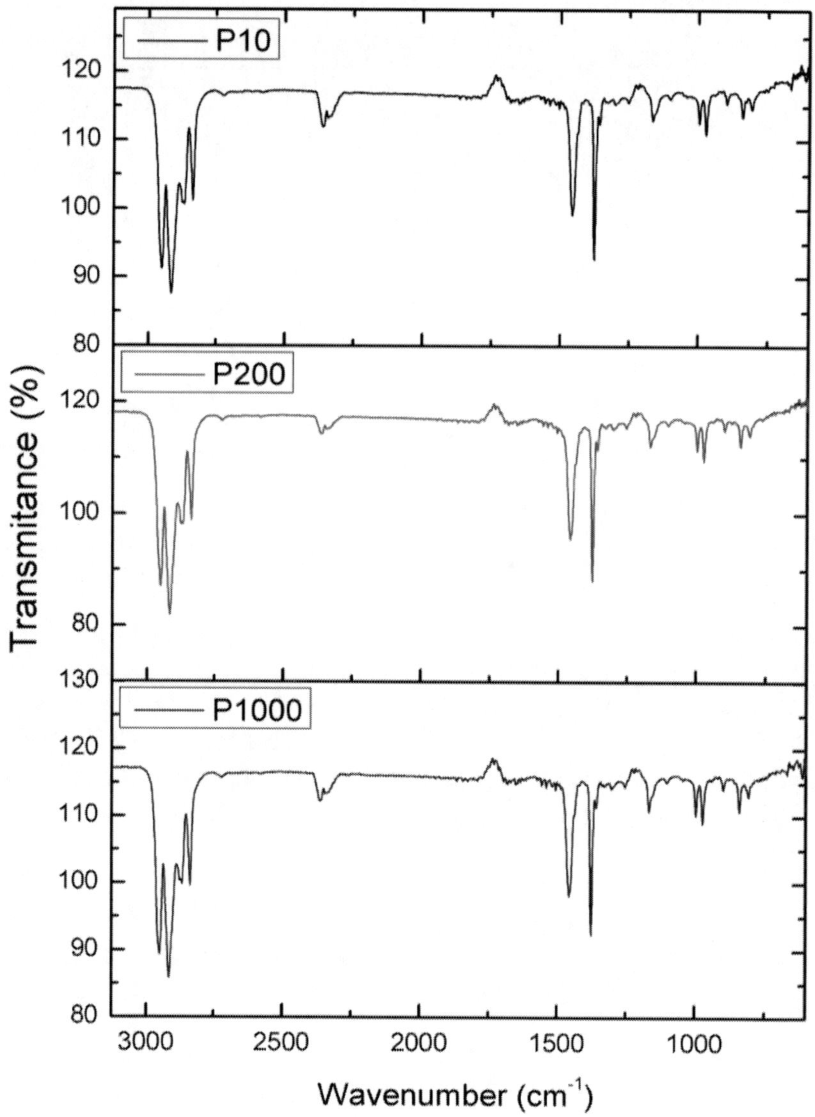

Figure 5. Analysis of Fourier transform infrared spectroscopy (FTIR) measured with attenuated total reflectance (ATR) for P10, P200, and P1000 PP tips. P10 represents internal PP tip surface results (0.5-10 μl micropipette), P200 refers to 20-200 μl micropipette, and P100 for 100-1000 μl micropipette.

Aiming to investigate the attractive and adhesives force behaviour of each PP tip surface, force spectroscopy (FS) experiments were performed.

FS experiments correspond to an AFM approach in which the probe is put onto the surface of each analyzed material. This approximation is mediated by piezoelectric ceramics, located in a scanner. Interactions between the AFM probe and the surface are measured by an integrated cantilever deflection.

Table 1. Main Fourier transform infrared spectroscopy (FTIR) assignments of crystalline PP

<table>
<tr><th>Frequency (cm^{-1})</th><th>Intensity*</th><th>Band analyzes</th></tr>
<tr><td>805</td><td>W</td><td rowspan="8">CH_2 rocking; C-C stretching; and CH_3 wagging and rocking modes</td></tr>
<tr><td>840</td><td>M</td></tr>
<tr><td>898</td><td>W</td></tr>
<tr><td>940</td><td>VW</td></tr>
<tr><td>972</td><td>M</td></tr>
<tr><td>995</td><td>M</td></tr>
<tr><td>1100</td><td>W</td></tr>
<tr><td>1166</td><td>M</td></tr>
<tr><td>1255</td><td>W</td><td>CH_2 twisting and CH bending</td></tr>
<tr><td>1374</td><td>S</td><td>Symmetric CH_3 bending</td></tr>
<tr><td>1455</td><td>S</td><td>CH_2 bending and antisymmetric CH_3 bending</td></tr>
<tr><td>2836</td><td>S</td><td>symmetric CH_2 stretching</td></tr>
<tr><td>2867</td><td>S</td><td rowspan="2">refers to symmetric CH_3 stretching, antisymmetric CH_2 stretching and CH stretching</td></tr>
<tr><td>2915</td><td>VS</td></tr>
<tr><td>2960</td><td>VS</td><td>antisymmetric CH_3 stretching</td></tr>
</table>

*Symbols meaning: *W* (weak); *M* (medium); *VW* (very weak); *S* (strong); *VS* (very strong)

The acquisition of force curves is based on a broad theory. It is related to Hooke's law applied to Hertz and Sneddon indentation models. Firstly, according to Domke and Radmacher, Hooke's law can connect cantilever deflection (d) with applied loading force (F) using force constant k of the cantilever, then loading force can be written as [17]:

$$F = kd \tag{3}$$

Hertz described theoretically the elastic deformation of two spherical surfaces touching under load [18]. Sneddon extended the calculation to other geometries, like a cone pushing onto a flat sample, as it was used here [19]. This model gives the following relation between indentation (δ) and loading force (F):

$$F = (2/\pi)[E/(1-\nu^2)]\delta^2 \tan(\alpha) \tag{4}$$

E represents elastic or Young's modulus, ν is the Poisson ratio of the sample, and α is the half opening angle of indenting cone. To calculate Young's modulus, a fit of Hertz (cone indentation) model was applied to the curve. Combining the two equations yields:

$$z - z_0 = d - d_0 + \sqrt{\frac{k(d - d_0)}{(2/\pi)[E(1-\nu^2)]\tan(\alpha)}} \tag{5}$$

Where z means the movement of the piezoelectric ceramic, zero deflection d_0 has to be determined in noncontact part of the force curve, and Poisson ratio assumed was 0.5. Pulling model automatically fitted to each detected unfolding event is based on the following formula:

$$F_{\mathrm{WLC}}(s) = \frac{k_B T}{l_p}\left[\frac{1}{4(1 - s/L_c)} + \frac{s}{L_c} - \frac{1}{4}\right] \tag{6}$$

$F_{WLC}(s)$ describes the force as function of separation s, k_B is Boltzmann's constant, T is the temperature in Kelvin, l_p represents the persistence length and L_c the contour length. In addition, peak force and the separation/unloading rate at rupture are reported.

In summary, when the AFM probe gets near to the sample, initially it is observed attractive interactions between atoms of the probe and surface. This event relates to Snap in parameter, which describes the time when the tip moves (jumps) towards the surface, due to attractive forces exceed the spring constant value of the cantilever (rod).

Figure 6 shows the patterns of interaction between the AFM probe and the PP tips surface.

Figure 6 shows Maximum Load, Maximum Pull, Snap In, and Detach force results related to the interactions between the probe and the PP tip surfaces. These results show high homogeneity between P10 and P200, corroborating the roughness outcomes. Statistic significant difference was found to Max. Pull and Detach force, indicating that P1000 tip presented reduced adhesion force to the probe when compared to the other tips, which suggests a possible lower tendency to retention or incrustation of undesirable materials. This found can be related to the smaller Ra and Rq roughness values presented by this sample when compared to the other tips (Figure 3).

Three types of PP microtubes (0.2, 0.6, and 1.5 ml) were selected to be investigated about some possible influence of the volume capacity over their internal surface properties. Figure 7 shows qualitative AFM results (three-dimensional images) of three representative sectioned areas along each sample.

To corroborate the qualitative analysis of Figure 7, quantitative roughness results presented in Figure 8 should be taken in to consideration.

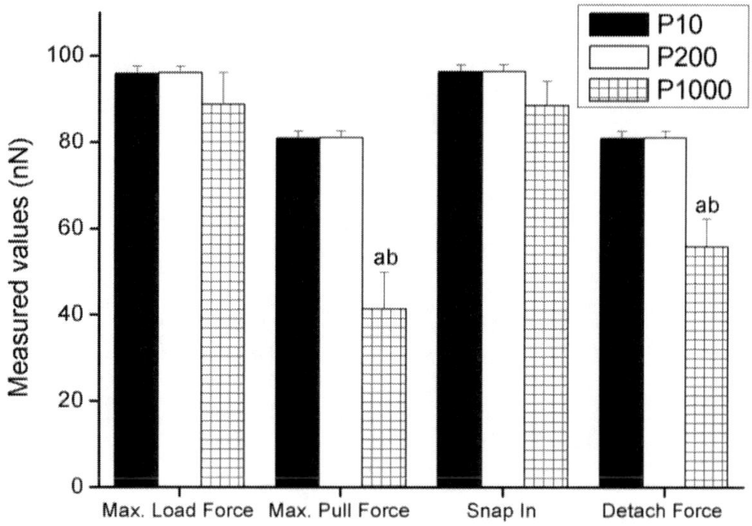

The symbol "a" means statistic significant difference of P10; and "b" from P200 ($p < 0.05$).

Figure 6. Force spectroscopy results obtained by AFM operating in contact mode. P10 represents the internal PP tip surface results for 0.5-10 μl micropipette, P200 to 20-200 μl micropipette, and P100 for 100-1000 μl micropipette.

PP Microtubes Nanoscale Characterization

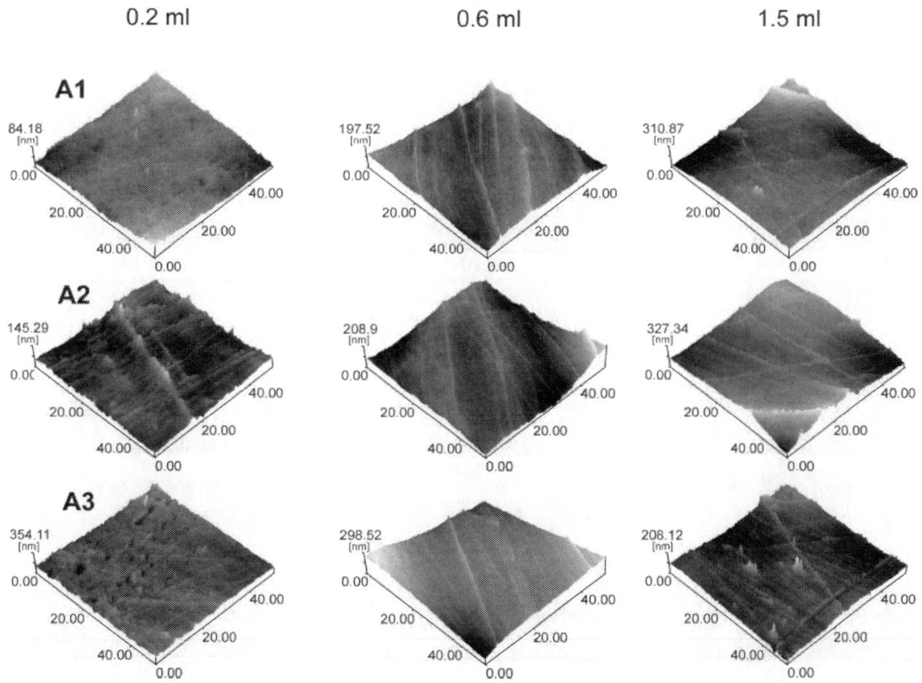

Figure 7. Representative three-dimensional images of PP microtubes obtained by atomic force microscopy analyses operating in dynamic mode. "A1, A2, and A3" represent the areas along each tube. Tube capacities were 0.2, 0.6, and 1.5 ml.

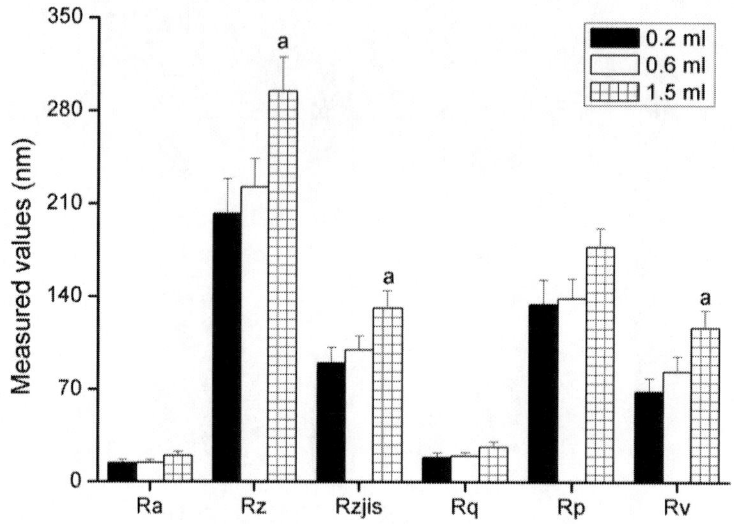

Presence of symbol "a" means statistic significant difference from 0.2 ml microtube ($p < 0.05$).

Figure 8. Nanoroughness results obtained by atomic force microscopy operating in dynamic mode. Analyzed microtubes capacity: 0.2, 0.6, and 1.5 ml.

Simple visual interpretation of Figure 7 might suggest that the analyzed surfaces could present high heterogeneity and disparity among the samples and the areas along each PP microtube. However, the quantitative nanoroughness results and statistical analysis pointed out that there was no statistically significant difference neither among the areas along each tube (data not showed) nor for Ra and Rq roughness results among the samples (Figure 8). These results showed considerable nanoroughness homogeneity among the tested PP microtubes. In spite of these facts, significant statistical differences were found to Rz, Rzjis, and Rv nanoroughness parameters, pointing the 1.5 ml tube as the most irregular material. This can be related to manufacturing steps associated with this tube size that evolves injection molding systems and metallic molds. An interesting study to corroborate this hypothesis is to investigate metallic molds surface morphology used to manufacture these tubes.

Under this perspective, it is important to highlight that nanoroughness results are directly dependent of scanning scale and conditions. According to Wong et al roughness measurements are scale dependent [20]. Then, it is not correct, for example, to establish any direct comparison between PP tips roughness results (scanned areas of 25 μm^2) with PP microtubes surface roughness results (scanned are of 2.500 μm^2). It is noteworthy that it was not possible to perform AFM scanning of similar areas for both PP materials due to specific sample macrostructural features. Rq (root mean square), for instance, is scale dependent through the power law $Rq = A.L^{3-D}$, where the coefficient, A, is the roughness at a scan size of 1 μm^2, D is the fractal dimension that relates the increase in roughness to the increase in scan size [6]. In this sense, comparisons of roughness measurements require lots of attention, requiring well established and constant scanned areas values.

To elucidate some possible differences concerning the chemical environment of each PP microtube type, FTIR analyses were performed, as showed in Figure 9. The results did not allow establishing any representative difference in chemical environment among the samples.

It showed the assignments related to PP, revealing its crystalline form, as evidenced by the same FITR assignments (Table 1).

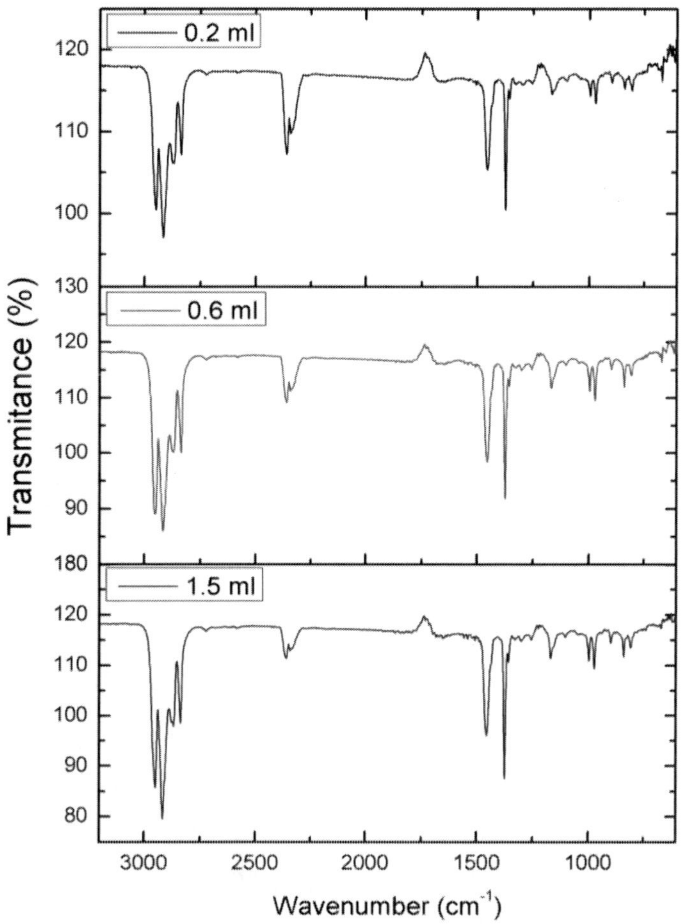

Figure 9. FTIR (Fourier transform infrared spectroscopy) analyses of PP microtubes internal surface. Measurements were performed with attenuated total reflectance (ATR). Tube capacities were: 0.2, 0.6, and 1.5 ml.

Lastly, FS experiments were performed to investigate some possible difference on the patterns of interaction between the internal PP microtube surfaces and AFM probe, as showed in Figure 10.

Force spectroscopy results indicate that there are significant differences concerning attraction, adhesion and stiffness features among the samples. As observed in Figure 10, 0.2 ml PP microtube showed a lower stiffness as assessed by its low Maximum Load force result. The 0.6 ml PP microtube presented stronger adhesive forces, indicated by higher Maximum Pull and Detach forces. The 1.5 ml microtube showed stronger attractive forces, since it presented highest Snap in force result. These strong differences concerning FS results could be related with undesirable adhesion and the lack of reproducibility involving PP microtubes applications, where the microtubes surface affect directly some experiments performance, as described by Mc Donald et al [6] and Watson et al [7].

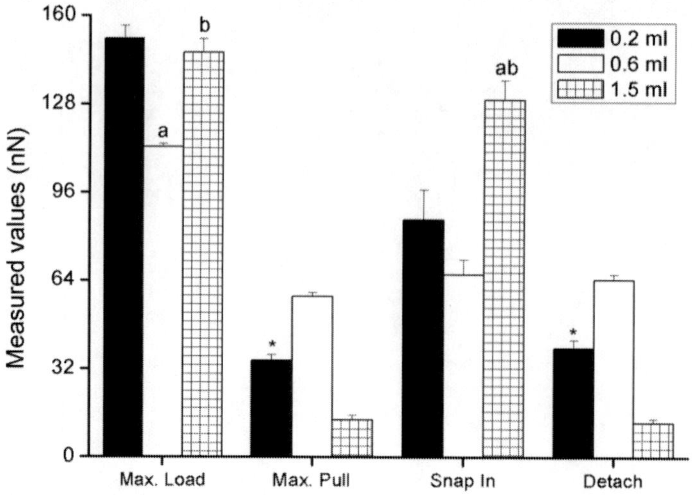

Asterisk symbol (*) means statistical significant difference among all other sample types ($p < 0.05$).

Figure 10. Force spectroscopy results obtained for PP microtubes by AFM operating in contact mode. Analyzed PP microtube capacities: 0.2, 0.6, and 1.5 ml. The symbol "a" means statistical significant difference of 0.2 ml; and "b" from 0.6 ml.

Conclusion

An alternative approach to investigate qualitative and quantitatively polypropylene plastic surfaces was performed. This insight includes AFM techniques to determinate nanoroughness and FS data. These parameters showed refined information about polypropylene surfaces nanometric organization allowing the evaluation of homogeneity and some interaction forces of each surface.

Nanoroughness and force spectroscopy results could be directly associated with the adhesion events onto polymer surface, since the surface area of contact can predicts tendencies of reactivity or leaching of different materials [8]. Coupled with other spectroscopic techniques, like FTIR, it was possible to evaluate the chemical environment of each surface with their morphological and structural aspects. Therefore, the experimental approach discussed in this chapter could be seen as a potential route for quality control, applicable for materials derived from polypropylene. Simultaneously, this information will be largely useful for further studies concerning enzyme immobilization on these surfaces.

Acknowledgments

The authors acknowledge the financial support from Brazilian agencies MCT/CNPq (processes no. 555.175/2005-7, 484.201/2007-7, 302.018/2008-5, 563.802/2010-3, 305.637/2011-8), FAPDF (processes no. 193.000.445/2008 and 193.000.429/2008), and CAPES (process no. 23038.019088/2009-58). The authors are also grateful to Laboratory of Environmental Analytical Chemistry of the Institute of Chemistry of the University of Brasilia.

The authors acknowledge Dr. Gerard Murray, Dr. Thales Lima Rocha, Ms. Rebecca Caldas Curley, and Mr. Pedro Henrique Cevallos Mijan by the English revision on this chapter.

REFERENCES

[1] J. Vlachopoulos, D. Strutt, *J. Mat. Sci. and Tech.* 19, 1161 (2003).

[2] M. Ahmed, Polypropylene fibers, science and technology. *Elsevier Sci. Pub. Co.* New York, 766 (1982).

[3] S. Zang, *Prog. Polym. Sci.* 28, 1517 (2003).

[4] C. Maier, T. Calafut, Polypropylene, the definitive guide and handbook. *Willian Andrew Publishing*, p. 32, (1998).

[5] Atwood et al. United Stated Patent Applications No. 10/929,961, Apr. 21, 2005.

[6] G. R. McDonald, A. L. Hudson, S. M. J. Dunn, H. You, G. B. Baker, R. M. Whittal, J. W. Martin, A. Jha, D. E. Edmondson, A. Holt, *Science* 322 (2008).

[7] J. Watson, E. B. Greenough, J. E. Leet, M. J. Ford, D. M. Drexler, J. V. Belcastro, J. J. Herbst, M. Chatterjee, M. Banks, *J. Biomol. Screening* 14 (2009).

[8] E. F. Barbosa, L. P. Silva, *Surf. Int. Anal.* 44, 389 (2012).

[9] Suovaniemi et al. United Stated Patent No. 4,151,750, May 1, 1979.

[10] Kelly et al. United Stated Patent No. 6,168,761, Jan 2, 2001.

[11] M. Benz, W. B. Euler and O.J. Gregory, *Langmuir* 17, 239 (2001).

[12] K. Feldman, T. Tervoort, P. Smith, N. D. Spencer, *Langmuir*, 14, 372 (1998).

[13] Y. Kumada, D. Kuroki, H. Yasui, T. Ohse, M. Kishimoto, *J. Biosci. Bioeng.* 109, 583 (2010).

[14] Y. Murakami, T. Endo, S. Yamamura, N. Nagatani, Y. Takamura, E. Tamiya, *Anal. Biochem.* 334, 111 (2004).

[15] E. F. Barbosa, L. P. Silva, *J. Memb. Sci.* Available online 17 March 2012.

[16] C. Y. Liang, F.G. Pearson, *J. Mol. Spec.* 5, 290 (1960).

[17] J. Domke, M. Radmacher, *Langmuir* 14, 3320 (1998).

[18] H. J. Hertz, *Reine Angew. Math.* 92, 156 (1882).N. Sneddon, *Int. J. Engng. Sci.* 3, 47 (1965).

[19] I. N. Sneddon, *Int. J. Engng. Sci.* **3**, 47 (1965).

[20] P.C.Y. Wong, Y.N. Kwon, C.S. Criddle, *J. Membr. Sci.* 340, 117 (2009).

INDEX

#

A

D

E

F

G

H

I

K

L

M

N

O

P

Q

R

S

T

U

V

W

X

Y

Z